AUSTRALIAN MATHEMATICAL SOCIETY LECTURE SERIES

Editor-in-Chief: Professor J.H. Loxton, School of Mathematics, Physics, Computing and Electronics, Macquarie University, NSW 2109, Australia

Editors:
Professor C.J. Thompson, Department of Mathematics, University of Melbourne, Parkville, Victoria 3052, Australia
Professor C.C. Heyde Department of Statistics, University of Melbourne, Parkville, Victoria 3052, Australia

1   Introduction to Linear and Convex Programming,   N. CAMERON
2   Manifolds and Mechanics,   A. JONES, A GRAY & R. HUTTON
3   Introduction to the Analysis of Metric Spaces,   J. R. GILES
4   An Introduction to Mathematical Physiology and Biology,   J. MAZUMDAR
5   2-Knots and their Groups,   J. HILLMAN
6   The Mathematics of Projectiles in Sport,   N. DE MESTRE
7   The Petersen Graph,   D. A. HOLTON & J. SHEEHAN
8   Low Rank Representations and Graphs for Sporadic Groups,
                                C. PRAEGER & L. SOICHER
9   Algebraic Groups and Lie Groups,   G. LEHRER (ed)

Australian Mathematical Society Lecture Series. 9

Algebraic Groups and Lie Groups

A volume of papers in honour of the late R. W. Richardson

Edited by

Gus Lehrer
University of Sydney

with

A.L. Carey
University of Adelaide

J.B. Carrell
University of British Columbia

M.K. Murray
University of Adelaide

T.A. Springer
University of Utrecht

CAMBRIDGE
UNIVERSITY PRESS

CAMBRIDGE UNIVERSITY PRESS
Cambridge, New York, Melbourne, Madrid, Cape Town, Singapore, São Paulo

Cambridge University Press
The Edinburgh Building, Cambridge CB2 2RU, UK

Published in the United States of America by Cambridge University Press, New York

www.cambridge.org
Information on this title: www.cambridge.org/9780521585323

First published 1997

A catalogue record for this publication is available from the British Library

ISBN-13 978-0-521-58532-3 paperback
ISBN-10 0-521-58532-5 paperback

Transferred to digital printing 2005

Table of Contents

ARMAND BOREL — Class functions, conjugacy classes and commutators in semisimple Lie groups ... 1

M. BRION — Curves and divisors in spherical varieties ... 21

A. L. CAREY, P. E. JUPP AND M. K. MURRAY — The phylon group and statistics ... 35

JAMES B. CARRELL — The span of the tangent cone of a Schubert variety ... 51

R. W. CARTER — Canonical bases, reduced words, and Lusztig's piecewise-linear function ... 61

W. A. CASSELMAN — Geometric rationality of Satake compactifications ... 81

EDWARD CLINE, BRIAN PARSHALL AND LEONARD SCOTT — Graded and non-graded Kazhdan-Lusztig theories ... 105

C. DE CONCINI AND C. PROCESI — Quantum Schubert cells and representations at roots of 1. ... 127

A. DIMCA AND G. I. LEHRER — Purity and equivariant weight polynomials ... 161

STEPHEN DONKIN — The restriction of the regular module for a quantum group ... 183

M. J. DYER — On coefficients of q in Kazhdan-Lusztig polynomials ... 189

A. F. M. TER ELST AND DEREK W. ROBINSON — Spectral estimates for positive Rockland operators ... 195

C. K. FAN AND G. LUSZTIG — Factorization of Certain Exponentials in Lie Groups ... 215

MICHAEL FIELD — Symmetry breaking for equivariant maps ... 219

JENS CARSTEN JANTZEN — Low dimensional representations of reductive groups are semisimple ... 255

D. LUNA — Grosses cellules pour les variétés sphériques ... 267

GEORGE LUSZTIG — Total positivity and canonical bases ... 281

VLADIMIR POPOV AND GERHARD RÖHRLE — On the Number of Orbits of a Parabolic Subgroup on its Unipotent Radical ... 297

GERALD W. SCHWARZ — On a homomorphism of Harish-Chandra ... 321

PETER SLODOWY — Two notes on a finiteness problem in the representation theory of finite groups ... 331

T. A. SPRINGER — A description of B-orbits on symmetric varieties. ... 349

ROBERT STEINBERG — Nagata's Example ... 375

PREFACE

This volume of research articles on the theme of group actions and geometry in a Lie theoretic context is dedicated to the memory of Roger Richardson, whose research illuminated the deep connections between these topics.

Roger Wolcott Richardson spent the last fifteen years of his life at the Australian National University. At the time of his death on 15th June, 1993 he had become a leading figure in Australian Mathematics, having founded (in 1989) the annual Australian Lie Groups Conference. After his death it was decided that a volume of research articles should be published by the Australian Mathematical Society as a tribute to Richardson. An editorial committee consisting of G. I. Lehrer (Chair), A. L. Carey, J. B. Carrell, M. K. Murray, and T. A. Springer decided to invite a group of his friends and colleagues to submit "research papers of the highest quality" for the volume. This volume, which aptly reflects his interests, is the result.

After initial training in physics and a long collaboration with Albert Nijenhuis on deformations of Lie algebras, Richardson's research was concerned with various aspects of actions of Lie groups and algebraic groups on manifolds and algebraic varieties. He wrote on algebraic, geometric, analytic and topological questions in this context and his work influenced the development of each aspect. The fact that many branches of mathematics are involved in this work reflects its central nature; most of these subjects are represented here.

Richardson's interests (and hence the contents of this volume) may be divided into six broad categories. They are: algebraic geometry (particularly geometric questions relating to algebraic group actions), algebraic and Lie groups (structure, properties, conjugacy), representation theory (of algebraic groups, Lie groups, reflection groups), invariant theory or the study of orbits of (particularly algebraic and Lie) group actions, symmetry breaking and analytical aspects of Lie groups.

Of course many of the articles in this volume belong to more than one category, particularly those which relate to representation theory, which is now closely linked with geometry in many ways. Particular aspects of this relationship which occur in this volume are Kazhdan-Lusztig theory (the papers of Dyer and Cline-Parshall-Scott) and canonical bases, addressed in the work of Carter, De Concini-Procesi, Lusztig and Fan-Lusztig, the last of which gives an independent application of ideas occurring in canonical basis theory. Representations and cohomology are directly addressed in Dimca-Lehrer and indirectly in several of the papers since Kazhdan-Lusztig polynomials may be thought of as Poincaré polynomials associated with certain perverse sheaves on Schubert varieties.

The area of algebraic geometry, or more particularly geometric aspects of algebraic group actions is represented by the papers of Brion and Luna on the theory of spherical varieties, which are far reaching generalisations of toric varieties, in the sense that they are varieties on which a reductive group acts so that a Borel subgroup has finitely many orbits, while toric varieties are those spherical varieties where the reductive group involved is a torus. Also in this area are the papers of Carrell on the geometry of Schubert varieties, Casselman on compactifications of symmetric spaces, Dimca-Lehrer on Hodge purity for algebraic varieties and representation theory and Springer on B-orbits in spherical symmetric varieties. The latter work particularly addresses the combinatorial structure of the set of orbits.

Algebraic groups and their associated geometries are the subject of the papers of Borel (conjugacy and coverings), De Concini-Procesi (quantum Schubert cells), Donkin (the regular module of a quantum group), Lusztig (total positivity) and Fan-Lusztig (factorization of exponentials). Some of the articles pertaining to representation theory have been mentioned above. Others are those of Jantzen on semisimplicity of low-dimensional representations and Slodowy on group representations into an algebraic group. In the field of orbit theory and invariant theory, the papers of Popov-Roehrle on the finiteness of the set of orbits of a parabolic subgroup on its unipotent radical and that of Springer (mentioned above) build on earlier joint work with Richardson. Also in the field are the papers of Schwarz on differential invariants and Steinberg on Nagata's counterexample.

Richardson's work on symmetry breaking was done in collaboration with Field whose contribution to this volume is also related to that collaboration. Finally, analytical aspects of Lie groups are addressed in the papers of Carey-Jupp-Murray (the phylon group) and ter Elst-Robinson (Rockland operators). Both of these represent work done in the context of projects sponsored by the Australian Research Council in which Richardson was involved.

G. I. Lehrer
The University of Sydney, December 1995

CLASS FUNCTIONS, CONJUGACY CLASSES AND COMMUTATORS IN SEMISIMPLE LIE GROUPS

ARMAND BOREL

To the memory of Roger Richardson

INTRODUCTION

Let G be a connected Lie group, $\pi : H \to G$ a covering of G. Let us say that the conjugacy class $C(g) = \{x.g.x^{-1} | x \in G\}$ of $g \in G$ splits in H if $\pi^{-1}(C(g))$ is the disjoint union of the conjugacy classes in H of the elements of $\pi^{-1}(g)$. J. Milnor pointed out that if $G = \mathbf{SL}_2(\mathbb{R})$ every conjugacy class splits completely in any covering. As a generalization and sharpening, we prove (see 1.2 for a slightly stronger statement and 1.3).

Theorem I. *Let G be a connected semisimple Lie group with finite center and $\pi : H \to G$ a covering.*

(i) *If $\pi_1(G)$ is torsion free, any conjugacy class in G splits in H.*

(ii) *If $\pi_1(G)$ has torsion and H is the universal covering of G, there is a conjugacy class in G which does not split.*

The first condition is fulfilled notably in the following case

(∗) G is linear, has no compact factor, its complexification is simply connected, and its quotient G/K by a maximal compact subgroup K is a bounded symmetric domain.

(see 1.4). This holds in particular for the symplectic group $\mathbf{Sp}_{2n}(\mathbb{R})$ $(n \geqq 1)$. In that case, Milnor had also established the splitting of conjugacy classes in the universal covering by constructing a continuous class function on $\mathbf{Sp}_{2n}(\mathbb{R})$ which extends the determinant function of $K = \mathbf{U}(n)$. This, and a similar construction by G. Lusztig on an indefinite special unitary group, led to ask whether any continuous class function on K extends to one on G. We shall prove it more generally for linear reductive groups of inner type (3.5).

Theorem II. *Let G be a linear real reductive group of inner type (see 2.1), K a maximal compact subgroup and M a topological space. Then any M-valued continuous class function f on K extends to one on G.*

This extension is not necessarily unique, but the proof will exhibit one constructed canonically, with the same set of values. To describe it, we use a slight sharpening of the (multiplicative) Jordan decomposition in a real linear algebraic group, discussed in §2, which states that any $g \in G$ can be written uniquely as a product of three commuting elements g_e, g_d and g_a where g_e is "elliptic" (i.e. contained in a compact subgroup), g_d is contained in the identity component of a

1

split torus, and g_u is unipotent. In fact, this is carried out slightly more generally for the class of "almost algebraic groups" (see 2.1), which includes the reductive groups of inner type.

The canonical extension F of f is then defined by $F(g) = f(C(g_e) \cap K)$. It is well defined since $C(g_e) \cap K$ is known to consist of exactly one conjugacy class in K (see 3.4) and the main point is to show that it is continuous. This follows directly from 3.3, which is the main technical result of this paper.

It is known that a connected semisimple linear Lie group G has finite *commutator length*, i.e. there exists an integer N such that any element in G is a product of at most N commutators (4.2). Milnor pointed out (see [20]) that this is not true anymore in the universal covering of $\mathbf{SL}_2(\mathbb{R})$. As a generalization, we shall show in §4 (see 4.2, 4.4):

Theorem III. *Let G be a connected semisimple Lie group. Then G has finite commutator length if and only if its center is finite.*

As can be seen from the above, the theorems proved here were mostly suggested by remarks of Milnor on $\mathbf{SL}_2(\mathbb{R})$ or $\mathbf{Sp}_{2n}(\mathbb{R})$. Those of §§1, 2, 3 were presented at an Oberwolfach meeting in 1976, in a somewhat less general form, but were not until now written up for publication.

0. NOTATION AND CONVENTIONS

0.1. Let G be a group. Then $\mathscr{D}G$ denotes its derived group and $\mathscr{C}G$ its center. If A is a subset of G, then $\mathscr{Z}_G(A)$ or $\mathscr{Z}(A)$ is its centralizer and $\mathscr{N}_G(A)$ or $\mathscr{N}(A)$ its normalizer:

$$\mathscr{Z}_G(A) = \{g \in G | g.a = a.g \ (a \in A)\}, \quad \mathscr{N}_G(A) = \{g \in G | g.A = A \cdot g\}.$$

The inner automorphism $x \mapsto g.x.g^{-1}$ of G defined by g is denoted Int g. We also write $^x g$ for $x.g.x^{-1}$ and, if $A \in G$, $^G A$ for $\{gag^{-1} | g \in G, a \in A\}$.

The conjugacy class of g in G is denoted $C(g)$ or $C_G(g)$.

0.2. If G is a Lie group, then G^o denotes its connected component of the identity. If G is an algebraic group over $\mathbb{C}$, it is also the identity component of G in the Zariski topology. In this paper, all algebraic groups are linear, over $\mathbb{C}$. If $G \subset \mathbf{GL}_n(\mathbb{R})$, we let G_c be the complexification of G, or, equivalently, the smallest complex Lie subgroup of $\mathbf{GL}_n(\mathbb{C})$ containing G.

The Lie algebra of a Lie group is usually denoted by the corresponding German letter, and similarly for an algebraic group.

As usual, Ad_G is the adjoint representation of G in its Lie algebra. It associates to g the differential at 1 of Int g.

0.3. Let G be a Lie group with finitely many connected components. We recall that any compact subgroup of G is contained in a maximal one, the maximal ones are conjugate under G^o and G is topologically the product of any one of them by a euclidean space [13].

1. Inverse Images of Conjugacy Classes

In this section, G and H are groups and $\pi : H \to G$ an epimorphism with kernel $N \subset \mathscr{C}H$.

Lemma 1.1. *Let $g \in G$ and $Q_g = \pi^{-1}(\mathscr{L}_G(g))$.*

(a)　*The following conditions are equivalent*

　　(i)　$Q_g = \mathscr{L}_H(h)$ *for some* $h \in \pi^{-1}(g)$.
　　(ii)　$Q_g = \mathscr{L}_H(h)$ *for all* $h \in \pi^{-1}(g)$.
　　(iii)　$C_G(g)$ *splits in* H.
　　(iv)　π *induces a bijection of* $C_H(h)$ *onto* $C_G(g)$ *for every* $h \in \pi^{-1}(g)$.

(b)　*Let $h \in \pi^{-1}(g)$. The restriction to Q_g of the commutator map*

$$c_h : x \mapsto x.h.x^{-1}.h^{-1}$$

is a homomorphism of Q_g into N, with kernel $\mathscr{L}_H(h)$.

Proof. Obviously, $\mathscr{L}_H(h) = \mathscr{L}_H(h.c)$ for all $h \in H$ and $c \in \mathscr{C}H$, hence (i) $\Leftrightarrow$ (ii) and

$$\mathscr{L}_H(h) \subset Q_g \quad (h \in \pi^{-1}(g)). \tag{1}$$

On the other hand

$$x.h.x^{-1} = h.n \quad (x \in H,\ h \in \pi^{-1}(g),\ n \in N) \Leftrightarrow x \in Q_g \tag{2}$$

and

$$n \neq 1 \text{ in (2)} \Leftrightarrow x \notin \mathscr{L}_H(h). \tag{3}$$

This shows that $(i) \Leftrightarrow (iii)$. Let $x, y \in H$ and $h \in \pi^{-1}(g)$. Then

$$x.h.x^{-1} = y.h.y^{-1} \Leftrightarrow x \in y.\mathscr{L}_H(h) \tag{4}$$

$$\pi(x.h.x^{-1}) = \pi(y.h.y^{-1}) \Leftrightarrow x \in y.Q_g, \tag{5}$$

whence $(i) \Leftrightarrow (iv)$. This proves (a).

(b) The map c_h is constant on the coset $x.\mathscr{L}_H(h)$. If $x \in Q_g$, then $c_h(x) \in N$ by (2) and then it is clear that c_h is a homomorphism of Q_g into N, with kernel $\mathscr{L}_H(h)$.

Theorem 1.2. *Assume that G and H are connected semisimple Lie groups and that $\mathscr{C}G$ is finite.*

(i)　*If N is torsion-free, then $C_G(g)$ splits in H for every $g \in G$.*
(ii)　*If N has torsion and the maximal compact subgroups of H are semisimple then there exists $g \in G$ such that $C_G(g)$ does not split in H.*

Proof. (i) Let $g \in G$, let $h \in \pi^{-1}(g)$ and $Q_g = \pi^{-1}(\mathscr{L}_G(g))$. By 1.1, it suffices to show that $Q_g/\mathscr{L}_H(h)$ is a finite group. Both groups contain N, hence $Q_g/\mathscr{L}_H(h) \cong \mathscr{L}_G(h)/\pi(Z_H(h))$. By 1.1(b), $Q_g^\circ = \mathscr{L}_H(h)^\circ$, hence $\big(\pi(\mathscr{L}_H(h))\big)^\circ = \mathscr{L}_G(g)^\circ$. Therefore $Q_g/\mathscr{L}_H(h)$ is isomorphic to a quotient of $\mathscr{L}_G(g)/\mathscr{L}_G(g)^\circ$ and it suffices to prove that $\mathscr{L}_G(g)$ has finitely many connected components.

Let G' be the adjoint group of G and $\sigma : G \to G'$ the canonical projection. Its kernel is $\mathscr{C}G$ and is finite by assumption. It suffices therefore to show that $\mathscr{Z}_{G'}(\sigma(g))$ has finitely many connected components. This reduces us to the case where G is linear. Let G_c be the complexification of G. It is a connected semisimple algebraic group and $G = G_c(\mathbb{R})^\circ$. The group $\mathscr{Z}_G(g)$ is equal to the group of real points of the algebraic group $\mathscr{Z}_{G_c}(g)$, hence has finitely many connected components ([19] or [3], 6.4). The same is then true of $\mathscr{Z}_{G_c}(g) \cap G = \mathscr{Z}_G(g)$.

(ii) Assume that N contains an element $z \neq 1$ of finite order. Any maximal compact subgroup K of H contains z. By assumption, K is semisimple. It is connected, since H is. By [8] there exists $a, b \in K$ such that $a.b.a^{-1}.b^{-1} = z$. We have then $z.b = a.b.a^{-1}$, which shows that $C_G(\pi(b))$ does not split in H (see 1.1).

1.3. Proof of Theorem I. Let $\tilde{\pi} : \tilde{G} \to G$ be the universal covering of G. Then $\pi_1(G) = \ker \tilde{\pi}$. If H is a covering of G, then it also has $\tilde{G}$ as a universal covering and $\pi_1(H)$ is a subgroup of $\pi_1(G)$. Assume the latter is torsion-free. Then so is $\pi_1(H)$. By 1.2(i) any conjugacy class in H or G is the bijective image of a conjugacy class in $\tilde{G}$ under the natural projection. The same is then true for $\pi : H \to G$. In view of 1.1, this implies (i). The maximal compact subgroups of $\tilde{G}$ are simply connected (0.3), in particular semisimple. Therefore (ii) is a special case of 1.2(ii).

Proposition 1.4. *Assume that G is a connected semisimple linear Lie group without compact factor, whose quotient by a maximal compact subgroup K is a bounded symmetric domain and that the complexification G_c of G is simply connected. Then $\pi_1(G)$ is torsion-free.*

Proof. Let S be the identity component of the center of K. The assumptions on G and G/K imply that $K = \mathscr{Z}_G(S)$. Furthermore, K is a maximal compact subgroup of the centralizer M of S in G_c. and $\mathscr{D}K$ a maximal compact subgroup of $\mathscr{D}M$. Since G_c is simply connected, it is known that $\mathscr{D}M$ is simply connected, hence so is $\mathscr{D}K$. But K is homeomorphic to the product of $\mathscr{D}K$ and of a torus, [1, 3.2], as already recalled. Hence $\pi_1(K)$ has no torsion.

2. JORDAN DECOMPOSITION IN ALMOST ALGEBRAIC LIE GROUPS

2.1. Let G be a Lie group with finitely many connected components. It is said to be *almost real algebraic* if there is given a linear algebraic $\mathbb{R}$-group H and a morphism of Lie groups $\sigma : G \to H(\mathbb{R})$ with finite central kernel, and image open in $H(\mathbb{R})$.

We view H and σ as part of the data, although the notation may not always indicate it. If (G', σ', H') is another such triple, a morphism $\varphi : G \to G'$ consists in fact of a morphism of Lie groups $\varphi : G \to G'$, and of a $\mathbb{R}$-morphism of algebraic groups $\psi : H \to H'$ such that $\sigma' \circ \varphi = \psi \circ \sigma$.

The Lie group G is *reductive* if it is almost algebraic and H° is a reductive algebraic group, i.e. has a unipotent radical reduced to $\{1\}$. It is *reductive of inner type* i.e. if $\mathrm{Ad}_H \sigma(G) \subset \mathrm{Ad}_H H^\circ$. The notion of reductive group of inner type is the same as that of ([18], 2.11, 2.28) and essentially equivalent to that of reductive group in [10, §3], also adopted in [17].

Our object of interest is G so, replacing H by a subgroup of finite index if needed, we may always assume $\sigma(G)$ to be Zariski dense in H. Note that, since H is of inner type, $\mathscr{C}G^\circ \subset \mathscr{C}G$ and $\mathscr{C}H^\circ \subset \mathscr{C}H$; moreover, $\mathscr{Z}_H(H^\circ)$ meets every connected component of H. The subgroup $\mathscr{Z}_H(H^\circ)$ is normal in H, defined over $\mathbb{R}$ and $H/\mathscr{Z}_H(H^\circ)$ is defined over $\mathbb{R}$, connected, semisimple of adjoint type. The group $\left(\mathscr{Z}_G(G^\circ)\right)^\circ$ is a real form of $\mathscr{Z}_H(H^\circ)^\circ$, the group $\sigma\left(\mathscr{Z}_G(G^\circ)\right)$ is of finite index in $\mathscr{Z}_H(H^\circ)(\mathbb{R})$, the map σ induces a morphism $\sigma' : G/\mathscr{Z}_G(G^\circ) \to H/\mathscr{Z}_H(H^\circ)$ and $\left(G/\mathscr{Z}_G(G^\circ),\, H/\mathscr{Z}_H(H^\circ),\, \sigma'\right)$ is an almost algebraic semisimple group of inner type.

2.2. Let G be almost algebraic and $g \in G$.

(a) A closed connected subgroup A is a *split toral subgroup* if $\sigma(A) = S(\mathbb{R})^\circ$, where S is a $\mathbb{R}$-split torus in H. Clearly, A is the identity component of $\sigma^{-1}\left(S(\mathbb{R})\right)^\circ$ and σ is an isomorphism of A onto its image $S(\mathbb{R})^\circ$.

(b) g is said to be *split positive* if it belongs to a split toral subgroup. Then g is the only split positive element in $\sigma^{-1}\left(\sigma(g)\right)$. The element g is split positive if and only if $\sigma(g)$ is diagonalisable over $\mathbb{R}$ and has strictly positive eigenvalues.

(c) g is *unipotent* if either $g = 1$ or $\sigma(g)$ is unipotent and $\neq 1$. Let X be the unique element in $\mathfrak{g}$ such that $\sigma(g) = \exp \sigma(X)$. Then $g = \exp X$ and is the unique unipotent element in $\sigma^{-1}\left(\sigma(g)\right)$.

(d) g is *elliptic* if it belongs to a compact subgroup of G, and semisimple if $\sigma(g)$ is so in H.

2.3. Many of the properties of semisimple groups (Cartan decomposition, Iwasawa decomposition etc.) extends to reductive groups of inner type, as is shown in those references given in 2.1. In particular we shall use the following:

The maximal compact subgroups of G are conjugate under G°. Let K be one. There is a unique involution θ_K of G with fixed point set K, called the Cartan involution of G with respect to K. Let $\mathfrak{p}$ be the (-1)-eigenspace of in $\mathfrak{g}$. Then $(k, X) \mapsto k.\exp X$ is an isomorphism of manifolds of $K \times \mathfrak{p}$ onto G. In particular $\exp \mathfrak{p} = P$ is closed. Any split toral subgroup of G is conjugate under K to a split toral subgroup contained in P. The maximal ones are conjugate under K. If G is commutative, then P is the biggest split toral subgroup of G and $G = K \times P$.

Let M be a subgroup of G which is stable under θ_K. Then M is reductive, and the restriction of θ_K to M is a Cartan involution of M. In particular $M \cap K$ is a maximal compact subgroup of M, and we have the Cartan decomposition $M = (M \cap K).\exp(\mathfrak{m} \cap \mathfrak{p})$. For this, see [11].

Proposition 2.4. *Let G be an almost algebraic group and $g \in G$.*

(i) *There exist unique elements g_s, g_u such that $g = g_s.g_u = g_u.g_s$ and g_s (resp. g_u) is semisimple (resp. unipotent).*

(ii) *Let g be semisimple. There exist unique elements g_e and g_d in G such that $g = g_e.g_d = g_d.g_e$ and g_e (resp. g_d) is elliptic (resp. split positive).*

If G is linear, g is elliptic (resp. split positive) if and only its eigenvalues are of modulus one (resp. real and > 0).

Proof. (i) Let $g' = \sigma(g)$. Then $g' = g'_s.g'_u$ with g'_u unipotent, g'_s semisimple and $g'_s g'_u = g'_u g'_s$ by the usual Jordan decomposition ([2], 4.4). If $g'_u \neq 1$, there exists a unique unipotent g_u in the inverse image of g'_u. It commutes with g. Then $g_s = g.g_u^{-1}$ commutes with g, and is semisimple since $\sigma(g_s) = g'_s$. This proves the existence of one decomposition. If $g = \tilde{g}_s.\tilde{g}_u$ is another one, then $\sigma(\tilde{g}_s) = g'_s$ and $\sigma(\tilde{g}_u) = g'_u$ by the uniqueness of the Jordan decomposition in H, whence $\tilde{g}_u = g_u$ and then $\tilde{g}_s = g_s$.

(ii) Let g be semisimple and let M be the smallest algebraic subgroup of H containing $\sigma(g)$. If $M(\mathbb{R})$ is compact, then so is its inverse image and g is elliptic. Assume it is not. It is defined over $\mathbb{R}$. and is of multiplicative type, i.e. it consists of semisimple elements and is commutative. Let S be the greatest $\mathbb{R}$-split torus contained in M and $A' = S(\mathbb{R})^\circ$. The group $M(\mathbb{R})$ is the direct product of A' and its greatest compact subgroup K'. Then $\sigma^{-1}\big(M(\mathbb{R})\big) = K \times A$, where K is compact, A is the identity component of $\sigma^{-1}(A')$ and is a split toral subgroup. Then $g = g_e.g_d$ with $g_e \in K$, hence elliptic, $g_d \in A$, hence split positive, and $g_e.g_d = g_d.g_e$. In view of its construction, $\sigma^{-1}\big(M(\mathbb{R})\big)$ is invariant under any automorphism of G leaving g and $\ker \sigma$ stable. In particular, any element of G commuting with g also commutes with g_e and g_d. Let now $g = g'_e.g'_d$ be another such decomposition. Then g_e, g_d, g'_e, g'_d commute pairwise. Assume first that $G \subset H(\mathbb{R})$. Then $g'_e g_e^{-1} = g_d g'^{-1}_d$. The right-hand side is diagonalisable over $\mathbb{R}$, with strictly positive eigenvalues, whereas the left-hand side is diagonalisable over $\mathbb{C}$, with eigenvalues of modulus one. Hence both sides are equal to 1, which proves the uniqueness of the decomposition in that case. Coming back to the general case, we see that $\sigma(g'_d) = \sigma(g_d)$, whence $g'_d = g_d$ (see 2.2 (a)) and the uniqueness. This proves the first assertion. If G is linear, then σ is the identity, and the second assertion follows, too.

Remark. A decomposition similar to (ii) is given in [17], part II, p. 41.

Corollary 2.5. *Given $g \in G$, there exist unique commuting elements g_e, g_d, g_u such that $g = g_e.g_d.g_u$, where g_e (resp. g_d, resp. g_u) is elliptic (resp. split positive, resp. unipotent).*

This decomposition will be referred to as the *refined Jordan decomposition*. It is invariant under morphisms (as defined in 2.1).

2.6. Let G be reductive of inner type. Let N be a maximal unipotent subgroup of G, necessarily contained in G°, A a maximal split toral subgroup normalizing N and K a maximal compact subgroup. Then $k \times a \times n \mapsto k.a.n$ is a diffeomorphism of $K \times A \times N$ onto G (Iwasawa decomposition, see e.g. [18], 2.1.8). Moreover $M.A.N$, where $M = \mathcal{Z}_K(A)$, is a minimal parabolic subgroup of G and any split toral subgroup (resp. unipotent subgroup) of G is conjugate to a subgroup of A (resp. N).

Let $g \in G$. As a consequence of the above, we see that if $g_e \in \mathscr{C}G$, then g is conjugate to an element $h \in M.A.N$ such that $h_e = g_e$, $h_d \in A$ and $h_u \in N$.

3. EXTENSION OF CERTAIN CLASS FUNCTIONS IN LINEAR REDUCTIVE GROUPS OF INNER TYPE

The following lemma is undoubtedly well-known. Not knowing of a reference, we give a proof for the sake of completeness.

Lemma 3.1. *Let L be a Lie group, $g \in L$ such that $\operatorname{Ad} g$ is semisimple, $Z = \mathscr{Z}_L(g)$ and $\mathfrak{m}$ a supplement to the Lie algebra $\mathfrak{z}$ of $\mathscr{Z}_L(g)$ in $\mathfrak{g}$ which is stable under $\operatorname{Ad} g$. Let $\tau : \mathfrak{m} \times Z \to G$ be the map defined by $\tau(X, z) = \operatorname{Ad} \exp X(z)$ ($X \in \mathfrak{m}, z \in Z$). Then its differential $d\tau_0$ at $(0, g)$ is invertible.*

Proof. The tangent space to $\mathfrak{m} \times Z$ (resp L) at $(0, g)$ (resp. g) is $\mathfrak{m} \oplus \mathfrak{z}.g$ (resp. $\mathfrak{g}.g$). Let $X \in \mathfrak{m}, Y \in \mathfrak{z}$. Then

$$\tau(sX, e^{tY}) = e^{sX}.g.e^{tY}.e^{-sX}.g^{-1}.g \qquad (s, t \in \mathbb{R}).$$

From

$$d\tau_0(X, Y) = \frac{d^2\tau}{ds\,dt} e^{sX}.e^{tY}.g e^{-sX}.g^{-1}.g \Big|_{s=t=0}$$

we get

$$d\tau_0(X, Y) = (I - \operatorname{Ad} g)X + Y.$$

Since $\operatorname{Ad} g - I$ is invertible on $\mathfrak{m}$, this proves the lemma.

Lemma 3.2. *Let G be a reductive linear group, $a \in \mathscr{C}G$ an elliptic element, $b \in G$ a semisimple element, $x_n, g_n \in G$ ($n = 1, 2...$) such that $x_{n,d} = 1$, $x_{n,e} \to a$, $g_n.x_n g_n^{-1} \to b$. Then $a = b$.*

Let K be a maximal compact subgroup of G containing a. The eigenvalues of b, in the given linear embedding of G, are limits of those of x_n, hence of $x_{n,e}$ and are therefore of modulus one. Consequently, b is elliptic (2.4) and there exists $g \in G$ such that $b' = g.b.g^{-1} \in K$. Let H be the complexification of G. Let L be a maximal compact subgroup of H containing K. It is a real form of H. By the Peter-Weyl theorem, the conjugacy classes in L are separated by the characters of the irreducible finite dimensional continuous (or real analytic) representations of L. By Tannaka duality (see [6], Chapter VI) the characters of the real analytic representations of L are the restrictions of linear combinations of those of the holomorphic or antiholomorphic finite dimensional irreducible representations of H. If σ is one and χ its character, then

$$\chi(b) = \lim_{n \to \infty} \chi(g_n.x_n.g_n^{-1}) = \lim_{n \to \infty} \chi(g_n.x_{n,e}.g_n^{-1}) = \lim_{n \to \infty} \chi(x_{n,e}) = \chi(a)$$

therefore a and b are conjugate in L. Since a is central also in H, this implies $b = a$.

The following theorem is the main technical result of this paper. Lemma 3.2 is a special case except for the fact that G is not assumed there to be of inner type.

Theorem 3.3. *Let G be a real linear reductive Lie group of inner type (see 2.1) and K a maximal compact subgroup of G. Let $a \in K$, $y \in G$ and $b = y_e$. Let g_n, x_n ($n = 1, 2...$) be sequences of elements in G such that $x_{n,e} \to a$ and $g_n.x_n.g_n^{-1} \to y$. Then a and b are conjugate in G.*

If G is compact, then $y = b$, and $x_n = x_{n,e}$. We may assume that $g_n \to g_o$ and get $b = g_o.a.g_o^{-1}$, which proves the theorem. From now on, G is assumed to be non compact. We let H be the complexification of G. Then G is Zariski dense in H and of finite index in $H(\mathbb{R})$.

The proof consists in a reduction in several steps to a special case where H is semisimple, connected, in which we can avail ourselves of a theorem of [9].

a) *We may assume* $b \in K$. Let $s \in G$ be such that $^s b \in K$. We have $(^s y)_e = {}^s y_e = {}^s b$. Moreover

$$^s y_n = (s g_n).x_n.(s g_n)^{-1} \to {}^s y.$$

The theorem, assumed to be proved if $b \in K$, shows that a is a conjugate to $^s b$, hence also to $^s b$.

b) *We may assume in addition that* $y = y_s$ *and* $g_n \in G^o$.

Let $\{V_j\}(j = 1, 2 \dots)$ be a decreasing sequence of neighborhoods of 1 converging to $\{1\}$. We claim we can find $z_j \in \mathscr{Z}_G(y_s)$ such that $z_j.y_u.z_j^{-1} \in V_j$ $(j = 1, 2 \dots)$.

If $y_u = 1$ there is nothing to prove. So assume $y_u \neq 1$. There exists then a unique $X \in \mathfrak{g}$ such that $\exp X = y_u$. The element X is nilpotent and belongs to the Lie algebra of the derived group of $\mathscr{Z}(g_s)$, which is semisimple. By the Jacobson-Morosow theorem, we can find in $\mathscr{D}\big(\mathscr{Z}(y_s)\big)$ a three-dimensional $\mathbb{R}$-split simple group containing $\exp \mathbb{R}X$ and a split toral subgroup D normalizing $\exp \mathbb{R}X$. The elements of D act by dilations on X under the adjoint representation. For any strictly positive real number r there is $d \in D$ such that $\mathrm{Ad}\,d$ is the dilation by r on X; the existence of the z_j's follows. Then $^{z_j}y \to y_s$. Choose a neighborhood W_j of y such that $^{z_j}W_j \subset y_s.V_j$. If $n(j)$ is such that

$$g_{n(j)}.x_{n(j)}\,g_{n(j)}^{-1} \in W_j.$$

Then

$$z_j.g_{n(j)}.x_{n(j)}.\,g_{n(j)}^{-1}.z_j^{-1} \to y_s.$$

Changing the notation, we may assume that our original sequence is this subsequence. So we have $g_n.x_n.g_n^{-1} \to y_s$. The group G has finitely many connected components. Passing again to a subsequence, we may assume that there exists $c \in G$ such that $g_n = c.g_n'$ with $g_n' \in G^o$. Then

$$g_n'.x_n.g_n'^{-1} \to c^{-1}.y_s.c.$$

The proposition, assumed to be proved under the assumption b), shows that a is conjugate to $c^{-1}.b.c$, hence also to b.

c) Assume G^o to be *commutative*. Then it is central in G, as already pointed out. Therefore we have $x_{n,e} \to a$, and $x_{n,d}.x_{n,e} \to y$. Also (see 2.3), G^o is the direct product of a compact subgroup K and a split toral subgroup A, therefore $x_{n,d} \to y_d$ and $x_{n,e} \to y_e = b$, whence $a = b$. In particular, this proves 3.3 when $\dim G \leqq 2$. From now on, we assume the theorem established in dimensions $< \dim G$.

d) *We may assume in addition that* $\mathscr{C}G$ *is compact.*

Assume $\mathscr{C}G$ is not compact. Since G is of inner type, $\mathscr{C}G^{\circ}$ and $\mathscr{C}G$ have the same connected component of the identity, and the same is true for $H(\mathbb{R})$ in the ordinary topology and for H in the Zariski topology.

Let S be the greatest split torus of the center of H. Since $\mathscr{C}H(\mathbb{R})$ is not compact, $\dim S \geq 1$. Then $A = S(\mathbb{R})$ is the greatest central split toral subgroup of G. We claim that there exists a normal subgroup G_1 of G such that $G = G_1 \times A$. Let L be the greatest compact subgroup of $\mathscr{C}G^{\circ}$. Then $\mathscr{C}G^{\circ} = L \times A$. The algebra $\mathfrak{g}$ being the direct sum of its derived algebra and of its center, we see that the group G_1° generated by L and $\mathscr{D}G^{\circ}$ is the almost direct product of these groups and $G^{\circ} = G_1^{\circ} \times A$. The group G_1° is invariant in G. We let $G_1 = K.G_1^{\circ}$; then G_1° is indeed the identity component of G_1. Since $\mathscr{C}G_1^{\circ} \subset \mathscr{C}G^{\circ}$ we have $\mathscr{C}G_1^{\circ} \subset \mathscr{C}G_1$. The group G_1° is normal in G°, hence its Zariski closure H_1° in H is normal in H. For $h \in H^{\circ}$ the restriction of $\mathrm{Ad}\, h$ to the Lie algebra of H_1 belongs to $\mathrm{Ad}\, H_1^{\circ}$. It follows that G_1 is reductive, of inner type, and of course linear. We can write uniquely $y = y' \cdot z$ with $y' \in G_1$ and $z \in A$. Then $y'_e = y_e = b$. Similarly we have

$$x_n = x'_n.z_n \qquad (x'_n \in G_1, z_n \in A, x'_{n,e} = x_{n,e}, n = 1, 2, \ldots)$$
$$g_n = g'_n.q_n \qquad (g'_n \in G_1, q_n \in A, n = 1, 2 \ldots)$$

hence

$$x'_{n,e} \to a, \quad g'_n.x'_n.g'^{-1}_n \to y',$$

so that the theorem, applied to G_1, as is allowed by our induction assumption, shows that a is conjugate to b already in G_1.

e) *We may assume in addition that* $x_{n,d} = 1$.

Assume that $x_{n,d} \neq 1$ for infinitely many n's, hence, going over to a subsequence, for all n's. Let S be a maximal $\mathbb{R}$-split torus of H. Since $\mathscr{C}G^{\circ}$ is compact, S belongs to $\mathscr{D}H^{\circ}$. Then $A = S(\mathbb{R})^{\circ}$ is a maximal split toral subgroup of G, belongs to $\mathscr{D}G^{\circ}$ and $A \cap \mathscr{C}G = \{1\}$.

Let θ be the Cartan involution of G with respect to K and $\mathfrak{p}$, P be as in 2.3. As recalled there we may, possibly after conjugation by an element of K, assume $A \subset P$. Let $Z_n = \mathscr{Z}_G(x_{n,e})$. It is stable under θ, since $x_{n,e} \in K$, reductive and contains x_n, $x_{n,d}$. The restriction of θ to Z_n is a Cartan involution and

$$Z_n = (K \cap Z_n).(P \cap Z_n)$$

is a Cartan decomposition, i.e. $K_n = K \cap Z_n$ is maximal compact in Z_n and $P \cap Z_n = \exp .\mathfrak{p}_n$, where $\mathfrak{p}_n = \mathfrak{p} \cap \mathfrak{z}_n$. There exists $u_n \in K_n$ such that $^{u_n}x_{n,d} \in P$ and $k_n \in K$ such that $k_n.u_n.x_{n,d}.u_n^{-1}.k_n^{-1} \in A$ (see 2.3).

Passing to a subsequence, we may assume that $k_n \to k_o \in K$. Let $g'_n = g_n.k_n^{-1}.u_n^{-1}$ and $x'_n = k_n.u_nx_n.u_n^{-1}.k_n^{-1}$. Then

$$x'_{n.e} = {}^{k_nu_n}x_{n,e} \to {}^{k_o}a, \quad x'_{n,d} = {}^{k_nu_n}x_{n,d} \in A.$$

If b is conjugate to ^{k_o}a, then it is conjugate to a as well. Therefore we are reduced to the case where $x_{n,d} \in A$ for all n's.

Let $\Phi = \Phi(S; H)$ be the set of roots of H° with respect to S. It may also be viewed as the set of roots of G with respect to A, or of $\mathfrak{g}$ with respect to $\mathfrak{a}$. For a subset Ψ of Φ, let A_Ψ be the intersection the kernels of the $\alpha \in \Psi$. There exists Ψ such that for infinitely many n's, $x_{n,d} \in A_\Psi$ and does not belong to $A_{\Psi'}$ for any bigger set Ψ'. Passing to a subsequence, we assume this is the case for all n's. Let $A' = A_\Psi$ and S' the Zariski closure of A' in H. It is a split torus contained in S. The group A belongs to $\mathscr{D}G$, as already remarked. Therefore the differentials of the restrictions of the roots to the Lie algebra $\mathfrak{a}'$ of A' span the dual of $\mathfrak{a}'$. As a consequence, A' is the split center of its centralizer and $\mathscr{Z}_G(A')$ is of finite index in its normalizer, which is equal to the normalizer $\mathscr{N}_G(A')$ of A'. In view of the construction, $\mathscr{Z}_G(x_{n,d}) = \mathscr{Z}_G(A')$, hence $x_n \in \mathscr{Z}_G(A')$ $(n = 1, 2, \dots)$.

Going over to a subsequence, we may assume that the $x_{n,d}$ belong to the same Weyl chamber in A', i.e. to the same connected component C of the complement of the union of the subgroups $\ker \alpha$ $(\alpha \in \Phi - \Psi)$. There exists then a parabolic subgroup $P \subset H^\circ$ with Levi subgroup $\mathscr{Z}_H(S')$ and unipotent radical $\mathscr{R}_u P$ corresponding to the positive roots for the order defined by C ([2], 20.4). Let $U = \mathscr{R}_u P(\mathbb{R})$ and $Q = \mathscr{Z}_G(A').U$.

Let $G' = Q/U$ and $\pi : G \to G'$ the canonical projection. The group G' is linear, reductive and of inner type ([18], 2.2.8).

We have $G = K.Q^\circ$ as follows from the Iwasawa decomposition (2.6), and can write $g_n = k_n.q_n$ with $k_n \in K$ and $q_n \in Q^\circ$. Passing to a subsequence, we may assume that $k_n \to k_o \in K$. Then

$$q_n.x_n.q_n^{-1} \to k_o^{-1}.y.k_o,$$

therefore $k_o^{-1}.y.k_o \in Q$. The induction assumption, applied to G', shows the existence of $v \in \mathscr{Z}_G(A')$ such that

$$^{\pi(v)}\pi(a) = \pi(k_o^{-1}bk_o).$$

The element $k_o^{-1}.b.k_o$ is semisimple, belongs to Q. Therefore ([12], §7) there exists $u \in U$ such that

$$u.k_o^{-1}.b.k_0.u^{-1} \in \mathscr{Z}_G(A').$$

This element has the same image, under π, as $k_o^{-1}.b.k_o$. Since π is an isomorphism of $\mathscr{Z}_G(A')$ onto G', we see that a is conjugate to $u.k_o^{-1}.b.k_o.u^{-1}$, hence to b. This proves e).

f) We have now reduced the proof of 3.3 to the following case:

($*$) $b \in K$, $y = y_s$, $\mathscr{C}G$ compact, $g_n \in G^\circ$ and $x_{n,d} = 1$, $(n = 1, 2 \dots)$.

We first claim that

$$y = y_e = b. \tag{1}$$

The kernel of Ad_G is $\mathscr{C}G$, hence is compact. It contains therefore no split positive element, except for the identity. Hence a semisimple element $x \in G$ is elliptic if and only if the eigenvalues of $\mathrm{Ad}_G x$ are of modulus one. The eigenvalues of $\mathrm{Ad}_G y$ are the limits of those of the elements $\mathrm{Ad}_G g_n.x_n.g_n^{-1}$, hence of $\mathrm{Ad}_G x_{n,e}$ since $x_{n,s} = x_{n,e}$ by ($*$), which are of modulus one. Therefore y, which is semisimple by ($*$), is elliptic, which proves (1).

The element y is then conjugate to one in K, so it is enough to consider the case where

$$y = y_e = b \in K. \qquad (2)$$

Let L be a maximal compact subgroup of H containing K. The argument in 3.2, based on the Peter-Weyl theorem and Tannaka duality, shows that a and b are conjugate in L. This implies first that either

$$G^{\circ} \subset \mathscr{Z}_G(a) \cap \mathscr{Z}_G(b) \qquad (3)$$

or

$$G^{\circ} \not\subset \mathscr{Z}_G(a) \qquad G^{\circ} \not\subset \mathscr{Z}_G(b). \qquad (4)$$

Assume (3). The element $x_{n,u}$ belongs to G°, hence $x_n \in G^{\circ}.a$ and then $b \in G^{\circ}.a$, too. We may replace G by the subgroup generated by G° and a, hence assume that $a \in \mathscr{C}(G)$. Then 3.2 implies that $a = b$, which proves the theorem in that case.

g) There remains to establish 3.3 under the assumptions $(*)$, (1) and (4). We first show that we can assume, in addition

$$x_{n,u} = 1 \qquad \text{for all } n\text{' s.} \qquad (5)$$

By 3.1, we can find sequences of elements $q_n \in G$ and $s_n \in \mathscr{Z}_G(b)$ such that

$$q_n \to 1, \; s_n \to b, \; g_n.x_n.g_n^{-1} = q_n.s_n.q_n^{-1} \qquad (n \geqq 0). \qquad (6)$$

The element b belongs to K, hence $\mathscr{Z}_G(b)$ is stable under the Cartan involution θ_K associated to K and (see 2.3) $K \cap \mathscr{Z}_G(b)$ is a maximal compact subgroup of $\mathscr{Z}_G(b)$. There exists therefore elements $r_n \in \mathscr{Z}_G(b)$ such that

$$r_n.s_{n,e}.r_n^{-1} \in K \cap \mathscr{Z}_G(b) \qquad (n \geqq n_o).$$

Passing to a subsequence, we may assume that $r_n.s_{n,e}.r_n^{-1} \to a' \in K \cap \mathscr{Z}_G(b)$. We now have

$$r_n^{-1}.(r_n.s_n.r_n^{-1})r_n = s_n \to b$$
$$(r_n s_n r_n^{-1})_e = r_n.s_{n,e}.r_n^{-1} \to a'.$$

Then 3.2 shows that $a' = b$. Now we have, taking (6) into account

$$r_n.q_n^{-1}.g_n.x_{n,e}.g_n^{-1}.q_n.r_n^{-1} = r_n.s_{n,e}.r_n^{-1} \to b$$

and of course still $x_{n,e} \to a$. This reduces us to the case where $x_n = x_{n,e} \, (n \geqq 1)$, i.e. $x_n \in K$, as claimed.

By 3.1, applied to K, there exist sequences $q_n \in K$, $s_n \in \mathscr{Z}_G(a) \cap K = \mathscr{Z}_K(a)$ such that

$$q_n \to 1, \; s_n \to a, \; x_n = q_n.s_n.q_n^{-1} \in K.$$

We have $s_n = s_{n,e}$ and

$$g_n.q_n^{-1}.s_n.q_n.g_n^{-1} = g_n.x_n g_n^{-1} \to b.$$

Replacing g_n by $g_n.q_n^{-1}$ and x_n by s_n, we are reduced to the case where

$$x_n = x_{n,e} \in \mathscr{Z}_G(a) \cap K = \mathscr{Z}_K(a).$$

Let T be a maximal torus of $\mathscr{Z}_K(a)^\circ \cap K$. Then any element of $\mathscr{Z}_K(a)^\circ.a$ is conjugate under $\mathscr{Z}_K(a)^\circ$ to an element of $T.a$. [If $x \in \mathscr{Z}_K(a)^\circ.a$, there exists $k \in \mathscr{Z}_K(a)^\circ$ such that $k.x.a^{-1}.k^{-1} \in T$, whence $k.x.k^{-1} \in T.a$.] Let then $k_n \in \mathscr{Z}_K(a)^\circ$ be such that $k_n.x_n.k_n^{-1} \in T.a$.

We may assume that $k_n \to k_o \in \mathscr{Z}_K(a)^\circ$. We have

$$k_n.x_n.k_n^{-1} \to k_o.a.k_o^{-1} = a$$

$$g_n.k_n^{-1}.(k_n.x_n.k_n^{-1}).(g_n.k_n^{-1})^{-1} = g_n.x_n.g_n^{-1} \to b.$$

Replacing x_n by $k_n.x_n.k_n^{-1}$ and g_n by $g_n.k_n^{-1}$, we see that we can arrange to have $x_n \in T.a$.

h) We may now assume (*), (1), (4) and

$$x_n \in T.a, \text{ where } T \text{ is a maximal torus of } \mathscr{Z}_G(a)^\circ \cap K. \tag{7}$$

We claim

(**) *there exists a compact set C in G such that we can write*

$$g_n = u_n.v_n \qquad (u_n \in C, \ v_n \in \mathscr{Z}_G(a), \ n = 1, \dots).$$

We first conclude the proof assuming (**). Passing to a subsequence, we may arrange that $u_n \to u$. Then

$$v_n.x_n.v_n^{-1} = u_n^{-1}.g_n.x_n.g_n^{-1}.u_n \to u^{-1}.b.u = b'$$

and therefore $b' \in \mathscr{Z}_G(a)$. Then $a = b'$ by 3.2, hence a is conjugate to b.

There remains to establish (**). We use the construction outlined at the end of 2.1, and write R for $\mathscr{Z}_G(G^\circ)$ and S for $\mathscr{Z}_H(H^\circ)$. Then R is of finite index in $S(\mathbb{R})$. We have $R^\circ = (\mathscr{C}G^\circ)^\circ$, which is compact under our present assumptions. Therefore R is compact. Let $\mu : G \to G' = G/R$ be the canonical projection and $a' = \mu(a)$.

Then $\mu(x_n) \to a'$, $\mu(g_n).\mu(x_n).\mu(g_n^{-1}) \to \mu(b)$. Assume that (**) holds in G'. We want to deduce (**) in G. There exists a compact set $C' \subset G'$ and elements $u_n' \in C'$, $v_n' \in \mathscr{Z}_{G'}(a')$ such that $\mu(g_n) = u_n'.v_n'$. Since R is compact, $\mu^{-1}(C') = C$ is compact. We may find $\tilde{u}_n \in C$ mapping onto u_n', hence such that $\tilde{v}_n = \tilde{u}_n^{-1}.g_n \in \mu^{-1}(\mathscr{Z}_G(a'))$. We claim that $\mathscr{Z}_G(a)$ has finite index in $\mu^{-1}(\mathscr{Z}_{G'}(a'))$. The Lie algebra $\mathfrak{z}$ of $\mathscr{Z}_G(a)$ is the 1-eigenspace of Ad a and contains the Lie algebra of R $\left(\text{since } \mathscr{C}G^\circ \subset \mathscr{C}G\right)$. The Lie algebra $\mathfrak{z}'$ of $\mathscr{Z}_{G'}(a')$ is the 1-eigenspace of $\mathrm{Ad}_{G'} a'$. Since these elements are semisimple,

$$d\mu(\mathfrak{z}) = \mathfrak{z}' \text{ and } d\mu^{-1}(\mathfrak{z}') = \mathfrak{z}$$

so that

$$\mu\left(\mathscr{Z}_G(a)^\circ\right) = \mathscr{Z}_{G'}(a')^\circ, \ \left(\mu^{-1}\left(\mathscr{Z}_{G'}(a)\right)\right)^\circ = \mathscr{Z}_G(a)^\circ.$$

Since these groups, being almost algebraic, have finitely many connected components, this proves our claim. Passing to a subsequence, we can find $c \in \mu^{-1}\big(\mathscr{Z}_{G'}(a')\big)$ such that $v'_n = c.v_n$ with $v_n \in \mathscr{Z}_G(a)$. We let then $u_n = \tilde{u}_n.c$. This proves $(\ast\ast)$ in G.

We may now replace G by G', hence H by $H' = H/S$. This reduces us to the case where H is connected semisimple, and even of adjoint type although we do not need that last fact. The normalizations (5), (7) still hold.

The centralizer $\mathscr{Z}_H(T)$ of T is connected ([2], 11.12) and contains a. There is a Cartan subgroup C of $\mathscr{Z}_H(T)$ containing T and a, and defined over $\mathbb{R}$. We now have

$$x_n \in C(\mathbb{R}) \cap \mathscr{Z}_G(a) \cap K, \qquad (n = 1, 2, \dots). \tag{8}$$

We want to deduce $(\ast\ast)$ from a theorem of [9], applied to H. To this effect, we need one further preliminary remark. Let $X = H/\mathscr{Z}_H(a)$ and $\nu : H \to X$ be the canonical projection. By 6.4 in [3], $X(\mathbb{R})$ is the union of finitely many orbits of $H(\mathbb{R})$, which are open and closed in $X(\mathbb{R})$, hence closed in X (in the ordinary topology). Since G has finite index in $H(\mathbb{R})$, this also holds for the orbits of G. In particular $\nu(G)$ is closed in X. The equality $G \cap \mathscr{Z}_H(a) = \mathscr{Z}_G(a)$ then shows that the projection μ induces an isomorphism of $G/\mathscr{Z}_G(a)$ onto a closed subset of X.

By Theorem 1 of [9], there exists a compact neighborhood B of a in C with the following property: given a compact subset $\omega \in H$, there is a compact subset $\omega^* \in X$ such that if $h \in H$, $x \in B$ are such that $h.x.h^{-1} \in \omega$, then $\nu(h) \in \omega^*$. We now apply this to our sequence. We may assume $x_n \in B$ and $g_n.x_n.g_n^{-1}$ in some compact neighborhood of b. Then the elements $\nu(g_n)$ belong to the intersection of $\nu(G)$ with a compact subset of X. The subset $\nu(G)$ is closed in X, isomorphic to $G/\mathscr{Z}_G(a)$, therefore the $\nu(g_n)$ belong to a compact subset of $G/\mathscr{Z}_G(a)$. Since any compact subset of $G/\mathscr{Z}_G(a)$ is the image of some compact subset of G, this proves $(\ast\ast)$.

Corollary 3.4. *Let $C(K)$ be the quotient of K be itself, acting by inner automorphisms and $\sigma_K : K \to C(K)$ the canonical projection. Then there exists a continuous map $\sigma_G : G \to C(K)$ which is constant on conjugacy classes and extends σ_K.*

It is known that two elements of K which are conjugate in G are already conjugate in K (see [2], 24.7, Prop. 2 for a more general statement). As a consequence, if $x \in G$ is elliptic, then $C_G(x) \cap K$ consists of one conjugacy class in K. We then define σ_G by $\sigma_G(g) = \sigma_K\big(C(g_e) \cap K\big)$. This is a map which extends σ_K and is constant on conjugacy classes. By definition then, $\sigma_G(g) = \sigma_G(g_e)$. There remains to show that σ is continuous.

Let $y \in G$ and $\{y_n\}$, $(n = 1, \dots,)$, a sequence of elements in G tending to y. We have to show that $\sigma_G(y_n) \to \sigma_G(y)$. There exists $g_n \in G$ such that

$$(g_n^{-1}.y_n.g_n)_e \in K.$$

Let $x_n = g_n^{-1}.y_n.g_n$. Then $x_{n,e} \in K$ and, passing to a subsequence, we may assume that $x_{n,e} \to a \in K$. Then

$$\sigma_G(y_n) = \sigma_G(x_n) = \sigma_K(x_{n,e}) \to \sigma_K(a) = \sigma_G(a). \tag{1}$$

On the other hand $\sigma_G(y) = \sigma_G(y_e)$ by definition. We have

$$x_{n,e} \in K, \ x_{n,e} \to a, \ y_n = g_n.x_n g_n^{-1} \to y.$$

By 3.3, a and y_e are conjugate in G, hence $\sigma_G(a) = \sigma_G(y_e) = \sigma_G(y)$. Together with (1), this shows that $\sigma_G(y_n) \to \sigma_G(y)$.

Corollary 3.5. *Let M be a topological space and $f : K \to M$ a continuous M-valued class function. Then f extends to a continuous M-valued class function on G.*

By definition, f can be written as $f = f' \circ \sigma_K$, with $f' : C(K) \to M$ continuous. Then $F' = f' \circ \sigma_G$ satisfies our conditions.

Corollary 3.6. *Let $u, v \in K$. Assume there exist sequences $x_n, y_n \in G$ such that $x_n \to u$, $y_n \to v$ and, for each n, x_n is conjugate to y_n in G. Then u and v are conjugate in K.*

For each n, $\sigma_G(x_n) = \sigma_G(y_n)$, hence $\sigma_G(u) = \sigma_G(v)$ and

$$\sigma_K(u) = \sigma_G(u) = \sigma_G(v) = \sigma_K(v)$$

which shows that u and v are conjugate in K.

3.7. Given f as in 3.5, we let F_f be the function constructed in proving 3.5, to be called the *canonical extension of f*.

Let S, T be tori, $\nu : S \to T$ and $\mu : K \to S$ continuous homomorphisms. Then ν and $\mu \circ \mu$ are continuous class functions on K. We claim that

$$F_{\nu \circ \mu} = \nu \circ F_\mu.$$

Let $g \in G$, and $k \in C(g_e) \cap K$. Let φ be either μ or $\nu \circ \mu$. By construction $F_\varphi(g) = \varphi(k)$. This implies our assertion.

3.8. We indicate briefly how to prove Theorem I(i) using 3.5 when G is linear. As in 1.3, it suffices to consider the case where H is the universal covering of G. Let K be a maximal compact subgroup of G. It is the semidirect product of its derived group $\mathscr{D}K$ by a torus S ([1], 3.2). Since $\pi_1(K)$ is isomorphic to $\pi_1(G)$, hence is torsion-free, the inclusion $S \to G$ induces an isomorphism $\pi_1(S) \tilde\to \pi_1(G)$. Let ν be the inverse isomorphism. We identify these groups to the kernels of the projections π and $\pi' : V \to S$, where V is the universal covering of S. Let f be the projection of K on S, with kernel $\mathscr{D}K$ and $F : G \to S$ its canonical extension. There is a unique lifting $\tilde F : \tilde G \to V$ of F which maps 1 onto 1 and satisfies the conditions $F(\pi(g)) = \pi'(\tilde F(g))$ and $\tilde F(g.n) = \tilde F(g).\nu(n)$ $(g \in G, n \in N)$. The first relation implies that $\tilde F$ is a class function. Since $\nu(N)$ acts freely on V, the second one shows that if $n \neq 1$ $\tilde F(g.n) \neq \tilde F(g)$, therefore g and $g.n$ are not conjugate. In view of 1.1, this proves I(i).

4. Commutator length in semisimple Lie groups

In this section, G is a connected semisimple Lie group.

4.1. Let L be a group which is equal to its derived group $\mathscr{D}L$. Then every element $g \in L$ is a product of finitely many commutators $(a, b) = a.b.a^{-1}.b^{-1} (a, b \in L)$. The commutator length $cl_L(g)$ or simply $cl(g)$ of g is the minimum of the number of factors over all such expressions. The *commutator length* $cl(L) \in \mathbb{N} \cup \infty$ of L is the lower upper bound of the $cl(g), (g \in L)$.

Let $\nu : L' \to L$ be a covering, and $N = \ker \nu$. It follows immediately from the definitions that

$$cl(L) \leqq cl(L') \leqq cl(L).\left(\sup{}_{n \in N} cl_L(n) \right). \tag{1}$$

In particular, if N is finite, $cl(L)$ is finite if and only if $cl(L')$ is so.

The following proposition is certainly known, even though I do not know of a reference for it.

Proposition 4.2. *Assume that G has a finite center. Then $cl(G)$ is finite.*

The Lie algebra $\mathfrak{g}$ is equal to its derived algebra, hence is spanned by elements $[x, y]$ $(x, y \in \mathfrak{g})$. This implies that the image of the map $G^{2n} \to G$ which associates to $(g_1, \ldots, g_{2n})$ the product of the commutators (g_i, g_{n+i}) $(i = 1, \ldots, n; n = \dim G)$ contains a neighborhood of the origin. Therefore $cl_G(g)$ has a finite upper bound if g runs through any compact subgroup. Let K be a maximal one and $G = K.P$ be the corresponding Cartan decomposition (2.3). It is known that any element in P is a commutator [7] (the argument is recalled below), whence the proposition.

4.3. Much more precise information is available in many cases. We review some of it, without claiming completeness. In [8] it is shown that if G is compact, then $cl(G) = 1$. To see this it suffices to prove that any element in a maximal torus T is a commutator. But if $w \in \mathscr{N}(T)$ represents an element of the Weyl group with finite fixed point set (e.g. a Coxeter transformation), then it is readily seen that $t \mapsto (w, t)$ is a surjective homomorphism of T onto itself.

Let now G be non-compact and $G = K.P$ as above. Let A be a maximal split toral subgroup contained in P. Then the same argument, using an element $w \in \mathscr{N}_G(A) \cap K$ representing a Coxeter transformation of the Weyl group of G with respect to A, proves that every element $a \in A$ is of the form $(w, b), (b \in A)$. By conjugation, this shows that any $p \in P$ is a commutator. Using that last fact and Goto's result, we see that if K is semisimple, then $cl(G) \leq 2$, as pointed out in [7]. On the other hand, if G is a complex semisimple group, then $cl(G) = 1$. This was proved first for $\mathbf{SL}_n$ over any algebraically closed field in [16], in general over $\mathbb{C}$ in [14] and in general over any algebraically closed groundfield in [15].

For further results on specific groups, see [7]. The precise value of $cl(G)$ for G of adjoint type, say, seems not to have been determined in general. I do not know of an example where it is > 2.

Proposition 4.4. *Assume G has an infinite center. Then $cl(G)$ is infinite.*

Proof. Let $G_o = G/N$ and $\tau : G \to G_o$ the canonical projection, where $N = \mathscr{C}G$. Then $\mathscr{C}G_o = \{1\}$. Let K_o be a maximal compact subgroup of G_o. The group N is commutative, finitely generated, hence product of a finite group N_t by a free commutative subgroup of some rank $d > 0$. It suffices to prove that $c\ell(G/N_t)$ is infinite and we may therefore assume N to be free abelian. It is contained in $\check{K} = \tau^{-1}(K_o)$. In view if the known structure of coverings of compact Lie groups (see e.g. [5], §1, n° 1, Prop. 5), $\check{K} = K' \times V$, where K' is a compact group on which τ is injective and V a vector group of rank d. Then K_o is the almost direct product of $\tau(K')$ and of a d-dimensional torus S_1. Let $S = K_o/\tau(K')$ and $f_o : K_o \to S$ the canonical projection. S is a d-dimensional torus, image of S_1. Let N' be the fundamental group of S and $\pi' : V' \to S$ the universal covering of S. Then f_o lifts to a surjective homomorphism $\rho : K' \times V \to V'$ with kernel K', which maps N onto a subgroup of finite index of N'. The map f_o may be viewed as a continuous S-valued class function on K_o. Let F_o be its canonical extension (3.7). It lifts uniquely to a continuous function $F = F_G : G \to V$ which maps 1 to 0 and satisfies the conditions

$$F(x.n) = F(x).\rho(n) \quad \pi'\big(F(x)\big) = F_o\big(\pi(x)\big) \quad (x \in G,\, n \in N).$$

In particular it is surjective.

For $m \in \mathbb{N}$, let us denote by $G_{(m)}$ the set of elements in G of commutator length $\leq m$. Then 4.4 is a consequence of

(∗) $F(G_{(m)})$ *is bounded in* V $(m \in \mathbb{N})$.

There remains to prove (∗). In order to avoid some technical complications in the last part of the proof we first show that we can assume G to be the quotient G_1 of its universal covering $\tilde{G}$ by the torsion subgroup of the center of $\tilde{G}$. The group K_o is the almost direct product of its derived group $\mathscr{D}K_o$ by the identity component C_o of its center. The latter has dimension c equal to the rank of $\mathscr{C}G_1$. The group $\mathscr{D}K_o$ belongs to the kernel of f_o and S_1 belongs to C_o. Therefore $f_o = \nu \circ f_1$, where f_1, ν

$$K_o \xrightarrow{f_1} K_o/\mathscr{D}K_o = C \xrightarrow{\nu} S$$

are the canonical projections. Let F_1 be the canonical extension of f_1 (3.7). It has a canonical lifting $F_{G_1} : G_1 \to V_C$ mapping 1 to 0.

Let $\tilde{\nu}$ be the map $V_C \to V$ covering ν. The map $\tilde{\nu} \circ F_{G_1} : G_1 \to V$ is constant on the left cosets of $\ker(G_1 \to G)$, hence defines a continuous map $G \to V$. We claim it is F_G. This follows from the fact that the diagram

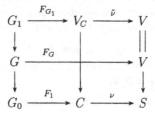

is commutative because $\tilde{\nu} \circ F_{G_1}$ is the lifting of F_o to G_1 by 3.7. It suffices then to prove that $F_{G_1}(G_{1(m)})$ is bounded in V_G. So we assume that $G = G_1$, hence that S is a quotient of $\mathscr{C}K_o$ by a finite group, under ν.

We consider the diagram

$$
\begin{array}{ccccccc}
G^{2m} & \xrightarrow{\ \gamma\ } & G^m & \xrightarrow{\ \mu\ } & G & \xrightarrow{\ F\ } & V \\
{\scriptstyle \sigma}\downarrow & & \downarrow & & \downarrow & & \downarrow{\scriptstyle \pi} \\
G_o^{2m} & \xrightarrow{\ \gamma_o\ } & G_o^m & \xrightarrow{\ \mu_o\ } & G_o & \xrightarrow{\ F_o\ } & S
\end{array}
\tag{1}
$$

where γ, γ_o are commutator maps, which assign to $(g_1, \dots, g_{2m})$ the element with components (g_i, g_{m+i}) $(i = 1, \dots, m)$, the maps μ, μ_o are product maps and the vertical maps natural projections. Let $\varphi = F_o \circ \mu_o \circ \gamma_o$. We claim

(∗∗) *there is a finite cover* $\mathscr{U} = \{U_i\}(i \in I)$ *of* S *by contractible closed subsets such that* $\varphi^{-1}(U_i)$ *has finitely many connected components* $(i \in I)$.

Assuming (∗∗) provisorily, we prove (∗).

The kernel of σ is N^{2m}. Since

$$(x.z, y.z') = (x, y) \quad (x, y \in G, \ z, z' \in N),$$

the map γ is constant of the cosets of N^{2m}, hence factors through σ, i.e. there exists a continuous map $\rho : G_o^{2m} \to G^m$ such that $\gamma = \rho \circ \sigma$. If we add ρ to (1), it clearly remains commutative. We have $\mu \circ \gamma(G^{2m}) = G_{(m)}$ and G^{2m} is the union of the $\sigma^{-1}\varphi^{-1}(U_i)$. Therefore $F(G_{(m)})$ is the union of the sets

$$V_i = (F \circ \mu \circ \rho)\left(\varphi^{-1}(U_i)\right) \quad (i \in I).$$

Let Z be a connected component of $\varphi^{-1}(U_i)$. Then $F.\mu.\rho(Z)$ is a *connected* subset of $\pi^{-1}(U_i)$. But U_i is closed, contractible, hence its inverse image is a disjoint union of compact subsets, all homeomorphic to U_i under π. Therefore $F.\mu.f(Z)$ is bounded, hence so is V_i and (∗) is proved.

We still have to establish (∗∗). We shall use the following fact: in ordinary topology, a real algebraic variety X and the complement $U = X - Z$ of a real algebraic subvariety Z in X have finitely many connected components (see [19], Theorems 3, 4). We shall call U a difference of real algebraic sets.

The group G_o is the product of simple groups with center reduced to $\{1\}$. A maximal compact subgroup of such a factor either is semisimple or has a one-dimensional center (in which case the corresponding symmetric space is isomorphic to a bounded symmetric domain). Let $G_{o,1}, \dots, G_{o,c}$ be the simple factors of G_o with non-semisimple maximal compact subgroup, H_o the product of the remaining factors and K_i a maximal compact subgroup of $G_{o,i}$. The group K_i is the almost direct product of its derived group $\mathscr{D}K_i$ and of a circle group C_i. The restriction f_i of f_o to K_i has kernel $\mathscr{D}K_i$ and induces an isomorphism of $C_i/C_i \cap \mathscr{D}K_i$ onto a direct factor S_i of S, and S is the direct product of the S_i's $(1 \leq i \leq c)$. The homomorphism f_o is the product of the f_i's by the trivial map of a maximal compact subgroup of H. Let F_i be the canonical extension of f_i to $G_{o,i}$. Then F_o is the product of the F_i's by the constant map of H_o into 1. The group G is the product of groups G_i $(i = 1, \ldots, c)$ by H, where G_i is a covering of $G_{o,i}$ with fundamental group $\mathbb{Z}$. The subset $G_{(m)}$ is contained in the product of the $G_{i(m)}$ by H and (1) may be written as a product of similar diagrams for the G_i's with F_o replaced by F_i, and F by the lifting of F_i mapping G_i to the universal covering V_i of S_i, and of a similar diagram for H where F and F_o are replaced by constant maps. From this it is clear that it suffices to prove $(**)$ for each G_i. We are therefore reduced to the case where G_o is simple, $(\mathscr{C}K_0)^o = C$ and S are one-dimensional.

Let $s \in S - \{1\}$ and $c \in C$ mapping onto S. Then $\mathscr{Z}_G(s) = K_o$, since, on one hand $\mathscr{Z}_G(s)$ contains K_o and on the other hand it is $\neq G_o$ and K_o is proper maximal in G_o. Therefore, if $F_o(g) = s\,(g \in G_o)$, then g is elliptic and contained in $^G(\mathscr{D}K_o.c)$.

As a consequence, if Z is a proper subset of $S - \{1\}$ and Z' its inverse image in C, then

$$F^{-1}(Z) = {}^G(\mathscr{D}K_o.Z'). \tag{2}$$

Let $\psi = \mu_o \circ \gamma_o$. It commutes with G_o acting on G_o by inner automorphisms and on G_o^{2m} by

$$^g(g_1, \ldots, g_{2m}) = ({}^g g_1, \ldots, {}^g g_{2m}) \qquad (g, g_i \in G_o; i = 1, \ldots, 2m) \tag{3}$$

$\psi^{-1}(K_o)$ and $\psi^{-1}(DK_o.c)\,(c \in C)$ are algebraic subsets in G_o^{2m}. Let E be a finite set of points on S containing 1 and $\{Z_j\}$ $(j \in J)$ the connected components of $S - E$. The set $\psi^{-1}(K_o) - \psi^{-1}(\mathscr{D}K_o.E)$ is a difference of algebraic sets, hence has finitely many connected components. It is the disjoint union of the $\mathscr{D}K_o.Z_j$, which are relatively open and closed subsets, hence consist each of finitely many connected components. Using (2) and (3) we see that

$(+)\varphi^{-1}(Z)$ *has finitely many connected components if Z is an interval in $S-\{1\}$, either open or closed in $S - \{1\}$.*

There remains to examine $F_o^{-1}(1)$. We want to show the existence of an algebraic subset $Y \subset G_o$ such that $F_o^{-1}(1) = {}^G Y$.

We note first that the set of subgroups $\mathscr{Z}_G(k)$, $k \in \mathscr{D}K_o$, form finitely many conjugacy classes in G. To see this we may assume that k is in a maximal torus T of $\mathscr{D}K_o$. Then the Lie algebra of $\mathscr{Z}_G(k)$ is described by the roots of G with respect to T which are equal to one on k, which gives only finitely many possibilities. In addition there are finitely many possibilities for the component group. Let $\{Q_i\}(i \in I)$ be a set of representatives of these conjugacy classes. For each Q_i, choose an Iwasawa decomposition $Q_i = K_i.A_i.N_i$, where $K_i = K_o \cap Q_i$ (see 2.3, 2.6). Let $Y_i = (\mathscr{D}K_o \cap \mathscr{C}Q_i).A_i.N_i$. The intersection $\mathscr{D}K_o \cap \mathscr{C}Q_i$ is a subgroup of $\mathscr{Z}_{Q_i}(A_i) \cap K_i$, hence normalizes $A_i.N_i$, so that Y_i is an algebraic subgroup.

Let Y be the union of the Y_i's. We want to show that $F_o^{-1}(1) = {}^G Y$. If $g \in Y_i$, then g_e is central in Y_i and belongs to $\mathscr{D}K_o$, hence $F_o(g) = 1$, which shows that $^G Y \subset F_o^{-1}(1)$. Conversely, let $g \in G$ be such that $F_o(g) = 1$. After conjugation, we may assume that $g_e \in K_o$ and therefore $g_e \in \mathscr{D}K_o$. The group $\mathscr{Z}_G(g_e)$ is conjugate to some Q_i, which reduces us to the case where $g \in Q_i$

and $Q_i = \mathscr{Z}_G(g_e)$. Then (see 2.6) g is conjugate to an element of Y_i, whence $F_o^{-1}(1) \subset {}^G Y$. Then $\psi^{-1}(Y)$ is algebraic and

$$\psi^{-1}\left(F_o^{-1}(1)\right) = {}^G\left(\psi^{-1}(Y)\right)$$

has finitely many connected components.

Now (**) follows readily. Take for $\mathscr{U}$ the union of two closed half circles covering S and intersecting only at their end points, one of which contains 1 in its interior. To the one not containing 1, we can apply (+). The other interval is the union of $\{1\}$, to which we apply the remark made above, and of two intervals for which (+) is valid.

REFERENCES

1. A. Borel, *Sous-groupes commutatifs et torsion des groupes de Lie compacts connexes*, Tôhoku Math. J. **13** (1961), no. 2, 216–240.
2. ———, *Linear algebraic groups*, Springer-Verlag, 1991, 2nd enlarged edition, GTM **126**.
3. A. Borel and J-P. Serre, *Théorèmes de finitude en cohomologie galoisienne*, Comment. Math. Helv. **39** (1964), 111–164.
4. A. Borel and J. Tits, *Groupes réductifs*, Publ. Math. I.H.E.S. **27** (1965), 55–150.
5. N. Bourbaki, *Groupes et algèbres de Lie*, Masson, Paris, 1982, Chap. 9.
6. C. Chevalley, *Theory of Lie groups I*, Princeton University Press, 1946.
7. D. Z. Djokovic, *On commutators in real semisimple Lie groups*, Osaka J. Math **23** (1986), 223–228.
8. M. Goto, *A theorem on compact semi-simple groups*, J. Math. Soc. Japan **1** (1949), 270–272.
9. Harish-Chandra, *A formula for semisimple Lie groups*, Amer. J. Math. **79** (1957), 733–760.
10. ———, *Harmonic analysis on real reductive groups I. the theory of the constant term*, J. Functional Analysis **19** (1975), 104–204.
11. G. D. Mostow, *Self-adjoint group*, Annals of Math. **62** (1955), 44–55.
12. ———, *Fully reducible subgroups of algebraic groups*, Amer. J. Math. **78** (1956), 200–221.
13. ———, *Covariant fiberings of Klein spaces, II*, Amer. J. Math. **84** (1962), 465–474.
14. S. Paciencer and H. C. Wang, *Commutators in a semi-simple Lie group*, Proc. A.M.S. **13** (1962), 907–913.
15. R. Ree, *Commutators in semi-simple algebraic groups*, Proc. A.M.S. **15** (1964), 457–460.
16. K. Shoda, *Einige Sätze über Matrizen*, Japanese J. Math. **13** (1936), 361–365.
17. V. S. Varadarajan, *Harmonic analysis on real reductive groups*, Springer-Verlag, 1977, LNM **576**.
18. N. Wallach, *Real reductive groups I*, Academic Press, 1988.
19. H. Whitney, *Elementary structure of real algebraic varieties*, Annals of Math. **66** (1957), 545–566.
20. J. Wood, *Bundles with totally disconnected structure group*, Comment. Math. Helv. **46** (1971), 257–273.

INSTITUTE FOR ADVANCED STUDY, OLDEN LANE, PRINCETON, NJ 08540, USA

CURVES AND DIVISORS IN SPHERICAL VARIETIES

M. BRION

Dedicated to the memory of Roger W. Richardson

INTRODUCTION

To any connected complex semisimple group G are associated the lattice of coroots and the dual lattice of weights. In geometric terms, the weight lattice is interpreted as the Picard group of the flag variety of G; then the fundamental weights correspond to Schubert varieties of codimension one. Moreover, the lattice of coroots is the Chow group of one-dimensional cycles in the flag variety. The simple coroots correspond to Schubert varieties of dimension one, and the duality between weights and coroots translates into the intersection pairing between divisors and curves.

In the present paper, we generalize this picture to spherical varieties, i.e., to normal algebraic varieties X where a connected reductive group G acts, and a Borel subgroup B of G acts with a dense orbit. To summarize our results, we assume for simplicity that X is a complete spherical variety which contains a unique closed G-orbit Y. Then it is known that $Pic(X)$ is freely generated by the classes of irreducible B-stable divisors which do not contain Y; see [3]. We show that the Chow group $A_1(X)$ of one-dimensional cycles is freely generated by the images of certain B-stable curves, termed basic curves; moreover, arbitrary B-stable curves in X are either basic, or rationally equivalent to a sum of two basic curves. Finally, these canonical bases of $Pic(X)$ and $A_1(X)$ are dual for the intersection pairing; see 3.3 below.

These results complete our previous work [5]. They are used in a description of the canonical divisor class of spherical varieties (see 4.2) and in the proof of a vanishing theorem for cohomology groups of line bundles over complete toroidal varieties (see 5.1; a spherical G-variety is toroidal if any G-orbit closure is an intersection of G-stable divisors). The latter result generalizes the vanishing statement in the theorem of Borel-Weil-Bott: $H^i(G/B, L_\lambda) = 0$ for $i \geq 1$ if $\langle \lambda, \check{\alpha} \rangle \geq -1$ for any simple coroot $\check{\alpha}$; here L_λ denotes the line bundle on G/B associated to the weight λ.

Notation. We consider algebraic groups and varieties defined over the field of complex numbers. Denote by G a connected reductive group; choose a Borel subgroup B of G and a maximal torus T of B. Let R (resp. R^+) be the set of roots of (G, T) (resp. (B, T)); let S be the set of simple roots. To any $I \subset S$ we associate the root system $R_I \subset R$ and the parabolic subgroup $P_I \supset B$ with its Levi subgroup $L_I \supset T$. Finally, H^u stands for the unipotent radical of the algebraic group H.

A G-variety X is an irreducible variety with an algebraic G-action. We will always assume that X is covered by G-stable open subsets which can be equivariantly embedded into projective spaces of G-modules; this assumption holds whenever X is a subvariety of a normal G-variety.

1. B-STABLE CURVES IN G-VARIETIES

1.1. Consider a G-variety X and a complete B-stable curve $C \subset X$; denote by $\tilde{C}$ the normalization of C. Then one of the following three cases occurs.

- Type 0: C is fixed pointwise by B; then $G \cdot C$ is a one-parameter family of projective G-orbits.
- Type I: B has two fixed points in $\tilde{C}$; then $\tilde{C}$ is isomorphic to the projective line where B acts via a character.
- ○ B has a unique fixed point in $\tilde{C}$; then C is isomorphic to the projective line where B acts by affine transformations. Moreover, the kernel of the B-action on C is the radical of a (unique) minimal parabolic subgroup $P(C)$.

This case subdivides into three subcases, according to the way $P(C)$ acts on C.

- Type II: C is stable by $P(C)$; then C is a one-dimensional Schubert variety in a projective G-orbit.
- Type III: The canonical morphism $\pi : P(C) \times_B C \to P(C) \cdot C$ is birational. Then π is the normalization of $P(C) \cdot C$, and, moreover, $P(C) \times_B C$ is isomorphic to $\mathbf{P}^1 \times \mathbf{P}^1$ where $P(C)$ acts diagonally via its quotient PGL(2). Moreover, C identifies with $\{\infty\} \times \mathbf{P}^1$.
- Type IV: The morphism π (defined above) has degree two. Then $P(C) \cdot C$ is isomorphic to $\mathbf{P}^2$ where $P(C)$ acts via its quotient PGL(2); here PGL(2) acts on $\mathbf{P}^2$ via the projectivization of the adjoint representation.

We refer to [5] §2 for the proofs of these results; we complete them in the next subsections, by making more precise the connections between curves of type III (resp. IV), and their "models" in $\mathbf{P}^1 \times \mathbf{P}^1$ (resp. $\mathbf{P}^2$).

1.2. For a B-stable curve C of type III or IV, we denote by $Q(C)$ the set of all $s \in G$ such that $s \cdot P(C) \cdot C \subset P(C) \cdot C$. Then $Q(C)$ is a subgroup of G which contains $P(C)$.

Proposition. *For any curve C of type III or IV in a G-variety X, the canonical morphism $\varphi : G \times_{Q(C)} P(C) \cdot C \to G \cdot C$ is birational. If moreover X is toroidal, then φ is an isomorphism.*

Proof. Because B acts on C as the group of affine transformations of the projective line, there exists a unique point $x \in C$ which is fixed by T and not by B. Set $H := G_x$ and denote by α the simple root such that $P(C) = P_\alpha$. Then $B \cap H$ is equal to $P_\alpha^u T$ and hence the Lie algebra of H decomposes as

$$\mathscr{H} = \mathscr{T} \oplus \bigoplus_{\beta \in R^+ \setminus \{\alpha\}} \mathscr{G}_\beta \oplus \bigoplus_{\gamma \in E} \mathscr{G}_\gamma$$

for some subset $E \subset R^-$. Observe that $\gamma' + \gamma'' \in E$ whenever γ', γ'' are in E and $\gamma' + \gamma'' \in R^-$.

We claim that all indecomposable elements of E are negatives of simple roots which are orthogonal to α. Namely, consider an indecomposable γ in E. Assume that $\gamma = \gamma' + \gamma''$ for some negative roots γ' and γ''. If $\gamma' \neq -\alpha \neq \gamma''$, then $\mathscr{G}_{-\gamma'} \subset \mathscr{H}$ and hence $\mathscr{G}_{\gamma''} = [\mathscr{G}_{-\gamma'}, \mathscr{G}_\gamma]$ is contained in $\mathscr{H}$, a contradiction. Therefore, we have $\gamma' = -\alpha \neq \gamma''$. Now $\mathscr{G}_{-\alpha} = [\mathscr{G}_\gamma, \mathscr{G}_{-\gamma''}]$ is contained in $\mathscr{H}$, and therefore H contains the Borel subgroup $s_\alpha(B)$. It follows that C is contained in the projective orbit G/H; but this contradicts the assumption that C has type III or IV. So γ is indecomposable in R^-, and hence $-\gamma$ is simple. If γ is not orthogonal to α, then $-\gamma + \alpha \in R^+ \setminus \{\alpha\}$ and hence $\mathscr{G}_\alpha = [\mathscr{G}_{-\gamma+\alpha}, \mathscr{G}_\gamma]$ is contained in $\mathscr{H}$, a contradiction. This proves our claim.

Now set $I := \{\beta \in S| -\beta \in E\}$, $J := I \cup \{\alpha\}$ and denote by R_I, R_J the corresponding root subsystems of R. Then we have $R_J = R_I \cup \{-\alpha, \alpha\}$ and moreover

$$\mathscr{H} = \mathscr{T} \oplus \bigoplus_{\beta \in R^+ \setminus R_J} \mathscr{G}_\beta \oplus \bigoplus_{\beta \in R_I} \mathscr{G}_{-\beta}.$$

So $H^u = P_J^u$ and hence H is contained in P_J. Moreover, we have $P_\alpha H = P_J$. It follows that $P_J \cdot x \subset P_\alpha \cdot x$, whence $P_J \cdot C \subset P_\alpha \cdot C$ and $P_J \subset Q(C)$. On the other hand, we have $Q(C) \cdot x \subset (P_\alpha \cdot C) \cap G \cdot x = P_\alpha \cdot x$. Thus, $Q(C) \subset P_\alpha H$; since $P_\alpha H = P_J$, we conclude that $H \subset Q(C) = P_J$. Therefore, the restriction $G \times_{Q(C)} P_\alpha \cdot x \to G \cdot x$ is bijective, i.e. φ is birational.

If, moreover, X is toroidal, then $G \cdot C$ is toroidal, too (this follows, e.g., from [7] 3.4). Denote by y the fixed point of B in C; then $G \cdot C = G \cdot x \cup G \cdot y$ and $G \cdot y$ is a divisor in $G \cdot C$. On the other hand, $G \times_{Q(C)} P_\alpha \cdot y$ is a divisor in $G \times_{Q(C)} P_\alpha \cdot C$, and hence the restriction $\psi : G \times_{Q(C)} P_\alpha \cdot y \to G \cdot y$ has finite fibers. Therefore, φ is an isomorphism by Zariski's main theorem.

1.3. For a curve C of type III in a G-variety X, it may happen that $P(C) \cdot C$ is singular, see [5] 2.5. However, it follows from 1.2 that $P(C) \cdot C$ is smooth whenever X is toroidal. More generally, we have the following

Proposition. *If C is a curve of type III in a spherical variety, then $P(C) \cdot C$ is isomorphic to $\mathbf{P}^1 \times \mathbf{P}^1$.*

Proof. By [7] 3.5, the G-variety $G \cdot C$ is spherical, and hence we may assume that $X = G \cdot C$. Denote by α the simple root such that $P(C) = P_\alpha$. Choose a one-parameter subgroup λ of T such that $\langle \lambda, \alpha \rangle = 0$ and such that $\langle \lambda, \beta \rangle < 0$ for all $\beta \in S \setminus \{\alpha\}$. Then the centralizer of λ is L_α. By [17] Theorem A, the orbit $L_\alpha \cdot x$ is an irreducible component of the fixed point set $(G \cdot x)^\lambda$. On the other hand, by [15] 1.3, the L_α-variety X^λ is spherical. It follows that $X^\lambda = \overline{L_\alpha \cdot x} = P_\alpha \cdot C$ is the unique normal completion of L_α/T, i.e., $\mathbf{P}^1 \times \mathbf{P}^1$.

1.4. We describe the behavior of B-stable curves under morphisms.

Proposition. *Let $\varphi : X \to X'$ be a proper G-morphism between two G-varieties; let $C \subset X$ be a B-stable curve which is not contracted by φ.*

   (i) *If C has type 0, I, II or IV, then $\varphi(C)$ has the same type.*
   (ii) *If C has type III, then $\varphi(C)$ has type II, III or IV.*

(iii) *If C has type II, III or IV, then φ induces an isomorphism $C \to \varphi(C)$.*

(iv) *If X' is toroidal and φ is birational, then $\varphi(C)$ has the same type as C.*

Proof. (i) is clear for types 0, I and II. If C has type IV, then the restriction of φ to $P(C) \cdot C$ identifies with a PSL(2)-morphism with source $\mathbf{P}^2$; but such a morphism is either bijective or trivial.

(ii) is obvious; observe that all cases can occur.

(iii) By (i) and (ii), the type of $\varphi(C)$ is II, III or IV, and hence $\varphi(C)$ is isomorphic to $\mathbf{P}^1$. On the other hand, $\varphi|_C$ is birational onto its image, because the isotropy group of the open B-orbit in $\varphi(C)$ is connected.

(iv) It suffices to show that $\varphi|_{G \cdot C}$ is an isomorphism onto its image, for any curve $C \subset X$ of type III. Consider X and X' as embeddings of some spherical homogeneous space; denote by $\mathbf{F}$, $\mathbf{F}'$ the corresponding subdivisions of the valuation cone $\mathscr{V}$, see [12] §4. Then the G-stable subvariety $G \cdot C \subset X$ corresponds to a cone of codimension one in $\mathbf{F}$. Moreover, this cone is the codimension one face of $\mathscr{V}$ given by $\alpha = 0$, where α is the simple root associated with $P(C)$ (this follows from 1.2 and [7] 3.6). Therefore, this cone must belong to $\mathbf{F}'$, and this implies our claim.

2. ELEMENTARY EQUIVALENCES IN SPHERICAL VARIETIES

2.1. Define an *elementary equivalence* between B-stable curves in a G-variety X, as a rational equivalence relation $[div(f)] = 0$ where f is a B-semi-invariant rational function on a B-stable surface in X. This definition is motivated by the following result: the Chow group of one-dimensional cycles $A_1(X)$ is generated by B-stable curves, and its relations are generated by elementary equivalences (see [10] Théorème 1, [9] Theorem 1).

Each curve C of type III gives rise to an elementary equivalence $C + C' - C'' = 0$ where C' (resp. C'') is a B-stable curve of type III (resp. II) in $P(C) \cdot C$; namely, in $\mathbf{P}^1 \times \mathbf{P}^1$, the sum $(\{\infty\} \times \mathbf{P}^1) + (\mathbf{P}^1 \times \{\infty\})$ is rationally equivalent to the diagonal. Similarly, for each curve C of type IV, there is an elementary equivalence $2C - C' = 0$ where C' is the B-stable curve of type II in $P(C) \cdot C$. Finally, for any curve C of type I and any minimal parabolic subgroup P which moves C, we have an elementary equivalence $P \cdot x - P \cdot y - nC = 0$ where x, y are the B-fixed points in C and n is a non-zero integer; see [5] 2.4. Here we set $P \cdot x = 0$ if P fixes x; otherwise, $P \cdot x$ is a curve of type II.

The elementary equivalences obtained above will be called *standard*. The following statement, a refinement of Proposition 3.8 in [5], shows that the non-standard elementary equivalences are very restricted in a toroidal variety. Its proof involves some properties of solvable spherical subgroups in reductive groups of low rank; we refer to [14] for a description of all solvable spherical subgroups.

Theorem. *In a toroidal variety, any elementary equivalence between B-stable curves is standard, or it involves only curves of type I, or it can be written $C_1 - C_2 = \gamma$ where γ is an integral combination of curves of type I, and where one of the following cases occurs:*

(i) *C_1, C_2 are disjoint curves of type III with $P(C_1) = P(C_2)$.*

(ii) C_1, C_2 are curves of type III with $P(C_1) \neq P(C_2)$.

(iii) C_1, C_2 are curves of type II with $P(C_1)$, $P(C_2)$ associated with orthogonal simple roots.

Proof. Let $\Sigma \subset X$ be a B-stable surface, together with a non-constant rational function $f \in k(\Sigma)^{(B)}$. Choose $x \in \Sigma$ such that $B \cdot x$ is open in Σ. Then $\overline{G \cdot x}$ is toroidal (see [7] 3.4) and hence we may assume that $G \cdot x$ is open in X. We set: $H := B_x$. Then $H \subset B$ is a subgroup of codimension two, and, moreover, H contains no maximal torus of B (otherwise $k(\Sigma)^{(B)} \subset k(B/H)^U = k$ where $U = B^U$).

If H contains U, then U acts trivially on Σ, and therefore Σ contains only curves of type I. Otherwise we can assume that $H = U_H \cdot T_H$ where U_H (resp. T_H) is a subgroup of codimension one in U (resp. T). Then T_H is the kernel of a non-trivial character χ of T, and χ is uniquely determined up to sign. It follows that the abelian group $k(\Sigma)^{(B)}/k^*$ is free of rank one; hence the elementary equivalence defined by Σ is unique up to sign.

Observe that U_H contains the derived subgroup of U. Therefore, we have

$$Lie(U_H) = V \oplus \bigoplus_{\beta \in R^+ \backslash S} \mathscr{G}_\beta$$

where V is a T_H-stable hyperplane in $\oplus_{\alpha \in S} \mathscr{G}_\alpha$. Then two cases can occur.
Case 1: V is T-stable. Then there exists $\alpha \in S$ such that

$$Lie(U_H) = \bigoplus_{\beta \in R^+ \backslash \{\alpha\}} \mathscr{G}_\beta \ .$$

Observe that $P_\alpha^u \subset H$; hence P_α^u acts trivially on Σ. So we have $P_\alpha \cdot \Sigma = L_\alpha \cdot \Sigma$ and any Borel subgroup of L has a dense orbit in $L_\alpha \cdot \Sigma$. If H contains the connected center of L_α, then L_α acts on $L_\alpha \cdot \Sigma$ via its quotient SL(2) or PSL(2). In this case, the elementary equivalence given by Σ is standard.

So we may assume that H does not contain the connected center of L_α. Then χ is not a multiple of α. We claim that there exists a set I of simple roots which are orthogonal to α, such that

$$Lie(G_x) = Lie(T_H) \oplus \bigoplus_{\beta \in R^+ \backslash R_J} \mathscr{G}_\beta \oplus \bigoplus_{\beta \in R_I} \mathscr{G}_\beta$$

where $J := I \cup \{\alpha\}$. Namely, choose a finite-dimensional rational G-module V and a point $v \in V$ such that G_x is the isotropy group of the line kv. Decompose V as a direct sum of simple G-modules V_λ with highest weight λ. Observe that the set of P_α^u-fixed points in each V_λ is a simple L_α-module with highest weight λ; therefore, its weights are $\lambda, \lambda - \alpha, \ldots, \lambda - \langle \lambda, \check{\alpha} \rangle \alpha$ with multiplicity one. Because χ is not proportional to α, each eigenspace of $P^u T_H$ in V_λ is one-dimensional and T-stable. Therefore, T normalizes H; now the claim follows as in the proof of 1.2.

Now observe that $G_x^u = P_J^u$, whence $P_J^u \subset G_x \subset P_J$. Because X is toroidal, it follows that the canonical map $G \times_{P_J} \overline{P_J \cdot x} \to X$ is an isomorphism. Therefore, $\overline{P_J \cdot x} = \overline{L \cdot x} = L \cdot \Sigma$ is a toroidal L-variety. So we may assume that G has semisimple rank one; then $H \subset T$ is the kernel of χ.

Because H is contained in B, the canonical map $G \times_B \overline{B \cdot x} \to X$ is an isomorphism. Therefore, $\Sigma = \overline{B \cdot x}$ is normal, and $X = G \times_B \Sigma$. Then X contains exactly two L-orbits of codimension one isomorphic to L/T, and their closures are disjoint. Moreover, the isotropy groups of all other codimension one L-orbits contain a conjugate of U. Hence the complement of the dense B-orbit in Σ is the union of two B-stable curves C_1, C_2 of type III, together with curves of type I.

To conclude the proof in this case, we claim that a relation $C_1 - C_2 = \gamma$ holds in the divisor class group of Σ, where γ is an integral combination of curves of type I. Namely, Σ contains $B/H \simeq U_\alpha \times T/ker(\chi)$ as a T-stable open subset, and hence the generic isotropy group of T in Σ is $ker(\chi) \cap ker(\alpha)$. We may assume that this group is trivial, i.e., that χ and α are a basis of the character group of T. Now χ can be considered as a rational function on Σ, with zeroes and poles outside B/H. Moreover, because C_1 contains $B/T \simeq U_\alpha$ as a T-stable open subset, the generic isotropy group of T in C_1 is $ker(\alpha)$, and the same holds for C_2. Therefore, the fan of the toric surface Σ contains two opposite half-lines corresponding to C_1 and C_2, and the values at χ of the associated one-parameter subgroups are 1 and -1. So the divisor of χ in Σ can be written $C_1 - C_2 - \gamma$.

(2) V is not stable by T. Then there exist distinct simple roots α_1, α_2 and a positive integer n such that $n\chi = \alpha_1 - \alpha_2$. Moreover, we have

$$Lie(U_H) = \ell \oplus \bigoplus_{\beta \in R^+ \setminus \{\alpha_1, \alpha_2\}} \mathscr{G}_\beta$$

where ℓ is a line in $\mathscr{G}_{\alpha_1} \oplus \mathscr{G}_{\alpha_2}$. Denote by P the parabolic subgroup associated to $\{\alpha_1, \alpha_2\}$ and by L the Levi subgroup of P which contains T. Observe that H contains $R(P)$. Therefore, $\overline{L \cdot x} = L \cdot \Sigma$ is fixed pointwise by $R(P)$. Choose a one-parameter subgroup λ of T such that $\langle \lambda, \alpha_1 \rangle = \langle \lambda, \alpha_2 \rangle = 0$ and that $\langle \lambda, \beta \rangle < 0$ for all $\beta \in S \setminus \{\alpha_1, \alpha_2\}$. Then the centralizer of λ is L, and λ fixes x. Moreover, all weights of λ in $Lie(G)/Lie(G_x)$ are non-negative. It follows that $\overline{L \cdot x}$ is the sink of the Bialynicki-Birula decomposition of X with respect to λ (see [1], [2]). Then the L-variety $\overline{L \cdot x}$ is toroidal by [15] 1.6. On the other hand, the connected center of L acts trivially on $\overline{L \cdot x}$. Therefore, we may assume that G is semisimple of rank two. Now two subcases can occur.

(2') The isotropy group G_x is contained in B. Then, arguing as in case (1), we obtain a relation of type (ii).

(2") The isotropy group G_x is not contained in B. Then it is easy to check that G/G_x is the quotient of $SL(2) \times SL(2)$ or of $PSL(2) \times PSL(2)$ by its diagonal subgroup, and that we obtain a relation of type (iii).

2.2. Recall that any spherical variety is dominated by a toroidal variety. Using 1.4 and 2.1, this leads to the following

Corollary. *In a spherical variety, any elementary equivalence is standard, or it involves only curves of type I, or it can be written $n_1 C_1 - n_2 C_2 = \gamma$ where the values of n_1, n_2 are 0 or 1, C_1, C_2 are curves of type II, III or IV, and γ is an integral combination of curves of type I.*

3. THE DUALITY BETWEEN CURVES AND DIVISORS

3.1. For any scheme X and any integer $k \geq 0$, one defines the Chow group $A_k(X)$ of k-cycles in X modulo rational equivalence, see [8]. If, moreover, X is complete, then there is a canonical pairing between the Picard group $Pic(X)$ and the first Chow group $A_1(X)$, and hence a map $\alpha_X : Pic(X) \to Hom(A_1(X), \mathbf{Z})$.

Proposition. *For any complete spherical variety X, the canonical map α_X : $Pic(X) \to Hom(A_1(X), \mathbf{Z})$ is an isomorphism.*

Proof. First we consider the case where X is projective and smooth. Then, because T acts on X with finitely many fixed points, X has a cellular decomposition; see [1], [2]. It follows that the maps $Pic(X) \to H^2(X, \mathbf{Z})$ and $A_1(X) \to H_2(X, \mathbf{Z})$ are isomorphisms; moreover, both groups are torsion-free. Then the map $H^2(X, \mathbf{Z}) \to Hom(H_2(X, \mathbf{Z}), \mathbf{Z})$ is an isomorphism, and our statement follows.

In the general case, we can find a smooth, projective, toroidal variety $\tilde{X}$ and a birational G-morphism $\varphi : \tilde{X} \to X$. Then the induced map $\varphi_* : A_1(\tilde{X}) \to A_1(X)$ is surjective over $\mathbf{Q}$ (it can be shown that $\varphi_* : A_*(\tilde{X}) \to A_*(X)$ is surjective, but we will not need this result). Consider the commutative square

$$
\begin{array}{ccc}
Pic(X) & \to & Hom(A_1(X), \mathbf{Z}) \\
\downarrow & & \downarrow \\
Pic(\tilde{X}) & \to & Hom(A_1(\tilde{X}), \mathbf{Z})
\end{array}
$$

with vertical maps φ^* and φ_*^t (the transpose of φ_*). Then φ_*^t is injective and $\alpha_{\tilde{X}}$ is an isomorphism, hence α_X is injective.

To check the surjectivity of α_X, choose $u \in Hom(A_1(X), \mathbf{Z})$. Then we can find $\tilde{\delta} \in Pic(\tilde{X})$ such that $\alpha_{\tilde{X}}(\tilde{\delta}) = \varphi_*^t(u)$. Moreover, $\tilde{\delta}$ vanishes on all curves which are contracted by φ. We claim that such a $\tilde{\delta}$ is in $\varphi^* Pic(X)$. Because φ_*^t is injective, the proposition follows from this claim.

Now we prove the claim, first in the case where X is toroidal. Consider X and $\tilde{X}$ as embeddings of a spherical homogeneous space, and denote by $\mathbf{F}$ and $\tilde{\mathbf{F}}$ the corresponding fans, see [12] §4. Then $\tilde{\mathbf{F}}$ is a subdivision of $\mathbf{F}$. For each wall μ of $\tilde{\mathbf{F}}$ which is not a wall of $\mathbf{F}$, we have a curve $C_\mu \subset X$ of type I, which is contracted by φ; see [5] 3.3. Moreover, it follows from [3] 3.1 that $\varphi^* Pic(X) \subset Pic(\tilde{X})$ is defined by the equations $(\tilde{\delta} \cdot C_\mu) = 0$ for all μ as before.

For arbitrary X, there is a unique "decoloration" $\pi : X' \to X$, i.e., X' is toroidal, π is a birational G-morphism and π is minimal for these properties. For any closed G-orbit $Y \subset X$, there exists a unique closed G-orbit $Y' \subset X'$ such that $\pi(Y') = Y$. Moreover, π restricts to a proper morphism $\pi_Y : X'_{Y'} \to X_Y$ where X_Y denotes the unique G-stable neighborhood of Y such that Y is closed in X_Y. Choose a prime, B-stable divisor $D \subset X'$ which is moved by G, and such that $\pi(D)$ contains Y. To this D is associated a B-stable curve $C_{D,Y} \subset X'_{Y'}$ of type II, III or IV which is contracted by π; see [5] 3.6. Moreover, it follows from [3] 3.1 that $\pi^* Pic(X) \subset Pic(X')$ is defined by the equations $(\delta' \cdot C_{D,Y}) = 0$ for Y and D as before. This completes the proof of the claim.

3.2. For any complete scheme X, we denote by $N_1(X)$ (resp. $N^1(X)$) the group of one-dimensional cycles (resp. Cartier divisors) of X modulo numerical equivalence. If, moreover, X is spherical, then we denote by $N_1(X)_I$ the subgroup of $N_1(X)$ generated by images of curves of type I, and we set: $\overline{N_1(X)} = N_1(X)/N_1(X)_I$. We define $A_1(X)_I$ and $\overline{A_1(X)}$ in a similar way.

For a projective, spherical variety X, the natural maps $Pic(X) \to N^1(X)$ and $N_1(X) \otimes_{\mathbf{Z}} \mathbf{Q} \to A_1(X) \otimes_{\mathbf{Z}} \mathbf{Q}$ are isomorphisms; see [5] 1.4 and 3.10. A sharper version of this result is as follows.

Proposition. *For any complete spherical variety X, the map $Pic(X) \to N^1(X)$ is an isomorphism. Moreover, the sequence*

$$0 \to A_1(X)_{tors} \to A_1(X) \to N_1(X) \to 0$$

is exact, it induces an isomorphism $\overline{A_1(X)} \to \overline{N_1(X)}$, and this last group is torsion-free.

In particular, $A_1(X)$ is torsion-free whenever X is complete, spherical and contains a unique closed G-orbit. Moreover, $Pic(X)$ and $A_1(X)$ are dual in this case.

Proof. Observe that the isomorphism $\alpha_X : Pic(X) \to Hom(A_1(X), \mathbf{Z})$ factors through

$$Pic(X) \to N^1(X) \to Hom(N_1(X), \mathbf{Z}) \to Hom(A_1(X), \mathbf{Z}) \ .$$

It follows that the map $Pic(X) \to N^1(X)$ is an isomorphism, and that the transpose of the map $A_1(X) \to N_1(X)$ is an isomorphism as well. Therefore, the kernel of this last map is the torsion subgroup of $A_1(X)$.

Recall that the group of rationally trivial one-dimensional cycles in X is generated by elementary equivalences. Now it follows from 2.2 that the quotient group $A_1(X)/A_1(X)_I$ is torsion-free. Therefore, $A_1(X)_{tors}$ is contained in $A_1(X)_I$.

Remark. For any complete scheme X, and for any integer $k \geq 0$, there is a pairing between the Chow group $A_k(X)$ and a "Chow cohomology" group $A^k(X)$; see [8] Chapter 17. Moreover, there is a canonical map $Pic(X) \to A^1(X)$ compatible with both pairings, but which is not always an isomorphism. By recent work of Fulton, MacPherson, Sottile and Sturmfels [9] Theorem 3, the canonical map $A^k(X) \to Hom(A_k(X), \mathbf{Z})$ is an isomorphism whenever a connected solvable linear algebraic group acts on X with finitely many orbits. Using the proposition above, we conclude that the map $Pic(X) \to A^1(X)$ is an isomorphism for complete, spherical X.

3.3. For a spherical variety X, we denote by $\mathscr{D}$ the set of all B-stable prime divisors in X which are not G-stable (this set depends only on the open G-orbit in X). To the G-orbit Y in X, we associate the set $\mathscr{D}_Y \subset \mathscr{D}$ consisting of divisors which contain Y; we set $\mathscr{D}_X := \cup_Y \mathscr{D}_Y$. Then X is toroidal if and only if $\mathscr{D}_X$ is empty.

Recall that any $D \in \mathscr{D} \setminus \mathscr{D}_X$ is a Cartier divisor without base point, see [3] 2.2 or [13] Lemma 2.2 . Moreover, by [5] 3.1, we have $(D \cdot C) = 0$ for any curve C of type I, and hence D defines an element $\overline{D}$ of $Hom(\overline{A_1(X)}, \mathbf{Z})$. We will show that the family $(\overline{D})_{D\in\mathscr{D}\backslash\mathscr{D}_X}$ is a basis of $Hom(\overline{A_1(X)},\mathbf{Z})$, and we will describe the dual basis of $\overline{A_1(X)}$. To this aim, we define the set of *basic curves*: it consists of all curves of type III or IV, and of those curves of type II which are not elementarily equivalent to an integral combination of curves or type I, or to a sum of two curves of type III, or to twice a curve of type IV.

Theorem. *For any complete spherical variety X, the family $(\overline{D})_{D\in\mathscr{D}\backslash\mathscr{D}_X}$ is a basis of $Hom(\overline{A_1(X)},\mathbf{Z})$, and the dual basis of $\overline{A_1(X)}$ consists of the images of basic curves. Moreover, the image in $\overline{A_1(X)}$ of the semigroup of effective cycles is generated by the images of basic curves.*

Proof. Denote by $NE(X)$ the convex cone in $N_1(X)_{\mathbf{Q}}$ generated by the images of curves. In [5] 3.2 and 3.3, it is shown that $NE(X)$ is generated by images of curves of type I, and by elements $C_{D,Y}$ (where $Y \subset X$ is a closed G-orbit, and $D \in \mathscr{D} \backslash \mathscr{D}_Y$) which take integral values on $N^1(X) = Pic(X)$. Moreover, the image of $C_{D,Y}$ in $\overline{N_1(X)}_{\mathbf{Q}}$ is independent of Y; denote it by $\overline{C_D}$. Finally, the $\overline{C_D}$ (where $D \in \mathscr{D} \backslash \mathscr{D}_X$) generate the cone $\overline{NE(X)}$, and we have $(\overline{D} \cdot \overline{C_E}) = \delta_{DE}$ for all D and E in $\mathscr{D} \backslash \mathscr{D}_X$.

By 3.1, the map $A_1(X) \to Hom(Pic(X),\mathbf{Z})$ is surjective. It follows that each $\overline{C_D}$ is in $\overline{A_1(X)}$ (identified with $\overline{N_1(X)}$). For any one-cycle γ, we have

$$\overline{\gamma} = \sum_{D\in\mathscr{D}\backslash\mathscr{D}_X} (\gamma \cdot D)\,\overline{C_D} \,.$$

Therefore, the $\overline{C_D}$'s are a basis of $\overline{A_1(X)}$, with the $\overline{D}$'s as the dual basis.

On the other hand, denote by $AE(X)$ the image in $A_1(X)$ of the semigroup of effective one-cycles. Recall that $AE(X)$ is generated by images of B-stable curves, and that the relations between these curves are generated by elementary equivalences. Denote by $\mathscr{B}$ the set of all basic curves; then $\overline{AE(X)}$ is generated by $\overline{\mathscr{B}}$. Moreover, it follows from 2.2 that $\overline{\mathscr{B}}$ is a basis of the free abelian group $\overline{A_1(X)}$. Because the cone $\overline{NE(X)}$ is generated by $\overline{AE(X)}$, any edge of $\overline{NE(X)}$ contains a unique element of $\overline{\mathscr{B}}$. Therefore, any $\overline{C_D}$ is in $\overline{\mathscr{B}}$.

4. THE CANONICAL DIVISOR OF A SPHERICAL VARIETY

4.1. Let X be a spherical variety. Because X is normal, we can define its canonical sheaf by $\omega_X = i_*\omega_{X^{reg}}$ where $i : X^{reg} \to X$ denotes the inclusion of the open subset of smooth points. Then ω_X is isomorphic to $\mathscr{O}_X(K_X)$ for some Weil divisor K_X on X. Any such K_X is called a canonical divisor of X.

We will express K_X as an integral combination of irreducible, B-stable divisors; we refer to [15] 3.6 for another proof of our result, in the case where X is smooth, toroidal and contains a unique closed G-orbit. We begin with the following

Proposition. *A canonical divisor for the spherical variety X can be written as $-\sum_{D\in\mathscr{D}} a_D\, D - \sum_{i=1}^{\ell} X_i$ where each a_D is a positive integer which depends only on D, and where $X_1,\dots,X_\ell$ are the G-stable prime divisors in X.*

Proof. By removing from X all G-orbits of codimension greater than two, we may assume that X consists of one open G-orbit Ω and a number of G-homogeneous divisors $X_1, \ldots, X_\ell$. Then X is smooth and toroidal. Denote by Ω^0 the open B-orbit in Ω. Set $P := \{s \in G \mid s \cdot \Omega^0 = \Omega^0\}$ and $X^0 := \{x \in X \mid B \cdot x$ is open in $G \cdot x\}$. Then, by [7] 3.4, X^0 is open in X and stable by P. Moreover, there exists a Levi subgroup L of P and a locally closed subvariety Z of X such that:

(i) Z is stable by L, and fixed pointwise by its derived subgroup $[L, L]$.
(ii) The map $P^u \times Z \to X^0 : (s, z) \to s \cdot z$ is an isomorphism.
(iii) Any G-orbit meets Z along a unique L-orbit.

It follows that Z is a toric variety for a quotient torus of T. Denote by T_0 the kernel of the T-action on Z, and choose a decomposition $T = T_0 T_1$ where T_1 is a subtorus of T such that $T_0 \cap T_1$ is finite. The the subgroup $P^u T_1 \subset B$ acts on X with an open orbit Ω^0, and its generic isotropy group is finite.

Denote by σ the exterior product of all members of a basis of $\mathscr{P}_u \oplus \mathscr{T}_1$. Then σ is a global section of the dual $\tilde{\omega}_X$ of the canonical sheaf. Moreover, the support of the zero set of σ is the complement of Ω^0, i.e., the union of the boundary divisors $X_1, \ldots, X_l$ and of the divisors in $\mathscr{D}$. Therefore, a canonical divisor of X can be written $-\sum_{D \in \mathscr{D}} a_D D - \sum_{i=1}^\ell a_i X_i$ for some positive integers $(a_D)_{D \in \mathscr{D}}$ and $a_1, \ldots, a_\ell$. We claim that $a_1 = \cdots = a_\ell = 1$, i.e., that σ vanishes to order one along each X_i. Namely, to check this, we may replace X by $X^0 \simeq P^u \times Z$. Decompose σ as $\sigma_u \wedge \sigma_1$ where σ_u (resp. σ_1) denotes the exterior product of all members of a basis of $\mathscr{P}_u$ (resp. $\mathscr{T}_1$). Then σ_1 vanishes to order one along each T_1-orbit of codimension one in Z, and this implies our claim.

4.2. To compute the coefficients a_D defined in the previous subsection, we may assume that X is toroidal. Then, by 3.3, we can associate to any $D \in \mathscr{D}$ a basic curve C_D such that $(D \cdot C_D) = 1$ and such that $(E \cdot C_D) = 0$ for all $E \in \mathscr{D} \setminus \{D\}$. Moreover, the image of C_D in $A_1(X)$ is uniquely determined. We say that D is *represented by the basic curve* C_D. Recall that C_D has type II, III or IV; moreover, it follows from 1.4 (iv) that this type depends only on D.

Theorem. *For any spherical variety X, with boundary divisors $X_1, \ldots, X_\ell$, a canonical divisor is given by $-\sum_{D \in \mathscr{D}} a_D D - \sum_{i=1}^\ell X_i$ where $a_D = 1$ if D is represented by a curve of type III or IV, and $a_D = 2\langle \rho - \rho_I, \check{\alpha}\rangle$ if D is represented by a curve C of type II, with $G \cdot C = G/P_I$ and $P(C) = P_\alpha$.*

Here ρ (resp. ρ_I) denotes the half-sum of positive roots in R (resp. R_I).

Proof. As before, we may assume that X is smooth and toroidal. Then we set $\tilde{\omega}_X := \omega_X(X_1 + \cdots + X_\ell)$. By 4.1, we have $\tilde{\omega}_X = \mathscr{O}_X(-\sum_{D \in \mathscr{D}} a_D D)$. Fix $D \in \mathscr{D}$ and represent D by a basic curve C; then we have $a_D = -(\tilde{\omega}_X \cdot C)$.

Recall that any G-stable subvariety $Y \subset X$ is smooth and toroidal; therefore, $\tilde{\omega}_Y$ makes sense. We claim that the restriction of $\tilde{\omega}_X$ to Y is isomorphic to $\tilde{\omega}_Y$. Namely, it follows from the local structure of toroidal varieties (see [7] 3.4) that Y is the transversal intersection of boundary divisors, say $X_1, \ldots, X_c$. Moreover, we may assume that the boundary divisors of Y are $Y \cap X_{c+1}, \ldots, Y \cap X_d$ and that the remaining X_j's do not meet Y. Then we have

$$\mathscr{O}_Y\Big(\sum_{i=c+1}^{d} Y \cap X_i \Big) \simeq \mathscr{O}_X\Big(\sum_{i=c+1}^{\ell} X_i \Big) | Y .$$

On the other hand, the adjunction formula implies $\omega_Y \simeq \omega_X(\sum_{i=1}^{c} X_i)|Y$ and our claim follows.

Now choose $Y = G \cdot C$; then the claim implies that $a_D = -(\tilde{\omega}_{G \cdot C} \cdot C)$. If C has type II with $G \cdot C \simeq G/P_I$, then $\tilde{\omega}_{G \cdot C} \simeq \omega_{G/P_I}$ is the G-linearized invertible sheaf associated with the character $-2(\rho - \rho_I)$ of P_I. Now our formula follows from [5] 1.6. On the other hand, if C has type III or IV, then the canonical map $\varphi : G \times_{Q(C)} P(C) \cdot C \to G \cdot C$ is an isomorphism by 1.2. It follows that $(\tilde{\omega}_{G \cdot C} \cdot C) = (\tilde{\omega}_{P(C) \cdot C} \cdot C)$. Now, for C of type III, we have $P(C) \cdot C \simeq \mathbf{P}^1 \times \mathbf{P}^1$ and hence $\tilde{\omega}_{P(C) \cdot C}$ is isomorphic to $\mathscr{O}(-2, -2) \otimes \mathscr{O}(1, 1) = \mathscr{O}(-1, -1)$. For C of type IV, we have $P(C) \cdot C \simeq \mathbf{P}^2$ and $\tilde{\omega}_{P(C) \cdot C} \simeq \mathscr{O}(-3) \otimes \mathscr{O}(2) = \mathscr{O}(-1)$. Therefore, $\tilde{\omega}_{P(C) \cdot C} = -1$ in both cases.

5. A VANISHING THEOREM FOR TOROIDAL VARIETIES

5.1. For any B-stable curve C in a G-variety X, we define a positive integer a_C by:

- $a_C = 1$ if the type of C is not II;
- $a_C = 2\langle \rho - \rho_I, \check{\alpha} \rangle$ if C has type II, where $G \cdot C = G/P_I$ and $P(C) = P_\alpha$.

Theorem. *Consider a toroidal variety X, a spherical variety Y and a proper G-morphism $\pi : X \to Y$. Let L be an invertible sheaf on X, such that $(L \cdot C) + a_C > 0$ for any B-stable curve $C \subset X$ such that $\pi(C)$ is a point. Then $R^i \pi_* L = 0$ for any $i \geq 1$.*

Proof. *Step 1:* We may assume that X is smooth and quasi-projective.

Namely, there exists a smooth, toroidal, quasi-projective G-variety $\hat{X}$ and a proper, birational G-morphism $p : \hat{X} \to X$. Set $\hat{\pi} := \pi \circ p$. Let $\hat{C}$ be a B-stable curve in $\hat{X}$ which is contracted by $\hat{\pi}$. We check that $(p^* L \cdot \hat{C}) + a_{\hat{C}} > 0$. This is clear if $\hat{C}$ is contracted by p, because $(p^* L \cdot \hat{C}) = 0$ in this case. Otherwise, set $C := p(\hat{C})$. Then C and $\hat{C}$ have the same type (see 1.4 (iv)), and hence $a_{\hat{C}} = a_C$. Moreover C is contracted by π, and therefore,

$$(p^* L \cdot \hat{C}) + a_{\hat{C}} = (L \cdot C) + a_C > 0 .$$

If the theorem holds for morphisms from smooth and quasi-projective varieties, then $R^i \hat{\pi}_*(p^* L) = 0$ for $i \geq 1$, and $R^i p_* \mathscr{O}_{\hat{X}} = 0$ as well. It follows that $R^i p_*(p^* L) = 0$ and hence that $R^i \pi_* L = R^i \hat{\pi}_*(p^* L) = 0$.

Step 2: We may assume that π is birational.

Namely we can choose an ample, G-linearized invertible sheaf M on X. Set

$$\hat{X} := \operatorname{Spec}_{\mathscr{O}_X}(\oplus_{n=0}^{\infty} M^n) \quad \text{and} \quad \hat{Y} = \operatorname{Spec}_{\mathscr{O}_Y}(\oplus_{n=0}^{\infty} \pi_* M^n)$$

with respective maps $p : \hat{X} \to X$ and $q : \hat{Y} \to Y$. Then we have a commutative square

$$
\begin{array}{ccc}
\hat{X} & \longrightarrow & \hat{Y} \\
\downarrow{\scriptstyle p} & & \downarrow{\scriptstyle q} \\
X & \longrightarrow & Y
\end{array}
$$

where the vertical arrows are affine morphisms. Set $\hat{G} := G \times \mathbf{G}_m$; then the G-action on X lifts to a $\hat{G}$-action on $\hat{X}$, and $\hat{X}$ is a smooth, toroidal, quasi-projective $\hat{G}$-variety. Moreover, $\hat{\pi}$ is a proper, birational $\hat{G}$-morphism, because M is ample and G-linearized.

Let $\hat{C} \subset \hat{X}$ be a complete curve which is stable by $\hat{B} := B \times \mathbf{G}_m$, and contracted by $\hat{\pi}$. Then $\hat{C}$ must be contained in the zero-section of the line bundle $p : \hat{X} \to X$. Therefore, the restriction of p to $\hat{C}$ is an isomorphism onto its image C. Moreover, C is contracted by π, because $\hat{C}$ is contracted by $q \circ \pi$. It follows that

$$
(p^* L \cdot \hat{C}) + a_{\hat{C}} = (L \cdot C) + a_C > 0 .
$$

If the theorem holds for $\hat{\pi}$, then $R^i \hat{\pi}_*(p^* L) = 0$ for $i \geq 1$. Moreover, because p and q are affine, we have

$$
q_*(R^i \hat{\pi}_* \mathscr{F}) = R^i \pi_*(p_* \mathscr{F})
$$

for any coherent sheaf $\mathscr{F}$ on $\hat{X}$. It follows that $(R^i \pi_*)(p_* p^* L) = 0$ for $i \geq 1$, and hence $R^i \pi_* L = 0$ because $p_*(p^* L) = \oplus_{n=0}^{\infty} L^{n+1}$.

Step 3: Consider a smooth, toroidal, quasi-projective G-variety X. Then we claim that there exists an effective divisor E supported on the whole boundary of X, such that $(E \cdot C) \geq 0$ for any curve $C \subset X$ of type I.

Namely, let $Z \subset X$ be a toric "slice" as in the proof of 4.1. Then Z is smooth and quasi-projective. Moreover, by [5] 3.3, any curve of type I in X is rationally equivalent to some curve of type I in Z. It follows that we can reduce to the case where X is a toric variety; then we may assume furthermore that X is projective. Now choose a very ample line bundle E_0 on X. Then E_0 is rationally equivalent to some effective divisor E_1 which is supported in the boundary of X. Let E_2 be any effective divisor with support on the whole boundary. Then we can take $E = n E_1 + E_2$ for n large enough.

Step 4: Completion of the proof.

By Steps 1 and 2, we may assume that X is smooth, toroidal and quasi-projective, and that π is birational. Denote by $X_1, \ldots, X_\ell$ the G-stable prime divisors in X. By Step 3, we can choose positive integers $a_1, \ldots, a_\ell$ such that $(a_1 X_1 + \cdots + a_\ell X_\ell \cdot C) \geq 0$ for any curve C of type I. For any integer $N \geq 1$, define $D_N \in Pic(X) \otimes_{\mathbf{Z}} \mathbf{Q}$ by

$$
D_N = -K_X + c_1(L) + \sum_{i=1}^{\ell} \left(\frac{a_i}{N} - 1 \right) X_i .
$$

Then we have by 4.2:

$$D_N = c_1(L) + \sum_{i=1}^{\ell} \frac{a_i}{N} X_i + \sum_{D \in \mathscr{D}} a_D D \ .$$

We check that for N large enough, we have $(D_N \cdot C) \geq 0$ for any curve $C \subset X$ which is contracted by π. For this, we may assume that C is B-stable (see [16] Lemma 6.1); then there are only finitely many such C's. If C has type I, then $(D \cdot C) = 0$ for any $D \in \mathscr{D}$, and hence we have

$$(D_N \cdot C) = (L \cdot C) + \frac{1}{N} \left(\sum_{i=1}^{\ell} a_i X_i \cdot C \right) \ .$$

If, moreover, C is contracted by π, then $(D_N \cdot C)$ is a sum of two non-negative numbers. On the other hand, for a curve C of type II, III or IV, the proof of Theorem 4.2 shows that $a_C = (\sum a_D D \cdot C)$. It follows that

$$(D_N \cdot C) = (L \cdot C) + \frac{1}{N} \left(\sum_{i=1}^{\ell} a_i X_i \cdot C \right) + a_C \ .$$

If, moreover, C is contracted by π, then $(L \cdot C) + a_C > 0$ and hence $(D_N \cdot C) \geq 0$ for N large enough.

For large N, the fractional part $\langle D_N \rangle = \sum_{i=1}^{\ell} \frac{a_i}{N} X_i$ of D_N is supported on the boundary, and the latter has normal crossings. Therefore, the Kawamata-Viehweg theorem (see [11] Theorem 1.2.3) applies to D_N, and yields

$$R^i \pi_* \omega_X(\lceil D_N \rceil) = 0$$

for all $i \geq 1$. Here $\lceil D_N \rceil$ is the round-up of D_N, i.e. $\lceil D_N \rceil = -K_X + c_1(L)$. So $\omega_X(\lceil D_N \rceil) = L$ and this completes the proof.

5.2. Theorem 5.1 implies readily the following result, first proved in [4] and [6] by different methods.

Corollary. *Let* $\pi : X \to Y$ *be a proper G-morphism between two spherical varieties. Let L be a line bundle over X, such that $(L \cdot C) \geq 0$ for any B-stable curve C which is contracted by π. Then $R^i \pi_* L = 0$ for $i \geq 1$.*

Proof. We can find a toroidal variety $\hat{X}$ and a proper, birational G-morphism $p : \hat{X} \to X$. Then Theorem 5.1 applies to the morphism $\pi \circ p : \hat{X} \to Y$ and the line bundle $p^* L$, to yield: $R^i(\pi \circ p)_*(p^* L) = 0$ for $i \geq 1$. Similarly, we obtain $R^i p_* \mathscr{O}_{\hat{X}} = 0$ for $i \geq 1$. Now the assertion follows from the Leray spectral sequence and the projection formula.

REFERENCES

1. A. Bialynicki-Birula, *Some theorems on actions of algebraic groups*, Ann. of Math. **98** (1973), 480–497.

2. _____, *Some properties of the decomposition of algebraic varieties determined by actions of a torus*, Bull. Acad. Sci. Séri. Sci. Math. Astronom. Phys. **24** (1976), 667–674.

3. M. Brion, *Groupe de Picard et nombres caractéristiques des variétés sphériques*, Duke Math. J. **58** (1989), 397–424.

4. _____, *Une extension du théorème de Borel-Weil*, Math. Ann. **286** (1990), 655–660.
5. _____, *Variétés sphériques et théorie de Mori*, Duke Math. J. **72** (1993), 369–404.
6. M. Brion and S. Inamdar, *Frobenius splitting of spherical varieties*, Algebraic groups and their generalizations (Rhode Island) (Haboush and Parshall, eds.), Providence, Rhode Island, 1994.
7. M. Brion and F. Pauer, *Valuations des espaces homogènes sphériques*, Comment. Math. Helv. **62** (1987), 265–285.
8. W. Fulton, *Intersection theory*, Springer-Verlag, Berlin, 1984.
9. W. Fulton, R. MacPherson, F. Sottile, and B. Sturmfels, *Intersection theory on spherical varieties*, J. of Alg. Geometry **4** (1995), 181–193.
10. A. Hirschowitz, *Le groupe de Chow équivariant*, C. R. Acad. Sci. Paris Série I Math. **298** (1984), 87–89.
11. Y. Kawamata, K. Matsuda, and K. Matsuki, *Introduction to the minimal model problem*, Algebraic geometry, Sendai 1985 (Kinokuniya) (Oda, ed.), Kinokuniya, 1987.
12. F. Knop, *The Luna-Vust theory of spherical embeddings*, Proceedings of the Hyderabad conference on algebraic groups (S. Ramanan, ed.), Manoj-Prakashan, 1991.
13. _____, *The asymptotic behavior of invariant collective motion*, Invent. math. **116** (1994), 309–328.
14. D. Luna, *Sous-groupes sphériques résolubles*, preprint 1995.
15. _____, *Grosses cellules des variétés sphériques*, this volume.
16. L. Moser-Jauslin, *The Chow rings of smooth, complete SL(2)-embeddings*, Compositio Math. **82** (1992), 67–106.
17. R. W. Richardson, *On orbits of algebraic groups and Lie groups*, Bull. Austral. Math. Soc. **25** (1982), 1–28.

THE PHYLON GROUP AND STATISTICS

A. L. CAREY, P. E. JUPP, AND M. K. MURRAY

In memory of Roger Richardson

1. INTRODUCTION

Statisticians are concerned with the question of the extent to which statistical inference is independent of the choice of parameters for a family of probability distributions. For example, of the three well-known forms of general hypothesis test, the likelihood ratio test and the score test are parameterisation-invariant but the Wald test is not. (See e.g., Cox and Hinkley, 1974, pp. 322–323.) It has been clear to statisticians for some time that differential geometry provides a convenient language for studying questions of parameterisation-independence and a fruitful interaction has arisen between the two subjects. In particular, statisticians have developed a theory of 'higher-order' tensors or 'strings' to study these questions. (See Barndorff-Nielsen *et al.*, 1994, for a review of string theory.) In Carey and Murray (1990) two of the present authors began to develop a mathematical framework for 'strings' involving an infinite-dimensional group. This group, later called the phylon group $\mathscr{P}(d)$, plays the same role in the theory of strings as $GL(d)$, the general linear group, plays in the theory of tensors. The theory of the phylon group and its representations was further developed in Barndorff-Nielsen *et al.* (1992). That work concentrated on the finite-dimensional representation theory of the phylon group, utilising in part results of Terng (1978). In the work presented here we consider two related extensions of this work: (i) twisted phylon representations, and (ii) infinite-dimensional phylon representations.

Twisted phylon representations arise naturally in statistical asymptotics and shed light on the relationship between the two senses in which 'order' is used in parametric statistics. These are: order in the sense of the order of a term in a Taylor expansion in some parameters, and order in the sense of an asymptotic expansion in powers of the sample size.

We shall see that twisted phylon representations are by their nature infinite-dimensional, so we are naturally led to study representations of the phylon group that are infinite-dimensional. In Barndorff-Nielsen *et al.* (1992) we defined a certain class of finite-dimensional representation of the phylon group which we called phylon representations. The statistical context and the fact that the phylon group has a Fréchet Lie group structure suggests that the appropriate class of infinite-dimensional phylon representations will have to be a subset of those representations arising from continuous actions on Fréchet spaces. We define this subset as follows. The phylon group $\mathscr{P}(d)$ contains a one-dimensional subgroup of dilations of $\mathbb{R}^d$, that is, maps of the form $x \mapsto \lambda x$ where λ is a non-zero real number. Following Terng (1978), we say that a vector v in a representation of the phylon group has homogeneous degree n if when we act on it by the dilation λ it is multiplied by λ^n. The subspace of all vectors of homogeneous degree n is called the homogeneous subspace of degree n. A representation arising from a continuous action of the phylon group on a Fréchet space is called a phylon representation if it decomposes into a direct product of finite-dimensional homogeneous subspaces whose degrees are bounded above. In Section 4 we prove that any phylon representation is a projective limit of finite-dimensional representations of finite-dimensional quotients $\mathscr{P}_k(d)$ of the phylon group $\mathscr{P}(d)$. The complete theory of the finite-dimensional algebraic representations of the finite-dimensional phylon groups $\mathscr{P}_k(d)$ has already been developed by Terng (1978). The primary example of an infinite-dimensional phylon representation is the adjoint representation of the phylon group $\mathscr{P}(d)$ on its Lie algebra. We shall see that the coadjoint action is not a phylon representation in our sense, illustrating the fact that infinite-dimensional strings and costrings are different.

2. FRÉCHET VECTOR SPACES

A Fréchet space is a complete, locally convex, metrisable, topological vector space. We shall be interested in a particular kind of Fréchet space, so we shall consider these rather than work with the full theory. Readers interested in the general framework should consult the book by Treves (1967).

Recall that if we have a countable collection $\{V_i \mid i = 1, 2, \dots\}$ of finite-dimensional vector spaces then we distinguish between the direct product

$$\prod_{i=1}^{\infty} V_i \qquad (2.1)$$

and the direct sum

$$\sum_{i=1}^{\infty} V_i, \qquad (2.2)$$

where the latter is the subset of the direct product consisting of all sequences $(v_1, v_2, \dots)$ with only a finite number of non-zero components. The direct product can be made into a Fréchet space by giving it the product topology, i.e. the topology of 'pointwise convergence'. Thus a sequence converges in this topology if and only if each of the sequences of components converges. Notice that there is no assumption of uniformity on the convergence of the components so, for example, $(a_1, 0, 0, \dots), (0, a_2, 0, \dots), (0, 0, a_3, \dots), \dots$ converges to $0 = (0, 0, 0, \dots)$ for any sequence $a_1, a_2, a_3, \dots$ of vectors. If $\sum_{i=1}^{\infty} V_i$ (with the product topology) were a Banach space then we could choose $a_1, a_2, a_3, \dots$ to get a sequence of elements of $\sum_{i=1}^{\infty} V_i$ with unit norm but converging to 0. Since this is impossible, $\sum_{i=1}^{\infty} V_i$ is not a Banach space.

If W_i is a subset of V_i let us denote by $W = W_1 \times W_2 \times W_3 \dots$ the subset of $\prod_{i=1}^{\infty} V_i$ defined by

$$W = \{(w_1, w_2, \dots) | w_i \in W_i \ \forall i = 1, 2, \dots\}. \qquad (2.3)$$

A basis of neighbourhoods of zero for the topology of pointwise convergence is given by sets of the form $U_1 \times U_2 \times U_3 \times \ldots$, where each of the U_i is an open neighbourhood of 0 in V_i and only a *finite number* of the U_i are not equal to V_i.

The direct sum $\sum_{i=1}^{\infty} V_i$ inherits a topology as a subspace of the direct product but is not complete. For this and other reasons the subspace topology is not the natural topology on $\sum_{i=1}^{\infty} V_i$ and instead we define the topology which has as a basis of neighbourhoods of the identity all sets of the form

$$(U_1 \times U_2 \times U_3 \times \ldots) \cap \sum_{i=1}^{\infty} V_i, \tag{2.4}$$

where each of the U_i is open in V_i. This is also the topology obtained by saying that a set is open if and only if its intersection with each of the subsets

$$V_1 \oplus 0 \oplus 0 \oplus \cdots \subset V_1 \oplus V_2 \oplus 0 \oplus \cdots \subset V_1 \oplus V_2 \oplus V_3 \oplus \cdots \subset \cdots \tag{2.5}$$

is open. As such, the direct sum of the V_i is the inductive limit in the category of topological vector spaces of a sequence of Fréchet spaces (Treves, 1967, p. 126).

The relationship between the direct sum and the direct product is best understood as one of 'duality'. More precisely, if $V = \prod V_i$ with the Fréchet topology just defined then the dual space V^* of continuous linear functionals is the direct sum $\sum V_i^*$. If the dual space is given the strong topology then this is an isomorphism of topological vector spaces. For example, let $\mathbb{R}[[x^1, \ldots, x^d]]$ be the space of all formal power series in d variables. Then this is a direct product of the kind we have been discussing if we let V_i be the space of all homogeneous polynomials of degree i. That is, $V_i = S^i(\mathbb{R}^d)^*$, the ith symmetric product of the dual of $\mathbb{R}^d$. The dual of $\mathbb{R}[[x^1, \ldots, x^d]]$ is the space of all polynomials, $\mathbb{R}[x^1, \ldots x^d]$, that is, formal power series of finite order. This is essentially shown in Treves (1967, pp. 227–228) and that proof can be adapted to the case at hand.

To understand this duality more geometrically, recall that the space of formal power series, $\mathbb{R}[[x^1, \ldots, x^d]]$, can be identified with $J^\infty(\mathbb{R}^d)_0$, the space of all infinite jets at the origin of smooth, real-valued functions on $\mathbb{R}^d$. By definition, the latter space is the quotient of the space $C^\infty(\mathbb{R}^d)$ of all smooth functions by the ideal of functions which have all their partial derivatives at the origin equal to zero. Then, by a theorem of Borel (Treves, 1967, p. 390), the map

$$C^\infty(\mathbb{R}^d) \to \mathbb{R}[[x^1, \ldots, x^d]] \tag{2.6}$$

which sends a smooth function to its Taylor expansion at the origin is a surjection. This surjection is continuous if we give $C^\infty(\mathbb{R}^d)$ the smooth (Fréchet) topology (Treves, 1967, p. 86) and hence induces an isomorphism from $J^\infty(\mathbb{R}^d)_0 \to \mathbb{R}[[x^1, \ldots, x^d]]$. Corresponding to the surjection in (2.6) there is an inclusion of the dual spaces. The dual of the space of all smooth functions with the smooth topology is just the space of all distributions $\mathscr{D}(\mathbb{R}^d)$. The dual of $J^\infty(\mathbb{R}^d)_0$ is the subspace of distributions that have support at the origin. As is well-known (Treves, 1967, p. 266), a distribution with support at the origin is a *finite* sum of derivatives of the Dirac delta distribution δ. This space is isomorphic to the space of polynomials. Indeed, we can map the polynomial p to the distribution $p(\partial/\partial x^1, \ldots, \partial/\partial x^d)\delta$. Notice that this isomorphism depends on choosing coordinates to define the partial derivative operators.

In the next section we study the phylon group as a Fréchet Lie group and, in particular, as a Fréchet manifold. The theory of finite-dimensional manifolds extends to manifolds modelled on any topological vector space. Manifolds modelled on Banach or Hilbert spaces behave in a very similar way to finite-dimensional manifolds. Manifolds modelled on Fréchet spaces, however, behave differently because two important theorems of multivariable calculus, the uniqueness of solutions to ordinary differential equations and the inverse function theorem, extend to Banach and Hilbert spaces but not to Fréchet spaces. The geometric consequence of this for the theory of Fréchet Lie groups is that there may not be an exponential map from the Lie algebra to the group and even if there is it need not be a local diffeomorphism. Useful references for the theory of Fréchet manifolds and Fréchet Lie groups are the articles by Hamilton (1982), Kobayashi et al. (1985) and Milnor (1984).

3. The infinite-dimensional phylon group

The infinite-dimensional phylon group $\mathscr{P}(d)$ is defined as either the set of infinite jets of germs of local diffeomorphisms of $\mathbb{R}^d$ which fix the origin or, equivalently, as all $\mathbb{R}^d$-valued formal power series in d variables with no constant term and invertible linear term. It is clear from these definitions that we can place on $\mathscr{P}(d)$ the topology of pointwise convergence of the coefficients of the power series. Inside the infinite phylon group we have a sequence of normal subgroups $\mathscr{P}^{(k)}(d)$, which are the sets of jets of diffeomorphisms that agree with the identity up to order k. The quotient groups $\mathscr{P}_k(d)$ are the groups of k-jets of local diffeomorphisms of $\mathbb{R}^d$ which fix the origin. For every k there is a projection map $\mathscr{P}(d) \to \mathscr{P}_k(d)$, which we shall denote by j^k. Moreover, for any $m \geq k$, there is a projection $\mathscr{P}_m(d) \to \mathscr{P}_k(d)$, which we shall also denote by j^k. The groups $\mathscr{P}_k(d)$ are finite-dimensional Lie groups. If $\mathscr{P}(d)$ is given the topology of pointwise convergence then each of the maps j^k is continuous and this topology is, in fact, the projective limit topology. That is, it is the coarsest topology for which each of the projections j^k is continuous.

With this topology the phylon group is a Fréchet Lie group. Indeed, we can say more than this; Omori (1980) shows that the phylon group is a regular Fréchet Lie group. The precise definition of a regular Fréchet Lie group is a little complicated and can be found, along with a survey of their properties, in Kobayashi et al. (1985). It suffices for our purposes to note that regular Fréchet Lie groups are a class of Fréchet Lie groups for which the exponential map from the Lie algebra to the group exists and is smooth. Recall (for example, from Kolář et al., 1993, p. 36) how we construct the exponential map $\exp : LG \to G$ for a finite dimensional Lie group G. Given an element X of the Lie algebra, we consider the integral curve γ of the left invariant vector field defined by X for which $\gamma(0)$ is the identity. The exponential map is then defined by $\exp(X) = \gamma(1)$ and the curve γ turns out to be a one-parameter subgroup, that is, $\gamma(t)\gamma(s) = \gamma(s+t)$, and satisfies $\gamma(t) = \exp(tX)$. The integral curve γ can be characterised by the fact that it is the unique one-parameter subgroup whose tangent at zero is X. As we have remarked at the end of Section 2, it is not always possible, even locally, to integrate vector fields on a Fréchet manifold, so this construction need not work for a Fréchet Lie group. Whether or not the exponential map exists for an arbitrary Fréchet Lie group is an open question.

The Lie algebra of $\mathscr{P}(d)$ is the tangent space at the identity and this is just the space of jets at the origin of vector fields on $\mathbb{R}^d$ which are zero at the origin. We denote this space by $L\mathscr{P}(d)$. An element X of $L\mathscr{P}(d)$ can be written as

$$X = \sum_i X^i \frac{\partial}{\partial x^i}, \tag{3.1}$$

where X^i is an element of $\mathbb{R}[[x^1, \ldots, x^d]]$ with constant term zero. The Lie bracket of any two elements X and Y of $L\mathscr{P}(d)$ is

$$[X, Y] = \sum_{i,j} \left(X^i \frac{\partial Y^j}{\partial x^i} - Y^i \frac{\partial X^j}{\partial x^i} \right) \frac{\partial}{\partial x^j}. \tag{3.2}$$

On a finite-dimensional Lie group we can prove that the exponential map is a diffeomorphism of a neighbourhood of the identity by using the inverse function theorem. The derivative of the exponential map is easily seen to be invertible. However, as we remarked at the end of Section 2, the inverse function theorem does not apply to Fréchet spaces and the exponential map on a Fréchet Lie group need not be a local diffeomorphism. See Milnor (1984) for a discussion of a counter-example. In the case of the phylon group $\mathscr{P}(d)$, we can exploit the fact that it is a projective limit of finite-dimensional Lie groups to show that exponential map $L\mathscr{P}^{(1)}(d) \to \mathscr{P}^{(1)}(d)$ is a smooth bijection, as follows.

Notice that $\mathscr{P}(d)/\mathscr{P}^{(1)}(d)$ is the group $GL(d)$ and, in fact, $\mathscr{P}(d)$ is the semi-direct product of $\mathscr{P}^{(1)}(d)$ and $GL(d)$. If we denote by $\mathscr{P}^{(m)}(d)$ the kernel of $\mathscr{P}_r(d) \to \mathscr{P}_m(d)$ then it is also true that $\mathscr{P}_k(d)$ is the semi-direct product of $\mathscr{P}_k^{(1)}(d)$ and $GL(d)$. It is pointed out by Kolář et al. (1993, p. 131) that the group $\mathscr{P}_k^{(1)}(d)$ is a connected, simply connected, nilpotent finite-dimensional Lie group and that therefore the exponential map $\exp: L\mathscr{P}_k^{(1)}(d) \to \mathscr{P}_k^{(1)}(d)$ is a diffeomorphism. In fact, the same is true of the groups $\mathscr{P}_k^{(r)}(d)$ for any $r \geq 1$, so the exponential maps $\exp: L\mathscr{P}_k^{(r)}(d) \to \mathscr{P}_k^{(r)}(d)$ are also diffeomorphisms. We shall use this fact to prove that the exponential map $L\mathscr{P}^{(r)}(d) \to \mathscr{P}^{(r)}(d)$ is a smooth bijection (for $r \geq 1$).

We have already noted that, because the phylon group $\mathscr{P}(d)$ is regular, it has a smooth exponential map. We have not yet considered the existence of the exponential map $L\mathscr{P}^{(r)}(d) \to \mathscr{P}^{(r)}(d)$. We would expect that it is the restriction of the exponential map on $L\mathscr{P}(d)$ and we show this as follows. Notice that the exponential map on $L\mathscr{P}(d)$ certainly restricts to a smooth map on $L\mathscr{P}^{(r)}(d)$. The question is: does its image lie in $\mathscr{P}^{(r)}(d)$? Let $X \in L\mathscr{P}^{(r)}(d)$ and consider $f(t) = \exp(tX)$. As we remarked earlier, this curve is a one-parameter subgroup whose tangent at zero is X. Because $j^r: \mathscr{P}(d) \to \mathscr{P}_r(d)$ is a group homomorphism,

$g(t) = j^r(f(t))$ is also a one-parameter subgroup and if we differentiate it at $t = 0$ we see that the tangent is the zero vector. The point is that the tangent map to j^r going from $L\mathscr{P}(d)$ to $L\mathscr{P}^{(r)}(d)$ is also the corresponding jet map j^r and $j^r(X) = 0$ by assumption. Standard facts about one-parameter subgroups in a finite-dimensional Lie group (as, for example, in Kolář et al., 1993, pp. 35–36) now apply and it follows that $g(t)$ is a constant function equal to the identity and hence $f(t)$ is in $\mathscr{P}^{(r)}(d)$ for all t, as required.

Because the map $j^k : \mathscr{P}^{(r)}(d) \to \mathscr{P}_k^{(r)}(d)$ is a group homomorphism for every k, its derivative is a homomorphism of the corresponding Lie algebras and this is, in fact, also the map j^k acting on jets of vector fields. Moreover, these two maps commute with the exponential map. Thus we have commutative diagrams

$$
\begin{array}{ccc}
L\mathscr{P}^{(r)}(d) & \xrightarrow{\ j^k\ } & L\mathscr{P}_k^{(r)}(d) \\
{\scriptstyle\exp}\Big\downarrow & & {\scriptstyle\exp_k}\Big\downarrow \\
\mathscr{P}^{(r)}(d) & \xrightarrow{\ j^k\ } & \mathscr{P}_k^{(r)}(d)
\end{array}
\tag{3.3}
$$

where we temporarily adopt the notation $\exp_k$ for the exponential map on $L\mathscr{P}_k^{(r)}(d)$. Let g be an element of $\mathscr{P}^{(r)}(d)$. Then, for every k, $j^k(g) = \exp_k(X_k)$ for some (unique) X_k in $L\mathscr{P}_k^{(r)}(d)$. If $m \geq k$ then $j^m(g) = \exp_m(X_m)$ and hence

$$
\begin{aligned}
j^k(g) &= j^k(j^m(g)) \\
&= j^k(\exp_m(X_m)) \\
&= \exp_k(j^k(X_m)).
\end{aligned}
\tag{3.4}
$$

Thus by the uniqueness of X_k we have that $X_k = j^k(X_m)$. Because $L\mathscr{P}^{(r)}(d)$ is the projective limit of the $L\mathscr{P}_k^{(r)}(d)$, it follows that there is a unique X in the Lie algebra of $\mathscr{P}^{(r)}(d)$ such that $j^k(X) = X_k$ and that $j^k(\exp(X)) = j^k(g)$, so that $\exp(X) = g$. Similarly, it is possible to show that exp is one to one.

Hence we have proved:

Proposition 3.1. *The exponential map* $\exp : L\mathscr{P}^{(r)}(d) \to \mathscr{P}^{(r)}(d)$ *is a smooth bijection.*

We can now prove the result that we need for the next section. Assume that we have a representation of $\mathscr{P}(d)$ on a Fréchet space V and that the map $\mathscr{P}(d) \times V \to V$ is smooth. By differentiating, it follows that we have a representation of the Lie algebra and that the map $L\mathscr{P}(d) \times V \to V$ is smooth. Let W be a closed subspace of finite codimension in V. We want to show that (i) if $L\mathscr{P}^{(k)}(d)$ stabilises W then so also does $\mathscr{P}^{(k)}(d)$, (ii) if $L\mathscr{P}^{(k)}(d)(V) \subset W$ then $\mathscr{P}^{(k)}(d)$ acts trivially on V/W. We just use the exponential map as we would in the finite-dimensional case. For X in $L\mathscr{P}^{(k)}(d)$ consider the curve $g_t = \exp(tX)$ and apply it to v in V. Differentiating gives

$$
\frac{d}{dt}(g_t . v) = X(g_t . v).
\tag{3.5}
$$

Let $\pi\colon V \to V/W$ be the projection. Then

$$\frac{d}{dt}\pi(g_t.v) = \pi X(g_t.v). \tag{3.6}$$

If either $L\mathscr{P}^{(k)}(d)$ stabilises W and $v \in W$, or $L\mathscr{P}^{(k)}(d)(V) \subset W$ then $\pi X(g_t.v) = 0$. Since $g_0 = 1$, uniqueness of solutions of first-order differential equations with values in V/W implies that $\pi(g_t.v) = \pi(v)$ for all t.

Hence we have proved:

Proposition 3.2. *Let W be a closed subspace of finite codimension in a vector space V on which the phylon group $\mathscr{P}(d)$ acts smoothly. Then, for $k = 1, 2, \ldots,$*

(i) *if $L\mathscr{P}^{(k)}(d)$ stabilises W then so does $\mathscr{P}^{(k)}(d)$,*
(ii) *if $L\mathscr{P}^{(k)}(d)(V) \subset W$ then the induced action of $\mathscr{P}^{(k)}(d)$ on V/W is trivial.*

4. INFINITE-DIMENSIONAL PHYLA

In Barndorff-Nielsen *et al.* (1992) we defined a phylon to be a certain type of finite-dimensional representation of the infinite-dimensional phylon group $\mathscr{P}(d)$. In this section we shall define a quite general class of infinite-dimensional representations V of $\mathscr{P}(d)$ to be phylon representations and show that they are actually projective limits of (finite-dimensional) representations V_k of $\mathscr{P}_k(d)$.

Before defining an infinite-dimensional phylon representation, let us note that the dilations of $\mathbb{R}^d$, i.e. the maps $x \mapsto \lambda x$ for λ a real non-zero number, form a subgroup of the phylon group isomorphic to the non-zero reals. Indeed, the dilations are actually the subgroup of $GL(d)$ of all non-zero multiples of the identity. Thus if the phylon group acts linearly on a vector space V then the dilations also act. We say that an element $v \in V$ has homogeneous degree n if $\lambda.1.v = \lambda^n.v$ (Terng, 1978). Then we have:

Definition 4.1. An (infinite-dimensional) phylon representation on a Fréchet space V is a group homomorphism

$$\mathscr{P}(d) \to GL(V) \tag{4.1}$$

such that:

(i) $V = \prod_{i=1}^{\infty} V_{n_i}$,
(ii) V_{n_i} is the subspace of all elements of homogeneous degree n_i, and $n_1 > n_2 > n_3 > \cdots$,
(iii) for $i = 1, 2, \ldots$, V_{n_i} is a finite-dimensional $GL(d)$-module,
(iv) the map

$$\mathscr{P}(d) \times V \to V \tag{4.2}$$

is smooth as a map between Fréchet manifolds.

We call the decomposition in (i) the homogenous decomposition of V. An important example of a phylon representation is the adjoint action of the phylon group $\mathscr{P}(d)$ on the Lie algebra $L\mathscr{P}(d)$, in which $\mathscr{P}(d)$ acts on $L\mathscr{P}(d)$ by conjugation. More precisely, for g in $\mathscr{P}(d)$ and X in $L\mathscr{P}(d)$, the adjoint action is

$$(g, X) \mapsto gXg^{-1}.$$

In this case the homogeneous decomposition is

$$\prod_{k=1}^{\infty} [\mathbb{R}^d \otimes S^k(\mathbb{R}^d)^*]. \tag{4.3}$$

The homogeneous degrees are $0, -1, \ldots$ with $L\mathscr{P}(d)_{-k} = \mathbb{R}^d \otimes S^{k+1}(\mathbb{R}^d)^*$ for all k. The vector space structure of the subalgebra $L\mathscr{P}^{(k)}(d)$ of $L\mathscr{P}(d)$ is

$$L\mathscr{P}^{(k)}(d) = \prod_{j \geq k} L\mathscr{P}(d)_{-j}. \tag{4.4}$$

This example is important in understanding the structure of general phylon representations, as we shall now see.

Proposition 4.1. *Every phylon representation is a projective limit of finite-dimensional representations of the phylon group.*

The proof of this result is quite straightforward. Consider the dilation $\lambda.1$. As the non-zero real number λ varies, this defines a curve in the phylon group $\mathscr{P}(d)$ and we can differentiate it at $\lambda = 1$ to obtain the element

$$Z = \sum_i x^i \frac{\partial}{\partial x^i} \tag{4.5}$$

in the Lie algebra $L\mathscr{P}(d)$.

Let v be an element of homogeneous degree n in a phylon representation. Then by definition we have

$$\lambda.v = \lambda^n v, \tag{4.6}$$

where we have written a dot on the left hand side to remind the reader that here λ is acting as an element of the phylon group, whereas on the right hand side λ^n is acting by scalar multiplication. Differentiating shows us that

$$Zv = nv. \tag{4.7}$$

In particular, if X is an element of $L\mathscr{P}(d)$ of homogeneous degree $-k$ then we have $[Z, X] = -kX$. Putting these facts together, we can show that Xv has a homogeneous degree and calculate it from

$$\begin{aligned}
Z(Xv) &= [Z, X]v + XZv \\
&= -kXv + nXv \\
&= (n - k)Xv.
\end{aligned} \tag{4.8}$$

In other words,

$$L\mathscr{P}(d)_{-k}V_n \subset V_{n-k}.\tag{4.9}$$

From the structure (4.4) of $L\mathscr{P}^{(k)}(d)$ it follows that

$$L\mathscr{P}^{(k)}(d)\left(\prod_{n<m}V_n\right) \subset \prod_{n<m-k}V_n.\tag{4.10}$$

By Proposition 3.2 (i) we therefore also have

$$\mathscr{P}^{(k)}(d)\left(\prod_{n<m}V_n\right) \subset \prod_{n<m}V_n.\tag{4.11}$$

Recalling that $n_1 > n_2 > n_3 > \ldots$ and using (4.10) and Proposition 3.2 (ii), we deduce that the quotient spaces

$$W_k = V/\prod_{n\leq n_1-k}V_n\tag{4.12}$$

are acted upon by $\mathscr{P}_k(d)$. Since V is the projective limit of the W_k, which are representations of the finite phylon groups $\mathscr{P}_k(d)$, this completes the proof.

5. JETS OF DENSITIES

In the next section we wish to work with the 'Taylor expansion' of a measure. For an arbitrary measure such a notion makes little sense. However, if the measure is a smooth function times Lebesgue measure then we can sensibly Taylor expand the smooth function. We shall assume, therefore, that the sample space Ω is a smooth manifold of dimension r, say an open subset of some $\mathbb{R}^r$. We are interested in the space $\Gamma(\Delta(\Omega))$ of smooth densities on Ω. A density on a finite-dimensional vector space V is a map ω from the set of all bases of V into $\mathbb{R}$ with the property that if $v = (v^1, \ldots, v^r)$ is a basis and $w = (w^1, \ldots, w^r)$ is another basis related by a linear transformation $X \in GL(r)$ then $\omega(v) = \omega(Xw) = |\det(X)|\omega(w)$. The set $\Delta(V)$ of all densities on V is a one-dimensional vector space. We can form a vector bundle $\Delta(\Omega)$ over Ω whose fibre at a point x in Ω is $\Delta(T_x\Omega)$, the space of all densities on the tangent space ot Ω at x. A (smooth) density on Ω is therefore a (smooth) choice of density on each tangent space, that is, an element of $\Gamma(\Delta(\Omega))$, the space of all smooth sections of $\Delta(\Omega)$. In the case that the sample space is an open subset of $\mathbb{R}^r$ this is just the space of all smooth functions times Lebesgue measure, i.e. of all measures of the form

$$f(x)dx^1 \ldots dx^r.\tag{5.1}$$

More generally, the densities will have such a representation locally. Because $\Delta(\Omega)$ is a vector bundle, we can form the space $J_x^\infty(\Delta(\Omega))$ of all infinite jets of its sections at any point x. Locally, the objects in here are expressions of the form (5.1) where $f(x)$ is now an infinite polynomial in $x^1 \ldots x^r$. More information on densities can be found in Section 6.9 of Murray and Rice (1993) The reader should note carefully that this use of the word 'density' is different from its use in 'probability density'.

Readers familiar with densities and measures will notice one immediate problem with this substitution. If $\phi \colon M \to N$ is a smooth map between manifolds of the same dimension then it induces both a push-out map ϕ_* which maps measures on M to measures on N and a pull-back map ϕ^* which maps densities on N to densities on M. Thus ϕ_* and ϕ^* behave quite differently. Moreover, whereas pulling back densities is something which can be defined at the level of jets, it is not clear that pushing out is. The solution to these dilemmas is the easily verified fact that if ω is a density on N then

$$\phi_* \phi^* \omega = \omega. \tag{5.2}$$

Thus if the map

$$j_m^\infty(\phi^*) \colon J_{\phi(m)}^\infty(\Delta(N)) \to J_m^\infty(\Delta(M)) \tag{5.3}$$

is invertible then we can just define $j_m^\infty(\phi_*)$ to be $j_m^\infty(\phi^*)^{-1}$. The condition for $j_m^\infty(\phi^*)$ to be invertible is that the derivative of ϕ, i.e. the tangent map

$$T_m(\phi) : T_m M \to T_{\phi(m)} N \tag{5.4}$$

is invertible. In coordinates this is just the condition that the Jacobian of ϕ at m be invertible. Moreover, we do not need to start with a function ϕ defined on all of M. It suffices to take some invertible element of

$$J^\infty(M, N)_m^{\phi(m)}, \tag{5.5}$$

the space of all infinite jets of maps from M to N that map m to $\phi(m)$. An invertible element of this space is an element that has its first order term invertible or equivalently, by the inverse function theorem, it is the jet of a local diffeomorphism.

Let P be a family of (mutually absolutely continuous) probability measures on the sample space Ω. Denote by $f(\omega; p)$ the value at ω of the Radon-Nikodym derivative of the probability measure p with respect to some reference measure. The maximum likelihood estimator based on random samples of size N is the function $M^N \colon \Omega^N \to P$ defined by $M^N(\omega_1, \ldots, \omega_N)$ being the element p (assumed unique) of P which maximises $\prod_{i=1}^N f(\omega_i; p)$. If p is any point in P then $M_*^N(p)$ is a measure on P and we are interested in its asymptotic behaviour as $N \to \infty$. It is a result due to Wald (1949) that $M_*^N(p)$ approaches a delta measure at p as $N \to \infty$. Of more interest is the measure

$$(N^{1/2}\phi)_* M_*^N(p), \tag{5.6}$$

where

$$\phi : P \to T_p P \tag{5.7}$$

is a map sending p to 0 which is a local diffeomorphism. This measure is asymptotically a normal distribution on the tangent space at p, with mean 0 and variance the dual of the Fisher information metric. (See, for instance, Cox and Hinkley (1974), pp. 294–295.) We want to understand the way in which the asymptotics of $(N^{1/2}\phi)_* M_*^N(p)$ depend on the choice of ϕ.

To do this, we assume that $M_*^N(p)$ is a density on P and consider its infinite jet

$$j_p^\infty(M_*^N(p)) \in J_p^\infty(\Delta(P)). \tag{5.8}$$

Then we choose a set of coordinates at p, or more precisely the infinite jet of a set of coordinates at p. This is an element

$$j_p^\infty(\phi) \in J_p^\infty(P, \mathbb{R}^d)_p^0, \tag{5.9}$$

which for convenience we shall denote by just ϕ. We can then define the element

$$j_0^\infty((N^{1/2}\phi)_* M_*^N(p)) \in J_0^\infty(\Delta(\mathbb{R}^d)). \tag{5.10}$$

Let us denote by $\mathcal{M}$ the space $J_0^\infty(\Delta(\mathbb{R}^d))$.

Thus $(N^{1/2}\phi)_* M_*^N(p)$ is a sequence of measures μ on $\mathbb{R}^d$ depending on N. Associated to this sequence of measures is a formal asymptotic series

$$\mu = \mu_0 + \frac{1}{N^{1/2}}\mu_1 + \frac{1}{N}\mu_2 + \ldots \tag{5.11}$$

where the μ_i are signed measures on $\mathbb{R}^d$.

If we change the choice of coordinates from ϕ to χ we have $\chi = g.\phi$ for g in the infinite phylon group $\mathscr{P}(d)$. The change in the asymptotic series can be computed from

$$\begin{aligned} (N^{1/2}\chi)_* M_*(p) &= (N^{1/2}g\phi)_* M_*(p) \\ &= (N^{1/2}g\frac{1}{N^{1/2}})_* N^{1/2}(\phi)_* M_*(p). \end{aligned} \tag{5.12}$$

Thus if μ is the asymptotic series of measures (5.11) induced by the ϕ coordinates, that induced by the χ coordinates is

$$(N^{1/2}g\frac{1}{N^{1/2}})_*\mu. \tag{5.13}$$

On passing to jets we obtain an action of $\mathscr{P}(d)$ on $\mathcal{M}$, which we shall call the *twisted phylon action*. This is to distinguish it from the action where g just acts on each μ_i in the natural way. Recall from the proof of Proposition 4.1 that every non-zero $t \in \mathbb{R}$ defines a dilation $t.1 \in GL(d) \subset \mathscr{P}(d)$. Thus $(N^{1/2}g\frac{1}{N^{1/2}})_*$ is the composition of three elements of $\mathscr{P}(d)$ acting on $\mathcal{M}$. To obtain further insight into this action it is useful to consider more general phylon representations and develop a theory of *twisted phylon representations*.

6. Twisted Phylon Representations

Let V be a phylon representation as in Section 4. Decompose it into subspaces V_i of homogeneous degree n_i

$$V = \prod_{i=1}^{\infty} V_{n_i} \tag{6.1}$$

As an example, let $\mathcal{M}$ be the space $J_0^\infty(\Delta(\mathbb{R}^d))$ of infinite jets of measures considered in the previous section. Thinking of these as infinite formal power series multiplied by Lebesgue measure, we see that they can be decomposed into sums of homogeneous terms of degree k for each $k \geq 0$. The subspaces C_k consisting of homogeneous functions of degree k are the irreducible $GL(d)$-factors of $\mathscr{M}$. Recalling that the action of an element $g : \mathbb{R}^d \to \mathbb{R}^d$ in the phylon group when applied to a function $f : \mathbb{R}^d \to \mathbb{R}$ is $f \circ g^{-1}$ and that the action of a dilation $\lambda.1$ on Lebesgue measure is to multiply it by λ^d, we see that the action of such a dilation on C_k is multiplication by λ^{d-k}, that is, $n_k = d - k$. This example will be our model.

Given the representation V, we can form the space $V[[t]]$ of all formal asymptotic power series in t:

$$v = v_0 + v_1 t + v_2 t^2 + \ldots, \qquad (6.2)$$

where each v_i is in V. As V is a direct product we see that $V[[t]]$ is, in fact, bi-graded as

$$V = \prod_{i \geq 1} \prod_{m \geq 0} V_{n_i} t^m. \qquad (6.3)$$

Thus any element $v \in V[[t]]$ can be written as

$$v = \sum_{i \geq 1} \sum_{m \geq 0} v_i t^m \qquad (6.4)$$

with $v_i \in V_{n_i}$. Given an element $w = v_i t^a$ with $v_i \in V_{n_i}$, let us call n_i the homogeneous degree of w and a the asymptotic degree of w.

We define the twisted action of the phylon group on $V[[t]]$ by

$$g \star v = t^{-1} g.t.v. \qquad (6.5)$$

To understand this action in more detail it is useful to consider the corresponding twisted Lie algebra action. Let $X \in L\mathscr{P}(d)_{-k}$ and $v \in V_{n_i}$. Then

$$X \star v t^a = t^{-1} X v t^{a+n_i} = (Xv) t^{a+k}, \qquad (6.6)$$

so that the homogeneous degree decreases and the asymptotic degree increases. The action on subspaces is therefore

$$L\mathscr{P}(d)_{-k} \star V_{n_i} t^m \subset V_{n_i-k} t^{m+k}. \qquad (6.7)$$

If we filter $V[[t]]$ by the subspaces $V(a) = V[[t]].t^a$ so that

$$V[[t]] = V(0) \supset V(1) \supset V(2) \supset \cdots \qquad (6.8)$$

then

$$L\mathscr{P}(d)_{-k} \star V(a) \subset V(a+k). \qquad (6.9)$$

It follows from Proposition 3.2 that there is an action of $\mathscr{P}(d)$ on the quotient spaces

$$V(a)/V(a+k) \qquad (6.10)$$

and that $L\mathscr{P}^{(k)}(d)$ acts trivially on $V(a)/V(a+k)$. Thus the action of $\mathscr{P}(d)$ on $V(a)/V(a+k)$ factors through an action of $\mathscr{P}_k(d)$.

In the application to the maximum likelihood estimator that we are interested in we just need to replace t by $1/N^{1/2}$. Then the action of $\mathscr{P}(d)$ on $\mathscr{M}(0)/\mathscr{M}(k)$ factors through an action of $\mathscr{P}_k(d)$. This says that if we want to understand the behaviour of the asymptotic expansion (5.11) up to and including order $1/N^{k/2}$ under change of parameterisation (i.e. under the action of the phylon group), we need only consider the Taylor expansion of the change of parameterisation up to order k. This fact is the connection between the two types of order in asymptotic expansions mentioned in the introduction. In particular, considering the action of $\mathscr{P}(d)$ on $\mathscr{M}(0)/\mathscr{M}(1)$, we deduce the familiar fact that the leading order term μ_0 in the asymptotic expansion (5.11) of $(N^{1/2}\phi)_* M_*^N(p)$ is acted on only by the matrix

$$\frac{\partial \chi^i}{\partial \phi^a}(p) \tag{6.11}$$

when we change ϕ to a new set of coordinates χ. Since this action is the usual action of $GL(d)$ on $\mathbb{R}^d$, this is another way of saying that μ_0 should be thought of as a measure on the tangent space to P at p.

7. THE COADJOINT ACTION

We conclude this paper with some remarks on the coadjoint action of $\mathscr{P}(d)$ on the dual $L\mathscr{P}(d)^*$ of the Lie algebra. For infinite-dimensional Lie groups it is known that the orbit theory of Kirillov (1976) often provides a guide to constructing an interesting class of representations. In the case of the phylon group we obtain yet another link with Terng's (1978) results.

Firstly, we note from the discussion in Section 2 that (i) because the Lie algebra of the phylon group is a direct product, the dual is a direct sum

$$L\mathscr{P}(d)^* = \oplus_{i\geq 1}[\mathbb{R}^d \otimes S^i(\mathbb{R}^d)^*], \tag{7.1}$$

(ii) $L\mathscr{P}(d)$ is not a Banach space. Since the dual of a Fréchet space which is not a Banach space is never a Fréchet space (Hamilton, 1982, p. 69), the coadjoint action is not a phylon representation in the sense of Definition 4.1. Clearly, this is a manifestation of the general fact that, in the language of string theory, infinite-dimensional strings and co-strings are quite different.

Let ξ be an element of $L\mathscr{P}(d)^*$. We want to consider the orbit of ξ under the coadjoint action of the phylon group. Let $\langle \ , \ \rangle$ denote the pairing of a vector space and its dual. Then under the coadjoint action g in $\mathscr{P}(d)$ maps ξ in $L\mathscr{P}(d)^*$ to $g^{-1}\xi g$, where

$$\langle g^{-1}\xi g, X \rangle = \langle \xi, g^{-1}Xg \rangle$$

for all X in $L\mathscr{P}(d)$. Recall that we have projections

$$j^k \colon L\mathscr{P}(d) \to L\mathscr{P}_k(d) \tag{7.2}$$

and therefore dual inclusions

$$j^{k*} \colon L\mathscr{P}_k(d)^* \to L\mathscr{P}(d)^*. \tag{7.3}$$

Because the dual of the Lie algebra is the direct sum (7.1), if $\xi \neq 0$ then for some k it has the form

$$\xi = (\xi_1, \xi_2, \xi_3, \ldots, \xi_k, 0, 0, \ldots) \tag{7.4}$$

with $\xi_i \in \mathbb{R}^d \otimes S^i(\mathbb{R}^d)^*$ and $\xi_k \neq 0$. It follows that $\xi = j^{k*}(\xi')$, where

$$\xi' = (\xi_1, \xi_2, \xi_3, \ldots, \xi_k) \in L\mathscr{P}_k(d)^*, \tag{7.5}$$

and that k is the smallest integer for which this is true. We call k the order of ξ.

For g in the phylon group and X in its Lie algebra we have

$$\begin{aligned}
\langle g^{-1}\xi g, X \rangle &= \langle g^{-1}j^{k*}(\xi')g, X \rangle \\
&= \langle \xi', j^k(g^{-1}Xg) \rangle \\
&= \langle \xi', j^k(g^{-1})j^k(X)j^k(g) \rangle \\
&= \langle j^{k*}(j^k(g)^{-1}\xi'j^k(g)), X \rangle.
\end{aligned} \tag{7.6}$$

Hence we have the identity:

$$g^{-1}\xi g = j^{k*}(j^k(g)^{-1}\xi'j^k(g)). \tag{7.7}$$

Thus if ξ is an element of order k then its stabiliser contains $\mathscr{P}^{(k)}(d)$ and hence the group $\mathscr{P}_k(d)$ acts transitively on its $\mathscr{P}(d)$-orbit, which must therefore be finite-dimensional. Moreover, the coadjoint orbit of ξ is the image under $j^{k*}: L\mathscr{P}_k(d)^* \to L\mathscr{P}(d)^*$ of the coadjoint orbit of ξ'. It follows that the representations of $\mathscr{P}(d)$ obtained from coadjoint orbits are actually representations of the finite-dimensional phylon groups $\mathscr{P}_k(d)$.

Note that we have not explored here the question of whether or not the representations of $\mathscr{P}_k(d)$ described by Terng (1978) arise from orbits of the coadjoint representation. All we have observed is that the orbit theory for the infinite phylon group is no more than the union of the theories for its finite-dimensional quotient groups. Thus the orbit theory will not give rise to the infinite-dimensional examples which we have been discussing.

REFERENCES

1. O. E. Barndorff-Nielsen, P. Blæsild, A. L. Carey, P. E. Jupp, M. Mora, and M. K. Murray, *Finite dimensional algebraic representations of the infinite phylon group*, Acta. Appl. Math. **28** (1992), 219–252.
2. O. E. Barndorff-Nielsen, P. E. Jupp, and W. S. Kendall, *Stochastic calculus, statistical asymptotics, Taylor strings and phyla*, Ann. Fac. Sci. Toulouse **III** (1994), no. 1, 5–62.
3. A. L. Carey and M. K. Murray, *Higher order tensors, strings and new tensors*, Proc. R. Soc. Lond. A. **430** (1990), 423–432.
4. D. R. Cox and D. V. Hinkley, *Theoretical statistics*, Chapman and Hall, London, 1974.
5. R. S. Hamilton, *The inverse function theorem of Nash and Moser*, Bull. Am. Math. Soc. **7** (1982), 65–222.
6. A. A. Kirillov, *Elements of the theory of representations*, Springer-Verlag, Berlin, 1976.
7. O. Kobayashi, A. Yoshioka, Y. Maeda, and H. Omori, *The theory of infinite-dimensional Lie groups and its applications*, Acta. Appl. Math. **3** (1985), 71–106.
8. I. Kolář, P.W. Michor, and J. Slovák, *Natural operations in differential geometry*, Springer-Verlag, London, 1993.

9. J. Milnor, *Remarks on infinite-dimensional Lie groups*, Les Houches, Session XL, 1983, Relativity, Groups and Topology II (B. S. DeWitt and R. Stora, eds.), Elsevier, 1984.
10. M. K. Murray and J. W. Rice, *Differential geometry and statistics*, Chapman and Hall, London, 1993.
11. H. Omori, *A method of classifying expansive singularities*, J. Diff. Geom. **15** (1980), no. 4, 493–512.
12. C. L. Terng, *Natural vector bundles and natural differential operators*, Amer. J. Math. **100** (1978), 775–828.
13. F. Treves, *Topological vector spaces, distributions and kernels*, Academic Press, New York, 1967.
14. A. Wald, *Note on the consistency of the maximum likelihood estimate*, Ann. Math. Stat. **20** (1949), 595–601.

THE SPAN OF THE TANGENT CONE
OF A SCHUBERT VARIETY

JAMES B. CARRELL

To the memory of my good friend and teacher, Roger Richardson

1. INTRODUCTION

Let G denote a simply connected, simple algebraic group over an algebraically closed field k, which we will assume has characteristic zero, although the arguments in the paper work for sufficiently large characteristics. Let B and T be respectively a Borel subgroup and a maximal torus contained in B, and let W be the Weyl group $N_G(T)/T$ with the natural Chevalley-Bruhat order $<$ determined by the B. The flag variety $G/B = \mathscr{B}$ of G is a projective G-variety admitting a locally closed decomposition into B-orbits. It is well known that each B-orbit contains a unique T-fixed point, *i.e.* a point of the form $w \cdot B = n_w B/B$, where $w = n_w T$ is an element of W and n_w is a fixed representative of w. The Schubert variety $X(w) \subseteq \mathscr{B}$ is the closure of the B-orbit $Bw \cdot B$. Fundamental relationships among the Bruhat graph of an interval $[e, w] = \{x \leq w\}$ in W, the set of T-invariant curves in the corresponding Schubert variety $X(w)$ and the triviality of the Kazhdan-Lusztig polynomials $P_{x,w}$ were established in [4]. Moreover, it was announced there without proof that if $G \neq G_2$, then the T-invariant curves in $X(w)$ containing $x \cdot B$ are in a canonical one to one correspondence with the T-invariant lines in the reduced tangent cone $\mathscr{T}_{x \cdot B}(X(w))$ of $X(w)$, and if G is simply laced, one can replace the tangent cone in this correspondence by its k-linear span $\theta_k(x, w)$ in the Zariski tangent space $T_{x \cdot B}(\mathscr{B})$ of $X(w)$ at $x \cdot B$. The purpose of this note is to prove a result about the set of weights of the T-module $\theta_k(x, w)$. In particular, we show that $\theta_k(e, w)$ is the B-module span of the set of T-invariant lines in $\mathscr{T}_{x \cdot B}(X(w))$. In addition, we will prove the result on T-invariant lines in $\mathscr{T}_{x \cdot B}(X(w))$ of $X(w)$ just mentioned.

2. STATEMENTS OF RESULTS

Let $\Phi \subset X(T)$ (*resp.* Φ^+) be the set of roots for (G, T) (*resp.* the set of roots determined by B). Let $R \subset W$ denote the set of all reflections and $S \subset R$ the set of simple reflections attached to the simple roots in Φ^+. For $\alpha \in \Phi$, r_α denotes the corresponding reflection, and similarly, for $r \in R$, $\alpha_r \in \Phi^+$ is the associated positive root. If $x \leq w$, put

$$\Phi(x, w) = \{\alpha \in \Phi \mid x^{-1}(\alpha) < 0 \text{ and } r_\alpha x \leq w\}.$$

1991 *Mathematics Subject Classification.* Primary 14M15, 20H15; Secondary 20F55, 51F15.
Partially supported by a grant of the Natural Sciences and Engineering research Council of Canada.

and

$$\Omega(x, w) = \{\alpha \in \Phi \mid \alpha \text{ occurs as a weight of the } T\text{-module } T_{x \cdot B}(X(w)\}.$$

Since $T_{x \cdot B}(X(w)) \subseteq T_{x \cdot B}(\mathscr{B})$, it is clear that $\Omega(x, w) \subseteq x\Phi^-$, where $\Phi^- = -\Phi^+$.

For $\alpha \in \Phi$, let U_α denote the one dimensional unipotent subgroup of G, normalized by T, such that $xU_\alpha x^{-1} = U_{x(\alpha)}$ for all $x \in W$. If $x(\alpha) < 0$, put $C_{\alpha, x} = \overline{U_{x(\alpha)}x \cdot B}$. Otherwise, put $C_{\alpha, x} = \overline{U_{-x(\alpha)}x \cdot B}$. These curves $C_{\alpha, x}$ pay a fundamental role in the geometry of Schubert varieties (cf. [4]).

Proposition 1. ([4]) $C_{\alpha, x}$ *is a nonsingular T-invariant curve in $\mathscr{B}$ such that* $C_{\alpha, x}^T = \{x \cdot B, r_\alpha x \cdot B\}$, *and every T-invariant curve in $\mathscr{B}$ is a $C_{\alpha, x}$ for suitable α and x. Moreover, $C_{\alpha, x} \subseteq X(w)$ if and only if $x, r_\alpha x \leq w$. Consequently, if $x \leq w$, then $\Phi(x, w) \subseteq \Omega(x, w)$.*

If $x, rx \leq w$, Proposition 1 implies that $T_{x \cdot B}(C_{\alpha_r, x})$ is a T-invariant line in $T_{x \cdot B}(X(w))$ associated to the one dimensional representation of T of weight α_r if $x^{-1}(\alpha_r) < 0$ and weight $-\alpha_r$ otherwise. Moreover, $T_{x \cdot B}(C_{\alpha_r, x}) \subseteq \mathscr{T}_{x \cdot B}(X(w))$. Our main theorem is inspired by the following result, announced in [4], which answers the question of which T-invariant lines occur in a $\mathscr{T}_{x \cdot B}(X(w))$, except in one case.

Theorem 1. *Suppose $G \neq G_2$, and $x \leq w$. Then every T-invariant line in $\mathscr{T}_{x \cdot B}(X(w))$ is $T_{x \cdot B}(C_{\alpha, x})$ for some $\alpha \in \Phi(x, w)$. In particular, the set of T-invariant lines in $\mathscr{T}_{x \cdot B}(X(w))$ is in one to one correspondence with set of the T-invariant curves in $X(w)$ containing $x \cdot B$. If G is simply laced, then the same is true if $\mathscr{T}_{x \cdot B}(X(w))$ is replaced by $\theta_k(x, w)$.*

We will give Dale Peterson's original proof of Theorem 1 in §4. In view of this result, it is natural to ask which T-invariant lines occur in the k-linear span $\theta_k(x, w)$ of $\mathscr{T}_{x \cdot B}(X(w))$ in $T_{x \cdot B}(X(w))$ when G is not simply laced. Thus let $\Theta(x, w)$ denote the set of weights of this T-module. Clearly $\Phi(x, w) \subseteq \Theta(x, w) \subseteq \Omega(x, w)$. Also, let $\mathscr{H}(x, w)$ denote the set of roots in the convex hull of $\Phi(x, w)$ in $X(T) \otimes \mathbb{R}$. We now state the main theorem of the paper.

Theorem 2. *Suppose $x \leq w$. Then $\Theta(x, w) \subseteq \mathscr{H}(x, w)$ and equality holds whenever G is simply laced. If $\gamma \in \mathscr{H}(x, w) \backslash \Theta(x, w)$, then γ is short, negative and may be expressed $\gamma = \beta + \epsilon\mu$, where $\beta \in \Phi(x, w)$, $\mu \in \Phi^+$, and $\epsilon \in \{1, 2\}$. Moreover, in any such expression, $x^{-1}(\mu) < 0$. If $G \neq G_2$, then β is long and $\epsilon = 1$.*

Several examples in rank two are given in §6. In particular, we give an example where $G = G_2$ in which every T-invariant line in $\theta_k(x, w)$ is contained in $\mathscr{T}_{x \cdot B}(X(w))$, but $\Phi(x, w) \neq \Theta(x, w)$. We also give an example in which $\Theta(x, w) \neq \mathscr{H}(x, w)$.

Corollary 1. *Suppose $w \in W$. Then $\Theta(e, w) = \mathscr{H}(e, w)$. That is, $\Theta(e, w)$ is the set of all roots in the convex hull of $\Phi(e, w)$.*

Proof. This is immediate from Theorem 2, since $x = e$.

Since $e \cdot B$ is also a B-fixed point, $\theta_k(e, w)$ and $T_{e \cdot B}(X(w))$ are B-modules. Let

$$M(w) := \bigoplus_{\alpha \in \Phi(e,w)} \mathrm{span}_k \, Be_\alpha$$

where $e_\alpha \in T_{e \cdot B}(X(w))$ has weight α. Thus $M(w)$ is the B-module generated by the tangent lines at $e \cdot B$ of T-invariant curves in $X(w)$ (except possibly when $G = G_2$).

Theorem 3. [1] *For any* $w \in W$, $\theta_k(e, w) = M(w)$. *Consequently,* $\theta_k(e, w)$ *is the B-module span of the set of T-invariant lines in* $\mathscr{T}_{x \cdot B}(X(w))$.

The proofs of Theorems 2 and 3 will be given in §6. In special cases, one can deduce some facts about $\Omega(x, w)$. In particular, it is shown in [7] that if G is of type C, then $\Theta(e, w) = \Omega(e, w)$. In addition, it is shown in [6], that in type A, $\Phi(x, w) = \Omega(x, w)$ as long as $x \le w$. From Theorem 2, we therefore obtain

Corollary 2. *If G is of type A or C, then* $\Omega(e, w) = \Theta(e, w) = \mathscr{H}(e, w)$. *Moreover, if G is of type A, then for all $x \le w$,* $\Phi(x, w) = \Theta(x, w) = \Omega(x, w) = \mathscr{H}(x, w)$.

Note that when G is simply laced, it is immediate that $\Phi(x, w) = \mathscr{H}(x, w)$ for all $x \le w$. What may not be obvious in general is that $\Theta(x, w) \subseteq \mathscr{H}(x, w)$. Also note that the convex cone generated by $\Phi(x, w)$ in general contains elements of Φ not in $\Phi(x, w)$. In [5], Dabrowski has shown that when $x = w$, this cone is spanned by the $\alpha \in \Phi$ such that $\ell(r_\alpha w) = \ell(w) - 1$, where $\ell(w)$ is the length of w. Since $w \cdot B$ is a non singular point of $X(w)$, it follows from [4] that $\Phi(w, w) = \Omega(w, w)$ and $|\Phi(w, w)| = \ell(w)$.

3. Some elementary results on torus actions

Let T be an algebraic torus over an algebraically closed field k with character group $X(T)$ and set of one parameter subgroups $Y(T)$. For $\lambda \in Y(T)$ and $\chi \in X(T)$, $\langle \lambda, \chi \rangle \in \mathbb{Z}$ is defined as usual by $\chi(\lambda(s)) = s^{\langle \lambda, \chi \rangle}$, where $s \in k^*$. Let V_χ denote the representation of T over k of weight χ and fixed dimension $d_\chi > 0$. For a finite subset $F \subset X(T)$, put $V_F = \bigoplus_{\chi \in F} V_\chi$. Let v_χ be the component of v in V_χ, and put $F(v) = \{\chi \in F | \, v_\chi \ne 0\}$. Finally, let $\mathscr{H}(v)$ denote the convex hull of $F(v)$ in $X(T) \otimes \mathbb{R}$.

Proposition 2. *Suppose Y is a Zariski closed T-invariant cone in V_F, and for $\alpha \in F$, let $Y_\alpha = Y \cap V_\alpha$. Then $Y_\alpha \ne 0$ if and only if α is an extreme point of $\mathscr{H}(y)$ for some $y \in Y$.*

Proof. The only if part of the first assertion follows imediately by taking $y \in Y_\alpha \backslash \{0\}$, since in that case, $\mathscr{H}(y) = \{\alpha\}$. For the sufficiency, assume α is an extreme point of $\mathscr{H}(y)$. Then, by definition, $y = \sum_{\chi \in F(y)} y_\chi$, where $y_\alpha \ne 0$. By

---

[1] Theorem 3 was originally conjectured by the author. As mentioned in [7], an idea for a proof, different from the one in this paper, was discussed in 1991 at the AMS symposium on algebraic groups.

assumption, there is a $\lambda \in Y(T)$ such that $\langle \lambda, \chi \rangle > \langle \lambda, \alpha \rangle$ for all $\chi \in \mathscr{H}(y)$ distinct from α. Then

$$s^{-\langle \lambda, \alpha \rangle} \lambda(s) \cdot y = \sum_{\chi \in F(y)} s^{\langle \lambda, \chi - \alpha \rangle} y_\chi.$$

Since Y is a closed T-stable cone, $\lim_{s \to 0} s^{-\langle \lambda, \alpha \rangle} \lambda(s) \cdot y = y_\alpha \in Y$. This shows $Y_\alpha \neq 0$.

If the condition that Y be a cone is dropped, the same conclusion holds as long as α is taken to be an extreme edge of $\mathscr{H}(y)$ for some $y \in Y$: that is, there exists a $\lambda \in Y(T)$ such that $\langle \lambda, \chi \rangle > 0$ for all $\chi \in F$ distinct from α, and $\langle \lambda, \alpha \rangle = 0$.

Next, put $F(Y) = \{\alpha \in F \mid Y_\alpha \neq 0\}$. We now want to relate $F(Y)$ to the k-span of Y in V_F, which will be denoted by $\theta_k(Y)$. Let

$$\Theta(Y) = \{\alpha \in F \mid V_\alpha \cap \theta_k(Y) \neq 0\},$$

and let $\mathscr{H}(Y)$ denote the convex hull of $F(Y)$.

Proposition 3. *If Y is a Zariski closed T-invariant cone in V_F, then*

$$\Theta(Y) = \bigcup_{y \in Y} F(y) \subseteq \bigcup_{y \in Y} \mathscr{H}(y) \subseteq \mathscr{H}(Y).$$

Proof. First note that for any $y \in Y$, $\mathscr{H}(y) \subseteq \mathscr{H}(Y)$ as is easily seen from the fact that $\mathscr{H}(y)$ is the convex hull of its extreme points, and by Proposition 2, every extreme point lies in $F(Y)$. It is clear that $\Theta(Y) \subseteq \cup_{y \in Y} F(y)$, so it remains to prove the opposite inclusion. But for this, it suffices to consider the case where $Y = \Theta(Y)$, and then it is obvious by complete reducibility of T.

We now give a basic application of this result. Let $\lambda \in X(T)$ be an integral dominant weight and $V(\lambda)$ the finite dimensional irreducible G-module of highest weight λ. For a weight $\mu \in X(T)$ of $V(\lambda)$, let $V_\mu \subseteq V(\lambda)$ denote the associated weight space and choose a non zero vector $v_\mu \in V_\mu$. In particular, v_λ denotes a highest weight vector. For $y \in W$, let $F_{y\lambda}$ be the B-module spanned by $v_{y\lambda}$ and let $\Omega_\lambda(y)$ be the set of weights of $F_{y\lambda}$.

Proposition 4. *Let λ be an integral dominant weight and suppose $y \in W$. Then $\Omega_\lambda(y) \subseteq \mathscr{H}(\{z\lambda \mid z \leq y\})$.*

Proof. Let P denote the stabilizer of v_λ let $\phi : G/P \longrightarrow \mathbb{P}(V(\lambda))$ be the natural T-equivariant closed immersion defined by $gP \to g[v_\lambda]$, where $[v] \in \mathbb{P}(V(\lambda))$ denotes the line spanned by a nonzero $v \in V(\lambda)$. Let $X_P(y) \subseteq G/P$ denote the Schubert variety $\overline{Bn_y P/P}$. By equivariance (see [2]),

$$\phi(X_P(y)^T) = \phi(X_P(y))^T = \{[v_{z\lambda}] \mid z \leq y\}.$$

Letting $\pi(v) = [v]$ if $v \neq 0$, it is clear that $Y := \overline{Bv_{y\lambda}} = \pi^{-1}(\phi(X_P(y)))$ is a closed T-invariant cone in $V(\lambda)$, so by the last observation, the set of T-invariant lines in Y is $\{[v_{z\lambda}] \mid z \leq y\}$ (due to the fact that the T-invariant lines in a cone Y are the T-fixed points in $\mathbb{P}(Y)$). The result therefore follows immediately from Proposition 3, since $F_{y\lambda} \subseteq \mathrm{span}_k(Y)$.

Proposition 4 is proved for regular dominant λ in [5]. The case in which $F_{y\lambda} = V(\lambda)$, i.e. y is the longest element of W, is proved in [8, p. 97].

4. SCHUBERT VARIETIES

The main purpose of this section is to prove Theorem 1. We begin by establishing the properties of T-invariant curves stated in Proposition 1. For $r \in R$, let $\mathscr{S}_r = \langle U_{\alpha_r}, U_{-\alpha_r} \rangle$.

Proposition 5. *Let $x \in W$, $r \in R$, and put $t = x^{-1}rx$. Then $C_{\alpha_t,x} = \mathscr{S}_r x \cdot B$. Moreover, $C_{\alpha_t,x} \subseteq X(w)$ if and only if x, $rx = xt \leq w$.*

Proof. The first statement is obvious since $x(\alpha_t) = \pm \alpha_r$. Since $x \cdot B$, $rx \cdot B \in X(w)$ if and only if x, $rx \leq w$, the necessity in the second statement is clear. To show the converse, suppose first that $x(\alpha_t) < 0$. Then $U_{-\alpha_t} \cdot B = x^{-1} U_{-x(\alpha_t)} x \cdot B \subset \overline{x^{-1}Bw \cdot B}$, so $x \mathscr{S}_t e \cdot B \subseteq \overline{Bw \cdot B}$. If $x(\alpha_t) > 0$, then $xt(\alpha_t) < 0$, so $x \mathscr{S}_t e \cdot B = xt \mathscr{S}_t e \cdot B \subseteq \overline{Bw \cdot B}$. In either case the conclusion holds.

The assertions of Proposition 1 are all now immediate. $C_{\alpha,x}$ is nonsingular since it is an orbit. The fact that every T-invariant curve is a $C_{\alpha,x}$ follows from the well known properties of the map Ψ defined in the next paragraph.

Let U^- be the unipotent radical of B^- and $\mathbf{u}^- = \mathrm{Lie}(U^-)$ the nilpotent radical with its usual T-module decomposition $\mathbf{u}^- = \sum_{\alpha>0} \mathbf{g}_{-\alpha}$. For any enumeration $\{\beta_1, \ldots, \beta_N\}$ of Φ^+, there is (see [2]) a T-equivariant isomorphism of algebraic varieties $\Psi : \mathbf{u}^- \longrightarrow U^-$ defined by setting

$$\Psi\Big(\sum_{i=1}^{N} u_i e_{-\beta_i} \Big) = \prod_{i=1}^{N} x_{-\beta_i}(u_i)$$

where $x_\alpha : k \longrightarrow U_\alpha$ is the T-equivariant additive isomorphism associated to the root α (i.e. $tx_\alpha(s)t^{-1} = x_\alpha(\alpha(t)s)$) and $e_{-\beta_i} \in \mathbf{g}_{-\beta_i}\backslash 0$. It is well known that $Z_{x,w} = \Psi^{-1}(x^{-1}BwB \cap U^-)$ is a closed T-stable subset of $\mathbf{u}^-$ isomorphic to a neighborhood of $x \cdot B$ in $X(w)$, and there are obvious one to one correspondences among the set of T-invariant lines in $\mathscr{T}_{x \cdot B}(X(w))$, $\mathscr{T}_1(\overline{x^{-1}BwB \cap U^-})$ and $\mathscr{T}_0(Z_{x,w})$. Hence it suffices to prove the following: If $G \neq G_2$, $x \leq w$ but $xt \not\leq w$, where $t \in R$, then $\mathbf{g}_{-\alpha_t} \not\subseteq \mathscr{T}_0(Z_{x,w})$. In other words, there is an $f \in k[\mathbf{u}^-]$ such that $f(Z_{x,w}) = 0$ whose lowest term does not vanish on $\mathbf{g}_{-\alpha_t}$. This shows, by Proposition 1, that if $\mathbf{g}_{-\alpha_t} \subseteq \mathscr{T}_0(Z_{x,w})$, then $C_{\alpha_r,x} \subseteq X(w)$, where $r = xtx^{-1}$.

Having chosen a highest weight vector $v_\lambda \in V(\lambda)$, fix v_λ^* of weight $-\lambda$ in the dual $V(\lambda)^*$ so that $\langle v_\lambda^*, v_\lambda \rangle = 1$. Define $\phi_\lambda \in k[G]$ by $\phi_\lambda(g) = \langle v_\lambda^*, g \cdot v_\lambda \rangle$. We need the following well known property.

Proposition 6. *If λ is a regular dominant weight and $y \not\leq w$, then $\phi_\lambda(y^{-1}BwB) = 0$.*

Proof. By [1], $y \not\leq w$ implies $yT \cdot v_\lambda \cap \mathrm{span}_k(BwB \cdot v_\lambda) = \phi$. Thus $kv_\lambda \cap \mathrm{span}_k(y^{-1}BwB \cdot v_\lambda) = \{0\}$, so the Proposition follows.

Now suppose $x \leq w$ but $rx = xt \not\leq w$, where $t = n_t T$. Let λ be a regular dominant weight so $\phi_\lambda(tx^{-1}BwB) = 0$. Thus $n_t^* \phi_\lambda(g) = \langle n_t \cdot v_\lambda^*, g \cdot v_\lambda \rangle = 0$ for $g \in x^{-1}BwB$. Put $f_\lambda := n_t^* \phi_\lambda \Psi$; then f_λ is in the ideal of $Z_{x,w}$ in $k[\mathbf{u}^-]$. It is easy to see that $\Psi(u_1, \dots, u_N) \cdot v_\lambda = \sum_I u^I v_I$ where the sum is over all $I = (i_1, \dots, i_n) \in \mathbb{Z}_+^N$ and v_I has weight $\lambda(I) := \lambda - \sum i_j \beta_j$. Thus

$$f_\lambda(u_1, \dots, u_N) = \sum_{\lambda(I)=t(\lambda)} a_I u^I$$

with $a_I = \langle n_t \cdot v_\lambda^*, v_I \rangle$. Since $t(\lambda) = \lambda - (\lambda, \alpha_t^\vee)\alpha_t$, where $(,)$ is a W-invariant inner product on $X(T) \otimes \mathbb{R}$, and $\alpha^\vee$ is the coroot associated to α, $\lambda(I) = t(\lambda)$ means

$$(\lambda, \alpha_t^\vee)\alpha_t = \sum i_j \beta_j.$$

Lemma 1. *Assume that $|(\alpha, \beta^\vee)| \leq 2$ for all $\alpha, \beta \in \Phi$. If $\alpha = \sum_{\beta > 0} k_\beta \beta$ where each $k_\beta \in \mathbb{R}_+$, then $\sum k_\beta \geq 1$.*

Proof. This is immediate from $(\alpha, \alpha^\vee) = 2 = \sum k_\beta(\beta, \alpha^\vee) \leq \sum k_\beta |(\beta, \alpha^\vee)|$.

When $G \neq G_2$, the hypotheses of the lemma are satisfied. It follows that if $\lambda(I) = t(\lambda)$, then $\sum i_j \geq (\lambda, \alpha_t^\vee)$. Thus the lowest term f_* of f_λ has degree at least $(\lambda, \alpha_t^\vee)$. If α_t is β_j in the ordering of Φ^+, then

$$f_*(0, \dots, u_j, \dots 0) = a u_j^{(\lambda, \alpha_t^\vee)}$$

for some $a \in k^*$, so $f_*(\mathbf{g}_{-\alpha_t}) \neq 0$ and therefore $g_{-\alpha_t} \not\subseteq \mathscr{I}_0(Z_{x,w})$, as asserted. To finish the proof, assume G is simply laced. Then $\Phi(x, w) = \mathscr{H}(x, w)$ by the remark following Theorem 3. Consequently, $\Phi(x, w) = \Theta(x, w)$ by Proposition 3.

5. A LEMMA ON ROOTS

If $\Psi \subseteq \Phi$, let

$$\mathscr{H}(\Psi) = \{\gamma \in \Phi |\ \gamma \text{ is in the convex hull of } \Psi\}.$$

Lemma 2. *Suppose Φ is irreducible and $\Psi \subseteq \Phi$. If $\gamma \in \mathscr{H}(\Psi)$ does not lie in Ψ, then:*

  (i) *Φ is not simply laced and γ is a short root;*
  (ii) *if Φ is not of type G_2, then in any expression $\gamma = \sum_{i=1}^r x_i \beta_i$, where all $\beta_i \in \Psi$, $x_i > 0$, and $\sum x_i = 1$, each β_i can be paired with a unique orthogonal $\beta_{i'}$ such that $\gamma = \frac{1}{2}(\beta_i + \beta_{i'})$; and*
  (iii) *in all cases there exist $\beta \in \Psi$ and $\mu \in \Phi$ such that $\gamma = \beta + \mu$ or $\gamma = \beta + 2\mu$ and the sign of μ is opposite to the sign of γ.*

Proof. Assume $\gamma \in \mathscr{H}(\Psi)$ does not lie in Ψ and let $\gamma = \sum_{i=1}^r x_i \beta_i$ be any expression as in (ii). Since long roots are extreme points of $\mathscr{H}(\Psi)$, hence already lie in Ψ, (i) is clear. Suppose Φ is not of type G_2. From $(\gamma, \gamma^\vee) = \sum x_i(\beta_i, \gamma^\vee) = 2$, we obtain that $(\beta_i, \gamma^\vee) = 2$ for all i. Thus all β_i are long, and the angle between γ and each β_i is $\frac{\pi}{4}$. This implies that the angle between any β_i and β_j lies in $[0, \frac{\pi}{2}]$, so every $(\beta_i, \beta_j^\vee) \in \{0, 1, 2\}$. Since $(\gamma, \beta_i^\vee) = 1$, it follows from the expression in (ii) that for any i there is an i' such that $(\beta_i, \beta_{i'}) = 0$, and that γ lies in the plane spanned by β_i and $\beta_{i'}$. (ii) now follows immediately from the orthogonality and $(\gamma, \beta_i^{\vee}) = 1$ for every i. If Φ is not of type G_2, choose any i and put $\beta_i = \beta$ and $\beta_{i'} = \beta'$. Then $\gamma - \beta = \frac{1}{2}(\beta' - \beta)$ is a root, and we have the two expressions

$$\gamma = \beta + \frac{1}{2}(\beta' - \beta) = \beta' + \frac{1}{2}(\beta - \beta'),$$

which verifies (iii) except for the case of G_2. Suppose Φ is of type G_2, and assume without loss of generality that $\gamma < 0$. Then the expression in (ii) implies that there is a $\beta_i = \beta$ such that the $\mathrm{ht}(\gamma - \beta) > 0$ (due to the fact that for each $i \in \{\pm 2, \pm 3, \pm 4, \pm 5\}$, there is only one $\mu \in \Phi$ with $\mathrm{ht}(\mu) = i$). Here ht denotes the height of a root. Now it is easy to check that $\gamma - \beta \in \Phi^+ \cup 2\Phi^+$, which completes the proof of Lemma 2.

The following result will be useful in the next section.

Corollary 3. *Let* $\mathbf{u}^-$ *be given the* B-*module structure of* $\mathbf{g}/\mathbf{b}$, *where* $\mathbf{g} = \mathrm{Lie}(G)$ *and* $\mathbf{b} = \mathrm{Lie}(B)$. *If* $\Psi \subseteq \Phi^-$, *let* $M(\Psi)$ *denote the* $\mathbf{b}$-*submodule of* $\mathbf{u}^-$ *generated by the* e_{α}, *where* $\alpha \in \Psi$. *If* Σ *is the set of weights of* $M(\Psi)$, *then* $\mathscr{H}(\Psi) \subseteq \Sigma$.

Proof. A necessary and sufficient condition for $\gamma \in \Sigma$ is that $\gamma < 0$ and $\gamma = \beta + \sum r_i \mu_i$, where $\beta \in \Phi(e, w)$, each $\mu_i \in \Phi^+$, and all $r_i \geq 0$. But Lemma 1 implies $\gamma \in \mathscr{H}(\Psi)$ has this form, so $\mathscr{H}(\Psi) \subseteq \Sigma$.

6. THE LINEAR SPAN OF THE TANGENT CONE

We now prove Theorem 2. First, $\Theta(x, w) \subseteq \mathscr{H}(x, w)$ follows immediately from Proposition 3. Assume $\gamma \in \mathscr{H}(x, w) \backslash \Theta(x, w)$. If γ is positive, then $r_\gamma x < x$, so $\gamma \in \Phi(x, w) \subseteq \Theta(x, w)$, which is a contradiction. Hence, γ is negative. By the previous lemma, γ is short, and may be expressed $\gamma = \beta + \epsilon\mu$, where $\beta \in \Phi(x, w)$, $\mu > 0$, and $\epsilon \in \{1, 2\}$. It remains to show that in any such expression for γ, $x^{-1}(\mu) < 0$. Suppose $x^{-1}(\mu) > 0$ for some $\mu > 0$ such that $\gamma = \beta + \epsilon\mu$ as above. Now U_μ stabilizes $X(w)$ and fixes $x \cdot B$. Indeed, $U_\mu x \cdot B = x U_{x^{-1}(\mu)} B = x \cdot B$. Therefore $\theta_k(x, w)$ is a U_μ-submodule of $T_{x \cdot B}(\mathscr{B})$. As $x^{-1}(\Phi(x, w)) \subseteq \Phi$, $x^{-1}(\gamma) < 0$. Consequently $x^{-1}(\beta + \mu) < 0$, irregardless of whether ϵ is 1 or 2 in the expression for γ. Since k has characteristic zero, this implies $[e_\beta, e_\mu]$ projects to a non zero element of $T_{x \cdot B}(\mathscr{B})$, so $\beta + \mu \in \Theta(x, w)$. If $\epsilon = 2$, the argument may be repeated, giving $\gamma \in \Theta(x, w)$ also. This is a contradiction, so $x^{-1}(\mu) < 0$.

Next, we prove Theorem 3. Recall that $e_\alpha \in T_{e \cdot B}(X(w)) \subseteq \mathbf{u}^-$ denotes a non zero vector of weight $\alpha < 0$, and

$$M(w) = \bigoplus_{\alpha \in \Phi(e, w)} \mathrm{span}_k Be_\alpha$$

is the B-module generated by the T-invariant lines in $\mathscr{T}_{x \cdot B}(X(w))$ (except possibly in the G_2 case). Let Σ be the set of weights of $M(w)$ as a B-module. Clearly, $M(w) \subseteq \theta_k(e, w)$ since $\theta_k(e, w)$ is B-stable, so $\Sigma \subseteq \Theta(e, w)$. We must show that $\Theta(e, w) = \Sigma$. But Corollary 3 implies $\mathscr{H}(e, w) \subseteq \Sigma$, and, by Corollary 1, $\Theta(e, w) = \mathscr{H}(e, w)$ so we are done.

7. EXAMPLES.

In this section, we will compute $\Phi(x,w)$, $\Theta(x,w)$ and $\mathscr{H}(x,w)$ for several Schubert varieties in $\mathscr{B}$, where G is of rank 2. Throughout, let α denote the simple short root and β the simple long root. Note that by [4], $|\Phi(x,w)| = \ell(w)$ for all $x \leq w$ in the rank two case.

Example 1. Suppose G is of type B_2, and let $w = r_\alpha r_\beta r_\alpha$.

(1) $x = e \implies \Phi(e,w) = \{-\alpha, -\beta, -(2\alpha+\beta)\}$, $\Theta(e,w) = \mathscr{H}(e,w) = \Phi(e,w) \cup \{-(\alpha+\beta)\}$;

(2) $x = r_\beta \implies \Phi(x,w) = \Theta(x,w) = \mathscr{H}(x,w) = \{-\alpha, \beta, -(\alpha+\beta)\}$;

(3) $x = r_\alpha \implies \Phi(x,w) = \{\alpha, -\beta, -(2\alpha+\beta)\}$, $\Theta(x,w) = \mathscr{H}(x,w) = \Phi(x,w) \cup \{-(\alpha+\beta)\}$;

(4) $x = r_\alpha r_\beta \implies \Phi(x,w) = \Theta(x,w) = \mathscr{H}(x,w) = \{\alpha, 2\alpha+\beta, -(\alpha+\beta)\}$;

(5) $x = r_\beta r_\alpha \implies \Phi(x,w) = \Theta(x,w) = \mathscr{H}(x,w) = \{-\alpha, \beta, \alpha+\beta\}$;

(6) $x = w \implies \Phi(x,w) = \Theta(x,w) = \mathscr{H}(x,w) = \{\alpha, \alpha+\beta, 2\alpha+\beta\}$.

In (3), $\mu = \alpha$. Since $r_\alpha(\mu) < 0$, Theorem 2 does not say whether or not $-(\alpha+\beta) \in \Theta(x,w)$. However, we can show that $\dim_k \theta_k(x,w) = 4$ using the following:

Proposition 7. *Suppose $r \in S$ has the property that $rw < w$. Then n_r stabilizes $X(w)$. Consequently, $|\Theta(x,w)| = |\Theta(rx,w)|$. In fact, $\Theta(rx,w) = r\Theta(x,w)$. In particular, $\dim_k \theta_k(x,w) = \dim_k \theta_k(rx,w)$.*

Proof. This is an immediate consequence of the fundamental property of the Bruhat decompositon that if $r \in S$, then $rBw \subset BwB \cup BrwB$.

Note that Proposition 7 also holds for the tangent spaces to $X(w)$ at $x \cdot B$ and $rx \cdot B$, e.g. $r\Omega(x,w) = \Omega(rx,w)$. In fact, $dn_r : T_{x \cdot B}(X(w)) \to T_{rx \cdot B}(X(w))$ is an isomorphism.

Example 2. Let G be of type B_2 as in Example 1. To obtain an example in which $\Theta(x,w) \neq \mathscr{H}(x,w)$, let $w = r_\beta r_\alpha r_\beta$ and $x = r_\beta r_\alpha$. In this case, $\Phi(x,w) = \{\beta, \alpha+\beta, -(2\alpha+\beta)\}$, and $\mathscr{H}(x,w) = \Phi(x,w) \cup \{-\alpha\}$. However, $x \cdot B$ is a smooth point of $X(w)$, due to the fact that every Schubert variety is non singular in codimension one. In fact, $X(w)$ is non singular. Thus $\Phi(x,w) = \Theta(x,w) = \Omega(x,w) \subset \mathscr{H}(x,w)$. Notice that the decomposition $\gamma = \beta + \mu$ of Theorem 2 is $-\alpha = -(2\alpha+\beta) + (\alpha+\beta)$, and $x^{-1}(\alpha+\beta) = -\alpha < 0$.

Combining Example 2 and Theorem 3 shows that there is no obvious relationship between $\mathscr{H}(x,w)$ and $\Omega(x,w)$.

Example 3. We now give an example where $G = G_2$, $\Phi(x,w) \subset \Theta(x,w) = \mathscr{H}(x,w)$, but every T-invariant line in $\Theta(x,w)$ is contained in $\mathscr{T}_{x \cdot B}(X(w))$. Thus Theorem 1 cannot hold in G_2. Let $w = r_\alpha r_\beta r_\alpha r_\beta r_\alpha = r_{\beta+2\alpha}$, let $x = r_\alpha r_\beta r_\alpha = r_{\beta+3\alpha}$ and put $\gamma = r_\beta(\alpha) = \alpha + \beta$. Then $\Phi(x,w) = \{\alpha, -\beta, 2\alpha+\beta, 3\alpha+\beta, -(3\alpha+2\beta)\}$, and by inspection, $\mathscr{H}(x,w) = \Phi(x,w) \cup \{-\gamma\}$. By Proposition 5, $C_{\gamma,x} \not\subseteq X(w)$. We claim that $T_{x \cdot B}(C_{\gamma,x}) \subseteq \mathscr{T}_{x \cdot B}(X(w))$. Clearly $x < w$, $xr_\gamma \not\leq w$. We will show that $e_{-x^{-1}(\gamma)} = e_{-\gamma} \in \mathscr{T}_0(Z_{x,w})$. Let λ_β denote the fundamental dominant weight for β. By [3, Appendix], the ideal of $Z_{x,w}$ is the principal ideal generated by $f_{\lambda_\beta} \in k[\mathbf{u}^-]$. Moreover, $\lambda_\beta - r_\gamma(\lambda) = 3\gamma = \lambda_\beta + \beta$, since λ_β is also the highest root. As $(\lambda_\beta, \gamma^\vee) = 3$, the leading term of f_{λ_β} is a non zero multiple of $u_\beta u_{\lambda_\beta}$. Thus $e_{-\gamma} \in \mathscr{T}_0(Z_{x,w})$, and the claim is proven.

Remark. These examples bring out a couple of points worth mentioning. First, Example 2 shows that the inequality $\mathscr{H}(x, w) \subseteq \Omega(x, w)$ cannot hold in general, although it holds in all the other examples. Moreover, Theorem 3 shows that the reverse inclusion in general fails when $x = e$. Secondly, we do not have any general results about the set $\Theta(x, w) \backslash \Phi(x, w)$ for $x \neq e$. Example 1, part 3, indicates that there is no analogue of Theorem 3 in the case $x \neq e$. Indeed, $-(\alpha + \beta) = -(2\alpha + \beta) + \alpha$, but $x^{-1}(\alpha) = -\alpha < 0$. That is, $\theta_k(x, w)$ is not the U_α-module span of the set of lines in $\mathscr{T}_{x \cdot B}(X(w))$. Essentially the same phenomenon occurs in Example 3.

The author heartily thanks the referee for finding some errors and for the suggestions for improving the exposition.

REFERENCES

1. I. N. Bernstein, I. M. Gelfand, and S. I. Gelfand, *Schubert cells and cohomology of the spaces G/P*, Russ. Math. Surveys **28** (1973), 1–26.
2. A. Borel, *Linear Algebraic Groups*, Graduate Texts in Mathematics, no. 126, Springer Verlag, New York, 1991.
3. J. B. Carrell, *Bruhat cells in the nilpotent variety and intersection rings of Schubert varieties*, J. Differential Geometry **37** (1993), 651–668.
4. _____, *The Bruhat graph of a Coxeter group, a conjecture of Deodhar, and rational smoothness of Schubert varieties*, Proc. Symp. in Pure Math. **56** (1994), 53–61, Part I.
5. R. Dabrowski, *A simple proof of a necessary and sufficient condition for the existence of global sections of a line bundle on a Schubert variety*, Kazhdan-Lusztig Theory and Related Topics, Contemporary Mathematics, no. 139, 1992, pp. 113–121.
6. V. Lakshmibai and C. S. Seshadri, *Singular locus of a Schubert variety*, Bull. Amer. Math. Soc. (N.S.) **11** (1984), 363–366.
7. P. Polo, *On Zariski tangent spaces of Schubert varieties and a proof of a conjecture of Deodhar*, Indag. Math. **5** (1994), 483–493.
8. H. Samelson, *Notes on Lie Algebras*, Springer Verlag, New York, 1990.

DEPARTMENT OF MATHEMATICS, UNIVERSITY OF BRITISH COLUMBIA, 121–1984 MATHEMATICS ROAD, VANCOUVER, B.C. V6T 1Z2, CANADA
E-mail address: carrell@math.ubc.ca

CANONICAL BASES, REDUCED WORDS, AND LUSZTIG'S PIECEWISE-LINEAR FUNCTION

R. W. CARTER

In memory of Roger Richardson

1. QUANTIZED ENVELOPING ALGEBRAS

Let g be a simple Lie algebra over $\mathbb{C}$. Then g is determined up to isomorphism by its Cartan matrix $A = (A_{ij})$. This is an $l \times l$ matrix where l is the rank of g. We shall assume that the Cartan matrix A is symmetric – this is equivalent to the condition that the Dynkin diagram of g has no multiple bonds.

Let $\mathscr{U}(g)$ be the universal enveloping algebra of g. This is an infinite dimensional associative algebra with the same representation theory as g. $\mathscr{U}(g)$ is the $\mathbb{C}$-algebra with 1 generated by elements

$$e_1, \ldots, e_l; \qquad f_1, \ldots, f_l; \qquad h_1, \ldots h_l$$

subject to a system of well known defining relations. Let $\mathscr{U}_\mathbb{Q}(g)$ be the $\mathbb{Q}$-algebra with 1 given by this system of generators and relations.

There is an important deformation of $\mathscr{U}_\mathbb{Q}(g)$ call the quantum group, or quantized enveloping algebra, of g. This is the $\mathbb{Q}(v)$-algebra with 1 generated by elements

$$E_1, \ldots, E_l, \qquad F_1, \ldots, F_l, \qquad K_1, \ldots, K_l, \qquad K_1^{-1}, \ldots, K_l^{-1}$$

subject to a system of defining relations which reduce to those of $\mathscr{U}_\mathbb{Q}(g)$ when the substitutions

$$E_i \to e_i, \qquad F_i \to f_i, \qquad K_i \to 1, \qquad v \to 1, \qquad \frac{K_i - K_i^{-1}}{v - v^{-1}} \to h_i$$

are made, a process known as specialisation.

Each of the above algebras has a triangular decomposition. Let n_+ be the Lie subalgebra of g generated by $e_1, \ldots e_l$; n_- the Lie subalgebra generated by $f_1, \ldots, f_l$; and h the Lie subalgebra generated by $h_1, \ldots, h_l$. Then we have

$$g = n_- \oplus h \oplus n_+$$

and h is a Cartan subalgebra of g. Let $\mathscr{U}^-, \mathscr{U}^0, \mathscr{U}^+$ be the enveloping algebras of n_-, h, n_+ respectively. These may be regarded as subalgebras of $\mathscr{U}$ and we have

$$\mathscr{U} = \mathscr{U}^- \otimes_\mathbb{C} \mathscr{U}^0 \otimes_\mathbb{C} \mathscr{U}^+.$$

The quantized enveloping algebra U also has a triangular decomposition

$$U = U^- \otimes_{\mathbb{Q}(v)} U^0 \otimes_{\mathbb{Q}(v)} U^+$$

where U^-, U^0, U^+ are the $\mathbb{Q}(v)$-subalgebras with 1 of U generated by $F_1, \ldots, F_l$; $K_1, \ldots, K_l,\ K_1^{-1}, \ldots, K_l^{-1}$; and $E_1, \ldots, E_l$ respectively.

We next recall some basic facts about the finite dimensional irreducible g-modules. Let $h^* = \operatorname{Hom}(h, \mathbb{C})$ be the dual space of h. The elements $h_1, \ldots, h_l$ form a basis for h and so the elements $\omega_1, \ldots, \omega_l \in h^*$ defined by

$$\omega_i(h_j) = \delta_{ij}$$

form a basis for h^*. These are called the fundamental weights. Each $\lambda \in h^*$ can be written in the form

$$\lambda = \lambda_1 \omega_1 + \cdots + \lambda_l \omega_l \qquad \lambda_i \in \mathbb{C}$$

Such an element λ is called integral if each $\lambda_i \in \mathbb{Z}$ and dominant integral if, in addition, each $\lambda_i \geq 0$. There is a bijection between finite dimensional irreducible g-modules, up to isomorphism, and dominant integral weights. The irreducible g-module corresponding to λ will be denoted $\mathscr{L}(\lambda)$. This module $\mathscr{L}(\lambda)$ has a unique 1-dimensional h-submodule $\mathscr{L}(\lambda)_\lambda$ giving the weight λ. $\mathscr{L}(\lambda)_\lambda$ is called the highest weight space of $\mathscr{L}(\lambda)$ and non-zero vectors in it highest weight vectors. The module $\mathscr{L}(\lambda)$ may also be regarded naturally as a $\mathscr{U}(g)$-module.

Lusztig [13] has shown how to obtain finite dimensional irreducible U-modules $L(\lambda)$ analogous to the $\mathscr{L}(\lambda)$, such that

$$\dim_{\mathbb{Q}(v)} L(\lambda) = \dim_{\mathbb{C}} \mathscr{L}(\lambda)$$

and such that $L(\lambda)$ specialises to $\mathscr{L}(\lambda)$ is a suitable sense. The module $L(\lambda)$ also has a 1-dimensional highest weight space $L(\lambda)_\lambda$ whose non-zero vectors are called highest weight vectors for $L(\lambda)$.

2. CANONICAL BASES

Kashiwara [12] and Lusztig [14] have independently, and by different methods, proved the existence of a basis B of U^- with very striking properties. This is called the canonical basis of U^-. Let λ be a dominant integral weight and v_λ be a highest weight vector in the finite dimensional irreducible U-module $L(\lambda)$. Then the elements $bv_\lambda \in L(\lambda)$ for $b \in B$ which are non-zero form a basis of $L(\lambda)$. Thus from the canonical basis B of U^- we obtain bases of all the modules $L(\lambda)$ simultaneously.

By specialising we obtain analogous results for g-modules. There is a canonical basis $\mathscr{B}$ for $\mathscr{U}^-$ such that, for all dominant integral weights λ, the elements $bv_\lambda \in \mathscr{L}(\lambda)$ for $b \in \mathscr{B}$ which are non-zero form a basis of $\mathscr{L}(\lambda)$. Thus we obtain bases for all finite dimensional irreducible g-modules simultaneously. This result was new, and most unexpected, even in the classical case.

Kashiwara and Lusztig have obtained algebraic proofs of this result, using rather different methods. Lusztig also has a geometric proof using methods of intersection cohomology. This geometric proof enables properties of the canonical basis to be proved which cannot be obtained by other methods. The most striking of these is the positivity property. This asserts that

$$bb' = \sum_{b'' \in B} c(b, b', b'') b'' \qquad b, b' \in B$$

where $c(b, b', b'') \in \mathbb{N}[v, v^{-1}]$. Thus the multiplication constants of U^- with respect to the canonical basis are Laurent polynomials in v with non-negative integer coefficients. The non-negative integers can be interpreted as the dimensions of certain intersection cohomology groups. By specialising we obtain

$$bb' = \sum_{b'' \in \mathscr{B}} c(b, b', b'') b'' \qquad b, b' \in \mathscr{B}$$

where $c(b, b', b'') \in \mathbb{N}$. Thus the multiplication constants of $\mathscr{U}^-$ with respect to the canonical basis are non-negative integers.

3. CANONICAL BASES AND LUSZTIG-DYER BASES

We shall now indicate how the canonical basis B can be obtained, using Lusztig's algebraic approach. We first recall some properties of the Weyl group and braid group of g.

The Weyl group W of g is the finite Coxeter group generated by elements $s_1, \ldots, s_l$ subject to relations

$$s_i^2 = 1$$
$$s_i s_j = s_j s_i \qquad \text{if } A_{ij} = 0$$
$$s_i s_j s_i = s_j s_i s_j \qquad \text{if } A_{ij} = -1.$$

(In the case we are considering we have $A_{ij} \in \{0, -1\}$ when $i \neq j$.)

The braid group T of g is the infinite group generated by elements $t_1, \ldots, t_l$ subject to relations

$$t_i t_j = t_j t_i \qquad \text{if } A_{ij} = 0$$
$$t_i t_j t_i = t_j t_i t_j \qquad \text{if } A_{ij} = -1$$

There is a standard action of the braid group T on g and on $\mathscr{U}(g)$.

Every element w in the Weyl group W can be expressed in the form

$$w = s_{i_k} \ldots s_{i_1}$$

and such an expression is called reduced if k is minimal. This minimal value of k is called $l(w)$. The Weyl group W has a unique element w_0 of maximal length.

We next recall the root system of g. We have the triangular decomposition

$$g = n_- \oplus h \oplus n_+$$

and n_+, n_- are h-submodules. Each of them decomposes into a direct sum of 1-dimensional h-submodules. Thus we have

$$n_+ = \Sigma \mathbb{C} e_\alpha \qquad \alpha \in h^*$$

where $[x e_\alpha] = \alpha(x) e_\alpha$ for all $x \in h$. The set of all α arising in this way is the set of positive roots, denoted by Φ^+. Similarly n_- gives rise to the set of negative roots Φ^-. We write $\Phi = \Phi^+ \cup \Phi^-$. The roots coming from the h-submodules $\mathbb{C}e_1, \ldots, \mathbb{C}e_l$ are called $\alpha_1, \ldots, \alpha_l$ respectively. We write $\Pi = \{\alpha_1, \ldots, \alpha_l\}$. This is the set of fundamental roots.

Now the Weyl group W acts on Φ, and the set Φ^+ can be obtained from Π in the following convenient manner. We begin with the longest element $w_0 \in W$ and let

$$w_0 = s_{i_N} \ldots s_{i_1}$$

be a reduced expression for w_0. Now the element w_0 has the property that $w_0(\Phi^+) = \Phi^-$. We have $|\Phi^+| = l(w_0) = N$ and we consider the positive roots made negative by the partial products

$$s_{i_1}, s_{i_2}s_{i_1}, \ldots, s_{i_N} \ldots s_{i_1} = w_0.$$

In fact just one additional positive root is made negative at each stage under such partial products. This gives us a total order on the set Φ^+ as follows:

$$\Phi^+ = \{\alpha_{i_1}, s_{i_1}(\alpha_{i_2}), s_{i_1}s_{i_2}(\alpha_{i_3}), \ldots, s_{i_1} \ldots s_{i_{N-1}}(\alpha_{i_N})\}.$$

Using a similar idea applied to negative roots we can obtain a basis for n_- given by

$$f_{i_1}, t_{i_1}(f_{i_2}), t_{i_1}t_{i_2}(f_{i_3}), \ldots, t_{i_1} \ldots t_{i_{N-1}}(f_{i_N}).$$

By the Poincaré-Birkhoff-Witt basis theorem the enveloping algebra $\mathscr{U}^- = \mathscr{U}(n_-)$ has a basis

$$\frac{f_{i_1}^{c_1}}{c_1!} \frac{t_{i_1}(f_{i_2})^{c_2}}{c_2!} \frac{t_{i_1}t_{i_2}(f_{i_3})^{c_3}}{c_3!} \cdots \frac{t_{i_1} \ldots t_{i_{N-1}}(f_{i_N})^{c_N}}{c_N!}$$

for all $c_i \geq 0$ in $\mathbb{Z}$.

Now Lusztig has defined an action of the braid group T on U which specialises to the action of T on $\mathscr{U}$. Using this action Lusztig and Dyer [16] have described a basis for U^- which specialises to the above PBW-basis of $\mathscr{U}^-$. As before we take a reduced expression

$$w_0 = s_{i_N} \ldots s_{i_1}.$$

We write $\mathbf{i} = (i_1, \ldots, i_N)$ and $\mathbf{c} = (c_1, \ldots, c_N)$ where $c_i \geq 0$ lie in $\mathbb{Z}$. Let $F_{\mathbf{i}}^{\mathbf{c}} \in U^-$ be the element defined by

$$F_{\mathbf{i}}^{\mathbf{c}} = \frac{F_{i_1}^{c_1}}{[c_1]!} \frac{t_{i_1}(F_{i_2})^{c_2}}{[c_2]!} \frac{t_{i_1}t_{i_2}(F_{i_3})^{c_3}}{[c_3]!} \cdots \frac{t_{i_1} \ldots t_{i_{N-1}}(F_{i_N})^{c_N}}{[c_N]!}.$$

Here $[c]! = [1][2] \ldots [c]$, where

$$[i] = \frac{v^i - v^{-i}}{v - v^{-1}} = v^{i-1} + v^{i-3} + \cdots + v^{-(i-1)}.$$

Also the action $t_{i_1}(F_{i_2})$ is that denoted by $\tilde{T}_{i_1}(F_{i_2})$ in Lusztig [14].

Then the elements $F_{\mathbf{i}}^{\mathbf{c}}$ for fixed $\mathbf{i}$ and variable $\mathbf{c}$ form a basis for U^-. We shall call it a LD-basis. We have one LD-basis for each reduced decomposition of w_0.

We now compare two different LD-bases $\{F_i^c\}$ and $\{F_{i'}^c\}$. It can be shown that

$$F_{i'}^{c'} = \sum_c \gamma_{i,\,i'}^{c,\,c'} F_i^c$$

where $\gamma_{i,\,i'}^{c,\,c'} \in \mathbb{Q}[v^{-1}]$. Moreover the constant term of $\gamma_{i,\,i'}^{c,\,c'}$ is 1 for one value of c and 0 for all the others. Let L be the $\mathbb{Q}[v^{-1}]$-submodule of U^- spanned by the elements F_i^c as c varies. We then see that the lattice L is independent of the choice of i. We also have a natural homomorphism

$$\theta : L \to L/v^{-1}L$$

where $L/v^{-1}L$ is a vector space over $\mathbb{Q}$. The elements $\theta(F_i^c)$ for fixed i as c varies from a $\mathbb{Q}$-basis of $L/v^{-1}L$ which is independent of i.

Now there is a map $\underset{u \to \bar{u}}{U \to U}$ which is a homomorphism of $\mathbb{Q}$-algebras and which satisfies

$$\bar{E}_i = E_i \quad \bar{F}_i = F_i, \qquad \bar{K}_i = K_i^{-1}, \qquad \bar{K}_i^{-1} = K_i,$$

$$\bar{v} = v^{-1}, \qquad \bar{v}^{-1} = v.$$

This map is an involution. Let $\bar{L}$ be the $\mathbb{Q}[v]$-submodule of U^- which is the image of L under this involution. We consider the restriction of θ to $L \cap \bar{L}$:

$$\theta : L \cap \bar{L} \to L/v^{-1}L.$$

It was shown by Lusztig that this map is bijective. Let B be the basis of $L \cap \bar{L}$ corresponding to the basis $\theta(F_i^c)$ of $L/v^{-1}L$ under this bijection. Then B is a $\mathbb{Q}$-basis of $L \cap \bar{L}$, a $\mathbb{Q}[v^{-1}]$-basis of L, a $\mathbb{Q}[v]$-basis of $\bar{L}$, and a $\mathbb{Q}(v)$-basis of U^-. This is the canonical basis of U^-.

We shall give two examples to illustrate this situation. Suppose first that g has type A_1. Then we have $W = \langle s_1; s_1^2 = 1 \rangle$ and $w_0 = s_1$. Thus there is a single reduced expression for w_0 and the corresponding LD-basis for U^- is $\frac{F^c}{[c]!}$ for $c \in \mathbb{Z}$, $c \geq 0$. The canonical basis B of U^- is the same as the LD-basis in this case.

Next we suppose that g has type A_2. Here the situation is somewhat more complicated. It is important to understand it because of its relevance in higher cases. We have

$$W = \langle s_1, s_2; s_1^2 = 1, s_2^2 = 1, s_1 s_2 s_1 = s_2 s_1 s_2 \rangle.$$

The largest element w_0 of W has two reduced expressions

$$w_0 = s_1 s_2 s_1$$

$$w_0 = s_2 s_1 s_2.$$

Let $\mathbf{i}' = (1\ 2\ 1)$ and $\mathbf{i}'' = (2\ 1\ 2)$. Then the corresponding LD-bases for U^- are

$$F_{\mathbf{i}'}^{\mathbf{c}} = \frac{F_1^a}{[a]!}\,\frac{t_1(F_2)^b}{[b]!}\,\frac{t_1t_2(F_1)^c}{[c]!} \qquad \mathbf{c} = (a\ b\ c)$$

$$F_{\mathbf{i}''}^{\mathbf{c}'} = \frac{F_2^{a'}}{[a']!}\,\frac{t_2(F_1)^{b'}}{[b']!}\,\frac{t_2t_1(F_2)^{c'}}{[c']!} \qquad \mathbf{c}' = (a'\ b'\ c')$$

The canonical basis B of U^- has the form

$$B = \left\{ \frac{F_1^m}{[m]!}\frac{F_2^{m'}}{[m']!}\frac{F_1^{m''}}{[m'']!}\ m' \geq m + m''; \ \frac{F_2^m}{[m]!}\frac{F_1^{m'}}{[m']!}\frac{F_2^{m''}}{[m'']!}\ m' > m + m'' \right\}$$

where we note that in the case $m' = m + m''$ we have the identity

$$\frac{F_1^m}{[m]!}\,\frac{F_2^{m+m''}}{[m+m'']!}\,\frac{F_1^{m''}}{[m'']!} = \frac{F_2^{m''}}{[m'']!}\,\frac{F_1^{m+m''}}{[m+m'']!}\,\frac{F_2^m}{[m]!}.$$

The bijective correspondence described above between the LD-basis $F_{\mathbf{i}'}^{\mathbf{c}}$ and the canonical basis B is given by

$$F_{\mathbf{i}'}^{(a\ b\ c)} \to \frac{F_1^a}{[a]!}\,\frac{F_2^{b+c}}{[b+c]!}\,\frac{F_1^b}{[b]!} \qquad \text{if } a \leq c$$

$$\to \frac{F_2^b}{[b]!}\,\frac{F_1^{a+b}}{[a+b]!}\,\frac{F_2^c}{[c]!} \qquad \text{if } a \geq c$$

On the other hand, the bijective correspondence between the LD-basis $F_{\mathbf{i}''}^{\mathbf{c}'}$ and the canonical basis B is given by

$$F_{\mathbf{i}''}^{(a'\ b'\ c')} \to \frac{F_1^{b'}}{[b']!}\,\frac{F_2^{a'+b'}}{[a'+b']!}\,\frac{F_1^{c'}}{[c']!} \qquad \text{if } a' \geq c'$$

$$\to \frac{F_2^{a'}}{[a']!}\,\frac{F_1^{b'+c'}}{[b'+c']!}\,\frac{F_2^{b'}}{[b']!} \qquad \text{if } a' \leq c'$$

By examining these bijections we see that $F_{\mathbf{i}'}^{(a\ b\ c)}$ gives rise to the same canonical basis element as $F_{\mathbf{i}''}^{(a'\ b'\ c')}$ if and only if

$$(a'\ b'\ c') = (b+c-a,a,b) \qquad \text{if } a \leq c$$
$$(b,c,a+b-c) \qquad \text{if } a \geq c.$$

4. Lusztig's piecewise-linear function

Let $N = l(w_0) = |\Phi^+|$. We shall now describe a piecewise-linear function $R : \mathbb{R}^N \to \mathbb{R}^N$ which was introduced by Lusztig and which is important in understanding the properties of the canonical basis. Let $\Delta = \{1,\dots,l\}$ be the set of vertices of the Dynkin diagram of g. Vertices i,j are joined in Δ if and only if $A_{ij} = -1$. Now there is just one way of decomposing Δ into two subsets

$\Delta' = \{1, \ldots, t\}$ and $\Delta'' = \{t+1, \ldots, l\}$ such that no two vertices in Δ' are joined and no two vertices in Δ'' are joined. With such a numbering of the vertices we define two reduced expressions for w_0:

$$\mathbf{i'} \qquad w_0 = s_1 s_2 \ldots s_l s_1 s_2 \ldots s_l \ldots$$

$$\mathbf{i''} \qquad w_0 = s_l s_{l-1} \ldots s_1 s_l s_{l-1} \ldots s_1 \ldots$$

where each expression has N terms.

Now it is possible to find a sequence of reduced expressions for w_0 beginning with $\mathbf{i'}$ and ending with $\mathbf{i''}$ such that consecutive expressions differ by a single braid relation. This will either have the form that $s_i s_j$ is replaced by $s_j s_i$ when $A_{ij} = 0$, or that $s_i s_j s_i$ is replaced by $s_j s_i s_j$ when $A_{ij} = -1$. We define a function $R : \mathbb{R}^N \to \mathbb{R}^N$ using this sequence of reduced expressions. We start with a vector $(c_1, \ldots, c_N) \in \mathbb{R}^N$. We modify this vector step by step. Whenever $s_i s_j$ is replaced by $s_j s_i$ we replace the corresponding pair of components a b of the vector by b a. Whenever $s_i s_j s_i$ is replaced by $s_j s_i s_j$ we replace the corresponding triple a b c by a' b' c' where

$$(a', b', c') = \begin{cases} (b + c - a, a, b) & \text{if } a \le c \\ (b, c, a + b - c) & \text{if } a \ge c \end{cases}.$$

The motivation for this definition comes from type A_2 discussed above. Suppose the final vector obtained from this process is $(c'_1, \ldots, c'_N)$. We define

$$R(c_1, \ldots, c_N) = (c'_1, \ldots, c'_N).$$

It is evident that the function R is piecewise-linear. Moreover it is independent of the sequence of steps used to pass from reduced expression $\mathbf{i'}$ to $\mathbf{i''}$. This is because the LD-basis element $F_{\mathbf{i'}}^{\mathbf{c}}$ gives rise to the same canonical basis element as the LD-basis element $F_{\mathbf{i''}}^{\mathbf{c'}}$ if and only if $\mathbf{c'} = R(\mathbf{c})$.

Thus each canonical basis element determines a vector $\mathbf{c}$ (using reduced expression $\mathbf{i'}$) and a vector $\mathbf{c'}$ (using reduced expression $\mathbf{i''}$) and these vectors are related by $R(\mathbf{c}) = \mathbf{c'}$. The importance of both vectors $\mathbf{c}$ and $\mathbf{c'}$ is illustrated in a striking way by a theorem of Lusztig [14] giving the condition for a canonical basis element $b \in B$ to give a non-zero image $bv_\lambda \in L(\lambda)$. We have $bv_\lambda \ne 0$ if and only if

$$c_i \le \lambda_i \qquad \text{for } i = 1, \ldots, t$$

$$c'_i \le \lambda_{l+1-i} \qquad \text{for } i = 1, \ldots, l - t.$$

Now the vector space $\mathbb{R}^N$ decomposes into the union of subsets which are the regions of linearity for the function R. It is conjectured that there is a bijective correspondence between the regions of linearity of the function R and different types of canonical basis element. This was illustrated in type A_2 above, where there are two regions of linearity for R and two types of canonical basis element.

5. SEQUENCES OF REDUCED WORDS IN TYPE A_l

We now suppose g has type A_l and that the vertices of the Dynkin diagram are numbered as shown

$$1 \qquad 2 \qquad 3 \qquad\qquad l-2 \quad l-1 \quad l$$

o————o————o—— $\cdots$ ——o————o————o

We define the two reduced expressions for w_0:

$$\mathbf{i'} \qquad w_0 = (s_1 s_3 s_5 \ldots)(s_2 s_4 s_6 \ldots)(s_1 s_3 s_5 \ldots)(s_2 s_4 s_6 \ldots)..$$
$$\mathbf{i''} \qquad w_0 = (s_2 s_4 s_6 \ldots)(s_1 s_3 s_5 \ldots)(s_2 s_4 s_6 \ldots)(s_1 s_3 s_5).. $$

where each expression has $l+1$ factors.

Each of these reduced expressions for w_0 determines a total order on the set Φ^+ of positive roots in the manner described in §3. These two orderings are actually opposite to each other, showing that expressions $\mathbf{i'}$ and $\mathbf{i''}$ for w_0 are as far apart from one another as possible. We consider a sequence of reduced expressions of w_0 from $\mathbf{i'}$ to $\mathbf{i''}$ of the type discussed above, and concentrate on the braid relations of the type $s_i s_j s_i \to s_j s_i s_j$ used in passing from one to the next. A relation $s_i s_{i+1} s_i \to s_{i+1} s_i s_{i+1}$ will be denoted by τ_i and a relation $s_{i+1} s_i s_{i+1} \to s_i s_{i+1} s_i$ by $\bar{\tau}_i$.

Suppose $l = 2$. Then we have

$$\mathbf{i'} \qquad s_1 s_2 s_1$$
$$\mathbf{i''} \qquad s_2 s_1 s_2$$

and the braid relation used is τ_1.

Next suppose $l = 3$. Then we have

$$\mathbf{i'} \qquad s_1 s_3 s_2 s_1 s_3 s_2$$
$$\mathbf{i''} \qquad s_2 s_1 s_3 s_2 s_1 s_3$$

and the following sequence of reduced words passes from $\mathbf{i'}$ to $\mathbf{i''}$.

1	3	2	1	3	2	
3	1	2	1	3	2	τ_1
3	2	1	2	3	2	τ_2
3	2	1	3	2	3	
3	2	3	1	2	3	$\bar{\tau}_2$
2	3	2	1	2	3	$\bar{\tau}_1$
2	3	1	2	1	3	
2	1	3	2	1	3	

The sequence of braid relations used (from right to left) is

$$\bar{\tau}_1 \bar{\tau}_2 \tau_2 \tau_1.$$

Next suppose $l = 4$. Then we have

$$\mathbf{i}' \qquad s_1 s_3 s_2 s_4 s_1 s_3 s_2 s_4 s_1 s_3$$

$$\mathbf{i}'' \qquad s_2 s_4 s_1 s_3 s_2 s_4 s_1 s_3 s_2 s_4$$

The reader might like to construct a sequence of reduced words passing from $\mathbf{i}'$ to $\mathbf{i}''$ in which the length 3 braid relations used (from right to left) are

$$\tau_3 \tau_2 \tau_1 \bar{\tau}_1 \bar{\tau}_2^2 \bar{\tau}_3 \tau_3 \tau_2 \tau_1$$

($\bar{\tau}_2^2$ means that relation $\bar{\tau}_2$ is applied in two different places in the given word).

We now describe the pattern for arbitrary l. This can be seen from the first few values.

$l = 2$ $\qquad\qquad\qquad\qquad\qquad \tau_1$

$l = 3$ $\qquad\qquad\qquad\qquad (\bar{\tau}_1 \bar{\tau}_2)(\tau_2 \tau_1)$

$l = 4$ $\qquad\qquad\qquad (\tau_3 \tau_2 \tau_1)(\bar{\tau}_1 \bar{\tau}_2^2 \bar{\tau}_3)(\tau_3 \tau_2 \tau_1)$

$l = 5$ $\qquad\qquad (\bar{\tau}_1 \bar{\tau}_2 \bar{\tau}_3 \bar{\tau}_4)(\tau_4 \tau_3^2 \tau_2^2 \tau_1)(\bar{\tau}_1 \bar{\tau}_2^2 \bar{\tau}_3^2 \bar{\tau}_4)(\tau_4 \tau_3 \tau_2 \tau_1)$

$l = 6$ $\quad (\tau_5 \tau_4 \tau_3 \tau_2 \tau_1)(\bar{\tau}_1 \bar{\tau}_2^2 \bar{\tau}_3^2 \bar{\tau}_4^2 \bar{\tau}_5)(\tau_5 \tau_4^2 \tau_3^3 \tau_2^2 \tau_1)(\bar{\tau}_1 \bar{\tau}_2^2 \bar{\tau}_3^2 \bar{\tau}_4^2 \bar{\tau}_5)(\tau_5 \tau_4 \tau_3 \tau_2 \tau_1)$

In general, when $l = 2m$, the sequence of length 3 braid relations has form

$$\theta_0 \varphi_1 \theta_1 \varphi_2 \ldots \varphi_2 \theta_1 \varphi_1 \theta_0 \qquad (2m - 1 \text{ factors})$$

where

$$\theta_i = \tau_{2m-1} \tau_{2m-2}^2 \cdots \tau_{2m-2i}^{2i} \tau_{2m-2i-1}^{2i+1} \cdots \tau_{2i+1}^{2i+1} \tau_{2i}^{2i} \cdots \tau_2^2 \tau_1$$

$$\varphi_1 = \bar{\tau}_1 \bar{\tau}_2^2 \cdots \bar{\tau}_{2i-1}^{2i-1} \bar{\tau}_{2i}^{2i} \cdots \bar{\tau}_{2m-2i}^{2i} \bar{\tau}_{2m-2i+1}^{2i-1} \cdots \bar{\tau}_{2m-2}^2 \bar{\tau}_{2m-1}$$

and, when $l = 2m - 1$, the sequence has form

$$\tilde{\varphi}_1 \tilde{\theta}_1 \tilde{\varphi}_2 \tilde{\theta}_2 \ldots \varphi_2 \theta_2 \varphi_1 \theta_1 \qquad (2m - 2 \text{ factors})$$

where

$$\theta_i = \tau_{2m-2} \tau_{2m-3}^2 \cdots \tau_{2m-2i+1}^{2i-2} \tau_{2m-2i}^{2m-1} \cdots \tau_{2i-1}^{2i-1} \tau_{2i-2}^{2i-2} \cdots \tau_2^2 \tau_1.$$

$$\varphi_i = \bar{\tau}_1 \bar{\tau}_2^2 \cdots \bar{\tau}_{2i-1}^{2i-1} \bar{\tau}_{2i}^{2i} \cdots \bar{\tau}_{2m-2i}^{2i} \bar{\tau}_{2m-2i+1}^{2i-1} \cdots \bar{\tau}_{2m-3}^2 \bar{\tau}_{2m-2}$$

$$\tilde{\theta}_i = \tau_{2m-2} \tau_{2m-3}^2 \cdots \tau_{2m-2i}^{2i-1} \tau_{2m-2i-1}^{2i} \cdots \tau_{2i}^{2i} \tau_{2i-1}^{2i-1} \cdots \tau_2^2 \tau_1$$

$$\tilde{\varphi}_i = \bar{\tau}_1 \bar{\tau}_2^2 \cdots \bar{\tau}_{2i-2}^{2i-2} \bar{\tau}_{2i-1}^{2i-1} \cdots \bar{\tau}_{2m-2i}^{2i-1} \bar{\tau}_{2m-2i+1}^{2i-2} \cdots \bar{\tau}_{2m-3}^2 \bar{\tau}_{2m-2}$$

We may define a graph in terms of this sequence of length 3 braid relations which has some useful properties. We call it the braid graph. We first note that each braid relation

$$\tau_i \quad \begin{matrix} i & i+1 & i \\ i+1 & i & i+1 \end{matrix} \qquad \bar{\tau}_i \quad \begin{matrix} i+1 & i & i+1 \\ i & i+1 & i \end{matrix}$$

has 3 initial letters and 3 terminal letters. The vertices of the braid graph are the length 3 braid relations τ_i or $\bar{\tau}_i$ occurring in the sequence. Two vertices of the braid graph are joined by an edge if and only if there is a terminal letter of one of the braid relations which, when followed through the sequence of reduced expressions, becomes an initial letter of the other.

The first few braid graphs are:

$\ell = 2$

$\ell = 3$

$\ell = 4$

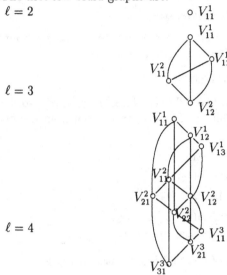

Let Γ be the set of vertices of the braid graph. Then we have

$$\Gamma = \Gamma_1 \,\dot\cup\, \Gamma_2 \,\dot\cup\, \ldots \,\dot\cup\, \Gamma_{l-1}$$

where Γ_i is the set of τ or $\bar\tau$ occurring in the ith factor of form $\theta, \varphi, \tilde\theta, \tilde\varphi$ in the sequence. We have

$$|\Gamma_1| = 1(l-1), \quad |\Gamma_2| = 2(l-2), \ldots, |\Gamma_{l-1}| = (l-1).1$$

and

$$|\Gamma| = \binom{l+1}{3}.$$

Consider the graph Γ_i whose edges are induced from those of Γ. Then Γ_i has a very simple form. It is a rectangle graph of shape $(l-i) \times i$.

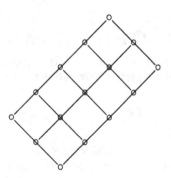

Its vertices are denoted by V_{jk}^i in a natural way.

It can be shown that any edge of the braid graph Γ is either an edge in one of the rectangle subgraphs Γ_i or is an edge joining vertices in two consecutive rectangle subgraphs Γ_i, Γ_{i+1}. The edges joining vertices in Γ_i to Γ_{i+1} can be readily described, but we shall not go into details here. See Carter [5],[2].

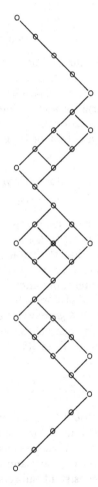

The braid graph of type A_6. (Not all edges included).

The sequence of reduced words passing from $\mathbf{i}'$ to $\mathbf{i}''$ described in this section is by no means the only possible one. However it appears that the braid graph obtained from this given sequence of reduced words has particularly favourable properties.

6. Regions of linearity of Lusztig's function in type A_l

We recall from §4 that Lusztig's piecewise-linear function R can be defined step by step, corresponding to a sequence of reduced expressions for w_0 proceeding from expression $\mathbf{i}'$ to expression $\mathbf{i}''$. Two consecutive reduced expressions differ by a braid relation. When $s_i s_j$ is replaced by $s_j s_i$ the corresponding pair of letters $a\ b$ is replaced by $b\ a$. When $s_i s_j s_i$ is replaced by $s_j s_i s_j$ the corresponding triple of letters $a\ b\ c$ is replaced by $a'\ b'\ c'$ where

$$(a'\ b'\ c') = (b + c - a, a, b) \qquad \text{if } a \le c$$
$$(b, c, a + b - c) \qquad \text{if } a \ge c.$$

Thus we have a choice of linear function for each application of a length 3 braid relation $s_i s_j s_i = s_j s_i s_j$.

Now the set of length 3 braid relations used are in bijective correspondence with the vertices of the braid graph. Thus for each vertex of the braid graph we have a choice between two linear functions. Let Γ be the set of vertices of the braid graph and S be a subset of Γ. Let C_S be the set of all vectors in $\mathbb{R}^N$ satisfying the inequalities

$$a < c \qquad \text{for each vertex in } S$$
$$a > c \qquad \text{for each vertex not in } S.$$

(a and c belong to the triple $a\ b\ c$ corresponding to the given vertex of Γ.) Then the restriction of the function R to C_S is linear.

It may appear from this that the total number of regions of linearity of the function R is $2^{|\Gamma|}$. This is not so, however, for some of the systems of inequalities defining C_S may be inconsistent, in which case C_S is empty. Moreover, even if we take two distinct subsets S, S' of Γ for which $C_S, C_{S'}$ are non-empty, the resulting linear functions R on C_S and $C_{S'}$ may be identical. We say that S is consistent if $C_S \ne \varphi$ and we obtain in this way an equivalence relation on the consistent subsets S of Γ. The regions of linearity of the function $R : \mathbb{R}^N \to \mathbb{R}^N$ have the form

$$\bigcup_S \bar{C}_S$$

taken over all consistent subsets S in an equivalence class.

Now it is possible to give a necessary and sufficient condition, in terms of the braid graph, for a subset $S \subset \Gamma$ to be consistent. See Carter [3], Corollary 25. It is also possible to give a necessary and sufficient condition, also in terms of the braid graph, for two consistent subsets S, S' to be equivalent. See Carter [3], Theorem 19. We shall not go into details here. By using these conditions it is possible to determine the number of regions of linearity of the function R for small values of l.

In type A_1 there is just one region of linearity, since R is a linear function. In type A_2 we have seen that there are two regions of linearity. The function R in type A_3 was investigated by Lusztig in [17]. In this case $|\Gamma| = 4$ and so there are 16 subsets $S \subset \Gamma$. It turns out that 11 of these subsets are consistent. Moreover

2 of these 11 consistent subsets are equivalent, and all others are inequivalent. Thus there are 10 regions of linearity of R in type A_3. In type A_4 it has been shown by Carter [4] that the function R has 144 regions of linearity. This is the highest case in which it seems possible to find the number of regions of linearity without using computational methods. In fact the conditions for consistency and for equivalence of subsets S are rather well adapted to the application of computational techniques. Using such methods Cockerton [8] has shown that in type A_5 the function R has 6608 regions of linearity. Thus we have:

Type	Number of regions of linearity
A_1	1
A_2	2
A_3	10
A_4	144
A_5	6608

7. THE SITUATION IN TYPE D_l

We now suppose g has type D_l and that the vertices of the Dynkin diagram are numbered as shown.

We again define two particular reduced expressions for the longest element w_0. These words differ when l is even and l is odd.

Suppose l is even. Define $\mathbf{i}', \mathbf{i}''$ by:

$\mathbf{i}'$ $\quad w_0 = (s_1 s_3 \ldots s_{l-3} s_{l-1} s_l s_2 s_4 \ldots s_{l-2})(s_1 s_3 \ldots s_{l-3} s_{l-1} s_l s_2 s_4 \ldots s_{l-2}) \ldots$

$\mathbf{i}''$ $\quad w_0 = (s_2 s_4 \ldots s_{l-2} s_1 s_3 \ldots s_{l-3} s_{l-1} s_l)(s_2 s_4 \ldots s_{l-2} s_1 s_3 \ldots s_{l-3} s_{l-1} s_l) \ldots$

In each expression there are $l-1$ repeated factors.
Now suppose l is odd. This time define $\mathbf{i}', \mathbf{i}''$ by:

$\mathbf{i}'$ $\quad w_0 = (s_1 s_3 \ldots s_{l-2} s_2 s_4 \ldots s_{l-3} s_{l-1} s_l)(s_1 s_3 \ldots s_{l-2} s_2 s_4 \ldots s_{l-3} s_{l-1} s_l) \ldots$

$\mathbf{i}''$ $\quad w_0 = (s_2 s_4 \ldots s_{l-3} s_{l-1} s_l s_1 s_3 \ldots s_{l-2})(s_2 s_4 \ldots s_{l-3} s_{l-1} s_l s_1 s_3 \ldots s_{l-2}) \ldots$

Again we have $l-1$ repeated factors in each expression.
We may again find a sequence of reduced expressions for w_0 passing from $\mathbf{i}'$ to $\mathbf{i}''$. This was done by Cockerton [11]. The length 3 braid relations used are:

τ_i	i	$i+1$	i	$\bar{\tau}_i$	$i+1$	i	$i+1$	$1 \leq i \leq l-2$
	$i+1$	i	$i+1$		i	$i+1$	i	

τ_{l-1}	$l-2$	l	$l-2$	$\bar{\tau}_{l-1}$	l	$l-2$	l
	l	$l-2$	l		$l-2$	l	$l-2$

The sequence of length 3 braid relations used to pass from $\mathbf{i}'$ to $\mathbf{i}''$ in low rank cases is as follows.

$$l = 3 \quad ([\tau_1 \bar{\tau}_2 \tau_2 \bar{\tau}_1])$$

$$l = 4 \quad (\bar{\tau}_1 [\tau_3 \bar{\tau}_2 \tau_2 \bar{\tau}_3 . \tau_2 \bar{\tau}_3 \tau_3 \bar{\tau}_2] \tau_1)(\bar{\tau}_1 [\tau_2 \bar{\tau}_3 \tau_3 \bar{\tau}_2] \tau_1)$$

$$l = 5 \quad (\bar{\tau}_1 \bar{\tau}_2^2 [\tau_4 \bar{\tau}_3 \tau_3 \bar{\tau}_4 . \tau_3 \bar{\tau}_4 \tau_4 \bar{\tau}_3 . \tau_4 \bar{\tau}_3 \tau_3 \bar{\tau}_4] \tau_2^2 \tau_1)(\bar{\tau}_1 \bar{\tau}_2^2 [\tau_4 \bar{\tau}_3 \tau_3 \bar{\tau}_4 . \tau_3 \bar{\tau}_4 \tau_4 \bar{\tau}_3] \tau_2^2 \tau_1)$$

$$(\bar{\tau}_1 \bar{\tau}_2 [\tau_3 \bar{\tau}_4 \tau_4 \bar{\tau}_3] \tau_2 \tau_1)$$

In general the sequence can be factorised naturally into $d - 2$ factors. Each factor has the form

$$\bar{\tau}_1 \bar{\tau}_2^2 \ldots \bar{\tau}_i^i \ldots \bar{\tau}_{l-3}^i [\quad] \tau_{l-3}^i \ldots \tau_i^i \ldots \tau_2^2 \tau_1$$

where the term in the middle bracket is a product of i or $i + 1$ quadruples of the form $\tau_{l-2} \bar{\tau}_{l-1} \tau_{l-1} \bar{\tau}_{l-2}$ or $\tau_{l-1} \bar{\tau}_{l-2} \tau_{l-2} \bar{\tau}_{l-1}$ taken alternately.

We may define the braid graph of type D_l in an analogous manner to our definition in type A_l. Let the set of vertices be Γ. This time we have

$$|\Gamma| = 4 \binom{l}{3}$$

and $\Gamma = \Gamma_1 \,\dot\cup\, \Gamma_2 \,\dot\cup\, \ldots \dot\cup\, \Gamma_{l-2}$ where Γ_i is the set of vertices in the ith factor above, starting from the right. The graphs $\Gamma_1, \Gamma_2, \ldots, \Gamma_{l-2}$ turn out to be what are called split rectangle graphs of shape $1 \times (2l - 5), 2 \times (2l - 6), \ldots, (l - 2) \times (l - 2)$ respectively. A split rectangle graph of shape 5×9 is shown in the figure. We

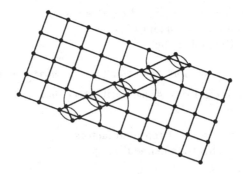

think of the rectangle as being split into two by a pair of tramlines across the middle. The vertices on the tramlines split into sets of 4, and each such set of 4 forms a braid graph of type $A_3 = D_3$. If each set of 4 is regarded as a single entity we recover the ordinary rectangle graph.

The way in which the split rectangle graphs Γ_i are joined together to form the braid graph Γ has been determined by Cockerton [7]. For each subset $S \subset \Gamma$ we have a region $C_S \subset \mathbb{R}^N$ as before. We can find a criterion in terms of the braid graph for a subset S to be consistent, i.e. $C_S \neq \emptyset$, and a criterion for two consistent subsets S, S' to be equivalent, i.e. R is given by the same linear function on C_S and $C_{S'}$.

Using this method Cockerton [9] has shown that Lusztig's piecewise-linear function $R : \mathbb{R}^{12} \to \mathbb{R}^{12}$ in type D_4 has 1204 regions of linearity.

8. THE NON-SIMPLY LACED CASES

The behaviour of Lusztig's piecewise-linear function in the non-simply laced cases can be deduced from its behaviour in the simply laced cases by making use of an appropriate graph automorphism.

We illustrate this by comparing types D_l and B_{l-1}.

The Dynkin diagrams of these Lie algebras are as shown. The diagram of type D_l has an automorphism σ which satisfies

$$\sigma(i) = i \text{ for } 1 \le i \le l - 2, \quad \sigma(l - 1) = l, \quad \sigma(l) = l - 1.$$

The orbits of σ on the Dynkin diagram of D_l are in bijective correspondence with the vertices of the Dynkin diagram of B_{l-1}. The last vertex of B_{l-1} will be denoted by $\overline{l-1}$. It corresponds to the pair of vertices $\{l - 1, l\}$ in type D_l.

We may obtain our two reduced expressions $\mathbf{i}'$ and $\mathbf{i}''$ for the longest element $w_0 \in W(B_{l-1})$ from the corresponding reduced expressions for $w_0 \in W(D_l)$. We note that in the reduced expressions $\mathbf{i}'$ and $\mathbf{i}''$ for $w_0 \in W(D_l)$ the fundamental reflections s_{l-1} and s_l always occur consecutively. Whenever they occur we replace $s_{l-1}s_l$ by $s_{\overline{l-1}} \in W(B_{l-1})$. Then we obtain the following reduced expressions for $w_0 \in W(B_{l-1})$. They differ when l is even and l is odd.

Suppose l is even. Define

$\mathbf{i}' \qquad w_0 = (s_1 s_3 \ldots s_{l-3} s_{\overline{l-1}} s_2 s_4 \ldots s_{l-2})(s_1 s_3 \ldots s_{l-3} s_{\overline{l-1}} s_2 s_4 \ldots s_{l-2}) \ldots$

$\mathbf{i}'' \qquad w_0 = (s_2 s_4 \ldots s_{l-2} s_1 s_3 \ldots s_{l-3} s_{\overline{l-1}})(s_2 s_4 \ldots s_{l-2} s_1 s_3 \ldots s_{l-3} s_{\overline{l-1}}) \ldots .$

In each expression there are $l - 1$ repeated factors.

Now suppose l is odd. Define

$\mathbf{i}' \qquad w_0 = (s_1 s_3 \ldots s_{l-2} s_2 s_4 \ldots s_{l-3} s_{\overline{l-1}})(s_1 s_3 \ldots s_{l-2} s_2 s_4 \ldots s_{l-3} s_{\overline{l-1}}) \ldots$

$\mathbf{i}'' \qquad w_0 = (s_2 s_4 \ldots s_{l-3} s_{\overline{l-1}} s_1 s_3 \ldots s_{l-2})(s_2 s_4 \ldots s_{l-3} s_{\overline{l-1}} s_1 s_3 \ldots s_{l-2}) \ldots .$

Again we have $l - 1$ repeated factors in each expression.

We may find a sequence of reduced expressions for w_0 passing from $\mathbf{i}'$ to $\mathbf{i}''$. As before, it is not necessary for our purposes to keep track of the length 2 braid relations used. However this time we have length 3 and length 4 braid relations. These are

$$
\begin{array}{llll}
\tau_i : & i & i+1 & i \\
& i+1 & i & i+1
\end{array}
\qquad
\begin{array}{lll}
\bar\tau_i : & i+1 & i & i+1 \\
& i & i+1 & i
\end{array}
\qquad 1 \le i \le l-2
$$

$$
\begin{array}{lllll}
\epsilon : & l-2 & \overline{l-1} & l-2 & \overline{l-1} \\
& \overline{l-1} & l-2 & \overline{l-1} & l-2
\end{array}
\qquad
\begin{array}{lllll}
\bar\epsilon : & \overline{l-1} & l-2 & \overline{l-1} & l-2 \\
& l-2 & \overline{l-1} & l-2 & \overline{l-1}
\end{array}
$$

The sequence of length 3 and 4 braid relations used to pass from **i′** to **i″** in low rank cases as follows.

$l = 3$ (Type B_2) ϵ

$l = 4$ (Type B_3) $(\bar\tau_1 \bar\epsilon^2 \tau_1)(\bar\tau_1 \epsilon \tau_1)$

$l = 5$ (Type B_4) $(\bar\tau_1 \bar\tau_2^2 \epsilon^3 \tau_2^2 \tau_1)(\bar\tau_1 \bar\tau_2^2 \bar\epsilon^2 \tau_2^2 \tau_1)(\bar\tau_1 \bar\tau_2 \epsilon \tau_2 \tau_1)$

$l = 6$ (Type B_5) $(\bar\tau_1 \bar\tau_2^2 \bar\tau_3^3 \bar\epsilon^4 \tau_3^3 \tau_2^2 \tau_1)(\bar\tau_1 \bar\tau_2^2 \bar\tau_3^3 \epsilon^3 \tau_3^3 \tau_2^2 \tau_1)(\bar\tau_1 \bar\tau_2^2 \bar\tau_3^2 \bar\epsilon^2 \tau_3^2 \tau_2^2 \tau_1)(\bar\tau_1 \bar\tau_2 \bar\tau_3 \epsilon \tau_3 \tau_2 \tau_1)$.

This sequence of relations is closely related to the corresponding sequence in type D_l. We recall that the sequence in type D_l factorised into $l - 2$ factors, each of which has the form

$$
\bar\tau_1 \bar\tau_2^2 \ldots \bar\tau_i^i \ldots \bar\tau_{l-3}^i [\quad] \tau_{l-3}^i \ldots \tau_i^i \ldots \tau_2^2 \tau_1
$$

where the term in the middle bracket is a product of i or $i + 1$ quadruples of the form $\tau_{l-2}\bar\tau_{l-1}\tau_{l-1}\bar\tau_{l-2}$ or $\tau_{l-1}\bar\tau_{l-2}\tau_{l-2}\bar\tau_{l-1}$. Each of these quadruples is replaced by a length 4 braid relation ϵ or $\bar\epsilon$ to give the corresponding sequence in type B_{l-1}. The terms τ_i and $\bar\tau_i$ are left unchanged.

We may define the braid graph of type B_{l-1} in an analogous manner to the previous cases. This time we have two different types of vertices – those corresponding to length 3 braid relations and those corresponding to length 4 relations. The former will be denoted as before, whereas the latter will be denoted by diamond symbols $\Diamond$.

Let Γ be the set of vertices of the braid graph of type B_{l-1}. Then we have

$$
|\Gamma| = 4\binom{l-1}{3} + \binom{l-1}{2}.
$$

There are $4\binom{l-1}{3}$ vertices corresponding to length 3 braid relations and $\binom{l-1}{2}$ vertices corresponding to length 4 braid relations. We have

$$
\Gamma = \Gamma_1 \dot\cup \Gamma_2 \dot\cup \ldots \dot\cup \Gamma_{l-2}
$$

where Γ_i is the set of vertices in the ith factor above, starting from the right. The graphs $\Gamma_1, \Gamma_2, \ldots \Gamma_{l-2}$ are rectangle graphs of type B and shape

$$
1 \times (2l - 5), \ 2 \times (2l - 6), \ 3 \times (2l - 7), \ \ldots (l - 2) \times (l - 2)
$$

respectively. A rectangle graph of type B and shape 5×9 is shown in the figure. We observe that it is obtained from the split rectangle graph of type D and shape 5×9 pictured in §7, by replacing each set of 4 tramline vertices forming an A_3 braid graph by a single vertex of type $\Diamond$ representing a length 4 braid relation.

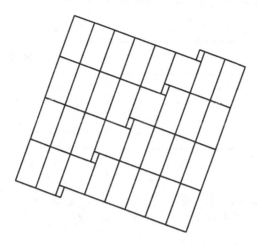

The way in which the rectangle graphs Γ_i of type B are joined together to form the braid graph Γ is closely related to the corresponding question in type D_l, and has been determined by Cockerton [10, 6].

The regions of linearity for Lusztig's piecewise-linear function R in type B_{l-1} are closely related to those in type D_l. Suppose $R(\mathbf{c}) = \mathbf{c}'$ where $\mathbf{c}, \mathbf{c}' \in \mathbb{R}^N$ and $N = l(w_0)$ where $w_0 \in W(D_l)$. We recall from §3 that the components of the vector $\mathbf{c}$ are in $1-1$ correspondence with the positive roots of D_l using the reduced decomposition $\mathbf{i}'$ of w_0. Similarly the components of the vector $\mathbf{c}'$ are also in $1-1$ correspondence with the positive roots of D_l, this time using the reduced decomposition $\mathbf{i}''$ of w_0. Now the symmetry σ of the Dynkin diagram of D_l extends naturally to an automorphism $\sigma : \Phi^+ \to \Phi^+$ of the set of positive roots of D_l. We say that a vector $\mathbf{c} \in \mathbb{R}^N$ is σ-stable if its components corresponding to positive roots in the same σ-orbit are equal. Similarly $\mathbf{c}' \in \mathbb{R}^N$ is called σ-stable if the same holds, this time using reduced expression $\mathbf{i}''$.

We now consider one of the linear functions f which make up the piecewise-linear function R. Thus R agrees with the linear function f on one of its regions of linearity in $\mathbb{R}^N$. We let $f(\mathbf{c}) = \mathbf{c}'$ where $\mathbf{c}, \mathbf{c}' \in \mathbb{R}^N$. We say f is σ-stable if $f(\mathbf{c})$ is σ-stable whenever $\mathbf{c}$ is σ-stable. Some but not all of the linear components of R will be σ-stable. There is a close connection between the regions of linearity of the function R in type B_{l-1} and the σ-stable regions of linearity of the function R in type D_l. For example, Lusztig showed [17] that, of the 10 regions of linearity in type $D_3 = A_3$, 4 are σ-stable, and these give the 4 regions of linearity in type B_2. Cockerton [9] has shown that, of the 1204 regions of linearity in type D_4, 140 are σ-stable. Each of these determines a region of linearity of the function R in type B_3. These regions of linearity are all distinct, and form a complete set. Thus the function R in type B_3 has 140 regions of linearity.

The other cases in which information in the non-simply laced cases can be obtained from that in simple laced cases are as follows.

$$A_{2\ell-1} \qquad \sigma(i) = 2\ell - i$$

$$C_\ell$$

For example, of the 6608 regions of linearity in type A_5, 140 are σ-stable. This leads to 140 regions of linearity in type C_3. (See [8].)

We also have

$$E_6 \qquad \begin{aligned} \sigma(1) &= 6 & \sigma(2) &= 5 \\ \sigma(3) &= 3 & \sigma(4) &= 4 \\ \sigma(5) &= 2 & \sigma(6) &= 1 \end{aligned}$$

$$F_4$$

Thus information about regions of linearity in type E_6 should yield corresponding information in type F_4.

Finally we can take $g = D_4$ and σ a graph automorphism of order 3.

$$D_4 \qquad \sigma(1) = 3 \quad \sigma(2) = 2 \quad \sigma(3) = 4 \quad \sigma(4) = 1$$

$$G_2$$

We can thus obtain information about regions of linearity in type G_2 from type D_4. Cockerton [9] has shown that, of the 1204 regions of linearity in type D_4, just 16 are σ-stable. These give rise to 16 regions of linearity in type G_2 which are all distinct and form a complete set. Lusztig's function R in type G_2 thus has 16 regions of linearity. The 16 corresponding linear functions can be written down explicitly. There are therefore conjectured to be 16 distinct types of canonical basis elements in type G_2.

REFERENCES

1. R. W. Carter, *The boundary walls of regions of linearity for Lusztig's piecewise-linear function*, Warwick preprint 12, 1994.
2. _____, *The braid graph*, Warwick preprint 9, 1994.
3. _____, *The braid graph and Lusztig's piecewise-linear function*, Warwick preprint 10, 1994.
4. _____, *On Lusztig's piecewise-linear function in types A_3 and A_4*, Warwick preprint 11, 1994.
5. _____, *Sequences of reduced words in symmetric groups*, Warwick preprint 8, 1994.
6. J. W. Cockerton, *The braid graph in type B_l*, Warwick preprint 70, 1994.
7. _____, *The braid graph in type D_l*, Warwick preprint 68, 1994.
8. _____, *Regions of linearity for Lusztig's piecewise-linear function in type A_5 and a counter example to a positivity conjecture*, Warwick preprint 72, 1994.
9. _____, *Regions of linearity for Lusztig's piecewise-linear function in types D_4, B_3, G_2*, Warwick preprint 71, 1994.
10. _____, *Sequences of reduced words in Coxeter groups of type B_l*, Warwick preprint 69, 1994.
11. _____, *Sequences of reduced words in Coxeter groups of type D_l*, Warwick preprint 67, 1994.
12. M. Kashiwara, *On crystal bases of the q-analogue of universal enveloping algebras*, Duke Math. Jour. **63** (1991), 465–516.
13. G. Lusztig, *Quantum deformations of certain simple modules over enveloping algebras*, Adv. in Math. **70** (1988), 237–249.
14. _____, *Canonical bases arising from quantized enveloping algebras*, Jour. Amer. Math. Soc. **3** (1990), 447–498.
15. _____, *Canonical bases arising from quantized enveloping algebras II*, Progress of Theor. Physics Suppl. **102** (1990), 175–201.
16. _____, *Quantum groups at roots of 1*, Geom. Dedicata **35** (1990), 89–114, (With appendix by M. Dyer).
17. _____, *Introduction to quantized enveloping algebras*, Progress in Mathematics **105** (1992), 49–65.
18. _____, *Introduction to Quantum Groups*, Birkhäuser, 1993.

MATHEMATICS INSTITUTE, UNIVERSITY OF WARWICK, COVENTRY CV4 7AL, UK

GEOMETRIC RATIONALITY OF SATAKE COMPACTIFICATIONS

W. A. CASSELMAN

Dedicated to the memory of Roger W. Richardson

INTRODUCTION

Throughout this paper, except in a few places, let

$G =$ the $\mathbb{R}$-rational points on a reductive group defined over $\mathbb{R}$

$K =$ a maximal compact subgroup

$X =$ the associated symmetric space

Let (π, V) be a finite-dimensional algebraic representation of G. Eventually G will be assumed semi-simple and π irreducible. If V contains a vector v with the property that K is the stabilizer of v, I will call (π, V, v) a *spherical representation*. Let $[v]$ be the image of v in $\mathbb{P}(V)$. The closure of the image of the G-orbit of $[v]$ in $\mathbb{P}(V)$ is then a G-covariant compactification $\overline{X}_\pi$ of the symmetric space X. Such compactifications were first systematically examined in Satake (1960a), and they are called *Satake compactifications*.

Suppose G to be defined over $\mathbb{Q}$ and Γ an arithmetic subgroup. In a second paper Satake (1960b) a procedure was given to obtain compactifications of the arithmetic quotient $\Gamma \backslash X$ from certain of these compactifications $\overline{X}_\pi$, by adjoining to X certain *rational boundary components* in $\overline{X}_\pi$. It was not clear at that time which (π, V, v) were *geometrically rational* in the sense that one could use them to construct such compactifications of $\Gamma \backslash X$, but Satake did formulate a useful criterion for geometric rationality in terms of the closures of Siegel sets in X, and verified that this criterion held for several classical arithmetical groups. Satake's geometrical conditions were reformulated more algebraically in Borel (1962), where it was shown very generally that (π, V, v) is geometrically rational if π is irreducible and $\mathbb{Q}$-rational. A little later it was shown in Baily-Borel (1966) that certain compactifications of Hermitian symmetric spaces X (those now usually called Baily-Borel compactifications) were also geometrically rational, even when not $\mathbb{Q}$-rational. It seems, however, to have remained an unsolved problem since then to formulate a simple necessary and sufficient criterion for geometrical rationality. In this paper I will give a reasonably useful result of this kind, involving the real and rational Galois indices of the group G, for the case of irreducible representations.

The literature on compactifications dealing with these questions, aside from the early papers already mentioned, is sparse. One exception is the book Ash et al. (1975), in which the complicated case-by-case argument of Baily and Borel was replaced by a more direct proof involving rational homogeneous cones. Another exception is Zucker (1983), in which the general subject of geometric rationality was broached perhaps for the first time since the original work. It was perhaps there that the question of whether or not one could find a criterion involving the Galois indices was first raised, although it was not a topic with which that paper was directly concerned. It will become apparent that I am indebted to Zucker's paper for several valuable suggestions. In order to avoid confusion, however, I should point out that Zucker's discussion of rationality matters (in §3 of Zucker (1983)) is somewhat obscure and, as far as I can tell, in error (particularly his Proposition (3.3)).

Aside from the criterion for geometric rationality, there are a few other points in this paper perhaps worth calling attention to. (1) Satake's original construction was for semi-simple groups G. It started with an arbitrary irreducible (π, V) (no K-fixed vector), and then considered the associated representation of G on the space of Hermitian forms on V, among which lies a positive definite one fixed by K. Things are in fact greatly simplified if one looks instead directly at spherical triples. It is possible also to use *projective spherical triples*, where v is only assumed to be an eigenvector of K, but this does not seem to add much, and makes things a bit more complicated. (2) The boundary components in Satake's compactifications are all symmetric spaces associated to semi-simple quotients of Levi factors of certain parabolic subgroups of G. In an initial version of this paper I looked at the case of reducible representations, when the boundary components possess abelian factors as well. Allowing this, one obtains a larger class of compactifications than the ones Satake found (the *reductive* as opposed to *semi-simple* ones). The most interesting example is probably the one used by Goresky-Harder-MacPherson (1994), obtained by collapsing the unipotent fibres at infinity in the Borel-Serre compactification. I have abandoned this idea in the present paper because I was not entirely satisfied with certain technically awkward points. I hope to return to it in a subsequent paper. (3) Another direction for generalization would be to consider certain compactification of other homogeneous quotients G/H. Among the most interesting of these are Oshima's $G \times G$-covariant compactification of G. (4) I am not sure that compactifications arising from finite-dimensional representations are ultimately the natural ones to consider. Satake's procedure for obtaining compactifications of symmetric spaces has much to recommend it, especially its simplicity. The topological structure of these compactifications is not difficult to understand. But in fact the compactifications one obtains are semi-algebraic spaces, and in various applications it this extra structure is useful. Moreover, one can also use the same technique to obtain $G(\mathbb{C})$-covariant completions of the algebraic varieties $G(\mathbb{C})/K(\mathbb{C})$. The problem is that as semi-algebraic or algebraic completions the varieties one gets by Satake's procedure leave much to be desired. This is pointed out in Vust (1990), where he constructs all normal G-covariant completions of $G(\mathbb{C})/K(\mathbb{C})$, and shows that Satake compactifications are not generally normal. If one wants compactifications of G/K with a structure finer than topological, one should probably look at real points on Vust's varieties, or at least use some of Vust's techniques.

The basic technique in this paper is to look at the image of split real tori in compactifications of X. These are looked at on their own in §§1–3. At the end of §3 there is an observation about the classification of regular polyhedra. In §4 I put results from §3 in a slightly more general context, with future work in mind. Finite dimensional representations of G and compactifications of X are examined in §§5–6. In §7 I introduce the Galois index of Borel and Tits, in §8 I formulate and prove the main results, and in §9 I look at a few examples.

Although the subject of this paper is not directly related to work of Roger Richardson's, it seems to me very likely that eventually the most elegant answer to questions raised here will in fact depend on the results of Richardson-Springer (1990) and Richardson-Springer (1993) on the structure of $B_{\mathbb{C}}\backslash G_{\mathbb{C}}/K_{\mathbb{C}}$, and for this reason it is perhaps not inappropriate to dedicate it to his memory. I also wish to thank I. Satake and A. Borel for helpful comments when I started out on this project.

1. Convex polyhedra

This section will formulate some simple results about the geometry of convex polyhedra. Proofs are straightforward and will be left as exercises. Suppose V to be a vector space over $\mathbb{R}$, V^* its dual. Suppose C to be a compact convex polyhedron in V^*.

If H is a hyperplane intersecting C and all of C lies on one side of H then $H \cap C$ is a closed face of C. If H is the hyperplane $\lambda = c$ and the side containing C is $\lambda \le c$ then this face is the subset of C where λ achieves its maximum value c on C. I call this the *maximum locus* of λ on C.

Conversely, if λ is any vector in V, its maximum locus is a closed face F of C. The value of λ on this face is a constant $\langle \lambda, F \rangle$, and on points not on F is strictly less.

Each closed face F of C therefore determines a subset σ_F of V, the set of λ such that F is the maximum locus of λ.

Lemma 1.1. *Suppose F to be a closed face of C.*

(a) *The vector λ lies in σ_F if and only if (i) λ takes a constant value $\langle \lambda, F \rangle$ on F, and (ii) for any x in C but not in F*

$$\langle \lambda, x \rangle < \langle \lambda, F \rangle.$$

(b) *The vector λ lies in the closure of σ_F if and only if (i) λ takes a constant value $\langle \lambda, F \rangle$ on F, and (ii) for any x in C*

$$\langle \lambda, x \rangle \le \langle \lambda, F \rangle.$$

The subset σ_F is conical, which is to say that λ lies in σ_F if and only if $c\lambda$ lies in it, for any $c > 0$. It is certainly convex. It is also polyhedral. We can describe its faces explicitly:

Proposition 1.2. *Suppose E and F are faces of C with F contained in the closure of E. For any u in the interior of E and v in the interior of F, the closure of σ_E is the maximum locus of $u - v$ in the closure of σ_F. The closure of σ_F is the union of these σ_E.*

A compact convex polyhedron C therefore determines a *dual* partition of V into conical convex polyhedral subsets σ_F, one for each face F of C. The face C itself corresponds to the linear space of functions constant on all of C, and vertices of C correspond to open polyhedral cones.

In the following picture, the edges of the σ_F are in light gray, the faces of C in black.

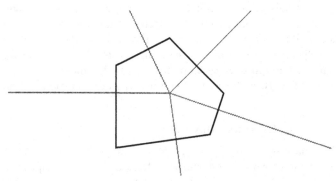

Several compact convex subsets C of V^* may determine the same partition of V. What in fact determines the partition is the local configuration at each of the vertices of C, not its global structure. In fact, a consequence of the previous result is that *if v is a vertex of C then the structure of the open cell σ_v is completely determined by the cone $c(u - v)$ as c ranges over all $c \geq 0$ and u ranges over all of C.* This cone will have one face for each closed face of C meeting v. Two compact convex polyhedra in V^* will determine the same partition of V precisely when the cones attached to each vertex are the same.

2. COMPACTIFICATIONS OF TORUS ORBITS

This section will formulate simple results relating homogeneous spaces of real algebaric tori and convex polyhedra. Again, proofs will generally be left as exercises.

Let A be the group of real points on a split torus defined over $\mathbb{R}$, A^{conn} its connected component. Let (π, V) be an algebraic representation of A and v a vector in V.

The vector v may be expressed as a sum of eigenvectors

$$v = \sum_{\alpha} v_{\alpha}$$

where the α are $\mathbb{R}$-rational characters of A. The representation π induces an action of A on $\mathbb{P}(V)$ as well. Let $[v]$ be the image of v in $\mathbb{P}(V)$, $A_{[v]}$ its isotropy subgroup—the subgroup of A acting by scalar multiplication on v. The closure of the A^{conn}-orbit of $[v]$ will be an A^{conn}-covariant compactification of $A^{\mathrm{conn}}/A_{[v]}^{\mathrm{conn}}$. In this section I shall describe the structure of this closure.

Define

$$\mathfrak{chars}(v) = \{\alpha \mid v_\alpha \neq 0\}$$
$$\mathbb{C} = \mathbb{C}(v) = \text{ the convex hull of } \mathfrak{chars}(v) \text{ in } X^*(A) \otimes \mathbb{R}$$

where $X^*(A)$ is the lattice $\text{Hom}(A, \mathbb{G}_m)$ of rational characters of A. Thus $\mathbb{C}$ is a compact convex polyhedron of dimension equal to that of $A/A_\mathfrak{c}$.

Recall that the lattice $X_*(A) = \text{Hom}(\mathbb{G}_m, A)$ is dual to $X^*(A)$, where the duality is expressed by the formula

$$t^{\langle \lambda, \alpha \rangle} = \alpha(\lambda(t))$$

for t in $\mathbb{R}^\times$. As suggested by the discussion in the previous section, to each closed face F of $\mathbb{C}$ associate the subset σ_F of all λ in $X_*(A) \otimes \mathbb{R}$ with the property that F is the maximum locus of λ. That is to say for each λ in σ_F the values

$$\langle \lambda, F \rangle = \langle \lambda, \alpha \rangle$$

are all the same for α in F, and if β is any other vertex of $\mathbb{C}$ then

$$\langle \lambda, \beta \rangle < \langle \lambda, F \rangle \, .$$

The σ_F make up the dual convex polyhedral decomposition of $X_*(A) \otimes \mathbb{R}$ described in §1. All the σ_F are invariant under translation by $X_*(A_\mathfrak{c}) \otimes \mathbb{R} = \sigma_\mathfrak{c}$. Whereas the λ in σ_F are characterized by the property that F is the maximum locus of λ, an element in the closure of the polyhedral cone σ_F will be constant on F, and this constant will be the maximum value on $\mathbb{C}$, but it may also have the same value on a larger face. From §1:

Lemma 2.1. *The closure of σ_F consists of all λ such that λ takes its maximum value $\max_\mathfrak{c} \lambda$ on F. If $E \subseteq F$ then $\sigma_F \subseteq \bar{\sigma}_E$.*

In other words, the partition of $X_*(A) \otimes \mathbb{R}$ is dual to that of $\mathbb{C}$ into faces. If F is any closed face of $\mathbb{C}$ then define

$$v_F = \sum_{\alpha \in F} v_\alpha \, .$$

Theorem 2.2. *The coweight λ lies in σ_F if and only if*

$$[\pi(\lambda(t))v] \to [v_F]$$

as $t \to \infty$.

Proof. This follows from the calculation

$$\pi(\lambda(t))v = \sum_{\alpha \in \mathfrak{chars}(v)} t^{\langle \lambda, \alpha \rangle} v_\alpha$$
$$= t^\ell \sum_{\alpha \in \mathfrak{chars}(v)} t^{\langle \lambda, \alpha \rangle - \ell} v_\alpha \, .$$

If $\ell = \langle \lambda, F \rangle$ then all terms vanish as $t \to \infty$ except the ones where $\langle \lambda, \alpha \rangle = \ell$.

In particular v_F lies in the closure of the A^{conn}-orbit of $[v]$. What does the whole closure look like?

Lemma 2.3. *The closure of the A^{conn}-orbit of $[v]$ is the union of the A^{conn}-orbits of the $[v_F]$ as F ranges over the closed faces of $\mathbb{C}$. The orbit of $[v_F]$ is contained in the closure of the orbit of $[v_E]$ if and only if F is contained in E.*

The orbit of $[v_F]$ is isomorphic to A/A_F where A_F is the subgroup of A with the property that all α in F are equal on A_F:

$$A_F = \cap_{\alpha,\beta \in F} \operatorname{Ker} \alpha\beta^{-1}$$

The subgroup A_F is also the *linear support* of σ_F, that is to say the smallest sub-torus of A such that $X_*(A_F) \otimes \mathbb{R}$ contains σ_F. Let γ_F be the character of A_F which is the common restriction to A_F of the characters in F. The vector space spanned by the v_α with α in F can be characterized as the eigenspace of V for the torus A_F with respect to the character γ_F.

Corresponding to the polyhedra σ_F in $X_*(A) \otimes \mathbb{R}$ are subsets of A^{conn} which are in some sense their *spans*. If F is a face of the cone $\mathbb{C}$ then define S_F to be the subset of a in A^{conn} such that all values of $\alpha(a)$ for α in F are equal to some real constant c_a, while $\alpha(a) < c_a$ for α not in F. Thus for λ in σ_F and $t > 1$, $\lambda(t)$ lies in S_F. Then $\pi(a^n)v \to v_F$ as $n \to \infty$. The sets S_F partition A^{conn}.

Proposition 2.4. *The closure of $S_F[v]$ in $\mathbb{P}(V)$ is the union of the $S_F[v_E]$ as E ranges over all faces of $\mathbb{C}$ containing F.*

Corollary 2.5. *The A^{conn}-orbits intersected by $S_F[v]$ in $\mathbb{P}(V)$ are those containing the limits*

$$\lim_{t \to \infty} [\pi(\lambda(t))v]$$

for λ in $\overline{\sigma}_F$.

3. THE CONVEX HULLS OF COXETER GROUP ORBITS

In this section, let (W, S) be an arbitrary finite Coxeter group. For each s, t in S let $m_{s,t}$ be the order of st.

Fix a realization (π, V) of (W, S). That is to say, π is a representation of W on the finite dimensional real vector space V in which elements of S act by reflections. There exists a positive definite metric invariant under W, and there exists a set of vectors $\{\alpha_s\}$ in V, indexed by S, such that the angle between α_s and α_t is $\pi - \pi/m_{s,t}$ for each s, t in S.

Let Δ be the set of α_s. The *Coxeter graph* of (W, S) is the graph with nodes indexed by S or equivalently Δ, in which α_s and α_t are linked by an edge labelled by $m_{s,t}$ when $m_{s,t} > 2$. (By default, an unlabelled edge has $m_{s,t} = 3$.)

In V the region

$$C^{++} = \{v \in V \mid \langle \alpha, v \rangle > 0 \text{ for all } \alpha \in \Delta\}$$

is an open fundamental domain for W. For each $\Theta \subseteq \Delta$ define

$$C_\Theta^{++} = \{v \mid \langle \alpha, v \rangle = 0 \text{ for } \alpha \in \Theta, \langle \alpha, v \rangle > 0 \text{ for } \alpha \notin \Theta\}.$$

Thus $C^{++} = C_\emptyset^{++}$, and the closure of C^{++} is the disjoint union of the C_Θ^{++}. We have similar sets $\hat{C}_\Theta^{++}$ in $\hat{V}$.

Fix for the moment χ in $\hat{V}$, lying in the closure of $\hat{C}^{++}$. It will lie in a unique $\hat{C}^{++}_\Theta$. Let $\delta = \delta_\chi$ be the complement of Θ in the set of nodes in the Coxeter graph. Equivalently

$$\delta_\chi = \Delta - \{\alpha_s \mid s\chi = \chi\}\,.$$

A subset $\kappa \subseteq \Delta$ is said to be δ-*connected* if every one of its nodes can be connected to an element of δ by a path inside itself. Given a δ-connected set κ, its δ-complement $\zeta(\kappa)$ is the set of α in the complement of κ which (a) are not in δ and (b) are not connected to κ by an edge in the Coxeter graph. Define its δ-*saturation*

$$\omega(\kappa) = \kappa \cup \zeta(\kappa)\,.$$

More generally, if θ is any set of nodes in the graph, define

$$\kappa(\theta) = \text{ the largest } \delta\text{-connected subset of } \theta$$
$$\zeta(\theta) = \zeta(\kappa(\theta))$$
$$\omega(\theta) = \kappa(\theta) \cup \zeta(\theta)$$

If $\kappa = \emptyset$ then ζ is the complement of δ. By construction, the two groups W_κ and W_ζ commute with each other.

If an element of θ has an edge linking it to $\kappa(\theta)$ then it lies in $\kappa(\theta)$ itself, and if doesn't then it lies in $\zeta(\theta)$. Hence

$$\kappa(\theta) \subseteq \theta \subseteq \omega(\theta)\,.$$

If κ is any δ-connected subset of Δ and ξ is any subset of Δ then $\kappa(\xi) = \kappa$ if and only if

$$\kappa \subseteq \xi \subseteq \omega(\kappa)\,.$$

The sets $\omega(\kappa)$ one gets as κ ranges over all δ-connected subsets of Δ I call the δ-*saturated* subsets of Δ. The correspondence between κ and ω is bijective.

The following is implicit in work of Satake and Borel-Tits (cf. Lemma 5 of Satake (1960a) and §12.16 of Borel-Tits (1965)). The proof is essentially theirs as well.

Theorem 3.1. *Suppose χ to lie in the closure of $\hat{C}^{++}$, and let $\delta = \delta_\chi$. The map taking κ to the convex hull F_κ of $W_\kappa \cdot \chi$ is a bijection between the δ-connected subsets of S and the faces F of the convex hull of $W \cdot \chi$ such that σ_F meets $\overline{C}^{++}$.*

Every face of the convex hull of $W \cdot \chi$ is W-conjugate to exactly one of these faces. The face corresponding to the empty set $\emptyset$ is χ itself. The dimension of the face F_κ is the cardinality of κ. An increasing chain of δ-connected subsets corresponds to an increasing chain of faces.

Proof. I begin with a simple observation. Suppose κ to be a δ-connected subset of Δ, $\omega = \omega(\kappa)$. Since s fixes χ and commutes with W_κ if s lies in $\zeta = \zeta(\kappa)$, the group W_ζ fixes all of $W_\kappa \cdot \chi$ and

$$W_\kappa \cdot \chi = W_\omega \cdot \chi$$

so that for any $\theta \subseteq \Delta$, the orbit $W_\theta \cdot \chi$ is the same as $W_{\kappa(\theta)} \cdot \chi$.

If φ lies in $\hat{V}$ then we can write it as $\sum c_\alpha \alpha$. The *support* of φ is the set $\mathrm{supp}(\varphi)$ of α with $c_\alpha \neq 0$.

For the duration of the proof, for each w in W let

$$\Sigma_w = \mathrm{supp}(\chi - w\chi) \, .$$

It is simple to calculate Σ_w in terms of a reduced expression for w. If $w = s$ lies in S we have (with $\alpha = \alpha_s$)

$$s\chi = \chi - \langle \chi, \alpha^\vee \rangle \alpha$$
$$\chi - s_\alpha \chi = \langle \chi, \alpha^\vee \rangle \alpha$$

and the coefficient on the right is always non-negative. It is positive if α lies in δ, otherwise 0. Therefore

$$\Sigma_s = \begin{cases} \{\alpha_s\} & \alpha_s \in \delta \\ \emptyset & \text{otherwise} \end{cases}$$

Inductively, we write

$$\chi - w\chi = \sum c_\gamma \gamma$$

with all $c_\gamma \geq 0$ and Σ_w a δ-connected subset of Δ. We then look at $s_\alpha w > w$. Since $w^{-1}\alpha > 0$

$$w\chi = \chi - \sum c_\gamma \gamma$$
$$s_\alpha w\chi = w\chi - \langle w\chi, \alpha^\vee \rangle \alpha$$
$$= \chi - \sum c_\gamma \gamma - \langle \chi, w^{-1}\alpha^\vee \rangle \alpha$$
$$\chi - s_\alpha w\chi = \sum c_\gamma \gamma + \langle \chi, w^{-1}\alpha^\vee \rangle \alpha$$

so that again the coefficients are always non-negative. Furthermore either (a) $\Sigma_{s_\alpha w} = \Sigma_w$ or (b) α does not lie in Σ_w and $\Sigma_{s_\alpha w}$ is the union of Σ_w and $\{\alpha\}$. In the second case

$$\langle w\chi, \alpha^\vee \rangle = \langle \chi, \alpha^\vee \rangle - \sum c_\gamma \langle \gamma, \alpha^\vee \rangle$$

where all the non-zero terms in the sum are non-positive. Since the total sum is non-zero, either α lies in δ or α is linked by an edge in the Coxeter diagram to an element in Σ_w. Either way we see that $\Sigma_{s_\alpha w}$ is again δ-connected.

Summarizing this argument:

Lemma 3.2. *The set Σ_w is always δ-connected. If $w = xy$ with $\ell(w) = \ell(x) + \ell(y)$ then $\Sigma_y \subseteq \Sigma_w$.*

The argument shows also that if w lies in W_κ then $\Sigma_w \subseteq \kappa$. The converse is also true:

Lemma 3.3. *If κ is a δ-connected subset of Δ then the character $w\chi$ lies in $W_\kappa \cdot \chi$ if and only if Σ_w is contained in κ.*

Proof. For the new half, let $w = s_n s_{n-1} \ldots s_1$ be a reduced expression for w. Let $w_i = s_i \ldots s_1$ for each i, $w_0 = 1$. If k is the smallest such that $w_k \chi$ doesn't lie in $W_\kappa \cdot \chi$, then $\epsilon = w_{k-1}\chi$ does lie in $W_\kappa \cdot \chi$, but $s_k \epsilon$ doesn't. Then s_k isn't in W_ω, and $s_k \epsilon \neq \epsilon$. Since

$$s_k \epsilon = \epsilon - n_{\alpha_k} \alpha_k$$

with $n_{\alpha_k} > 0$, the support of $\chi - w_k \chi$ contains α_k. The same is true of $\chi - w\chi$ since support increases along a chain.

Lemma 3.4. *Every δ-connected subset κ of Δ is the support of some $\chi - w\chi$.*

Proof. We prove this by induction on the size of the δ-connected subset. It is clear for single elements in δ.

Otherwise, suppose that κ is δ-connected and $\chi - x\chi$ has support κ. Suppose then that α does not lie in κ and that $\kappa \cup \alpha$ is δ-connected. Either there is an edge in the Coxeter graph connecting α to κ or α itself lies in δ. If $x\chi = \chi - \sum n_\beta \beta$ where all $n_\beta > 0$ for β in κ, then

$$s_\alpha x \chi = s_\alpha \chi - \sum n_\beta s_\alpha \beta$$
$$= s_\alpha \chi - \sum n_\beta [\beta - \langle \beta, \alpha^\vee \rangle \alpha]$$
$$s_\alpha x \chi - \chi = (s_\alpha \chi - \chi) - \sum n_\beta \beta + \sum n_\alpha \langle \beta, \alpha^\vee \rangle \alpha$$

and all of the coefficients in the last sum are non-positive. If α itself lies in δ then $s_\alpha \chi$ has support α, so the support of the whole expression is the union of κ and α. If not, then the first term vanishes, but at least one term in the sum is actually negative.

Lemma 3.5. *For λ in C_θ^{++} the subset of $W \cdot \chi$ where λ achieves its maximum is $W_{\kappa(\theta)} \cdot \chi$.*

This is immediate from Lemma 3.3. The main Theorem now follows from it and Lemma 3.4, which guarantees that the sets $W_\kappa \cdot \chi$ are all distinct.

The main Theorem of this section is curiously relevant to the classification of regular polyhedra. It follows by induction that the symmetry group of any regular polyhedron is a Coxeter group. Since all vertices are W-conjugate every regular polyhedron must be the convex hull of a single vector. Since all faces of a given dimension must also be W-conjugate, the δ-connected sets must form a single ascending chain. This means that δ itself must be a single node at one end of the Coxeter graph and that the Coxeter graph has no branching. Also, an automorphism of the Coxeter graph induces an isomorphism of corresponding polyhedra. Therefore *the regular polyhedra correspond to isomorphism classes of pairs Δ, α where Δ is the Coxeter graph of a finite Coxeter group, without branching, and α is an end-node in Δ.* This is of course the usual classification, which I exhibit in these terms in the following table. Keep in mind that the dimension in which the figure is embedded is the number of nodes. The marked end denotes the vertex. Thus, in the first line for H_3 we list the vertex, then the pentagonal face, on the dodecahedron, whereas in the next line, reading from right to left, we meet the vertex and then a triangular face.

System	Marked Coxeter diagram	Figure
A_n $(n \geq 2)$	●—○— ••• —○—○	tetrahedron
B_n $(n \geq 2)$	●—⁴—○— ••• —○—○	cube
	○—⁴—○— ••• —○—●	octahedron
F_4	●—○—⁴—○—○	
G_2	●—⁶—○	hexagon
H_3	●—⁵—○—○	dodecahedron
	○—⁵—○—●	icosahedron
H_4	●—⁵—○—○—○	
	○—⁵—○—○—●	
I_p $(p = 5, p \geq 7)$	●—ᵖ—○	polygon

4. SATAKE PARTITIONS

Continue the notation of the last section. In particular, χ is an element of V^* lying in the closure of $\widehat{C}^{++}$, and $\delta = \delta_\chi$. In this section I shall describe in more detail the partition of V dual to the convex hull of $W \cdot \chi$.

I shall begin in a little more generality. I define a *Satake partition* of V to be any partition of V into convex polyhedral cones σ which is geometric in the sense that any open face of one of these cones is also a subset of the partition, and which is also W-invariant. According to Proposition 1.2, the partition of V dual to the convex hull of any W-orbit of a finite set of points in V^* is a Satake partition.

Define a *Weyl cell* in V to be a W-transform of some C_θ^{++}. The partition of V into Weyl cells is a Satake partition I call the *Weyl partition*.

Suppose now and for a while that that σ is one of the cells in an arbitrary Satake partition. Since the partition is W-invariant, for any given w in W exactly one of two possibilities is true: (a) $w\sigma$ and σ are disjoint, or (b) $w\sigma = \sigma$. Define W_σ to be the group of w in W with $w\sigma = \sigma$. Define the *centre* of σ to be the elements of σ fixed by W_σ. For any x in σ define

$$\Pi_\sigma x = \frac{1}{\#W_\sigma} \sum_{W_\sigma} w\,x$$

to be the average of the transforms of x by elements of W_σ. Because of convexity, $\Pi_\sigma x$ lies also in σ, and in particular the *centre* C_σ of σ is not empty. If $wx = x$ for w in W and x in C_σ then $w\sigma \cap \sigma \neq \emptyset$ so w has to be in W_σ. In other words, for any x in C_σ

$$W_\sigma = \{w \in W \mid wx = x\}$$

which proves:

Lemma 4.1. *There exists a unique Weyl cell containing the centre of* σ.

For $\theta \subseteq \Delta$ define

$$\Pi_\theta x = \frac{1}{\#W_\theta} \sum_{W_\theta} w\,x$$

to be the average over W_θ.

Lemma 4.2. *For $\psi \subseteq \theta \subseteq \Delta$ and x in C_θ^{++} the projection $\Pi_\psi x$ lies in C_ψ^{++}.*

Proof. For x in C_θ^{++}

$$\Pi_\psi x = x - \sum_{\theta - \psi} c_\alpha \alpha^\vee$$

with $c_\alpha \geq 0$. But then for $\beta \in \Delta - \psi$

$$\langle \beta, \Pi_\psi x \rangle = \langle \beta, x \rangle - \sum c_\alpha \langle \beta, \alpha^\vee \rangle .$$

Since by assumption the first term is positive and the rest are non-negative, the sum is positive.

Theorem 4.3. *If C is any Weyl cell intersecting σ, then the closure of C contains the centre C_σ.*

Proof. The interior of the segment from x to $\Pi_\sigma x$ cannot cross any root hyperplane.

If σ intersects $\overline{C}^{++}$, suppose that κ is minimal such that σ intersects C_κ^{++}. This will be unique, since a segment from C_θ^{++} to C_ψ^{++} will cross $C_{\theta \cap \psi}^{++}$. I call it the *linear support* of σ. The centre of σ will intersect a unique Weyl cell, which according to Lemma 4.3 will be contained in C_κ^{++}. It will therefore be C_ω^{++} for some set $\omega \supseteq \kappa$ which I call the *central support* of σ.

Lemma 4.4. *Suppose that σ intersects $\overline{C}^{++}$. Let κ be the linear support of σ and ω its central support. Then*

$$\sigma \cap \overline{C}^{++} = \bigcup_{\kappa \subseteq \theta \subseteq \omega} \sigma \cap C_\theta^{++}$$

Proof. A simple consequence of Lemma 4.2.

Lemma 4.5. *Suppose that σ intersects $\overline{C}^{++}$ and let ω be its central support. then*

$$\sigma = W_\omega(\sigma \cap \overline{C}^{++}) .$$

Proof. If w in W takes C to C_θ^{++} then $\overline{C}$ and $\overline{C}_\theta^{++}$ must both contain C_ω^{++}. But then $wC_\omega^{++} = C_\omega^{++}$, which in turn means that w must lie in W_ω.

Suppose that κ is a δ-connected subset of Δ, and let F be the face of the convex hull of $W \cdot \chi$ spanned by $W_\kappa \cdot \chi$. We know from the previous section that the intersection of σ_F with $\overline{C}^{++}$ is the union of the C_θ^{++} with θ ranging over all subsets of Δ with $\kappa \subseteq \theta \subseteq \omega(\kappa)$. Lemma 4.5 implies:

Proposition 4.6. *If κ is a δ-connected subset of Δ, $\omega = \omega(\kappa)$, and F is the face of the convex hull of $W \cdot \chi$ spanned by $W_\kappa \cdot \chi$ then σ_F is the union of the transforms of the C_θ^{++} by elements of W_ω as θ ranges over all subsets of Δ such that*

$$\kappa \subseteq \theta \subseteq \omega .$$

It can also be characterized as the union of all Weyl cells C in the linear subspace spanned by the cell C_κ^{++} with $\overline{C}$ containing C_ω^{++}.

This result shows that the partition dual to a W-orbit has the property that all its cells are unions of Weyl cells. Suppose that, conversely, we are given a Satake partition $\{\sigma\}$ with this property. If σ is an open cell in this partition then it must contain at least one Weyl chamber as an open subset. We can transform this cell by an element of W so that this chamber is just C^{++}. Let δ be the central support of this σ. Lemma 4.4 then shows that the intersection of σ and $\overline{C}^{++}$ consists of all C_θ^{++} with $\theta \subseteq \delta$. In other words, the open cell of the partition containing C^{++} is the same as the open cell of the Satake partition associated to δ. The geometric condition implies that in fact the two partitions are the same.

Proposition 4.7. *If Σ is a Satake partition of V whose cells are unions of Weyl cells, then there exists a subset δ of Δ such that Σ is the Satake partition associated to δ.*

In other words, there is a natural bijection between Satake partitions of this sort and subsets of Δ. We know then that the structure of the partition can be described in terms of the apparatus of δ-connected sets. Can we understand arbitrary Satake partitions in terms of similar combinatoric data? Lemma 4.4, Lemma 4.5, and Proposition 4.7 suggest a start.

5. PARABOLIC SUBSPACES

In this section, let

$G =$ a complex reductive group

$B = AN =$ a Borel subgroup of G

$\Delta =$ the basic roots associated to the choice of B

For each $\theta \subseteq \Delta$ let $P_\theta = M_\theta N_\theta$ be the associated parabolic subgroup containing $B = P_\emptyset$, so that the Lie algebra $\mathfrak{n}_\theta$ is the sum of positive root spaces $\mathfrak{n}_\alpha$ with α not a linear combination of elements of θ. The split centre A_θ of M_θ is the intersection of the kernels $\ker(\alpha)$ as α ranges over θ. Let C_θ^{++} be the subset of $X_*(A_\theta)$ comprising λ with $\langle \alpha, \lambda \rangle = 0$ for α in θ, $\langle \alpha, \lambda \rangle > 0$ for α in $\Delta - \theta$. The Lie algebra of M_θ is spanned by $\mathfrak{a}_\theta$ and the root spaces $\mathfrak{g}_\alpha$ with $\langle \alpha, \lambda \rangle = 0$ for λ in C_θ^{++} and the Lie algebra of N_θ is spanned by the $\mathfrak{g}_\alpha$ with $\langle \alpha, \lambda \rangle > 0$ for λ in C_θ^{++}.

Let W be the Weyl group of G with respect to A, S the reflections associated to elements of Δ. The set Δ may be identified with the nodes of the Coxeter graph of (W, S).

Fix a finite-dimensional representation (π, V) of G. A *parabolic subspace* of V is one of the form $\mathrm{Fix}(N)$ where N is the unipotent radical of some parabolic subgroup of G. I recall that

$$\mathrm{Fix}(N) = \{v \in V \mid \pi(n)v = v \text{ for all } n \in N\}.$$

Suppose that π is irreducible with highest weight χ. The set of all weights of π is contained in the convex hull of the Weyl orbit $W \cdot \chi$. Let $\delta = \delta_\pi$ be the set of roots α in Δ such that $s_\alpha \chi \neq \chi$. Equivalently, δ_π is the complement in Δ of the set θ such that P_θ is the stabilizer of the line through the highest weight vector of π. For example, $\delta = \emptyset$ when π is the trivial representation of G, and $\delta = \Delta$ itself when the highest weight is regular. Recall that a δ-saturated subset ω is a subset of the nodes of the Dynkin diagram of G with this property: let κ be the union of connected components of ω containing elements of δ. Then the complement of ω in Δ is made up of exactly those nodes of Δ which do not lie in κ, and which either lie in δ or possess an edge in common with an element of κ.

Theorem 5.1. *Suppose (π, V) to be an irreducible finite dimensional represent-ation of G with highest weight χ. Suppose that χ lies in C_θ^{++} and let $\delta = \Delta - \theta$. Then for any δ-saturated subset ω of Δ the subspace $\mathrm{Fix}(N_\omega)$ is the sum of weight spaces with weights in the convex hull of $W_\omega \cdot \chi$. This space is the same as $\mathrm{Fix}(N_\theta)$ for any θ with $\omega(\theta) = \omega$. It is an irreducible representation of M_ω.*

Proof. Let V_ω be the direct sum of the weight spaces with weights in the convex hull of $W_\omega \cdot \chi$. It follows from the remarks above about C_θ^{++} and Lemma 2.1 that $\pi(\nu)v = 0$ if ν lies in $\mathfrak{n}_\theta$ and v in V_ω. Since $N_\omega \subseteq N_\theta$ we have

$$V_\omega \subseteq \mathrm{Fix}(N_\theta) \subseteq \mathrm{Fix}(N_\omega) \, .$$

The space V_ω is stable under P_ω, hence a representation of M_ω, as is $\mathrm{Fix}(N_\omega)$. Both have a unique highest weight vector, namely the highest weight vector of V itself, and are hence irreducible and identical. So the inclusions above are all equalities.

Another way of phrasing this is to define a *saturated parabolic subgroup* of G (with respect to π) to be a conjugate of some P_ω with ω saturated. The Theorem amounts to the assertion that *the map $P = MN \mapsto \mathrm{Fix}(N)$ is a bijection between saturated parabolic subgroups of G and parabolic subspaces of V.*

Recall that if A is a torus in G and λ in $X_*(A)$ then P_λ is the parabolic subgroup of G corresponding to the sum of eigenspaces $\mathfrak{g}_\alpha$ with $\langle \alpha, \lambda \rangle \geq 0$. Its unipotent radical N_λ has as Lie algebra the sum of eigenspaces with $\langle \alpha, \lambda \rangle > 0$. The proof above also leads to the following result.

Proposition 5.2. *Suppose A to be any torus in G, λ in $X_*(A)$. Let π be an irreducible representation of G, let F be the face of the convex hull of the weights of π restriced to A where λ takes its maximum value, and let V_F be the sum of weight spaces associated to F. Then $V_F = \mathrm{Fix}(N_\lambda)$. This is also $\mathrm{Fix}(N)$ if $P = MN$ is the minimal saturated subgroup containing P_λ.*

How does M_ω act on V_ω? The Dynkin diagram of M_ω is the sub-diagram of the Dynkin diagram of G corresponding to the nodes in ω. The group M_ω is isogeneous to a product of a torus and semi-simple groups G_κ, G_ζ whose diagrams are the sub-diagrams corresponding to κ and ζ. Since N_κ acts trivially on V_ω, so do all its conjugates in M_ω, and in particular by elements of W_ω. The group A_ω acts by scalars on V_ω. Therefore the kernel of the representation of M_ω on V_ω contains G_ζ, and this representation factors through a representation of G_κ.

6. Compactifications of G/K

Again suppose G to be the group of $\mathbb{R}$-valued points on a semi-simple group defined over $\mathbb{R}$. Let $P_\emptyset$ be a minimal real parabolic subgroup with unipotent radical $N_\emptyset$, $M_\emptyset$ the quotient $P_\emptyset/N_\emptyset$, $A_\emptyset$ the maximal split real torus in $P_\emptyset$ invariant under the involution associated to the maximal compact subgroup K.

A *spherical representation* of G is a triple (π, V, v) where v in V is fixed by K.

Irreducible spherical representations arise in a simple manner. If χ is a rational character of $P_\emptyset$ then the space

$$\operatorname{Ind}(\chi \mid P_\emptyset, G) = \{f \in C^\infty(G) \mid f(pg) = \chi(p)\,f(g) \text{ for all } p \in P_\emptyset\}$$

is a smooth representation of G with respect to the right regular action. If (π, V) is the irreducible finite dimensional representation with lowest weight χ then the map from V to $V/\mathfrak{n}_\emptyset V$ is χ-covariant, hence induces by Frobenius reciprocity a G-map from V to $\operatorname{Ind}(\chi \mid P_\emptyset, G)$. If χ is trivial on $P_\emptyset \cap K$ then since $G = PK$ the space of K-invariant functions in $\operatorname{Ind}(\chi \mid P_\emptyset, G)$ has dimension 1. It is a theorem of Helgason that the finite-dimensional representation contains the space of K-invariants. The converse is also true: any irreducible spherical representation has the property that its lowest weight is a character of $P_\emptyset$ trivial on $P_\emptyset \cap K$. (A strictly algebraic proof of all these assertions can be found in Vust (1974).)

To summarize:

Proposition 6.1. *The irreducible spherical representations of G are the irreducible finite dimensional representations with highest weights which are characters of $P_\emptyset$ trivial on $P_\emptyset \cap K$.*

Corollary 6.2. *If (π, V, v) is a spherical representation of G then it is defined over $\mathbb{R}$, and it possesses a highest weight whose stabilizer in G is a real parabolic subgroup of G.*

Proof. Since $P_\emptyset/N_\emptyset(P_\emptyset \cap K)$ is split over $\mathbb{R}$, its algebraic characters are all real. The stabilizer of a highest weight stabilized by $P_\emptyset$ must be defined over $\mathbb{R}$ and contain $P_\emptyset$, hence be a real parabolic subgroup of G.

In the terminology of Borel-Tits (1965), the irreducible spherical representations of G are *strongly rational* over $\mathbb{R}$.

Proposition 6.3. *If (π, V, v) is an irreducible spherical representation of G then the highest weight of v with respect to $A_\emptyset$ is the restriction to $A_\emptyset$ of the highest weight of V.*

Proof. Since π is irreducible, $V = U(\mathfrak{g})v$. Because $\mathfrak{g} = \mathfrak{n}_\emptyset + \mathfrak{a}_\emptyset + \mathfrak{k}$, $U(\mathfrak{g})v = U(\mathfrak{n}_\emptyset)U(\mathfrak{a}_\emptyset)v$.

Corollary 6.4. *If (π, V, v) is an irreducible spherical representation of G then the extremal weights of the restriction to $A_\emptyset$ of the weights of V are the same as the exremal $A_\emptyset$-weights of v.*

Suppose that (π, V, v) is a spherical triple, non-trivial on each simple factor of G. We can embed $X = G/K$ into $\mathbb{P}(V)$ according to the recipe

$$x \mapsto \iota(x) = \pi(g)v, \quad (x = gK) \, .$$

What does the closure $\overline{X}_\pi$ of $X_\pi = \iota(X)$ look like?

If P is a parabolic subgroup of G and A the maximal split torus in the centre of a Levi component of P, define A^{++} to be the subset of a in A with the property that all its eigenvalues on the Lie algebra of P are ≥ 1, and let $\overline{A}^{++}$ be the closure of A^{++}.

The Cartan decomposition asserts that $G = K\overline{A}_\emptyset^{++}K$, and implies that the closure $\overline{X}_\pi$ is the same as the K-orbit of the closure of $A_\emptyset^{++}[v]$, and also of the closure of $A_\emptyset[v]$. According to results of §2 this closure is obtained in the following way: for any face F of the convex hull of the $A_\emptyset$-weights of v, let v_F be the sum of the components of v corresponding to characters in F. The closure of $A_\emptyset[v]$ is the union of $A_\emptyset$-orbits of the $[v_F]$.

What are the faces of this convex hull?

It will be shown in the next section that the extremal characters of the restriction of π to $A_\emptyset$ are the $W_\mathbb{R}$-transforms of the restriction of the highest weight, and according to Proposition 6.3 this is also the set of extremal $A_\emptyset$-weights of v. It will also be shown in the next section that the weight space corresponding to a face of this hull is stablized by a π-saturated parabolic subgroups of G which is real.

Suppose P to be a real π-saturated parabolic subgroup of G. Let V_P be Fix(N_P), π_P the representation of P on π_P, which factors through P/N_P. If ι is the Cartan involution of G fixing the elements of K, let M_P be $P \cap P^\iota$, the unique Levi subgroup of P stable under ι. Let A_P be the maximal split torus contained in M_P, a in A_P^{++}. Then

$$v_P = \lim_{n \to \infty} \pi(a^n)v$$

lies in V_P, and since π_P is irreducible, (π_P, V_P, v_P) is a spherical triple for M_P. If P corresponds to the face F of the convex hull of the $A_\emptyset$-weights of v, then v_P is the same as v_F. The natural embedding of V_P into V induces an embedding of $\mathbb{P}(V_P)$ into $\mathbb{P}(V)$. Let L_P be the projective kernel of the representation of M_P on V_P, G_P the quotient M_P/L_P. The embedding of $\mathbb{P}(V_P)$ into $\mathbb{P}(V)$ induces one of $X_P = G_P/K_P$ into X, where K_P is the image of $K \cap P$ in G_P. The symmetric space X_P is contained in the closure of X, and called a *boundary component* of X. The transverse structure of a neighbourhood of X_P in $\overline{X}$ is related to the group L_P, which I call the *link* group.

Proposition 6.5. *The closure of X in $\mathbb{P}(V)$ is the union of its boundary components. If A is a split torus in G, λ in $X_*(A)$, then the limit*

$$\lim_{t \to \infty} \lambda(t)x_0$$

lies in X_P if P is the smallest saturated parabolic subgroup of G containing P_λ.

The parabolic subgroup P_λ is the one with the property that its Lie algebra is spanned by the eigenvectors of $\lambda(t)$ $(t > 1)$ with eigenvalues ≤ 1.

7. GALOIS INDICES

Let k be a subfield of $\mathbb{C}$, G a semi-simple group defined over k. Let $P_{\emptyset,k}$ be a minimal parabolic subgroup of G, A_k a maximal k-split torus in $P_{\emptyset,k}$.

Choose a Borel subgroup $B_\mathbb{C}$ in $G_\mathbb{C}$ and let $A_\mathbb{C}$ be a maximal torus contained in it. Let $\Delta_\mathbb{C}$ be the corresponding basis of roots of the pair $\mathfrak{g}_\mathbb{C}$, $\mathfrak{a}_\mathbb{C}$. The maximal parabolic subgroups of $G_\mathbb{C}$ containing $B_\mathbb{C}$ are parametrized by maximal proper subsets of $\Delta_\mathbb{C}$, or equivalently by their complements, which are singletons. If τ is an automorphism of $\mathbb{C}/k$ and P is a maximal proper parabolic subgroup of $G_\mathbb{C}$ then P^τ is conjugate to a unique maximal proper parabolic subgroup of G containing B. This induces an action of $\mathrm{Aut}(\mathbb{C}/k)$ on the complex Dynkin diagram which Borel-Tits (1965) call the $*$ action (see also §2.5 of Tits (1966)). It factors through the Galois group of the algebraic closure of k in $\mathbb{C}$, and does not depend on the particular choice of data.

We may assume that $B_\mathbb{C} \subseteq P_{\emptyset,k}$, $A_k \subseteq A_\mathbb{C}$. Restriction of roots determines a map

$$\rho_{\mathbb{C}/k} : \Delta_\mathbb{C} \to \Delta_k \cup \{0\} \, .$$

The *anisotropic kernel* of $\rho_{\mathbb{C}/k}$ is the inverse image $\Delta^0_{\mathbb{C}/k}$ of $\{0\}$. It is stable under $\mathrm{Aut}(\mathbb{C}/k)$, as is its complement. Galois orbits in the complement are exactly the inverse images in $\Delta_\mathbb{C}$ of single elements of Δ_k. The Galois action and the anisotropic kernel together make up the *index* of G_k. Tits explains in §2.5 of Tits (1966) how the relative root system can be reconstructed from the index.

If $k = \mathbb{R}$, there is a standard way of coding the index in the complex Dynkin diagram. The nodes of the anisotropic kernel are coloured black, and the remaining white nodes interchanged by conjugation are linked together. We shall see some examples later on.

For a subset $\theta \subseteq \Delta_k$ define

$$\epsilon_{\mathbb{C}/k}(\theta) = \rho^{-1}(\theta) \cup \Delta^0_{\mathbb{C}/k} \, .$$

Such sets are exactly the subsets of $\Delta_{\mathbb{C}/k}$ stable under $\mathrm{Aut}(\mathbb{C}/k)$ containing $\Delta^0_{\mathbb{C}/k}$, and the parabolic subgroups they parametrize are the k-rational parabolic subgroups containing $P_{\emptyset,k}$.

For any subset θ of Δ_k let $C^{++}_{k,\theta}$ be the corresponding wall of the closed positive chamber in $X_*(A_k)$. The definitions imply immediately:

Proposition 7.1. *If θ is a subset of Δ_k and the element λ of $X_*(A_k)$ lies in $C^{++}_{k,\theta}$ then its image in $X_*(A_\mathbb{C})$ lies in $C^{++}_{\mathbb{C},\psi}$ where $\psi = \epsilon_{\mathbb{C},k}(\theta)$.*

If we are given a Satake partition of $X_*(A_\mathbb{C})$ with the property that its cells are unions of Weyl cells, then the partition of A_k induced from inclusion in $A_\mathbb{C}$ will also be such a Satake partition. According to Proposition 4.7 it is therefore determined by a subset δ_k of Δ_k. The significance of this is that $C^{++}_{k,\theta}$ is in the same cell of the partition of $X_*(A_{\mathbb{C}})$ as $C_{k,\kappa}^{++}$ if κ is the union of connected components in θ containing elements of δ_k. There is a simple criterion to determine the set δ_k in terms of the index. Let $\kappa^0 = \kappa_{\mathbb{C}/k}^0$ be $\kappa(\Delta_{\mathbb{C}/k}^0)$, and let

$\delta_{\mathbb{C}/k}$ = the nodes of $\Delta - \Delta_{\mathbb{C}/k}^0$ which are either in $\delta_{\mathbb{C}}$ or connected by an edge to an element of $\kappa_{\mathbb{C}/k}^0$.

Corollary 7.2. *A node α in Δ_k lies in δ_k if and only if it is the image under restriction of an element of $\delta_{\mathbb{C}/k}$.*

Proof. The set δ_k is the complement of $\omega_k(\emptyset)$. This is the largest set θ such that $C_{k,\theta}^{++}$ lies in the same partition cell as $C_{k,\emptyset}^{++}$. By the previous result, this is the partition cell containing $C_{\Delta^0}^{++}$, which contains, in addition to $\Delta^0 = \Delta_{\mathbb{C}/k}^0$, all nodes not in $\delta_{\mathbb{C}}$ and not connected to κ^0 by an edge.

A strongly k-rational representation of G is one where the stabilizer of some highest weight is a k-rational parabolic subgroup. When $\delta_{\mathbb{C}}$ arises from a strongly k-rational representation of G then it has no nodes inside $\Delta_{\mathbb{C}/\mathbb{R}}^0$, and it is invariant under the Galois group. It is therefore equal to the inverse image of δ_k. In these circumstances, if θ is any subset of Δ_k then the smallest saturated subset of Δ containing it is the same as the inverse image of the smallest saturated subset of Δ_k with respect to the subset δ_k.

Suppose that π is strongly k-rational. It follows from the remarks above and from Proposition 5.2 that the convex hull of the weights of π restricted to A_k has its vertices the W_k-transforma of the highest weight, and that the stabilizers of its faces are the parabolic subgroups of G associated to saturated subsets of Δ_k. (This observation was required in the previous section for $k = \mathbb{R}$.)

For some groups G, the symmetric space $X = G/K$ possesses a G-invariant Hermitian structure (this is explained nicely in §1.2 of Deligne (1979)). In these cases, X may be embedded as a bounded symmetric domain in a complex vector space, and its closure in this vector space will be a G-stable compactification of X. It is called a *Baily-Borel* compactification of X. It is not associated to a spherical triple (not even for $PGL_2(\mathbb{R})$), but it is topologically equivalent to one which is. In the following pictures, I exhibit the index data for Baily-Borel compactifications (compare the diagrams in Deligne (1979)). Nodes in δ are dotted, compact nodes (making up the anisotropic kernel of $\mathbb{C}/\mathbb{R}$) are black.

System   Type   Marked index diagram

$AIII$ $SU(p,q)$ $(q \geq p + 2)$

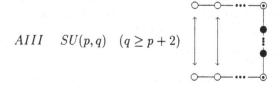

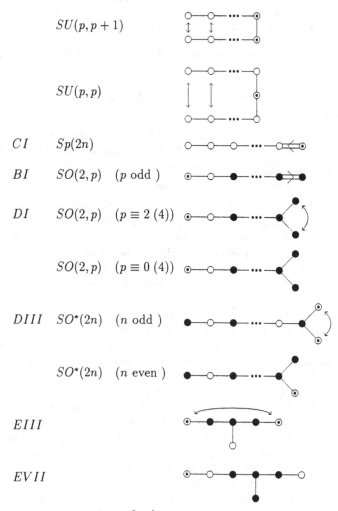

$$SU(p, p+1)$$

$$SU(p, p)$$

CI $Sp(2n)$

BI $SO(2, p)$ $(p$ odd $)$

DI $SO(2, p)$ $(p \equiv 2\ (4))$

       $SO(2, p)$ $(p \equiv 0\ (4))$

$DIII$ $SO^*(2n)$ $(n$ odd $)$

       $SO^*(2n)$ $(n$ even $)$

$EIII$

$EVII$

8. Arithmetic quotients

Suppose G to be defined over $\mathbb{Q}$. Let x_0 be the point in X fixed by K. If P is a parabolic subgroup of G and A a maximally split torus in the centre of a Levi component then I shall call (P, A) a *parabolic pair*. If (P, A) is a parabolic pair, let Σ_P be the eigencharacters of A acting on the Lie algebra of the unipotent radical N of P, and define for $T > 0$ the set $A^{++}(T)$ to be

$$A^{++}(T) = \left\{ a \in A \,\middle|\, |\alpha(a)| > T \text{ for all } \alpha \in \Sigma_P \right\}.$$

If $(P < A)$ is a $\mathbb{Q}$-rational parabolic pair, Ω is a compact subset of P, and $T > 0$, the *Siegel set* associated to these data is the image of

$$\mathfrak{S}(P, A, \Omega, T) = \Omega A^{++}(T)$$

in $X = G/K$. The main result of reduction theory, in its simplest form, is that (a) the arithmetic quotient $\Gamma \backslash X$ is covered by a finite number of Siegel sets with respect to minimal $\mathbb{Q}$-rational parabolic subgroups, and (b) the covering of X by the Γ-transforms of this finite collection of Siegel sets is locally finite. In other words, we can assemble a sort of fundamental domain for Γ from Siegel sets. We can choose one set for each Γ-conjugacy class of minimal $\mathbb{Q}$-rational parabolic subgroups.

Fix a spherical triple (π, V, v), and let $\overline{X} = \overline{X}_\pi$ be the corresponding Satake compactification of X. For brevity let $\Delta^0 = \Delta^0_{C/\mathbb{Q}}$ and similarly for κ^0.

A boundary component X_P of $\overline{X}$ is called *geometrically rational* if it satisfies these two conditions:

(GR1) Its stabilizer P is a $\mathbb{Q}$-rational parabolic subgroup of G.

(GR2) Its link group L_P is isogeneous to the product of a rational group and a compact one.

These conditions guarantee that the image of $\Gamma \cap P$ in M_P is arithmetic, and acts discretely on X_P. I shall say that the compactification $X \hookrightarrow \overline{X}$ itself is geometrically rational if the boundary components met by the closure of every Siegel set are all geometrically rational.

If $X \hookrightarrow \overline{X}$ is geometrically rational, the *rational boundary components* are those intersected by the closure of a Siegel set. In these circumstances, define X^* to be the union of X and all of the rational boundary components. The conditions guarantee that each of these components X_P is a symmetric space of some semi-simple group defined over $\mathbb{Q}$, which will be the product of G_P and some possibly trivial compact factor. Assign to X^* the topology characterized by the condition that a set is open if and only if its intersection with the closure of every Siegel set in $\overline{X}$ is open. Combining the main result of Satake (1960b) with a result of Borel (1962), we see that Γ acts discretely on X^* and that the quotient $\Gamma \backslash X^*$ is both compact and Hausdorff.

What properties of π determine whether or not $X \hookrightarrow \overline{X}_\pi$ is geometrically rational or not? The first step is to decide what boundary components are intersected by Siegel sets. Suppose (P, A) is a $\mathbb{Q}$-rational parabolic pair with P minimal $\mathbb{Q}$-rational. Choose parabolic pairs $(P_{\mathbb{C}}, A_{\mathbb{C}})$, $(P_{\mathbb{R}}, A_{\mathbb{R}})$ with

$$P \subseteq P_{\mathbb{R}} \subseteq P_{\mathbb{C}}, \quad A \supseteq A_{\mathbb{R}} \supseteq A_{\mathbb{C}}.$$

Let $\delta = \delta_\pi$ in $\Delta_{\mathbb{C}}$. If θ is a subset of $\Delta_{\mathbb{Q}}$ then for each λ in $X^{++}(A_\theta)$ the limit of $\lambda(t)x_0$ lies in X_{P_ω} where $\omega = \omega(\epsilon_{C/\mathbb{Q}}(\theta))$. According to Proposition 5.2 each of the boundary components intersected by the closure of $A^{++}(T)$ is of this form. Since each one of the saturated parabolic subgroups contains P, these are also the boundary components intersected by any Siegel set for P.

Lemma 8.1. *If (P, A) is a $\mathbb{Q}$-rational parabolic pair and P is a minimal $\mathbb{Q}$-rational parabolic subgroup of G then the boundary components met by the closure of a Siegel set $\mathfrak{S}(P, A, \Omega, T)$ are those whose stabilizer is P_ω, where ω is of the form $\omega(\epsilon_{C/\mathbb{Q}}(\theta))$.*

Therefore, in order that (GR1) hold it is necessary and sufficient that

($\bullet$) *for every* $\theta \subseteq \Delta_{\mathbb{Q}}$ *the group* P_ω *is* $\mathbb{Q}$-*rational, where* $\omega = \omega(\epsilon_{\mathbb{C}/\mathbb{Q}}(\theta))$.

This condition amounts to a finite number of conditions, but the number of conditions grows rapidly with the $\mathbb{Q}$-rank, so that it is not entirely practical.

What does condition (GR1) amount to for $\theta = \emptyset$? Since $\epsilon_{\mathbb{C}/\mathbb{Q}}(\emptyset)$ is just $\Delta^0_{\mathbb{C}/\mathbb{Q}}$, ω in this case will be the union of $\Delta^0_{\mathbb{C}/\mathbb{Q}}$ and those roots outside $\Delta^0_{\mathbb{C}/\mathbb{Q}}$ which are not in $\Delta_{\mathbb{C},\delta}$ and not connected to $\kappa^0_{\mathbb{C}/\mathbb{Q}}$ by a single edge. Since this contains $\Delta^0_{\mathbb{C}/\mathbb{Q}}$, it will parametrize a $\mathbb{Q}$-rational parabolic subgroup if and only if the elements in it which are not in $\Delta^0_{\mathbb{C}/\mathbb{Q}}$ are Galois invariant, or equivalently if and only if its complement $\delta_{\mathbb{C}/\mathbb{Q}}$ in $\Delta_{\mathbb{C}} - \Delta^0_{\mathbb{C}/\mathbb{Q}}$ is Galois invariant. I recall that this complement consists of elements of $\Delta_{\mathbb{C}} - \Delta^0_{\mathbb{C}/\mathbb{Q}}$ which are either in the set $\delta_{\mathbb{C}}$ or connected to $\kappa^0_{\mathbb{C}/\mathbb{Q}}$ by a single edge. In order that the stabilizer of every minimal rational boundary component be $\mathbb{Q}$-rational, it is thus necessary and sufficient that $\delta_{\mathbb{C}/\mathbb{Q}}$ (defined in §7 by these two conditions) be Galois invariant. In fact:

Theorem 8.2. *Condition (GR1) holds if and only if* $\delta_{\mathbb{C}/\mathbb{Q}}$ *is Galois invariant.*

Proof. We need to show sufficiency of this condition. It remains to show that under the assumption, if θ is a Galois invariant subset of $\Delta_{\mathbb{C}}$ containing $\Delta^0_{\mathbb{C}/\mathbb{Q}}$, then $\omega(\theta)$ is Galois invariant.

Suppose that α lies in $\kappa(\theta)$. There exists a path inside θ connecting α to an element of δ. Either this path lies entirely inside $\Delta^0_{\mathbb{C}/\mathbb{Q}}$, or it doesn't. In the first case, α lies in $\kappa^0_{\mathbb{C}/\mathbb{Q}}$. In the second, either the path goes through an element of $\Delta - \Delta^0$ which lies in δ, or it passes back into Δ^0 without meeting such an element. In the second case it passes through an element of $\Delta - \Delta^0$ which is connected to κ^0 by a single edge. In either of the last two cases, it is connected to an element of $\delta_{\mathbb{C}/\mathbb{Q}}$ inside θ. Conversely, if it is connected inside θ to an element of $\delta_{\mathbb{C}/\mathbb{Q}}$, then it lies in $\kappa(\theta)$. As a consequence

($\bullet$) *the set* $\kappa(\theta) - \kappa^0$ *is Galois invariant.*

We shall use this observation a bit later on.

The set $\zeta(\theta)$ is made up of the elements of $\Delta_{\mathbb{C}}$ which are not connected to $\kappa(\theta)$ by a single edge, and $\omega(\theta)$ is the union of $\kappa(\theta)$ and $\zeta(\theta)$. Since θ contains Δ^0 it only has to be shown that if α lies in $\omega(\theta)$ and the complement of Δ^0 then any Galois transform also lies in $\omega(\theta)$.

If α lies in $\kappa(\theta) - \Delta^0$ then we have just seen that any Galois transform lies also in the same set. So it remains to be seen that if α lies in $\zeta(\theta)$ but not in Δ^0 then any Galois transform lies in the same set. If β is a transform of α which does not lie in $\zeta(\theta)$ then there exists an edge from β to an element of $\kappa(\theta)$. If this edge leads to an element of $\kappa(\theta) - \kappa^0$ we get a contradiction, while if it leads to an element of κ^0 then β lies in $\delta_{\mathbb{C}/\mathbb{Q}}$ and we again get a contradiction.

Corollary 8.3. *If* G *is quasi-split, then a compactification*

$$X \hookrightarrow \overline{X}_\pi$$

is geometrically rational precisely when it is rational.

Proof. In this case the anisotropic kernel is empty and condition (GR2) is automatic.

The situation for condition (GR2) is similar—that is to say, the condition for it to hold with respect to minimal boundary components, which is clearly necessary, is also sufficient for it to hold in general. For the minimal boundary component, the factor G_P has Dynkin diagram κ^0. Condition (GR2) means that this must be a rational group up to a compact factor, which means in turn that the Galois orbit of κ^0 must be the union of κ^0 and a set of compact nodes.

Theorem 8.4. *Assuming that condition (GR1) is valid, condition (GR2) holds if and only if the Galois orbit of κ^0 in the index is the union of κ^0 and a set of compact nodes.*

Proof. This is clear, since condition (GR1) implies, as the proof of the last result showed, that for any θ parametrizing a $\mathbb{Q}$-rational subgroup, $\kappa(\theta)$ is the union of a Galois invariant subset and κ^0.

If $X \hookrightarrow \overline{X}$ is a geometrically rational compactification, then the rational boundary components coincide with those whose stabilizers are $\mathbb{Q}$-rational, and the stabilizers in this case coincide with the rational parabolic subgroups parametrized by saturated subsets of $\Delta_{\mathbb{Q}}$.

9. Some examples

Fix a real quadratic extension F of $\mathbb{Q}$.

• Let G be the unitary group of an isotropic Hermitian form of dimension three over F, which is a $\mathbb{Q}$-rational group isomorphic over F to $SL_3(F)$. Over $\mathbb{Q}$ it is quasi-split. Its Dynkin diagram and index are

Over $\mathbb{R}$ the group is split. If we pick as δ either one of the two nodes (say the one marked by a dot in the centre), condition (GR1) is not satisfied. On the other hand this example satisfies Assumption 1 (quasi-rationality) of §3.3 in Zucker (1983), and this example seems to show that his Proposition 3.3.(ii) is false. It is not clear to me what the significance of Zucker's notion of quasi-rationality is.

• Let Q be an anisotropic quadratic form over F of dimension four, arising from a quaternion algebra over F which is split at both real embeddings of F into $\mathbb{R}$. Let G be the orthogonal group of the quadratic form $H \oplus Q$, where H is hyperbolic space. The group G may be considered by restriction of scalars as a group over $\mathbb{Q}$. The index of G is shown by the following figure:

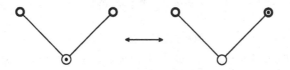

Here the anisotropic kernel of G over $\mathbb{Q}$ consists of the top four nodes. Conjugation over $\mathbb{Q}$ just interchanges the two components. If δ is made up of the left bottom node and the right top node, then (GR1) is valid but (GR2) is not.

• Let K be a quadratic imaginary extension of F obtained by adjunction from one over $\mathbb{Q}$. Let Q be a Hermitian form of dimension three over K, anisotropic at one real embedding of F and of signature $(2,1)$ at the other. Let G be the unitary group of of $H \oplus Q$, where H is the hyperbolic Hermitian form, considered as a group over $\mathbb{Q}$. Its index is

where the anisotropic kernel over $\mathbb{Q}$ consists of the four bottom nodes. Conjugation over $\mathbb{Q}$ interchanges the two components, and complex conjugation interchanges the branches inside each component.

Let δ be the marked nodes. Then the compactification of each real factor of G is its Baily-Borel compactification, and conditions (GR1) and (GR2) are both valid. All other Baily-Borel compactifications can be dealt with in a similar case by case analysis, somewhat different from the one in Baily-Borel (1966). What one needs for this are the diagrams in §7 and the observation in §3.1.2 of Tits (1966) about induced indices.

This compactification does not arise from a rational representation, and the boundary components have a different structure from those of the compactification associated to the similar rational representation whose figure is

although the saturated parabolic subgroups are the same for both cases. In the second case, the minimal boundary components are just points, while in the first they are non-trivial symmetric spaces.

I have not attempted a classification of all semi-simple groups over $\mathbb{Q}$ in order to try to find all geometrically rational Satake compactifications. This example illustrates that the first step is probably to break them up into families, each family having the same set of saturated $\mathbb{Q}$-rational parabolic subgroups, but differing in the structure of its boundary components.

References

1. A. Ash, D. Mumford, M. Rapoport, and Y. Tai, *Smooth compactifications of locally symmetric varieties*, Math. Sci. Press, Brookline, 1975.

2. W. Baily and A. Borel, *Compactifications of arithmetic quotients of bounded symmetric domains*, Ann. Math. **84** (1966), 442–528.
3. A. Borel, *Ensembles fondamentaux pour les groupes arithmétiques*, Colloque sur la Théorie des Groupes Algébriques (Bruxelles), CBRM, Bruxelles, 1962, pp. 23–40.
4. A. Borel and J. Tits, *Groupes réductifs*, Publ. Math. I.H.E.S. **27** (1965), 55–151.
5. P. Deligne, *Variétés de Shimura*, vol. 2 of Proc. Symp. Pure Math. XXXIII (Providence) (A. Borel and W. Casselman, eds.), Amer. Math. Soc., Providence, 1979, pp. 247–290.
6. M. Goresky, G. Harder, and R. MacPherson, *Weighted cohomology*, Inv. Math. **116** (1994), 139–213.
7. G. Kempf, F. Knudsen, D. Mumford, and B. Saint-Donat, *Toroidal embeddings I.*, Springer-Verlag, Berlin, 1973, Lect. Notes Math. 339.
8. R. Richardson and T. Springer, *The Bruhat order on symmetric varieties*, Geom. Ded. **35** (1990), 389–436.
9. _____, *Combinatorics and geometry of k-orbits on the flag manifold*, Contemp. Math. **153** (1993), 109–142.
10. I. Satake, *On representations and compactifications of symmetric riemannian symmetric spaces*, Ann. Math. **71** (1960), 77–110.
11. _____, *On compactifications of the quotient spaces for arithmetically defined discontinuous groups*, Ann. Math. **72** (1960), 555–580.
12. J. Tits, *Classification of algebraic semi-simple groups*, Proc. Symp. Pure Math. IX (Providence) (A. Borel and G. D. Mostow, eds.), Amer. Math. Soc., Providence, 1966, pp. 33–62.
13. Th. Vust, *Opérations de groupes réductifs dans un type de cônes presque homogènes*, Bull. Soc. Math. France **102** (1974), 317–333.
14. _____, *Plongements d'espaces symétriques algébriques: une classification*, Annali Sc. Norm. Pisa **17** (1990), 165–195.
15. S. Zucker, *Satake compactifications*, Comm. Math. Helv. **78** (1983), 312–343.

GRADED AND NON-GRADED KAZHDAN-LUSZTIG THEORIES

EDWARD CLINE, BRIAN PARSHALL, AND LEONARD SCOTT

We dedicate this paper to the memory of Roger Richardson.

ABSTRACT. Let $\mathscr{C}_A$ be the category of finite dimensional right modules for a quasi-hereditary algebra A. In the context of various types of Kazhdan-Lusztig theories, we study both the homological dual

$$A^! = \operatorname{Ext}^{\bullet}_{\mathscr{C}_A}(A/\operatorname{rad}(A), A/\operatorname{rad}(A))$$

and the graded algebra

$$\operatorname{gr} A = \bigoplus_j \operatorname{rad}(A)^j / \operatorname{rad}(A)^{j+1}.$$

For example, we investigate a condition introduced in [6], and here called (SKL'), which guarantees that $A^{!!} \cong \operatorname{gr} A$. A strengthening of (SKL') leads to the notion of a *strong Kazhdan-Lusztig theory* (SKL) for $\mathscr{C}_A$. We show that (SKL) behaves well with respect to recollement and we relate this notion to that of a graded Kazhdan-Lusztig theory. In particular, we prove that the category $\mathscr{O}_{\operatorname{triv}}$ associated to a complex semisimple Lie algebra satisfies the (SKL) condition. Finally, we present an example showing that, even in very favorable circumstances, strong properties of the algebras $A^!$ and $\operatorname{gr} A$ may not imply similar properties for A.

Let A be a quasi-hereditary algebra over an algebraically closed field k. Let $\mathscr{C}_A$ be the category of finite dimensional right A-modules. In general, the homological dual $A^! = \operatorname{Ext}^{\bullet}_{\mathscr{C}_A}(A/\operatorname{rad}(A), A/\operatorname{rad}(A))$ is *not* a quasi-hereditary algebra. However, when $\mathscr{C}_A$ has a Kazhdan-Lusztig theory in the sense of [5], a main result of [6] established that $A^!$ is quasi-hereditary. Another main result in [6] presented a strengthening of the condition for a Kazhdan-Lusztig theory. This new condition, here called (SKL'), is a parity requirement, involving, for example, $\operatorname{Ext}^{\bullet}_{\mathscr{C}_A}(-, L)$, for L an irreducible A-module, as applied to the radical series of standard modules. The (SKL') condition suffices to guarantee that the category $\mathscr{C}^{\operatorname{gr}}_{A^!}$ of finite dimensional graded right $A^!$-modules has a *graded* Kazhdan-Lusztig theory (and, consequently, $A^!$ is Koszul). As remarked in [6, (3.10)] (see also [14]), the stronger (SKL') condition does hold in at least the classical case when $\mathscr{C}_A \cong \mathscr{O}_{\operatorname{triv}}$, the principal block for the Bernstein-Gelfand-Gelfand category $\mathscr{O}$ associated with a complex semisimple Lie algebra. Thus, (SKL') illustrates a

1991 *Mathematics Subject Classification.* 20G05.
Research supported in part by the National Science Foundation

new property for $\mathscr{O}_{\text{triv}}$ discovered in the context of quasi-hereditary algebras and abstract highest weight category theory.

This paper continues the study of the homological dual begun in [6] and of graded algebras in [4, 14]. One general theme is that homological properties of $A^!$ are often reflected in terms of properties of the graded algebra $\text{gr } A = \bigoplus_j \text{rad}(A)^j / \text{rad}(A)^{j+1}$ obtained from A using its radical filtration. This fact is clear and well-known in case A has a Koszul grading, since then $A^{!!} \cong A \cong \text{gr } A$. (Here $A^{!!} = (A^!)^!$ is the homological dual of $A^!$.) In §1, we discuss some general results concerning quadratic and Koszul algebras. In particular, Corollary (1.3.2) shows that if A is a finite dimensional algebra such that $A^!$ is Koszul (a property that always holds if A is Koszul), then the graded algebra $\text{gr } A$ is naturally a homomorphic image of $A^{!!}$.

We apply these results to quasi-hereditary algebras in §2. Let A be a quasi-hereditary algebra such that $\mathscr{C}_A$ satisfies the (SKL') condition. Then Theorem 2.2.1 demonstrates that $\text{gr } A$ is also quasi-hereditary. Actually, we show that $\text{gr } A \cong A^{!!}$, so, by results of [6], the graded highest weight category $\mathscr{C}^{\text{gr}}_{\text{gr } A}$ of finite dimensional graded right $\text{gr } A$-modules has a graded Kazhdan-Lusztig theory. In particular, the algebra $\text{gr } A$ is quasi-hereditary as well as Koszul, both properties difficult to prove *a priori*. We further show that $\mathscr{C}_A$ and $\mathscr{C}_{\text{gr } A}$ have the same Kazhdan-Lusztig polynomials.

This paper also presents a stronger version of (SKL'), called (SKL). As we prove in Theorem 2.4.2, when A is a graded quasi-hereditary algebra such that $\mathscr{C}^{\text{gr}}_A$ has a graded Kazhdan-Lusztig theory, then $\mathscr{C}_A$ satisfies (SKL). (In particular, (SKL) holds for the categories $\mathscr{O}_{\text{triv}}$.) For a general quasi-hereditary algebra A, we prove in Theorem 2.3.3 that the (SKL) condition behaves well with respect to recollement (as studied for highest weight categories in [3]). It turns out that (SKL) represents the "universal" version of (SKL'): (SKL) is equivalent to (SKL') holding in each of the quotient categories $\mathscr{C}_A(\Omega)$, $\Omega \subset \Lambda$, a coideal; see §2.4. (SKL) has other advantages involving certain derived category constructions; see (2.1.6), for example. Encouraged by these results, we call (SKL) the *strong Kazhdan-Lusztig* condition for $\mathscr{C}_A$.

As indicated in several (largely unpublished) talks by the authors, the existence of a sufficiently rich graded structure on the relevant quasi-hereditary algebras suffices to establish the truth of the Lusztig conjecture for semisimple algebraic groups G in positive characteristic. For a version of this argument, see (2.3.5). Evidence involving the category $\mathscr{O}_{\text{triv}}$ ([2], [14]) as well as categories associated with the restricted enveloping algebra of G ([1]) suggests the graded algebras involved are actually Koszul. For these reasons, it is important to understand when a quasi-hereditary algebra A possesses a (non-trivial) positive grading, and even a Koszul structure. For example, let σ be an automorphism of an algebra A. In [14, (1.16)], the second two authors proved that A is Koszul provided the eigenvalue structure for the induced action of σ on $A^!$ determines the natural grading of $A^!$. They then used these results, together with calculations by Kazhdan-Lusztig of eigenvalues of the Frobenius morphism on cohomology stalks of perverse sheaves, to give a proof of the results mentioned above on $\mathscr{O}_{\text{triv}}$ (results also proved in [2]).

In view of the previous paragraph as well as the results given in §2 of this paper, it is reasonable to ask if properties of either its homological dual $A^!$ or of gr A suffice to determine if a quasi-hereditary algebra A is Koszul. (When gr A is Koszul, the issue becomes: "Is $A \cong$ gr A?") In §3, we present an example which indicates a negative answer to this question even when $\mathscr{C}_A$ has the strong Kazhdan-Lusztig condition (SKL).

1. QUADRATIC AND KOSZUL ALGEBRAS

Throughout this paper, k denotes a fixed algebraically closed field which serves as the base field for the algebras under consideration. Given an algebra A, let $\mathscr{C}_A$ denote the category of finitely generated right A-modules.

By a graded algebra A, we mean that A is *positively graded*: $A = \bigoplus_{n \in \mathbb{Z}^+} A_n$, $A_i A_j \subset A_{i+j}$. *In this paper, it will always be the case that A_0 is a finite dimensional semisimple algebra, while A_1 is finite dimensional.* If A is a graded algebra, let $\mathscr{C}_A^{\mathrm{gr}}$ denote the category of finitely generated graded right A-modules. For $M = \bigoplus_{n \in \mathbb{Z}} M_n \in \mathrm{Ob}(\mathscr{C}_A^{\mathrm{gr}})$ and for an integer i, let $M(i) \in \mathrm{Ob}(\mathscr{C}_A^{\mathrm{gr}})$ be defined by $M(i)_n = M_{n-i}$. If L is an irreducible right A_0-module, we regard it as an object in $\mathscr{C}_A^{\mathrm{gr}}$, by making it into a graded A-module (concentrated in degree 0) via the homomorphism $A \to A/\bigoplus_{n>0} A_n \cong A_0$. If Λ is a set indexing the distinct isomorphism classes of irreducible A_0-modules (with $\lambda \in \Lambda$ corresponding to the irreducible A_0-module $L(\lambda)$), then the modules $L(\lambda)(i)$, $\lambda \in \Lambda$, $i \in \mathbb{Z}$, index the distinct isomorphism classes of irreducible graded A-modules.

If A is a finite dimensional algebra having radical $\mathrm{rad}(A)$, let gr A denote the associated graded algebra

$$\mathrm{gr}\ A = \bigoplus_j \mathrm{rad}(A)^j / \mathrm{rad}(A)^{j+1}.$$

The reader should keep in mind that, by our notational conventions, $\mathscr{C}_{\mathrm{gr}\ A}$ denotes the category of finite dimensional right gr A-modules, while $\mathscr{C}_{\mathrm{gr}\ A}^{\mathrm{gr}}$ denotes the category of finite dimensional *graded* right gr A-modules. If $M \in \mathrm{Ob}(\mathscr{C}_A)$, let

$$0 = \mathrm{soc}^0(M) \subset \mathrm{soc}^{-1}(M) \subset \mathrm{soc}^{-2}(M) \subset \cdots$$

denote the socle series of M (note the negative indices). In addition, we will consider the radical series

$$M = \mathrm{rad}^0(M) \supset \mathrm{rad}^1(M) \supset \mathrm{rad}^2(M) \supset \cdots$$

of M. (Thus, $\mathrm{rad}^j(M) = M \mathrm{rad}(A)^j$ and $\mathrm{rad}^j(A) = \mathrm{rad}(A)^j$.) Let $t(M) \geq 0$ be the smallest integer such that $\mathrm{soc}^{-t(M)}(M) = M$. Of course, $t(M)$ is also the smallest non-negative integer for which $\mathrm{rad}^{t(M)}(M) = 0$, and is called the Loewy length of M. Put

$$\mathrm{gr}'\ M = \bigoplus_i \mathrm{soc}^{i-1}(M)/\mathrm{soc}^i(M), \quad \mathrm{gr}\ M = \bigoplus_j \mathrm{rad}^j(M)/\mathrm{rad}^{j+1}(M). \quad (1.0.1)$$

Both gr$'$ M and gr M are graded gr A-modules, taking

$$(\mathrm{gr}'\ M)_i = \mathrm{soc}^{i-1}(M)/\mathrm{soc}^i(M) \quad \text{and} \quad (\mathrm{gr}\ M)_i = \mathrm{rad}^i(M)/\mathrm{rad}^{i+1}(M).$$

Finally, recall that a finite dimensional algebra A is *tightly graded* if there is an algebra isomorphism $A \cong \operatorname{gr} A$. If A is already a *graded* algebra (from context), we only use this terminology if the isomorphism is an isomorphism of graded algebras.[1]

1.1. Quadratic algebras.

The definitions and most of the discussion of this subsection are standard, though notation varies. Let us first recall the notion of a *quadratic algebra* over k. Let $A = A_0 \oplus A_1 \oplus \cdots$ be a graded algebra. Then A is quadratic provided A is isomorphic to a quotient $T_{A_0}(A_1)/\mathfrak{J}$ of the tensor algebra $T_{A_0}(A_1)$ of the (A_0, A_0)-bimodule A_1 by a homogeneous ideal $\mathfrak{J}$ which is generated by its term $\mathfrak{J}_2 = A_2 \cap \mathfrak{J}$ in grade 2.

There are three possible "dual" bimodules associated to an (A_0, A_0)-bimodule V, defined as follows: Let V^* (respectively, ${}^{\mathrm{tr}}V$, V^{tr}) denote the k-linear dual $\operatorname{Hom}_k(V, k)$ (respectively, $\operatorname{Hom}_{A_0}({}_{A_0}V, {}_{A_0}A_0)$, $\operatorname{Hom}_{A_0}(V_{A_0}, A_{0A_0})$). Each of these vector spaces is naturally an (A_0, A_0)-bimodule. We have canonical isomorphisms

$$ {}^{\mathrm{tr}}V \cong V^* \cong V^{\mathrm{tr}}. \tag{1.1.1} $$

To prove this, observe the isomorphism

$$ V^* \xrightarrow{\sim} \operatorname{Hom}_{\mathscr{C}_{A_0}}(V_{A_0}, A^*_{0A_0}), \; \phi \mapsto [v \mapsto [a \mapsto \phi(va)]]. $$

Since A_0 is semisimple, the bimodules A_0 and A^*_0 are isomorphic. (This isomorphism can be taken to be canonical by using the trace form.) Hence, $V^* \cong V^{\mathrm{tr}}$. Similarly, $V^* \cong {}^{\mathrm{tr}}V$.

If V, W are two finite dimensional (A_0, A_0)-bimodules, there is a natural bimodule isomorphism

$$ V^{\mathrm{tr}} \otimes_{A_0} W^{\mathrm{tr}} \xrightarrow{\sim} (W \otimes_{A_0} V)^{\mathrm{tr}} \tag{1.1.2} $$

defined by mapping a generator $\phi \otimes \psi \in V^{\mathrm{tr}} \otimes_{A_0} W^{\mathrm{tr}}$ to the element of $(W \otimes_{A_0} V)^{\mathrm{tr}}$ defined by the rule $(\phi \otimes \psi)(w \otimes v) = \phi(\psi(w)v)$, $v \in V, w \in W$. For an (A_0, A_0)-subbimodule $I \subset W \otimes_{A_0} V$, let $I' \subset (W \otimes_{A_0} V)^{\mathrm{tr}}$ be the annihilator of I, and let $I^\perp \subset V^{\mathrm{tr}} \otimes_{A_0} W^{\mathrm{tr}}$ denote the corresponding subbimodule under the isomorphism (1.1.2).

For a quadratic algebra $A = T_{A_0}(A_1)/\mathfrak{J}$ as above, the *quadratic dual* is defined by

$$ A^\perp = T_{A_0}(A_1^{\mathrm{tr}})/(\mathfrak{J}_2^\perp). $$

By construction, $A^\perp$ is again a quadratic algebra satisfying $A^{\perp\perp} \cong A$.

---

[1]More generally, we say that a (not necessarily finite dimensional) graded algebra $A = \bigoplus_i A_i$ is *tightly graded* if $A_i \cdot A_j = A_{i+j}$ for all i, j.

1.2. A surjectivity criterion. Let A be a finite dimensional algebra over k. Let $I = \mathrm{rad}(A)$ be the radical of A, and let A_0 be a fixed Wedderburn complement. Thus, we have a vector space decomposition $A = A_0 \oplus I$. Multiplication determines the following commutative diagram having exact rows:

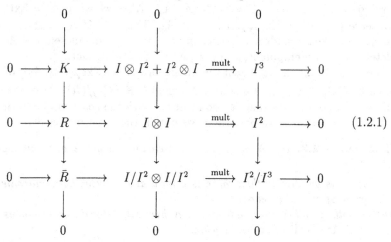

$$
\begin{array}{ccccccccc}
& & 0 & & 0 & & 0 & & \\
& & \downarrow & & \downarrow & & \downarrow & & \\
0 & \longrightarrow & K & \longrightarrow & I \otimes I^2 + I^2 \otimes I & \xrightarrow{\text{mult}} & I^3 & \longrightarrow & 0 \\
& & \downarrow & & \downarrow & & \downarrow & & \\
0 & \longrightarrow & R & \longrightarrow & I \otimes I & \xrightarrow{\text{mult}} & I^2 & \longrightarrow & 0 \quad (1.2.1) \\
& & \downarrow & & \downarrow & & \downarrow & & \\
0 & \longrightarrow & \bar{R} & \longrightarrow & I/I^2 \otimes I/I^2 & \xrightarrow{\text{mult}} & I^2/I^3 & \longrightarrow & 0 \\
& & \downarrow & & \downarrow & & \downarrow & & \\
& & 0 & & 0 & & 0 & &
\end{array}
$$

Thus, $K, R, \bar{R}$ are defined by the diagram. Also, we have written $\otimes$ for $\otimes_{A_0}$; unless otherwise noted, we follow this convention below. In addition to exact rows, the second and third columns are exact. By the snake lemma, the first column is also exact.

Let $V = I/I^2$. If we view V as an (A_0, A_0)-bimodule and A_0 as a right A-module by means of the quotient map $\pi : A \to A_0 \cong A/I$, we have $V^{\mathrm{tr}} \cong \mathrm{Ext}^1_{\mathscr{C}_A}(A_0, A_0)$. We have the following result concerning the product structure in the Ext-algebra. This result is essentially contained in Priddy [15, p. 42] (see also Löfwall [10, Cor. 1.1]). For completeness, we give a simple proof.

Lemma 1.2.2. *There is a natural isomorphism from $\bar{R}^{\mathrm{tr}}$ to the image of the Yoneda product map*

$$
V^{\mathrm{tr}} \otimes V^{\mathrm{tr}} = \mathrm{Ext}^1_{\mathscr{C}_A}(A_0, A_0) \otimes \mathrm{Ext}^1_A(A_0, A_0) \xrightarrow{\text{mult}} \mathrm{Ext}^2_{\mathscr{C}_A}(A_0, A_0).
$$

Proof. The inclusion $\bar{R} \subset V \otimes V$ induces a surjective homomorphism $\pi : V^{\mathrm{tr}} \otimes V^{\mathrm{tr}} \to \bar{R}^{\mathrm{tr}}$. The proof consists of showing that the Yoneda product map $V^{\mathrm{tr}} \otimes V^{\mathrm{tr}} \xrightarrow{\rho} \mathrm{Ext}^2_{\mathscr{C}_A}(A_0, A_0)$ factors through π via an injective map $\bar{\rho} : \bar{R}^{\mathrm{tr}} \to \mathrm{Ext}^2_{\mathscr{C}_A}(A_0, A_0)$ (so $\rho = \bar{\rho} \circ \pi$).

Consider the resolution $0 \to I \otimes I \to I \otimes A \xrightarrow{\text{mult}} A \to A/I \to 0$. Here the map on $I \otimes I$ is given by $x \otimes y \mapsto x \otimes y - xy \otimes 1$. First, observe that any A-module homomorphism $\bar{R} \to A/I$ extends to an A-module homomorphism from the semisimple module $I/I^2 \otimes I/I^2$ to A/I, and thus defines, by composition, a map $I \otimes I \to A/I$. Also, no homomorphism $I \otimes I \to A/I$ which has a nonzero restriction to R factors through $I \otimes A$, since the image of R in the latter module is contained in its radical $I \otimes I$. This gives an injective map $\bar{\rho} : \bar{R}^{\mathrm{tr}} \to \mathrm{Ext}^2_{\mathscr{C}_A}(A/I, A/I)$. The reader may directly verify that $\rho = \bar{\rho} \circ \pi$.

We now consider another type of dual associated to a finite dimensional algebra A—its *homological dual* $A^!$. Again, view A_0 as a right A-module through the quotient map $A \to A/I$. Form the algebra $A^! = \text{Ext}^{\bullet}_{\mathscr{C}_A}(A_0, A_0)$ (in which the product structure is defined by using Yoneda composition).[2] We give $A^!$ its natural graded structure: $A^!_n = \text{Ext}^n_{\mathscr{C}_A}(A_0, A_0)$. Also, we write $A^{!!} = \text{Ext}^{\bullet}_{A^!}(A^!_0, A^!_0)$. Observe that $A^!_0 = \text{Hom}_{\mathscr{C}_A}(A_0, A_0) \cong A_0$. Thus, $A^{!!}_0 \cong A^!_0 \cong A_0$. In what follows, we make these identifications; in particular, an idempotent $e \in A_0$ naturally determines an idempotent, still denoted e, in both $A^!_0$ and $A^{!!}_0$.

Lemma (1.2.2) shows that if the subalgebra $A' = \langle \text{Ext}^1_{\mathscr{C}_A}(A_0, A_0) \rangle$ of $A^!$ generated by $\text{Ext}^1_{\mathscr{C}_A}(A_0, A_0)$ is quadratic, then $A'^{\perp} \cong T(V)/(\bar{R})$. Also, the exactness of (1.2.1) above shows that $\bar{R}$ also identifies with the space of quadratic relations for the algebra gr A. In particular, we obtain the following result.

Theorem 1.2.3. *Let A be a finite dimensional algebra over an algebraically closed field k.*

(a) *If $A^!$ is quadratic, then there is a natural surjective homomorphism $A^{!\perp} \to$ gr A of graded algebras.*

(b) *If both A and $A^!$ are quadratic, then there are natural isomorphisms $A^{!\perp} \xrightarrow{\sim} A$ and $A^{\perp} \xrightarrow{\sim} A^!$ of graded algebras.*

1.3. Koszul algebras. Let $B = B_0 \oplus B_1 \oplus \cdots$ be a graded algebra. By definition, B is *Koszul* provided that for simple B_0-modules L, L', and integers m, n, p, we have

$$\text{Ext}^p_{\mathscr{C}^{gr}_B}(L(m), L'(n)) \neq 0 \implies n - m = p. \tag{1.3.1}$$

If B is Koszul, (1.3.1) implies that there can be no non-split extension between simple (graded) B-modules $L(m), L'(n)$ unless their grades differ by 1. In particular, this implies easily that B is generated as an algebra by B_0 and B_1, i. e., B is tightly graded.

If B is Koszul, then it is known that B is a quadratic algebra. Also, the algebra $B^!$ is again quadratic and $B^{!!} \cong B$. (See [10] in the local case, [7] in the basic case, and [2] in the general case.) Thus, we obtain the following result

Corollary 1.3.2. *Let A be a finite dimensional algebra. Assume that $A^!$ is Koszul. Then there exists a natural surjective homomorphism $A^{!!} \cong A^{!\perp} \to$ gr A of graded algebras.*

---

[2]Our notation differs from that used in [6]. Let $L = \bigoplus_{\lambda \in \Lambda} L(\lambda)$ and form the algebra $A^{\dagger} = \text{Ext}^{\bullet}_{\mathscr{C}_A}(L, L)$. In [6], we most often worked with the algebra $A^{\dagger}$, which was denoted there by the symbol $A^!$. In this paper, we prefer the larger version of $A^!$, with possibly multiple copies of $L(\lambda)$ in L. This causes no problems, since the two algebras are Morita equivalent. ($A^{\dagger} \cong eA^!e$ for an idempotent $e \in A^!_0$ which has a nonzero projection into each simple factor of $A^!/\text{rad}(A^!)$.) Thus, $A^{\dagger}$ is quasi-hereditary if and only if $A^!$ is quasi-hereditary, the module categories $\mathscr{C}_{A^!}$ and $\mathscr{C}_{A^{\dagger}}$ are equivalent (as are the corresponding graded module categories), etc. Also, in contrast to [6], we do not assume our highest weight categories possess a duality. As remarked in [6, p. 310], the assumption of a duality was largely a matter of convenience. We continue to use the results of [6], making appropriate remarks and minor adjustments.

GRADED AND NON-GRADED KAZHDAN-LUSZTIG THEORIES   111

2. Applications to Quasi-Hereditary Algebras

For the theory of quasi-hereditary algebras, we refer to [3, 4, 5]. Let A be a quasi-hereditary algebra; thus, the module category $\mathscr{C}_A$ is a highest weight category relative to some partial ordering $\leq$ on a set Λ indexing the irreducible A-modules. (When speaking of $\mathscr{C}_A$ as a highest weight category, we will assume some fixed poset structure on Λ is given—we call Λ the weight poset of $\mathscr{C}_A$.) For the theory of highest weight categories, see also [3]. In general, for a highest weight category $\mathscr{C}$ with weight poset Λ, let $L(\mathscr{C}, \lambda) = L(\lambda)$, $\Delta(\mathscr{C}, \lambda) = \Delta(\lambda)$, $\nabla(\mathscr{C}, \lambda) = \nabla(\lambda)$ denote the irreducible, standard, and costandard objects, respectively, corresponding to λ.[3]

2.1. Kazhdan-Lusztig theories. We will make use of several families of (Laurent) polynomials in a variable t which are associated to the highest weight category $\mathscr{C}_A$ with weight poset Λ. Let $l : \Lambda \to \mathbb{Z}$ be some fixed function.

For $\lambda, \nu \in \Lambda$, the associated (left) Kazhdan-Lusztig polynomial $P_{\nu,\lambda}$ is defined by

$$P_{\nu,\lambda} = t^{l(\lambda)-l(\nu)} \sum_n \dim \operatorname{Ext}^n_{\mathscr{C}_A}(L(\lambda), \nabla(\nu))t^{-n}. \tag{2.1.1}$$

The *right* Kazhdan-Lusztig polynomials $P^R_{\lambda,\nu}$ are defined by a similar formula:

$$P^R_{\nu,\lambda} = t^{l(\lambda)-l(\nu)} \sum_n \dim \operatorname{Ext}^n_{\mathscr{C}_A}(\Delta(\nu), L(\lambda))t^{-n}. \tag{2.1.2}$$

Recall that $\mathscr{C}_A$ has a Kazhdan-Lusztig theory relative to l provided that, for all $\lambda, \nu \in \Lambda$, the polynomials $P_{\nu,\lambda}$, $P^R_{\nu,\lambda}$ have only *even* powers of t with nonzero coefficients. If $\mathscr{C}_A$ has a Kazhdan-Lusztig theory relative to l, then $\mathscr{C}^!_A \overset{\text{def}}{=} \mathscr{C}_{A^!}$ is a highest weight category. The weight poset for $\mathscr{C}_{A^!}$ can be taken to be the opposite poset Λ^{op}, obtained from Λ by reversing the relations. See [6, (2.1)].

As suggested by work of Irving [8], *filtered* (left and right) Kazhdan-Lusztig polynomials are defined as follows. For $\lambda, \nu \in \Lambda$, define

$$\begin{cases} F_{\nu,\lambda} = t^{l(\nu)-l(\lambda)} \sum_n [\operatorname{gr} \Delta(\nu) : L(\lambda)(n)](-t)^{-n}, \\ F^R_{\nu,\lambda} = t^{l(\nu)-l(\lambda)} \sum_n [\operatorname{gr}' \nabla(\nu) : L(\lambda)(n)](-t)^{-n} \end{cases} \tag{2.1.3}$$

For more details, see [6, (1.2.6)].

In case A is a *graded* quasi-hereditary algebra, the category $\mathscr{C}^{\mathrm{gr}}_A$ carries the structure of a graded highest weight category; see [5, 6] for a discussion. The category $\mathscr{C}^{\mathrm{gr}}_A$ has a *graded* Kazhdan-Lusztig theory relative to $l : \Lambda \to \mathbb{Z}$ provided the non-vanishing of either $\operatorname{Ext}^n_{\mathscr{C}^{\mathrm{gr}}_A}(\Delta(\nu), L(\lambda)(m))$ or $\operatorname{Ext}^n_{\mathscr{C}^{\mathrm{gr}}_A}(L(\lambda), \nabla(\nu)(m))$ implies that $m = n \equiv l(\lambda) - l(\nu) \pmod 2$. In case, $\mathscr{C}^{\mathrm{gr}}_A$ has a graded Kazhdan-Lusztig

---

[3]Thus, we depart from [3] in our notation. There, by analogy with the representation theory of algebraic groups, we called $\nabla(\lambda)$ (respectively, $\Delta(\lambda)$) the "induced object" (respectively, "Weyl object") corresponding to a weight $\lambda \in \Lambda$. When dealing with highest weight categories from the theory of algebraic groups, we will continue to use the terms "induced module" and "Weyl module" (or "Verma module") when this terminology agrees with the commonly accepted one. It also seems good in these contexts to conform to standard notations for Weyl or Verma modules. The ∇-, Δ-notation was proposed by Ringel, who thought it visually suggestive.

theory, it follows that the Kazhdan-Lusztig polynomials for the ungraded module category $\mathscr{C}_A$ take the form

$$P_{\nu,\lambda} = t^{l(\lambda)-l(\nu)} \sum_n \dim \operatorname{Ext}^n_{\mathscr{C}_A^{\mathrm{gr}}}(L(\lambda), \nabla(\nu)(n))t^{-n}.$$

This is because

$$\operatorname{Ext}^n_{\mathscr{C}_A}(L(\lambda), \nabla(\nu)) \cong \bigoplus_m \operatorname{Ext}^n_{\mathscr{C}_A^{\mathrm{gr}}}(L(\lambda), \nabla(\nu)(m))$$

$$\cong \operatorname{Ext}^n_{\mathscr{C}_A^{\mathrm{gr}}}(L(\lambda), \nabla(\nu)(n))$$

in the presence of a graded Kazhdan-Lusztig theory. A similar formula exists for the polynomials $P^R_{\nu,\lambda}$. We also speak of the $P_{\nu,\lambda}$, $P^R_{\nu,\lambda}$ as the Kazhdan-Lusztig polynomials of $\mathscr{C}_A^{\mathrm{gr}}$. By [6, (3.9)], $\mathscr{C}_A^{\mathrm{gr}}$ has a graded Kazhdan-Lusztig theory relative to l if and only if $\mathscr{C}_A$ has a Kazhdan-Lusztig theory relative to l and A is a Koszul algebra.

Now let $\mathscr{C}_A$ be a highest weight category with weight poset Λ. We do not assume that A is necessarily graded. For $\lambda \in \Lambda$, put

$$\nabla^i(\lambda) = \begin{cases} \nabla(\lambda)/\operatorname{soc}^i(\nabla(\lambda)) & \text{if } -t(\nabla(\lambda)) \leq i \leq 0; \\ 0 & \text{if } i > 0 \text{ or } i < -t(\nabla(\lambda)). \end{cases}$$

(Recall that $t(M)$ is the Loewy length of a module M.) Thus, $\nabla^0(\lambda) = \nabla(\lambda)$, $\nabla^{-1}(\lambda) = \nabla(\lambda)/\operatorname{soc}(\nabla(\lambda)) \cong \nabla(\lambda)/L(\lambda)$, etc. Dually, we put

$$\Delta^i(\lambda) = \begin{cases} \operatorname{rad}^i(\Delta(\lambda)) & \text{if } 0 \leq i \leq t(\Delta(\lambda)); \\ 0 & \text{if } i < 0 \text{ or } i > t(\Delta(\lambda)). \end{cases}$$

Thus, $\Delta^0(\lambda) = \Delta(\lambda)$, $\Delta^1(\lambda) = \operatorname{rad}(\Delta(\lambda))$, etc.

We will say that $\mathscr{C}_A$ satisfies the *strong Kazhdan-Lusztig* (SKL) condition relative to $l : \Lambda \to \mathbb{Z}$ provided the following condition holds for all $\lambda, \nu \in \Lambda$ and $n, i \in \mathbb{Z}$:

$$\begin{cases} \operatorname{Ext}^n_{\mathscr{C}_A}(\Delta(\nu), \nabla^i(\lambda)) \neq 0 \implies n \equiv l(\nu) - l(\lambda) + i \mod 2; \text{ and} \\ \operatorname{Ext}^n_{\mathscr{C}_A}(\Delta^i(\nu), \nabla(\lambda)) \neq 0 \implies n \equiv l(\nu) - l(\lambda) + i \mod 2. \end{cases} \quad \text{(SKL)}$$

We will see in (2.1.5(d)) below that this condition implies very strong properties of $A^!$; in particular, it means that $A^!$ is both a quasi-hereditary and a Koszul algebra. While (SKL) appears technical, it is natural that properties of radical filtrations, or of gr A, should appear as hypotheses guaranteeing homological conditions on $A^!$. (Also, examining homological properties of A-module radical and socle series is at least less foreboding than passing to $A^!$ and testing hypotheses there.)

In the following result, $\mathscr{E}^L = \mathscr{E}^L(\mathscr{C}_A)$ and $\mathscr{E}^R = \mathscr{E}^R(\mathscr{C}_A)$ are the full subcategories of the bounded derived category $D^b(\mathscr{C}_A)$ defined and studied in [5, §2]. (Here $\mathscr{C}_A$ is a highest weight category with weight poset Λ and we fix a function $l : \Lambda \to \mathbb{Z}$.) We don't repeat the precise definition here, but, roughly speaking, $\mathscr{E}^L$ (respectively, $\mathscr{E}^R$) consists of objects $X \in \operatorname{Ob}(D^b(\mathscr{C}_A))$ which have a "filtration" with sections of the form $\Delta(\lambda)[r]$ (respectively, $\nabla(\lambda)[r]$) for $\lambda \in \Lambda$ and $r \equiv l(\lambda) \mod 2$. The Recognition Theorem [5, (2.4)] provides a homological

GRADED AND NON-GRADED KAZHDAN-LUSZTIG THEORIES   113

criterion for membership in the categories $\mathscr{E}^L$ and $\mathscr{E}^R$. Thus, $X \in \mathrm{Ob}(\mathscr{E}^L)$ if and only if, for $\tau \in \Lambda$, $\mathrm{Hom}_{D^b(\mathscr{C}_A)}^n(X, \nabla(\tau)) \neq 0$ implies that $n \equiv l(\tau) \bmod 2$. A dual criterion applies for $\mathscr{E}^R$. In particular, the highest weight category $\mathscr{C}_A$ has a Kazhdan-Lusztig theory relative to l if and only if $L(\lambda)[-l(\lambda)]$ belongs to both $\mathscr{E}^L$ and $\mathscr{E}^R$ for all $\lambda \in \Lambda$.

Lemma 2.1.4. *Let $\mathscr{C}_A$ be a highest weight category with weight poset Λ, and let $l : \Lambda \to \mathbb{Z}$. Then:-*

(a) *$\mathscr{C}_A$ satisfies the (SKL) condition relative to l if and only if*

$$\Delta^i(\lambda)[-l(\lambda) + i] \in \mathrm{Ob}(\mathscr{E}^L) \quad \text{and} \quad \nabla^i(\lambda)[-l(\lambda) + i] \in \mathrm{Ob}(\mathscr{E}^R) \quad (2.1.4.1)$$

for all $\lambda \in \Lambda$ and all integers i.

(b) *Assume that $\mathscr{C}_A$ satisfies the (SKL) condition relative to l. Given $\lambda \in \Lambda$, for any positive odd integer $j = 2m + 1$, $m \geq 0$, we have*

$$\begin{cases} (1) \quad \Delta^i(\lambda)/\Delta^{i+j}(\lambda)[-l(\lambda) + i] \in \mathrm{Ob}(\mathscr{E}^L) \quad (i \geq 0), \\ (2) \quad \mathrm{soc}^{i-j}(\nabla(\lambda))/\mathrm{soc}^i(\nabla(\lambda))[-l(\lambda) + i] \in \mathrm{Ob}(\mathscr{E}^R) \quad (i \leq 0). \end{cases} \quad (2.1.4.2)$$

In particular, $\mathscr{C}_A$ has a Kazhdan-Lusztig theory relative to l.

Proof. Part (a) follows immediately from the Recognition Theorem [5, (2.4)]. Next, we prove (b). Shifting the short exact sequence

$$0 \to \Delta^{i+j}(\lambda) \to \Delta^i(\lambda) \to \Delta^i(\lambda)/\Delta^{i+j}(\lambda) \to 0$$

gives the distinguished triangle

$$\Delta^i(\lambda)[-l(\lambda) + i] \to \Delta^i(\lambda)/\Delta^{i+j}(\lambda)[-l(\lambda) + i] \to \Delta^{i+j}(\lambda)[-l(\lambda) + i + 1] \to$$

in $D^b(\mathscr{C}_A)$. Since $1 + j \equiv i + j \bmod 2$,

$$\Delta^{i+j}(\lambda)[-l(\lambda) + i + j] \in \mathrm{Ob}(\mathscr{E}^L) \implies \Delta^{i+j}(\lambda)[-l(\lambda) + i + 1] \in \mathrm{Ob}(\mathscr{E}^L).$$

Since $\mathscr{C}_A$ satisfies the (SKL) condition relative to l, this proves (1), using (a) and the definition of $\mathscr{E}^L$. A dual argument establishes assertion (2).

Finally, taking $i = 0$ and $j = 1$ in (2.1.4.2), we conclude that

$$L(\lambda)[-l(\lambda)] \in \mathrm{Ob}(\mathscr{E}^L) \cap \mathrm{Ob}(\mathscr{E}^R)$$

for all $\lambda \in \Lambda$. By [5, (2.4)] again, $\mathscr{C}_A$ has a Kazhdan-Lusztig theory relative to l.

In [6, §3] a somewhat weaker condition than (SKL) was used. We say that $\mathscr{C}_A$ satisfies the (SKL′) condition relative to $l : \Lambda \to \mathbb{Z}$ provided that for all $\lambda, \nu \in \Lambda$ and $n, i \in \mathbb{Z}$, we have:[4]

$$\begin{cases} \mathrm{Ext}^n_{\mathscr{C}_A}(L(\nu), \nabla^i(\lambda)) \neq 0 \implies n \equiv l(\nu) - l(\lambda) + i \mod 2; \text{ and} \\ \mathrm{Ext}^n_{\mathscr{C}_A}(\Delta^i(\nu), L(\lambda)) \neq 0 \implies n \equiv l(\nu) - l(\lambda) + i \mod 2. \end{cases} \quad (\mathrm{SKL}')$$

Our next result recapitulates, in part, some of the conclusions of [6]. In Theorem (2.4.2) below, we will improve upon part (d) by showing that if $\mathscr{C}_A^{\mathrm{gr}}$ has a graded Kazhdan-Lusztig theory then the (SKL) condition holds relative to l.

---

[4]For other equivalent formulations, see [6, (3.6)]

Lemma 2.1.5. (a) *If a highest weight category $\mathscr{C}_A$ satisfies the (SKL′) condition relative to l, then it has a Kazhdan-Lusztig theory relative to l.*

(b) *Suppose that the highest weight category $\mathscr{C}_A$ satisfies the (SKL) condition relative to l. Then $\mathscr{C}_A$ also satisfies the (SKL′) condition relative to l. In particular, $\mathscr{C}_A$ has a Kazhdan-Lusztig theory relative to l.*

(c) *If the highest weight category $\mathscr{C}_A$ satisfies the (SKL′) condition relative to l, then $\mathscr{C}_{A^!}^{\mathrm{gr}}$ has a graded Kazhdan-Lusztig theory relative to $l^{\mathrm{op}} \overset{\text{def}}{=} -l : \Lambda^{\mathrm{op}} \to \mathbb{Z}$. The (left and right) Kazhdan-Lusztig polynomials for $\mathscr{C}_{A^!}$ relative to l^{op} identify with the (left and right) filtered Kazhdan-Lusztig polynomials for $\mathscr{C}_A$ relative to l.*

(d) *Let A be a graded quasi-hereditary algebra and assume that $\mathscr{C}_A^{\mathrm{gr}}$ has a graded Kazhdan-Lusztig theory relative to l. Then the (SKL′) condition relative to l holds. Also, $\mathscr{C}_{A^!}^{\mathrm{gr}}$ has a graded Kazhdan-Lusztig theory relative to l^{op}.*

Proof. Part (a) follows by taking $i = 0$ in (SKL′). Now consider (b). Since any $L(\tau)[-l(\tau)]$ lies in $\mathscr{E}^L$ because $\mathscr{C}_A$ has a Kazhdan-Lusztig theory by (2.1.4(b)), it follows that

$$0 \neq \mathrm{Ext}^n_{\mathscr{C}_A}(L(\tau), \nabla^i(\lambda)) \cong \mathrm{Hom}^{n-l(\tau)+l(\lambda)-i}_{D^b(\mathscr{C}_A)}(L(\tau)[-l(\tau)], \nabla^i(\lambda)[-l(\lambda)+i])$$

implies $n \equiv l(\lambda) - l(\tau) + i \bmod 2$ by [5, (2.4)]. The dual condition in (SKL′) follows similarly.

(b,c,d) follow from [6, (3.8), (3.9)].[5]

We conclude this subsection with some further remarks on polynomials.

Lemma 2.1.6. *Let $\mathscr{C}_A$ and $\mathscr{C}_{\tilde{A}}$ be highest weight categories having the same weight poset Λ. Suppose that $\mathscr{C}_A$ and $\mathscr{C}_{\tilde{A}}$ both have a Kazhdan-Lusztig theory relative to $l : \Lambda \to \mathbb{Z}$ and that $A^! \cong \tilde{A}^!$ as graded algebras. Then $\mathscr{C}_A$ and $\mathscr{C}_{\tilde{A}}$ have the same (left and right) Kazhdan-Lusztig polynomials.*

Proof. The hypothesis implies that $A/\mathrm{rad}(A) \cong \tilde{A}/\mathrm{rad}(\tilde{A})$, so we identify a Wedderburn complement A_0 of A with one of $\tilde{A}$. Hence, we also identify the irreducible modules $L(\mathscr{C}_A, \lambda)$ and $L(\mathscr{C}_{\tilde{A}}, \lambda)$ for all $\lambda \in \Lambda$.

Since $\mathscr{C}_A$ and $\mathscr{C}_{\tilde{A}}$ both have a Kazhdan-Lusztig theory relative to l, both $\mathscr{C}_{A^!}$ and $\mathscr{C}_{\tilde{A}^!}$ are highest weight categories with weight poset Λ^{op} by [6, (2.1)]. Thus, [6, (2.1.11b)] gives two descriptions of the $\Delta(\mathscr{C}_{A^!}, \lambda)$ as a graded module:

$$\Delta(\mathscr{C}_{A^!}, \lambda)_n \cong \mathrm{Ext}^n_{\mathscr{C}_A}(A_0, \nabla(\mathscr{C}_A, \lambda)) \cong \mathrm{Ext}^n_{\mathscr{C}_{\tilde{A}}}(A_0, \nabla(\mathscr{C}_{\tilde{A}}, \lambda)).$$

Let $\nu \in \Lambda$. Multiplying both sides by an appropriate central idempotent (corresponding to $L(\lambda)$), taking dimensions, and dividing by $\dim L(\nu)$ yields that

$$\dim \mathrm{Ext}^n_{\mathscr{C}_A}(L(\nu), \nabla(\mathscr{C}_A, \lambda)) = \dim \mathrm{Ext}^n_{\mathscr{C}_{\tilde{A}}}(L(\nu), \nabla(\mathscr{C}_{\tilde{A}}, \lambda)).$$

---

[5]In [6, (3.8)], we assumed for convenience that $\mathscr{C}_A$ had a duality D. This is unnecessary in the present case, because our formulation of (SKL′) contains both left-handed and right-handed versions of the hypotheses of [6, (3.8)]. Observe that when $\mathscr{C}_A$ has a duality D, then $D(\nabla^{-i}(\lambda)) \cong \Delta^i(\lambda)$ for all i.

It follows from (2.1.1) that the (left) Kazhdan-Lusztig polynomials $P_{\nu,\lambda}$ for $\mathscr{C}_A$ and $\mathscr{C}_{\tilde{A}}$ are equal.

A dual argument, using the remark immediately preceding [6, (2.2)], shows that $\mathscr{C}_A$ and $\mathscr{C}_{\tilde{A}}$ have the same right Kazhdan-Lusztig polynomials.

Given a highest weight category $\mathscr{C}_A$ and a length function $l : \Lambda \to \mathbb{Z}$, the (left and right) *inverse Kazhdan-Lusztig polynomials* $Q_{\nu,\lambda}, Q_{\nu,\lambda}^R$ ($\lambda, \nu \in \Lambda$) are formally (and uniquely) determined by the relations:[6]

$$\begin{cases} \sum_{\lambda \leq \nu \leq \lambda'}(-1)^{l(\lambda')-l(\nu)}P_{\nu,\lambda'}Q_{\nu,\lambda} = \delta_{\lambda',\lambda} \\ \sum_{\lambda \leq \nu \leq \lambda'}(-1)^{l(\lambda')-l(\nu)}P_{\nu,\lambda'}^R Q_{\nu,\lambda}^R = \delta_{\lambda',\lambda}. \end{cases} \qquad (2.1.7)$$

When A is a graded quasi-hereditary algebra and $\mathscr{C}_A^{\mathrm{gr}}$ has a graded Kazhdan-Lusztig theory relative to l, then we have an identification $F_{\nu,\lambda} = Q_{\nu,\lambda}$ and $F_{\nu,\lambda}^R = Q_{\nu,\lambda}^R$ of polynomials for all $\lambda, \nu \in \Lambda$ [5, §3, appendix]. More generally, we have the following result:

Corollary 2.1.8. *Let $\mathscr{C}_A$ be a highest weight category satisfying the (SKL) condition relative to l. Then, for $\nu, \lambda \in \Lambda$, the left and right filtered Kazhdan-Lusztig polynomials $F_{\nu,\lambda}, F_{\nu,\lambda}^R$ defined in (2.1.3) identify with the left and right inverse Kazhdan-Lusztig polynomials $Q_{\nu,\lambda}, Q_{\nu,\lambda}^R$, respectively, defined by (2.1.7).*

Proof. By (2.1.5), $\mathscr{C}_{A^!}^{\mathrm{gr}}$ has a graded Kazhdan-Lusztig theory relative to $-l$: $\Lambda^{\mathrm{op}} \to \mathbb{Z}$ and $\mathscr{C}_{A^{!!!}}^{\mathrm{gr}}$ also has a graded Kazhdan-Lusztig theory relative to l (and, in particular, a Kazhdan-Lusztig theory). By [6, (3.9)], $A^!$ is a Koszul algebra, so $A^! \cong A^{!!!}$. Thus, (2.1.6) implies that $\mathscr{C}_{A^{!!}}$ and $\mathscr{C}_A$ have the same (left and right) Kazhdan-Lusztig polynomials. By [6, (3.13)], the (left and right) Kazhdan-Lusztig polynomials for $\mathscr{C}_{A^{!!!}} \cong \mathscr{C}_{A^!}$ identify with the (left and right) inverse Kazhdan-Lusztig polynomials for $\mathscr{C}_{A^{!!}}$, while by (2.1.5(c)), the (left and right) Kazhdan-Lusztig polynomials for $\mathscr{C}_{A^!}$ identify with the (left and right) filtered Kazhdan-Lusztig polynomials for $\mathscr{C}_A$.

Finally, we note the following result on dimensions of quasi-hereditary algebras.

Lemma 2.1.9. *Let $\mathscr{C}_A$ and $\mathscr{C}_{\tilde{A}}$ be highest weight categories having the same weight poset Λ and the same (left and right) filtered Kazhdan-Lusztig polynomials as defined in (2.1.3) for some function $l : \Lambda \to \mathbb{Z}$. Suppose*

$$\dim L(\mathscr{C}_A, \lambda) = \dim L(\mathscr{C}_{\tilde{A}}, \lambda)$$

for all $\lambda \in \Lambda$. Then

$$\dim A = \dim \tilde{A}.$$

---

[6]By [3, (3.2)] (and its dual version), if $P_{\nu,\lambda}$ or $P_{\nu,\lambda}^R \neq 0$, then $\nu \leq \lambda$. Also, $P_{\lambda,\lambda} = P_{\lambda,\lambda}^R = 1$. Thus, the inverse polynomials $Q_{\nu,\lambda}, Q_{\nu,\lambda}^R$ are uniquely determined. As with the $P_{\nu,\lambda}, P_{\nu,\lambda}^R$ and the $F_{\nu,\lambda}, F_{\nu,\lambda}^R$, the $Q_{\nu,\lambda}, Q_{\nu,\lambda}^R$ depend on the choice of the function l. Also, they are Laurent polynomials in t. In [5, 6], the inverse Kazhdan-Lusztig polynomials are defined differently—however, it can be shown that the definition agrees with the current one in the presence of a graded Kazhdan-Lusztig theory; see [5, (3A.3), p. 525]. When quoting [5, 6] for facts about these polynomials, we will be in the situation of a graded Kazhdan-Lusztig theory. Further, the polynomials $Q_{\nu,\lambda}, Q_{\nu,\lambda}^R$ play a minor role in this paper, arising only in (2.1.8).

Proof. For $\lambda \in \Lambda$, let $P(\lambda) = P(\mathscr{C}_A, \lambda)$ denote the projective cover in $\mathscr{C}_A$ of $L(\lambda)$. Then $P(\lambda)$ has a filtration with sections of the form $\Delta(\nu)$ for certain $\nu \geq \lambda$. By Brauer-Humphreys reciprocity [3, (3.11)], the number of times a given $\Delta(\nu)$ appears as a section in $P(\lambda)$ equals the multiplicity $[\nabla(\nu) : L(\lambda)]$ of $L(\lambda)$ as a composition factor of $\nabla(\nu)$. By (2.1.3),

$$[\nabla(\nu) : L(\lambda)] = (-1)^{l(\nu)-l(\lambda)} F^R_{\nu,\lambda}(-1).$$

Also, (2.1.3) implies that

$$\dim \Delta(\nu) = \sum_\tau (-1)^{l(\tau)-l(\nu)} F_{\nu,\tau}(-1) \dim L(\tau).$$

Thus, putting everything together, we obtain

$$\dim A = \sum_\lambda \dim P(\lambda) \dim L(\lambda)$$

$$= \sum_{\lambda,\nu,\tau} (-1)^{l(\tau)-l(\lambda)} F^R_{\nu,\lambda}(-1) F_{\nu,\tau}(-1) \dim L(\tau) \dim L(\lambda).$$

A similar formula holds for $\dim \tilde{A}$. Because $\mathscr{C}_A$ and $\mathscr{C}_{\tilde{A}}$ have the same (left and right) filtered Kazhdan-Lusztig polynomials, we conclude that $\dim A = \dim \tilde{A}$.

2.2. An application. The hypotheses of the following theorem are very close to, but slightly weaker than, those of [6, (3.8)].[7] There we concluded that $\mathscr{C}^{gr}_{A^!}$ has a graded Kazhdan-Lusztig theory (and, in particular, that $A^!$ is a Koszul algebra). We now establish that the present hypotheses, and thus those of [6, (3.8)], guarantee even more, viz., gr A is Koszul and $\mathscr{C}^{gr}_{gr\,A}$ has a graded Kazhdan-Lusztig theory. We had showed in [6] that the hypotheses do indeed hold when A itself is graded and $\mathscr{C}^{gr}_A$ has a graded Kazhdan-Lusztig theory.

Theorem 2.2.1. *Let A be a quasi-hereditary algebra such that $\mathscr{C}_A$ satisfies the (SKL') condition relative to l. Then:-*

(a) *There is a natural isomorphism $A^{!!} \xrightarrow{\sim} gr\,A$ of graded algebras.*
(b) *The graded algebra gr A is quasi-hereditary. The corresponding graded module category $\mathscr{C}^{gr}_{gr\,A}$ is a graded highest weight category having a graded Kazhdan-Lusztig theory relative to l. Also, the (left and right) Kazhdan-Lusztig polynomials of $\mathscr{C}_A$ and $\mathscr{C}_{gr\,A}$ agree.*
(c) *The algebra gr A is a Koszul algebra.*

Proof. By [6, (3.8)], the algebra $A^!$ is Koszul. Therefore, by (1.3.2), there is a surjective homomorphism $A^{!!} \to gr\,A$. The proof of (a) is completed by showing that A and $A^{!!}$ have the same dimensions. By construction, $\dim L(\mathscr{C}_A, \lambda) = \dim L(\mathscr{C}_{A^{!!}}, \lambda)$ if $\lambda \in \Lambda$. Also, the poset Λ can serve as the weight poset for both the highest weight categories $\mathscr{C}_A$ and $\mathscr{C}_{A^{!!}}$.

Since the (SKL') condition holds relative to l for $\mathscr{C}_A$, $\mathscr{C}^{gr}_{A^!}$ has, by (2.1.5(c)), a graded Kazhdan-Lusztig theory with respect to $l^{op} = -l : \Lambda^{op} \to \mathbb{Z}$. For

---

[7]Besides the assumption of a duality (an unnecessarily strong assumption, as already discussed in footnote 5 above), we assumed that $\mathscr{C}_A$ had a Kazhdan-Lusztig theory. However, as observed in (2.1.5(a)), this assumption follows from (SKL').

$\nu, \lambda \in \Lambda^{\mathrm{op}}$, let $P^!_{\nu,\lambda}$ denote the corresponding (left) Kazhdan-Lusztig polynomial for $\mathscr{C}^{\mathrm{gr}}_{A^!}$. By (2.1.5(c)) again, we have $P^!_{\nu,\lambda} = F_{\nu,\lambda}$. Since $\mathscr{C}_{A'''} \cong \mathscr{C}_{A^!}$, the (left and right) Kazhdan-Lusztig polynomials for $\mathscr{C}_{A^!}$ also identify with the (left and right) filtered Kazhdan-Lusztig polynomials $F^{!!}_{\nu,\lambda}$ for $\mathscr{C}^{\mathrm{gr}}_{A^{!!}}$. Thus, we see that $F^{!!}_{\nu,\lambda} = F_{\nu,\lambda}$ for all $\lambda, \nu \in \Lambda$. A dual argument establishes that $\mathscr{C}_A$ and $\mathscr{C}_{A^{!!}}$ have the same right filtered Kazhdan-Lusztig polynomials. Hence, (2.1.9) implies that $\dim A = \dim A^{!!}$, as required to complete the proof of (a).

By (a) and (2.1.5(d)), $\mathscr{C}^{\mathrm{gr}}_{\mathrm{gr}\,A} \cong \mathscr{C}^{\mathrm{gr}}_{A^{!!}}$ can be assumed to have weight poset Λ and also has a graded Kazhdan-Lusztig theory relative to l. We next show that the Kazhdan-Lusztig polynomials of $\mathscr{C}_A$ and $\mathscr{C}_{\mathrm{gr}\,A}$ agree. By (a) again, $(\mathrm{gr}\,A)^! \cong A^{!!!} \cong A^!$ as graded algebras. By (2.1.6), $\mathscr{C}_{\mathrm{gr}\,A}$ and $\mathscr{C}_A$ have the same (left and right)] Kazhdan-Lusztig polynomials. This completes the proof of (b).

Finally, (c) follows from (b) and [6, (3.9)].

Remarks 2.2.2. (a) By (2.1.5(b)), we see that (2.2.1) remains valid provided (SKL') is replaced by (SKL) in the hypothesis.

(b) Under the hypotheses of (2.2.1),

$$\Delta(\mathscr{C}_{\mathrm{gr}\,A}, \lambda) \cong \mathrm{gr}\,\Delta(\mathscr{C}_A, \lambda), \qquad \nabla(\mathscr{C}_{\mathrm{gr}\,A}, \lambda) \cong \mathrm{gr}'\,\nabla(\mathscr{C}_A, \lambda)$$

for $\lambda \in \Lambda$.

2.3. Recollement and the strong Kazhdan-Lusztig condition. In this subsection, we investigate the behavior of (SKL) under the recollement set-up for highest weight categories. It will be convenient to review briefly how this machinery works.

Let A be a quasi-hereditary algebra and let $0 = J_0 \subset J_1 \subset \cdots \subset J_t = A$ be a defining sequence in A, in the sense of [3] (see also [6, (1.1)]). Write $J_i = Ae_iA$ for an idempotent $e_i \in A$. We can assume that for $i \leq j$, $e_ie_j = e_je_i = e_i$, and that the defining sequence is compatible with a poset structure on Λ making $\mathscr{C}_A$ into a highest weight category in the sense that $t = |\Lambda|$ and $L(\lambda_j)e_i \neq 0 \implies j \leq i$.

Fix some integer j, and write $J = J_j$ and $e_j = e$. By [3], the quotient algebra A/J is again quasi-hereditary as is the centralizer algebra $eAe \cong \mathrm{End}_{\mathscr{C}_A}(eA)$. In fact, for some ideal $\Gamma \subset \Lambda$, $\mathscr{C}_{A/J}$ is equivalent to the full subcategory $\mathscr{C}_A[\Gamma]$ of $\mathscr{C}_A$ having objects with composition factors $L(\gamma)$, $\gamma \in \Gamma$. Also, letting Ω denote the coideal $\Lambda \backslash \Gamma$, $\mathscr{C}_{eAe}$ is equivalent to the quotient category $\mathscr{C}_A(\Omega) = \mathscr{C}_A/\mathscr{C}_A[\Gamma]$.

Let $i_* : \mathscr{C}_A[\Gamma] \to \mathscr{C}_A$ be the natural full embedding and let $j^* : \mathscr{C}_A \to \mathscr{C}_A(\Omega)$ be the (exact) quotient morphism. Then i_* admits a left adjoint i^* and a right adjoint $i^!$. Similarly, the quotient morphism j^* admits a left adjoint $j_!$ and a right adjoint j_*. These morphisms fit into a standard "recollement" diagram

$$\mathscr{C}_A[\Gamma] \overset{i_*}{\underset{}{\to}} \mathscr{C}_A \overset{j^*}{\underset{}{\to}} \mathscr{C}_A(\Omega). \tag{2.3.1}$$

(See [11, §2].) At the level of bounded derived categories, there is also a recollement diagram:

$$D^b(\mathscr{C}_A[\Gamma]) \underset{\substack{\longleftarrow \\ \longleftarrow}}{\overset{i_*}{\longrightarrow}} D^b(\mathscr{C}_A) \underset{\substack{\longleftarrow \\ \longleftarrow}}{\overset{j^*}{\longrightarrow}} D^b(\mathscr{C}_A(\Omega)). \qquad (2.3.2)$$

(See [3].) Here we follow the convention that if $t : \mathscr{C} \to \mathscr{C}'$ denotes a (left or right exact) additive functor of abelian categories, then (boldface) $\mathbf{t} : D^*(\mathscr{C}) \to D^*(\mathscr{C}')$ denotes the corresponding (left or right) derived functor. Thus, for example, $\mathbf{j}_!$ is the left derived functor of $j_!$. Also, $\mathbf{i}_*, \mathbf{j}_!, \mathbf{j}_*$ are full embeddings of derived categories.

We have the following result:

Theorem 2.3.3. *Assume that the highest weight category $\mathscr{C}_A$ satisfies the (SKL) condition relative to l. Let $J = AeA$, $e^2 = e$, belong to the defining sequence $0 = J_0 \subset J_1 \subset \cdots \subset J_t = A$ chosen as above. Then both $\mathscr{C}_A[\Gamma] \cong \mathscr{C}_{A/J}$ and $\mathscr{C}_A(\Omega) \cong \mathscr{C}_{eAe}$ satisfy the (SKL) condition relative to $l|_\Gamma$ and $l|_\Omega$, respectively.*

Proof. The functor i_* takes ∇, Δ and simple objects in $\mathscr{C}_A[\Gamma]$ to the corresponding objects in $\mathscr{C}_A$. Since i_* is the inflation functor defined by the quotient homomorphism $A \to A/J$, it preserves radical and socle series. Thus, for any $\lambda \in \Gamma$ and any integer i, we have $i_*\Delta^i(\mathscr{C}_A[\Gamma], \lambda) \cong \Delta^i(\mathscr{C}_A, \lambda)$ and $i_*\nabla^i(\mathscr{C}_A[\Gamma], \lambda) \cong \nabla^i(\mathscr{C}_A, \lambda)$. Since $\mathbf{i}_*$ is a full embedding, $\mathrm{Ext}^\bullet_{\mathscr{C}_A[\Gamma]}(M, N) \cong \mathrm{Ext}^\bullet_{\mathscr{C}_A}(i_*M, i_*N)$ for $M, N \in \mathrm{Ob}(\mathscr{C}_A[\Gamma])$. Hence, $\mathscr{C}_A[\Gamma]$ satisfies the (SKL) condition relative to $l|_\Gamma$.

Now we show that $\mathscr{C}_{eAe}$ satisfies (SKL) relative to $l|_\Omega$. We first show that the (SKL') condition relative to l for $\mathscr{C}_A$ implies that

$$j^*\Delta^i(\mathscr{C}_A, \lambda) \cong \Delta^i(\mathscr{C}_{eAe}, \lambda), \qquad j^*\nabla^i(\mathscr{C}_A, \lambda) \cong \nabla^i(\mathscr{C}_{eAe}, \lambda) \qquad (2.3.3.1)$$

for all $\lambda \in \Omega$ and all integers i. By [6, (2.3)], $\mathscr{C}_{(eAe)^!} \cong \mathscr{C}_{A^!}[\Omega^{\mathrm{op}}]$. Since $\mathscr{C}^{\mathrm{gr}}_{A^!}$ has a graded Kazhdan-Lusztig theory (relative to $-l : \Lambda^{\mathrm{op}} \to \mathbb{Z}$), $\mathscr{C}^{\mathrm{gr}}_{(eAe)^!}$ also has a graded Kazhdan-Lusztig theory (relative to $-l|_{\Omega^{\mathrm{op}}}$). Also, for $\nu, \lambda \in \Omega$, the (left and right) Kazhdan-Lusztig polynomials for $\mathscr{C}_{(eAe)^!}$ are the filtered polynomials $F_{\nu,\lambda}, F^R_{\nu,\lambda}$ (2.1.3) for $\mathscr{C}_A$, using (2.1.5(c)) and [5, (3.10)]. Since $(eAe)^{!!!} \cong (eAe)^!$, we can apply (2.1.5(c)) again to conclude that these polynomials identify with the filtered Kazhdan-Lusztig polynomials for $\mathscr{C}_{(eAe)^{!!}}$. Considering the natural isomorphisms $(eAe)/\mathrm{rad}(eAe) \cong ((eAe)^!)_0 \cong ((eAe)^{!!})_0$, we have $\dim L(\mathscr{C}_{eAe}, \nu) = \dim L(\mathscr{C}_{(eAe)^{!!}}, \nu)$ for $\mu \in \Omega$. Because $\mathscr{C}_{eAe}$ and $\mathscr{C}_{(eAe)^{!!}}$ have the same weight poset Ω, (2.1.9) implies that $\dim eAe = \dim (eAe)^{!!}$. Thus, by (1.2.3), this proves that $(eAe)^{!!} \cong \mathrm{gr}(eAe)$.

Let e also denote the image of e in $(\mathrm{gr}\, A)_0 = A/\mathrm{rad}(A)$ under the quotient map $A \to A/\mathrm{gr}\, A$. Since $(\mathrm{gr}\, A)_0 \cong (A^!)_0 \cong (A^{!!})_0$, we can view e as an idempotent in each of the three algebras $\mathrm{gr}\, A$, $A^!$ and $A^{!!}$. Using [6, (2.3.3)], $\mathscr{C}_{(eAe)^{!!}} \cong \mathscr{C}_{eA^{!!}e}$. Since $\mathrm{gr}(eAe) \cong (eAe)^{!!}$ and, by (2.2.1), $eA^{!!}e \cong e(\mathrm{gr}\, A)e$, the algebras $\mathrm{gr}(eAe)$ and $e(\mathrm{gr}\, A)e$ are Morita equivalent. However, since they have isomorphic

Wedderburn complements, it follows that

$$\text{gr}(eAe) \cong e(\text{gr } A)e. \tag{2.3.3.2}$$

Observe that both algebras above are Koszul, and hence tightly graded. Thus, (2.3.3.2) can be chosen to be an isomorphism of graded algebras. If $I = \text{rad}(A)$, it follows for any integer $i \geq 0$ that $eI^i e = (eIe)^i$.

Let $\Gamma' \supset \Gamma$ be an ideal in Λ containing λ as a maximal element. The equivalence

$$\mathscr{C}_A[\Gamma'](\Omega \cap \Gamma') \cong \mathscr{C}_A(\Omega)[\Omega \cap \Gamma']$$

shows it is enough to prove (2.3.3.1) under the assumption that $\lambda \in \Omega$ is maximal. But now $\Delta(\mathscr{C}_A, \lambda) \cong e'A$ for some idempotent $e' \in A$ satisfying $e'e = ee' = e'$. Also, $\Delta(\mathscr{C}_A(\Omega), \lambda) \cong e'Ae$. Thus,

$$\begin{aligned}
j^* \Delta^i(\mathscr{C}_A, \lambda) &= j^* \text{rad}^i(\Delta(\mathscr{C}_A, \lambda)) \\
&= e'AI^i e = e'eI^i e = e'(eIe)^i = (e'Ae)(eIe)^i \\
&= \text{rad}^i(\Delta(\mathscr{C}_A(\Omega), \lambda)) = \Delta^i(\mathscr{C}_A(\Omega), \lambda).
\end{aligned}$$

A dual argument proves the claim concerning $\nabla^i(\mathscr{C}_A, \lambda)$. Hence, (SKL') $\implies$ (2.3.3.1).

By (1.2.4), each $\Delta^i(\mathscr{C}_A, \lambda)[-l(\lambda) + i] \in \text{Ob}(\mathscr{E}^L(\mathscr{C}_A))$ for $\lambda \in \Omega$. But we have

$$j^*(\mathscr{E}^L(\mathscr{C}_A)) \subset \mathscr{E}^L(\mathscr{C}_A(\Omega)),$$

arguing as in [5, (3.10)]. Thus, since j^* is exact, (2.3.3.1) implies

$$\begin{aligned}
j^*(\Delta^i(\mathscr{C}_A, \lambda)[-l(\lambda) + i]) &\cong (j^*\Delta^i(\mathscr{C}_A, \lambda))[-l(\lambda) + i] \\
&\cong \Delta^i(\mathscr{C}_A(\Omega), \lambda)[-l(\lambda) + i].
\end{aligned}$$

We conclude that $\Delta^i(\mathscr{C}_A(\Omega), \lambda)[-l(\lambda) + i] \in \text{Ob}(\mathscr{E}^L(\mathscr{C}_A(\Omega)))$ for all $\lambda \in \Omega$. A dual argument establishes that $\nabla^i(\mathscr{C}_A(\Omega), \lambda)[-l(\lambda)+i] \in \text{Ob}(\mathscr{E}^R(\mathscr{C}_A(\Omega)))$. Thus, $\mathscr{C}_A(\Omega)$ satisfies the (SKL) condition relative to $l|_\Omega$.

We can draw two important corollaries from the proof of (2.3.3) for highest weight categories $\mathscr{C}_A$ which satisfy the weaker condition (SKL').

Corollary 2.3.4. *Suppose the highest weight category $\mathscr{C}_A$ satisfies the (SKL') condition relative to l. Let $J = AeA$ be as in the statement of (2.3.3). Then:-*
(a) *For $\lambda \in \Omega$, $i \in \mathbb{Z}$,*

$$j^*\Delta^i(\mathscr{C}_A, \lambda) = \Delta^i(\mathscr{C}_{eAe}, \lambda), \qquad j^*\nabla^i(\mathscr{C}_A, \lambda) = \nabla^i(\mathscr{C}_{eAe}, \lambda).$$

(b) *There is an isomorphism*

$$\text{gr}(eAe) \cong e(\text{gr } A)e$$

of graded algebras[8] [9] .

Remark 2.3.5. Let G be a semisimple, simply connected algebraic group defined over an algebraically closed field k of positive characteristic p. Assume that $p \geq h$, the Coxeter number of G. Let Λ denote the poset of dominant weights $\lambda = w \cdot 0$ lying in the Janzten region. (Here $w \in W_p$, the affine Weyl group associated to G, and the poset structure is induced by $\uparrow$ [9, §II(6.4)].) The full subcategory of finite dimensional rational G-modules having composition factors $L(\lambda)$, $\lambda \in \Lambda$, has the form $\mathscr{C}_A$ for a quasi-hereditary algebra A. A main result [5, (5.3)] showed that the Lusztig conjecture for G is equivalent to the simple assertion that

$$\operatorname{Ext}^1_G(L(\lambda), L(\lambda')) \neq 0 \qquad (2.3.5.1)$$

for all $\lambda > \lambda'$ in Λ which are mirror images of each other in adjacent p-alcoves. Observe that (2.3.5.1) holds provided for any coideal $\Omega \subset \Lambda$, the conclusion of (2.3.4(a)) holds for the $\Delta^i(\lambda)$, $\lambda \in \Omega$, $i = 1, 2$. To see this, let $\lambda > \lambda'$ be as in (2.3.5.1). Replacing Λ by a suitable ideal to assume that λ is maximal, let $\Omega = \{\lambda, \lambda'\}$. (Using [9, §II(6.6)], it is easy to see that Ω is a coideal.) The quotient functor $j^* : \mathscr{C}_A \to \mathscr{C}_{eAe}$ kills the simple objects in $\mathscr{C}_A$ except for $L(\lambda), L(\lambda')$. Since $[\Delta(\lambda) : L(\lambda')] = 1$ (see [9, §II(7.13)]) and $j^*\Delta(\lambda) = \Delta(\mathscr{C}_{eAe}, \lambda)$, the isomorphisms $j^*\Delta^i(\lambda) = \Delta^i(\mathscr{C}_{eAe}, \lambda)$, $i = 1, 2$, easily imply that (2.3.5.1) holds.

One can also establish (2.3.5.1) by a similar argument, but under the hypotheses that A is tightly graded, that the idempotent $e \in A$ above is chosen homogenous, and that the graded algebra eAe (whose grading is induced from that of A) is isomorphic to gr eAe (i. e., eAe is also tightly graded).

2.4. Graded Kazhdan-Lusztig theories and the strong Kazhdan-Lusztig condition. The following result justifies our claim in the introduction that (SKL) is a "universal" version of (SKL').

Lemma 2.4.1. *A highest weight category $\mathscr{C}_A$ satisfies the (SKL) condition relative to l if and only if, for all coideals $\Omega \subset \Lambda$, the quotient category $\mathscr{C}_A(\Omega)$ satisfies the (SKL') condition relative to $l|_\Omega$.*

Proof. By (2.1.5(b)), (SKL) $\Longrightarrow$ (SKL'), so (2.3.3) implies that if $\mathscr{C}_A$ satisfies (SKL) relative to l, then each quotient category $\mathscr{C}_A(\Omega)$ satisfies (SKL') relative to $l|_\Omega$.

Conversely, assume that each $\mathscr{C}_A(\Omega)$ satisfies (SKL') relative to $l|_\Omega$. We will show that the (SKL) condition relative to l holds for $\mathscr{C}_A$. Suppose that

$$\operatorname{Ext}^n_{\mathscr{C}_A}(\Delta(\lambda), \nabla^i(\nu)) \neq 0 \quad \text{for some } n \in \mathbb{Z}^+, i \in \mathbb{Z}, \lambda, \nu \in \Lambda.$$

---

[8]In general, it is not difficult to find examples of algebras A containing idempotents e for which such isomorphisms fail.

[9]There is a "dual" version of this isomorphism, also obtained using [6, (2.3.3)]. If $J^{\mathrm{gr}} = (\mathrm{gr}\ A)e(\mathrm{gr}\ A)$, then

$$\mathrm{gr}(A/J) \cong (\mathrm{gr}\ A)/J^{\mathrm{gr}}.$$

We omit further details.

Choose a coideal $\Omega \subset \Lambda$ which contains λ as a *minimal* element. Since

$$\mathbf{j}^* \nabla(\mathscr{C}_A, \tau) \cong \begin{cases} \nabla(\mathscr{C}_A(\Omega), \tau), & \tau \in \Omega \\ 0, & \tau \notin \Omega, \end{cases}$$

the adjointness properties of the pair $(\mathbf{j}_!, \mathbf{j}^*)$, the fact that

$$L(\mathscr{C}_A(\Omega), \lambda) \cong \Delta(\mathscr{C}_A(\Omega), \lambda)$$

(because λ is minimal in Ω), and [5, (2.2)] imply that

$$\dim \operatorname{Hom}^n_{D^b(\mathscr{C}_A)}(\mathbf{j}_! L(\mathscr{C}_A(\Omega), \lambda), \nabla(\mathscr{C}_A, \tau)) = \delta_{n,0}\delta_{\lambda,\tau}.$$

The Recognition Theorem [5, (2.4)] implies that

$$\mathbf{j}_! L(\mathscr{C}_A(\Omega), \lambda)[-l(\lambda)] \in \operatorname{Ob}(\mathscr{E}^L(\mathscr{C}_A)).$$

It now follows immediately from [5, (2.3)] that $\mathbf{j}_! L(\mathscr{C}_A(\Omega), \lambda) \cong \Delta(\mathscr{C}_A, \lambda)$. Thus,

$$\begin{aligned} 0 \neq \operatorname{Ext}^n_{\mathscr{C}_A}(\Delta(\mathscr{C}_A, \lambda), \nabla^i(\mathscr{C}_A, \nu)) \\ \cong \operatorname{Hom}^n_{D^b(\mathscr{C}_A)}(\mathbf{j}_! \Delta(\mathscr{C}_A(\Omega), \lambda), \nabla^i(\mathscr{C}_A, \nu)) \\ \cong \operatorname{Hom}^n_{D^b(\mathscr{C}_A(\Omega))}(\Delta(\mathscr{C}_A(\Omega), \lambda), \mathbf{j}^* \nabla^i(\mathscr{C}_A, \nu)) \\ \cong \operatorname{Hom}^n_{D^b(\mathscr{C}_A(\Omega))}(L(\mathscr{C}_A(\Omega), \lambda), \mathbf{j}^* \nabla^i(\mathscr{C}_A, \nu)) \\ \cong \operatorname{Ext}^n_{\mathscr{C}_A(\Omega)}(L(\mathscr{C}_A(\Omega), \lambda), \mathbf{j}^* \nabla^i(\mathscr{C}_A, \nu)). \end{aligned} \qquad (2.4.1.1)$$

In particular, $\nu \in \Omega$. Our hypothesis implies that $\mathscr{C}_A$ satisfies (SKL$'$) relative to l, so that (2.3.4) implies that

$$j^* \nabla^i(\mathscr{C}_A, \nu) \cong \nabla^i(\mathscr{C}_A(\Omega), \nu).$$

Hence, since (SKL$'$) holds relative to $l|_\Omega$, (2.4.1.1) now implies that $n \equiv l(\lambda) - l(\nu) + i \mod 2$.

A dual argument shows that

$$\operatorname{Ext}^n_{\mathscr{C}_A}(\Delta^i(\nu), \nabla(\lambda)) \neq 0 \implies n \equiv l(\nu) - l(\lambda) + i \mod 2.$$

Thus, the (SKL) condition relative to l holds for $\mathscr{C}_A$.

Using this result, we obtain the following:

Theorem 2.4.2. *Suppose that A is a graded quasi-hereditary algebra so that $\mathscr{C}_A$ is a graded highest weight category. Suppose that $\mathscr{C}_A^{gr}$ has a graded Kazhdan-Lusztig theory relative to l. Then $\mathscr{C}_A$ satisfies the strong Kazhdan-Lusztig condition (SKL) relative to l.*

Proof. Let $\Omega \subset \Lambda$ be a coideal. Then $\mathscr{C}_A(\Omega) \cong \mathscr{C}_{eAe}$ for some idempotent $e \in A_0$ [4, (5.4)]. By [6, (3.11)], $\mathscr{C}_{eAe}^{gr}$ has a graded Kazhdan-Lusztig theory relative to l. By (2.1.5(d)), the (SKL$'$) condition relative to $l|_\Omega$ holds for $\mathscr{C}_{eAe}$. Thus, by (2.4.1), the (SKL) condition relative to l holds for $\mathscr{C}_A$.

This theorem and [6, (3.9)] (see (2.1.5)) yield the following characterization of graded Kazhdan-Lusztig theories.

Corollary 2.4.3. *Let A be a graded quasi-hereditary algebra and consider the graded highest weight category $\mathscr{C}_A^{\mathrm{gr}}$. The following two conditions are equivalent:*

(1) *$\mathscr{C}_A^{\mathrm{gr}}$ has a graded Kazhdan-Lusztig theory relative to $l : \Lambda \to \mathbb{Z}$.*

(2) *A is tightly graded and $\mathscr{C}_A$ satisfies the (SKL) condition relative to $l : \Lambda \to \mathbb{Z}$.*

Finally, consider the Bernstein-Gelfand-Gelfand category $\mathcal{O}$ associated to a complex semisimple Lie algebra $\mathfrak{g}$. The principal block $\mathcal{O}_{\mathrm{triv}}$ is a highest weight category with weight poset the Weyl group W (regarded as a poset by means of the Bruhat-Chevalley partial ordering). If $l : W \to \mathbb{Z}$ is the usual length function on W, it is known that $\mathcal{O}_{\mathrm{triv}}$ is Koszul and has a Kazhdan-Lusztig theory relative to l. (See [2], [14], [6].) Thus, $\mathcal{O}_{\mathrm{triv}}$ has a graded Kazhdan-Lusztig theory relative to l. Hence, using (2.4.2), we obtain the following new result on $\mathcal{O}_{\mathrm{triv}}$:

Corollary 2.4.4. *For a complex semisimple Lie algebra $\mathfrak{g}$, the highest weight category $\mathcal{O}_{\mathrm{triv}}$ defined above satisfies the strong Kazhdan-Lusztig condition (SKL) relative to the classical length function $l : W \to \mathbb{Z}$.*

3. An example

In this section, we address two questions:

(1) In [14, (1.16)], it was shown that a finite dimensional algebra A is Koszul whenever the natural grading on $A^!$ is induced by the eigenvalue structure of a suitable automorphism $\sigma \in \mathrm{Aut}(A)$. This suggests the question: *can it be determined if A is Koszul entirely from knowledge of $A^!$ (or perhaps $A^!$ and $A^{!!}$)?*

(2) In his study of the category $\mathcal{O}$, Irving [8] obtained many properties of $\mathrm{gr}\, A$, where A is the basic algebra satisfying $\mathscr{C}_A \cong \mathcal{O}_{\mathrm{triv}}$. The algebra A was later shown [2] to be Koszul; in particular, $A \cong \mathrm{gr}\, A$. This suggests the question: *does the Koszul property for $\mathrm{gr}\, A$ imply the same property for A?*

Both of these questions have negative answers even when A is quasi-hereditary, and even when the highest weight category $\mathscr{C}_A$ has a Kazhdan-Lusztig theory. This section gives an example of a 21-dimensional quasi-hereditary algebra A such that $\mathrm{gr}\, A$ is a Koszul algebra. In contrast, $A \not\cong \mathrm{gr}\, A$, and, in particular, A is not Koszul. However, the associated highest weight category $\mathscr{C}_A$ satisfies the strong Kazhdan-Lusztig condition (SKL) relative to a suitable l. Thus, the algebra $A^!$ is Koszul and the algebra $A^{!!} \cong \mathrm{gr}\, A$ is quasi-hereditary. The highest weight categories $\mathscr{C}_{A^!}^{\mathrm{gr}}$ and $\mathscr{C}_{A^{!!}}^{\mathrm{gr}} \cong \mathscr{C}_{\mathrm{gr}\, A}^{\mathrm{gr}}$ have graded Kazhdan-Lusztig theories (and thus $\mathscr{C}_{A^{!!}}$ and $\mathscr{C}_{\mathrm{gr}\, A}$ satisfy the (SKL) condition relative to a suitable l).

3.1. The "Moose" construction. The algebra A will be constructed using the recursive "Moose" construction for quasi-hereditary algebras given in [13]. We briefly review this construction. We consider data

$$\xi = (B, M_B, {}_B N, \gamma), \tag{3.1.1}$$

consisting of a quasi-hereditary algebra B, a left (respectively, right) finite dimensional B-module N (respectively, M), and a Hochschild cohomology element

$\gamma \in H^2(B, N \otimes_k M)$. (Here we regard $N \otimes_k M$ as a (B, B)-bimodule in the natural way.) Let $\tilde{B}$ denote the extension algebra (unique up to isomorphism) of B by $N \otimes_k M$ defined by γ, so $N \otimes_k M$ identifies with an ideal of $\tilde{B}$ of square 0 and $B \cong \tilde{B}/(N \otimes_k M)$. We extend the action of B on M and N to an action of $\tilde{B}$ by inflation, and we form the "matrix algebra" [10]

$$A = A(\xi) = \begin{pmatrix} \tilde{B} & N \\ M & k \end{pmatrix}. \qquad (3.1.2)$$

If $e = \begin{pmatrix} 0 & 0 \\ 0 & 1 \end{pmatrix}$, then $J = AeA = \begin{pmatrix} N \otimes_k M & N \\ M & k \end{pmatrix}$ satisfies $A/J \cong B$. Then A is quasi-hereditary (and all quasi-hereditary algebras arise, up to Morita equivalence, in this way).

Let Γ be a weight poset for the highest weight category $\mathscr{C}_B$. The poset $\Lambda = \Gamma \cup \{\lambda\}$, such that $\lambda > \gamma$ for all $\gamma \in \Gamma$, serves as a weight poset for $\mathscr{C}_A$. For $\tau \in \Gamma$, $\Delta(\mathscr{C}_A, \tau)$ is obtained from $\Delta(\mathscr{C}_B, \tau)$ by inflation through the quotient map $A \to B$; a similar assertion holds for $\nabla(\mathscr{C}_A, \tau)$. Finally, $\Delta(\mathscr{C}_A, \lambda)$ and $\nabla(\mathscr{C}_A, \lambda)$ are defined by short exact sequences $0 \to M \to \Delta(\mathscr{C}_A, \lambda) \to L(\lambda) \to 0$ and $0 \to L(\lambda) \to \nabla(\mathscr{C}_A, \lambda) \to N^* \to 0$. (See [12].) Diagram chasing establishes:

Lemma 3.1.3. *With the above notation, suppose that $\mathscr{C}_B$ has a Kazhdan-Lusztig theory relative to $l : \Gamma \to \mathbb{Z}$. Then:–*

(a) *$\mathscr{C}_A$ has a Kazhdan-Lusztig theory relative to a suitable extension $\tilde{l} : \Lambda \to \mathbb{Z}$ of l if and only if there exists an integer ϵ such that for all $\nu \in \Gamma$ and all $n \in \mathbb{Z}^+$:*

$$\begin{cases} \mathrm{Ext}^n_{\mathscr{C}_B}(M, \nabla(\nu)) \neq 0 \implies n \equiv l(\nu) + \epsilon \mod 2 \\ \mathrm{Ext}^n_{\mathscr{C}_B}(\Delta(\nu), N^*) \neq 0 \implies n \equiv l(\nu) + \epsilon \mod 2. \end{cases}$$

(b) *Suppose that $\mathscr{C}_A$ has a Kazhdan-Lusztig theory relative to an extension $\tilde{l} : \Lambda \to \mathbb{Z}$ of l. Let $N^{*i} = N^*/\mathrm{soc}^i(N^*)$ and $M^i = \mathrm{rad}^i(M)$. Then $\mathscr{C}$ satisfies the (SKL) condition relative to $\tilde{l}$ if and only if for all $n \in \mathbb{Z}^+$:*

$$\begin{cases} \mathrm{Ext}^n_{\mathscr{C}_B}(M^i, \nabla(\nu)) \neq 0 \implies n \equiv \tilde{l}(\lambda) - l(\nu) + i - 1 \mod 2 \ (i \geq 0) \\ \mathrm{Ext}^n_{\mathscr{C}_B}(\Delta(\nu), N^{*i}) \neq 0 \implies n \equiv \tilde{l}(\lambda) - l(\nu) + i - 1 \mod 2 \ (i \leq 0). \end{cases}$$

---

[10]The "matrix multiplication" in A is subject to the conventions: (i) in calculating the $(1, 1)$ position in a product, the product $n \cdot m$ ($m \in M, n \in N$) is identified with $n \otimes m \in N \otimes M \subset \tilde{B}$; (ii) in calculating the $(1, 2)$ or $(2, 1)$ positions, N and M are regarded as $(\tilde{B}, k)$ and $(k, \tilde{B})$ bimodules, respectively; and (iii) in calculating the $(2, 2)$ position, a product $m \cdot n$ is set equal to 0, $m \in M, n \in N$.

3.2. The example. We wish to apply the above construction of quasi-hereditary algebras. Let B be the basic algebra with quiver and defining relations

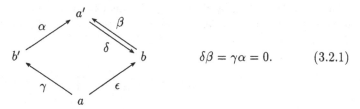

$$\delta\beta = \gamma\alpha = 0. \qquad (3.2.1)$$

Thus, B has simple (right or left) modules denoted a, a', b, b'; let e_x, $x = a, a', b, b'$, denote the corresponding primitive idempotents.

The defining relations (3.2.1) are homogeneous, so B is a tightly graded algebra. Giving $\Gamma = \{a, a', b, b'\}$ the poset structure generated by the relations $a' > b, b'$ and $b, b' > a$, $\mathscr{C}_B$ is a highest weight category, and B is quasi-hereditary. Also, $\mathscr{C}_B$ has a Kazhdan-Lusztig theory relative to the function $l : \Gamma \to \mathbb{Z}$ in which $l(a), l(a')$ are odd, while $l(b), l(b')$ are even.

Continuing with (3.1), let M be the irreducible right B-module a' and let N be the left B-module $Be_{a'}/b'$. We regard M as graded in degree 1 and N^* as graded with socle in grade -1. In $\mathscr{C}_B^{\mathrm{gr}}$, form the minimal resolution

$$0 \to e_{a'}B(-1) \oplus e_{a'}B(1) \to e_{b'}B(-2) \oplus e_b B(0) \to e_a B(-3) \to N^* \to 0.$$

Let $\gamma \in H^2(B, N \otimes M) \cong \mathrm{Ext}^2_{\mathscr{C}_B}(N^*, M)$ have nonzero projection into exactly the summands $\mathrm{Ext}^2_{\mathscr{C}_B^{\mathrm{gr}}}(N^*, M(-2))$, $\mathrm{Ext}^2_{\mathscr{C}_B^{\mathrm{gr}}}(N^*, M(0))$ of $\mathrm{Ext}^2_{\mathscr{C}_B}(N^*, M) \cong \bigoplus_i \mathrm{Ext}^{\bullet}_{\mathscr{C}_B^{\mathrm{gr}}}(N^*, M(i))$.

Let A be the quasi-hereditary algebra which is obtained in (3.1.2) using the data (B, M, N, γ). Assume that A is tightly graded. Then the idempotent ideal J is generated by a homogeneous idempotent e' of degree 0 by [4, (1.2), (4.2)]. Hence, the grading on A induces, by means of the quotient map $A \to B$, a grading on B which must be the tight grading. Since B is tightly graded, it follows that the map $A \to B$ is a graded homomorphism. Hence, $\tilde{B} = (1 - e')A(1 - e')$ is graded and the natural map $\pi : \tilde{B} \to B$ is a graded homomorphism. Hence, $\mathrm{Ker}(\pi) = N \otimes_k M$ is graded compatibly with the grading given on M and N, and γ can be constructed from a transversal $B \to \tilde{B}$ which is a mapping of graded vector spaces. The resulting cocycle is homogeneous of degree 0. This contradicts the choice of γ, so A is not tightly graded.

Next, using (3.1.3), a direct verification shows that $\mathscr{C}_A$ satisfies the (SKL) condition relative to l. By (2.2.1), $\mathrm{gr}\, A \cong A^{!!}$ and $A^!$ is Koszul.

Remark 3.2.4. Although constructed using the recursive construction [13], the algebra A above can easily be described by generators and defining relations. Attach an extra node (labeled c and maximal in the poset structure) to the quiver in (3.2.1) by the scheme: $c \overset{\varsigma}{\underset{\xi}{\rightleftarrows}} a' \cdots$. Then A is given by this quiver with the defining relations $\alpha\xi = \zeta\xi = \zeta\delta = \gamma\alpha - \epsilon\beta\xi\zeta = \delta\beta - \xi\zeta = 0$. A tedious, but direct, calculation, shows that $\mathscr{C}_A$ satisfies the (SKL) condition relative to any length function l which assigns irreducible modules labeled by vowels (respectively, consonants) odd (respectively, even) parity. The algebra gr A has the same quiver as does A, but defining relations $\alpha\xi = \zeta\xi = \zeta\delta = \delta\beta - \xi\zeta = 0$.

REFERENCES

1. H. Andersen, J. Janzten, and W. Soergel, *Representations of quantum groups at a pth root of unity and of semisimple groups in characteristic p: independence of p*, Astérisque **220** (1994).

2. A. Beilinson, V. Ginzburg, and W. Soergel, *Koszul duality patterns in representation theory*, preprint.

3. E. Cline, B. Parshall, and L. Scott, *Finite dimensional algebras and highest weight categories*, J. reine angew. Math. **391** (1988), 85–99.

4. _____, *Integral and graded quasi-hereditary algebras, I*, J. Algebra **131** (1990), 126–160.

5. _____, *Abstract Kazhdan-Lusztig theories*, Tôhoku Math. J. **45** (1993), 511–534.

6. _____, *The homological dual of a highest weight category*, Proc. London Math. Soc. **68** (1994), 294–316.

7. E. Green and R. Martinez Villa, *Koszul and Yoneda algebras*, Proc. ICRA Conference, to appear.

8. R. Irving, *A filtered category $\mathscr{O}_S$*, Memoirs Amer. Math. Soc. **419** (1990).

9. J. Jantzen, *Representations of algebraic groups*, Academic Press, 1987.

10. C. Löfwall, *On the subalgebra generated by the one-dimensional elements in the Yoneda Ext-algebra*, Springer-Verlag LNM **1183** (1983), 291–338.

11. B. Parshall, *Finite dimensional algebras and algebraic groups*, Contemp. Math. **82** (1989), 97–114.

12. _____, *Hyperalgebras, highest weight categories, and finite dimensional algebras*, Contemp. Math. **110** (1990), 203–215.

13. B. Parshall and L. Scott, *Derived categories, quasi-hereditary algebras, and algebraic groups*, Math. Lecture Notes Series, vol. 3, Carleton University, 1988, pp. 1–105.

14. _____, *Koszul algebras and the Frobenius endomorphism*, Quarterly J. Math. **46** (1995), 345–384.

15. S. Priddy, *Koszul resolutions*, Trans. Amer. Math. Soc. **152** (1970), 39–60.

DEPARTMENT OF MATHEMATICS, UNIVERSITY OF OKLAHOMA, NORMAN, OK 73019-0001, USA

DEPARTMENT OF MATHEMATICS, UNIVERSITY OF VIRGINIA, CHARLOTTESVILLE, VA 22903-3199, USA

DEPARTMENT OF MATHEMATICS, UNIVERSITY OF VIRGINIA, CHARLOTTESVILLE, VA 22903-3199, USA

QUANTUM SCHUBERT CELLS AND REPRESENTATIONS AT ROOTS OF 1.

C. DE CONCINI AND C. PROCESI

February 1994

To the memory of Roger Richardson good mathematician and friend

INTRODUCTION

One of the several Hopf algebras recently introduced as *Quantum groups* (cf. the address of Drinfeld in the I.C.M. of Berkeley [12] and [14],[23]) is the q-analogue of the coordinate ring of a semisimple algebraic group G. It is possible to define in a coherent way a specialization of this algebra $R_q[G]$ when $q = \varepsilon$ is a root of 1 (cf. [9]). The resulting algebra $F_\varepsilon[G]$ is a finite module over the coordinate ring Z_0 of G, (which also is central). From the non commutative nature of $R_q[G]$ one inherits a Poisson structure on G and there is a compatibility between this structure and the non commutative algebra $F_\varepsilon[G]$ (which in [10] we have axiomatized into the notion of Hamiltonian algebra, here we will not stress this point but only some of its consequences).

We are concerned with the study of the representation theory of $F_\varepsilon[G]$. By general facts it is natural to analyze irreducible representations by first specifying their central character on Z_0. The symmetries appearing in this picture suggest to stratify the spectrum of Z_0 according to the intersections $X_{w_1,w_2} := U^- n_{w_1} B^- \cap U^+ n_{w_2} B^+$ of double cosets.

Define $\tilde{R}_{w_1,w_2}$ as $F_\varepsilon[G]$ restricted to X_{w_1,w_2} modulo its nilpotent radical. Our main results are:

(i)   The algebra $\tilde{R}_{w_1,w_2}$ is an Azumaya algebra of rank $\ell^{\frac{1}{2}(\ell(w_1)+\ell(w_2)+\mathrm{rank}(w_1-w_2))}$.

(ii) The spectrum of the center of $\tilde{R}_{w_1,w_2}$ is a Galois covering of X_{w_1,w_2} with Galois group the ℓ torsion subgroup of the torus $T/S_{w_1,w_2}$ where $S_{w_1,w_2} := \{t \in T | t^{w_1} = t^{w_2}\}$.

Remark. (a) $\ell(w_1)+\ell(w_2)+\mathrm{rank}(w_1-w_2)$ is the dimension of a symplectic leaf in X_{w_1,w_2}.

(b) The degree of the Galois covering (equal to the number of irreducible representations with given central character in X_{w_1,w_2}) is $\ell^{n-\mathrm{rank}(w_1-w_2)}$ and $n - \mathrm{rank}(w_1 - w_2)$ is the codimension in X_{w_1,w_2} of a symplectic leaf.

Similar results are obtained for the quantized Borel algebras, in particular considering the strata $X_{w,e}$ (e the identity element) of B^+. These results are

---

partially supported by M.U.R.S.T. 40%

127

on one hand a special case of those for G, on the other a necessary step for the general theorem.

Particular importance have the strata $X_{s_i,e}$, X_{e,s_i} where s_i is a simple reflection. In this case the irreducible representations are ℓ dimensional and can be viewed as building blocks (by tensor products) of essentially all other representations.

This result should be compared with the results of Soibelman-Vaksman [29] and Levendorski -Soibelman [17] in the case of q a positive real number and the ones of Joseph [15] and Hodge-Levasseur [13] for a general q not a root of 1. Also Lakshmibai and Reshetikhin [16] study quantum Schubert cells but their treatment is not directly connected with our point of view, essentially they study a q−analogue of the projective coordinate ring while we take an affine point of view and study the q−analogues of the intersection of Schubert varieties with the opposite big cell.

Our techniques consist in studying such algebras as if they were restriction of functions on a Schubert cell. They are partly inspired and based on the work of the previously mentioned authors. The main difference is our stressing of finite dimensional algebras and their representations and Poisson geometry.

The paper is organized as follows: in §1 we recall basic notations and the results and definitions of the theory of Quantum groups and non commutative algebra which are used in this paper. In doing this we have been necessarily somewhat sketchy but we refer to the original papers where the topics here recalled are treated in detail. In §2 we introduce and study certain ideals in the algebras $F_q[G]$, $F_q[B^{\pm}]$ which should be thought as defining some (affine cone over the) *quantum Schubert cell*. In §3 we prove the existence of a birational isomorphism between the coordinate ring of the quantum Schubert cell and another algebra which is a q−analogue of the coordinate ring of a suitable root subgroup. For this last algebra we already know the degree by [6]. In §4 we use the information previously obtained to prove our main theorem. Finally in §5 we discuss the explicit construction of irreducible representations as tensor products of elementary ones.

1. NOTATIONS AND RECOLLECTIONS

1.1. In this paragraph we wish to recall several well known facts in the theory of semisimple groups over $\mathbb{C}$ (see for example [5]).

Consider a finite Cartan matrix $A = (a_{ij})$ of rank n. To A we associate a root system with positive roots Φ^+, simple roots $\Delta := \{\alpha_1, \alpha_2, \ldots, \alpha_n\}$, a semisimple Lie algebra $\mathfrak{g}$ and a simply connected semisimple group G all defined over $\mathbb{C}$.

$\mathfrak{g}$ has the decomposition into root spaces $\mathfrak{g} = \mathfrak{t} \oplus_{\alpha \in \Phi} \mathfrak{g}_\alpha$, where $\mathfrak{t}$ denotes a fixed Cartan subalgebra.

Let P, P^+ denote the integral, resp. dominant integral, weights. Q the root lattice W the Weyl group, s_i the simple reflections associated to the simple roots α_i (given by $s_i(\alpha_j) = \alpha_j - a_{ij}\alpha_i$), ω_i the fundamental weights and $\rho = \sum_{i=1}^{n} \omega_i$.

In W we have the length function. If $w = s_{i_1} s_{i_2} \ldots s_{i_k}$ is a minimal expression we set $k = \ell(w)$.

Given $w \in W$ set $\Phi_w := \{\alpha \in \Phi^+ | w^{-1}(\alpha) \in \Phi^-\} = \Phi^+ \cap w(\Phi^-)$, then $\ell(w)$ equals the cardinality of Φ_w.

If $w = ab$ we say that this is a *reduced decomposition* if $\ell(w) = \ell(a) + \ell(b)$.

In W there is a unique element w_0 of maximal length, it plays a very important role in the theory. It is characterized by the property $w_0(\Phi^+) = \Phi^-$. We have $w_0^2 = 1$ and its length equals the number of positive roots, which will always be denoted by N in this paper.

If $w \in W$ and $w_0 = wa$, this is a reduced decomposition and

$$\Phi^+ = \Phi_w \cup w(\Phi_a), \quad \text{disjoint union.} \tag{1.1.1}$$

$Q \subset P$ is contained in a Euclidean space where the scalar product is denoted by $(\alpha|\beta)$, A is symmetrizable, i.e. there is a diagonal matrix D with entries d_i and DA is the matrix of the scalar products $d_i a_{ij} = (\alpha_i|\alpha_j)$.

Let T, B^+, B^-, U^+, U^- be a maximal torus, the positive, resp. negative Borel subgroups and their unipotent radicals in G (relative to the choices made) and:

$$\mathfrak{t}, \ \mathfrak{b}^+ := \mathfrak{t} \oplus_{\alpha \in \Phi^+} \mathfrak{g}_\alpha, \ \mathfrak{b}^- := \mathfrak{t} \oplus_{\alpha \in \Phi^-} \mathfrak{g}_\alpha, \ \mathfrak{u}^+ := \oplus_{\alpha \in \Phi^+} \mathfrak{g}_\alpha, \ \mathfrak{u}^- := \oplus_{\alpha \in \Phi^-} \mathfrak{g}_\alpha \tag{1.1.2}$$

be the Lie algebras of these groups.

1.2. We introduce now the fundamental objects of study of this paper namely the quantum function algebra $F_q[G]$, its integral form and its specialization at roots of one.

For all the notations relative to q-numbers, q-factorials etc. the reader can refer, for example, to [10].

First of all let us briefly recall the definition of the algebra $F_q[G]$. One starts with the quantized universal enveloping algebra $U_q(\mathfrak{g})$, of Drinfeld-Jimbo, which is generated over the field $\mathbb{C}(q)$ by elements E_i, F_i and $K_i^{\pm 1}$ satisfying suitable q-analogues of Serre relations (see [20]) and which is a Hopf algebra (since this algebra and especially the definition of its comultiplication appear in various slightly different forms in the literature, we will adhere to the definition and notations given in [20]. One should be particularly careful when one uses results from [9]).

One then takes $F_q[G]$ as the Hopf algebra consisting of those linear functions f on $U_q(\mathfrak{g})$ satisfying the following properties:

(a)   There exists a finite codimensional ideal $I \subset U_q(\mathfrak{g})$ such that $I \subset Ker(f)$.
(b)   For all $i = 1, \ldots, r$ there are integers $n_{i1}, \ldots, n_{ik}$ such that $f(\prod_{s=1}^k (K_i - q^{n_{is}})) = 0$.

From this definition one sees that, in order to get elements of $F_q[G]$ one can take matrix coefficients of finite dimensional representations on which the spectrum of the K_i's consists of powers of q. Let us denote this class of representations by $\mathscr{C}$. To be more precise, given a representation $V \in \mathscr{C}$, a vector $v \in V$ and a linear form $\varphi \in V^*$ one defines the element $c_{\varphi,v} \in F_q[G]$, by the formula

$$c_{\varphi,v}(x) = \varphi(xv) \qquad \forall x \in U_q(\mathfrak{g}).$$

Furthermore one easily sees that $F_q[G]$ is linearly spanned by matrix coefficients and, by the complete reducibility of finite dimensional representations of $U_q(\mathfrak{g})$ ([20], [25]), one gets that one can restrict oneself to take V to be irreducible.

From the theory of the R−matrix one has an immediate implication on the commutation rules among the elements $c_{\phi,v}$. Assume that v, w have weights μ_1, μ_2 and that ϕ, ψ have weights ν_1, ν_2 with respect to the action of the elements K_i cf. [17]. Then:

$$c_{\phi,v} c_{\psi,w} = q^{-(\mu_1|\mu_2)+(\nu_1|\nu_2)} c_{\psi,w} c_{\phi,v} + \sum c_{\psi_i,w_i} c_{\phi_i,v_i} \qquad (1.2.1)$$

where

$$\psi_i \otimes \phi_i = p_i(q)(M_i(E) \otimes M_i(F))\psi \otimes \phi, \ \ w_i \otimes v_i = p_i'(q)(M_i'(E) \otimes M_i'(F))w \otimes v \qquad (1.2.2)$$

where the p_i, p_i' are in $\mathbb{C}(q)$ and M_i, M_i' are monomials of which at least one is not constant.

Let us now recall the classification of the irreducible elements in $\mathscr{C}$.

There is a bijective correspondence between dominant weights and isomorphism classes of irreducible left $U_q(\mathfrak{g})$ modules in $\mathscr{C}$. Given $\lambda \in P^+$ the corresponding module V_λ is characterized as follows. It contains a unique line L_λ such that, if $v \in L_\lambda$, one has $E_i v = 0$ and $K_i v = q^{(\lambda,\alpha_i)}v$ for all $i = 1,\dots,n$. Furthermore Weyl character formula holds for V_λ [20].

Let us now remark that we can consider $F_q[G]$ as a left $U_q(\mathfrak{g}) \otimes U_q(\mathfrak{g})$ module. If S denotes the antipode this action is defined by

$$(a \otimes bf)(c) = f(S(a)cb) \qquad \forall a,b,c \in U_q(\mathfrak{g}), f \in F_q[G].$$

We shall often call the action of $U_q(\mathfrak{g}) \otimes 1$ (resp. $1 \otimes U_q(\mathfrak{g})$) the left (resp. right) action of $U_q(\mathfrak{g})$.

The maps $V_\lambda^* \otimes V_\lambda \to F_q[G]$ defined above (by taking matrix coefficients) give an isomorphism of left $U_q(\mathfrak{g}) \otimes U_q(\mathfrak{g})$-modules

$$\delta : \bigoplus_{\lambda \in P^+} V_\lambda^* \otimes V_\lambda \longrightarrow F_q[G] \qquad (1.2.3)$$

Let us consider now the generalized braid group $\mathscr{B}$ associated to our root system [31]. Let us denote by $t_1,\dots,t_n$ its canonical generators. By $p : \mathscr{B} \to W$ the homomorphism taking t_i to the simple reflection s_i. One can define an action ([20], [17]) of $\mathscr{B}$ by algebra automorphisms on $U_q(\mathfrak{g})$ and an action of $\mathscr{B} \times \mathscr{B}$ on $F_q[G]$ by linear isomorphisms. These actions satisfy the following compatibility condition:

$$(h_1,h_2)(a \otimes b \, f) = h_1 a \otimes h_2 b \, ((h_1,h_2)(f))$$

for any $h_1, h_2 \in \mathscr{B}$, $a,b \in U_q(\mathfrak{g})$, $f \in F_q[G]$.

One then easily verifies that the decomposition (1.2.3) is preserved by the $\mathscr{B} \times \mathscr{B}$ action and one gets an action of $\mathscr{B}$ on each irreducible module V_λ compatible with the action of $\mathscr{B}$ on $U_q(\mathfrak{g})$.

Of this action we will systematically use one simple property which we wish to stress.

Let i be a node in the Dynkin diagram, the elements $E_i, F_i, K_i^{\pm 1}$ generate a subalgebra which is the extension to $\mathbb{C}(q)$ of a copy of the algebra $U_{q_i}(sl_2)$, with $q_i = q^{d_i}$. The element $t_i \in \mathscr{B}$ induces on this subalgebra the corresponding Braid automorphism. If we consider a representation V_λ this will decompose under $U_{q_i}(sl_2)$ into irreducible representations which are stable under t_i. Each of these is isomorphic, with the action of t_i to a highest weight representation of $U_{q_i}(sl_2)$. For the representation of highest weight m of the algebra $U_q(sl_2)$, if v_0 is a highest weight vector then $v_m := F^{(m)}v_0$ is a lowest weight vector and moreover (cf. [9]

$$t(v_0) = (-q)^m v_m, \ t(v_m) = v_0, \ E^{(m)}v_m = v_0. \tag{1.2.4}$$

We shall often use the previous remarks in the following form:

Lemma. *If v is a vector of weight λ in a representation in $\mathscr{C}$ and $E_i v = 0$ then setting $m = (\lambda|\alpha_i)/d_i$, $v' := F_i^{(m)}v$ we have:*

$$t_i(v) = (-q_i)^m v', \ t_i(v') = v, \ E_i^{(m)}v' = v. \tag{1.2.5}$$

Similarly for a weight vector killed by F_i.

Proof. The hypothesis implies that v is a highest weight vector for a representation of highest weight m of the corresponding $U_q(sl_2)$.

Given $w \in W$, let $w = s_{i_1} \cdots s_{i_r}$ be a reduced expression. One knows that the element $t_w = t_{i_1} \cdots t_{i_r} \in \mathscr{B}$ is independent from the choice of the reduced expression [21]. This allows us to define some distinguished elements in $F_q[G]$. Fix $\lambda \in P^+$, consider a lowest weight vector $v_{-\lambda} \in V_{-w_0\lambda}$ ($v_{-\lambda}$ is uniquely defined up to multiplication by a non zero constant). Let $\phi_\lambda \in V^*_{-w_0\lambda} \cong V_\lambda$ be a highest weight vector normalized in such a way that $\phi_\lambda(v_{-\lambda}) = 1$. We then set

$$z_w^\lambda = t_w(\phi^\lambda) \otimes v_{-\lambda} \qquad \zeta_w^\lambda = \phi^\lambda \otimes t_{w^{-1}}(v_{-\lambda}). \tag{1.2.6}$$

(we shall often write z_w (resp. ζ_w) for z_w^ρ (resp. ζ_w^ρ, notice also that $z_e^\lambda = \phi^\lambda \otimes v_{-\lambda}$).

1.3. Let us consider now the three subalgebras U^0, $U_q(\mathfrak{b}^+)$ and $U_q(\mathfrak{b}^-)$ in $U_q(\mathfrak{g})$ generated respectively by the elements K_i, by the elements E_i and K_i and finally by the F_i and K_i. These are Hopf subalgebras and U^0 can be identified with the group algebra of the root lattice Q, where the K_i correspond to the simple roots. The braid group action restricts to U^0 inducing the Weyl group action on Q.

Choose a reduced expression $w_0 = s_{i_1}s_{i_2} \ldots s_{i_k}s_{i_{k+1}} \ldots s_{i_N}$, of the longest element w_0 in the Weyl group. From this deduce a convex ordering of the positive roots

$$\beta_1, \beta_2, \ldots, \beta_k, \ldots, \beta_N$$

and corresponding root vectors

$$F_{\beta_1}, \ldots, F_{\beta_N} \in U_q(\mathfrak{b}^-); \ E_{\beta_1}, \ldots, E_{\beta_N} \in U_q(\mathfrak{b}^+)$$

defined as:

$$\beta_k := s_{i_1} s_{i_2} \ldots s_{i_{k-1}}(\alpha_{i_k}), \ F_{\beta_k} := t_{i_1} t_{i_2} \ldots t_{i_{k-1}}(F_{i_k}), \ E_{\beta_k} := t_{i_1} t_{i_2} \ldots t_{i_{k-1}}(E_{i_k}). \tag{1.3.1}$$

Each F_β is a polynomial in the F_i with coefficients in U^0 and no constant term (similarly for the E_β). One has that the elements:

$$\prod_{i=1}^{N} F_{\beta_i}^{h_i} K_\alpha := F_{\beta_N}^{h_N} \ldots F_{\beta_k}^{h_k} \ldots F_{\beta_2}^{h_2} F_{\beta_1}^{h_1} K_\alpha, \ \alpha \in Q, \ h_i \geq 0 \tag{1.3.2}$$

are a basis of $U_q(\mathfrak{b}^-)$.

Similarly for the corresponding monomials in $U_q(\mathfrak{b}^+)$. One could also use the automorphisms t_i^{-1} and obtain parallel results (see [20]).

There are basic commutation relations between the elements E_β (or the F_β) which can be found in [17] (cf. also [24]) and we will often use referring to them as the L-S relations.

We can now define three quotient Hopf algebras of $F_q[G]$. For this consider the restriction of the linear functions in $F_q[G]$ to the subalgebras U^0, $U_q(\mathfrak{b}^+)$ and $U_q(\mathfrak{b}^-)$. Since these are Hopf subalgebras the kernels of these restrictions are Hopf ideals and we thus get quotient Hopf algebras which we denote by $F[T]$, $F_q[B^+]$ and $F_q[B^-]$. Notice that $F[T]$ is the classical coordinate ring of the torus having Q as character group, while the other two algebras are genuine quantizations of function algebras.

There is a different very important way of presenting the algebras $F_q[B^+]$ and $F_q[B^-]$ by means of the so called Drinfeld pairing (see [30]). Remark first of all that one can define (see [6]) a *simply connected* form $U_q^s(\mathfrak{g})$ by extending the indices α in the K from the root to the weight lattice. Correspondingly we also have algebras $U_q^s(\mathfrak{b}^\pm)$. We have a Hopf algebra pairing between $U_q^s(\mathfrak{b}^-)$ and $U_q(\mathfrak{b}^+)_{op}$ where the subscript op reminds us that we have to take the opposite multiplication. Similarly we have a pairing between $U_q(\mathfrak{b}^-)$ and $U_q^s(\mathfrak{b}^+)_{op}$.

Under the canonical pairing we have:

$$\left(\prod_{i=N}^{1} F_{\beta_i}^{h_i} K_\alpha, \prod_{i=N}^{1} E_{\beta_i}^{h_i} K_\beta \right) = q^{-(\alpha|\beta)} \prod_{i=1}^{N} (h_i)_{q_{\beta_i}^2}! (q_{\beta_i}^{-1} - q_{\beta_i})^{-h_i} \tag{1.3.3}$$

(where $q_\beta = q^{\frac{(\beta,\beta)}{2}}$ and $(h)_q = \frac{q^h-1}{q-1}$) and 0 otherwise.

The algebra $U_q^s(\mathfrak{b}^-)$ under this pairing is identified to $F_q[B^+]$ (similarly for $^-$).

Remark. The $U_q(\mathfrak{g}) \otimes U_q(\mathfrak{g})$ action on $F_q[G]$ induces a $U_q(\mathfrak{b}^+) \otimes U_q(\mathfrak{b}^+)$ action on $F_q[B^+]$.

Given a left ideal $I \subset U_q(\mathfrak{b}^+)$ and an element $f \in F_q[B^+]$, we have that f is orthogonal in the Drinfeld pairing to the right ideal $S(I)$ if and only if it is killed by I under the left action.

Proof. $(a \otimes 1)f(x) = f(S(a)x) = (f, S(a)x)$.

1.4. We now set $R = \mathbb{C}[q, q^{-1}]$; in [9] one defines a form $R_q[G]$ of $F_q[G]$, i.e. a free R Hopf subalgebra such that the obvious homomorphism $\mathbb{C}(q) \otimes_R R_q[G] \to F_q[G]$ is an isomorphism. We want to recall a few properties of $R_q[G]$. First of all $R_q[G]$ is stable under the $\mathscr{B} \times \mathscr{B}$ action and the elements $z_w^\lambda, \zeta_w^\lambda \in R_q[G]$.

Under the restriction homomorphisms $r_\pm : F_q[G] \to F_q[B^\pm] \cong U_q^s(\mathfrak{b}^\mp)$, the image of $R_q[G]$ equals the subalgebra generated over R by the elements K_λ, $\overline{F}_\beta = (q_\beta - q_\beta^{-1})F_\beta$ (resp. K_λ, $\overline{E}_\beta = (q_\beta - q_\beta^{-1})E_\beta$). Call these algebras $R_q[B^\pm]$.

Furthermore consider the homomorphism

$$\gamma : R_q[G] \xrightarrow{\Delta} R_q[G] \otimes R_q[G] \to R_q[B^+] \otimes R_q[B^-],$$

in [9] it is proved that γ is injective and $\gamma(z_e^\lambda) = K_\lambda \otimes K_{-\lambda}$ ($e \in W$ being the identity element), so that $\gamma(z_e^\lambda)$ is invertible in $R_q[B^+] \otimes R_q[B^-]$. Furthermore if we set $d = z_e^\rho$, then the subalgebra $\gamma(R_q[G])[d^{-1}]$ of $R_q[B^+] \otimes R_q[B^-]$ coincides with the one generated by the elements $K_\lambda \otimes K_{-\lambda}$, $\overline{F}_\beta \otimes 1$ and $1 \otimes \overline{E}_\beta$ (see Theorem 4.6 in [9]).

The properties of γ allow us to show (cf. 1.2.6):

Proposition. *Let* λ, $\mu \in P^+$, $w \in W$ *then* $z_w^\lambda z_w^\mu = z_w^{\lambda+\mu}$ *and* $\zeta_w^\lambda \zeta_w^\mu = \zeta_w^{\lambda+\mu}$.

Proof. We prove the proposition for the z_w^λ's the proof for the ζ_w^λ's being the same.

We proceed by induction on the length of w. If $w = e$ then everything follows from the fact that $\gamma(z_e^\lambda) = K_\lambda \otimes K_{-\lambda}$ and the injectivity of γ. Now write $w = s_i w'$ with $\ell(w') = \ell(w) - 1$. By induction we have $z_{w'}^\lambda z_{w'}^\mu = z_{w'}^{\lambda+\mu}$. Also by [17] (see the formula for $\Delta(t_i)$), we have that

$$z_w^{\lambda+\mu} = (t_i, e) z_{w'}^{\lambda+\mu} = (t_i, e)(z_{w'}^\lambda z_{w'}^\mu)$$

$$= \sum_{n \geq 0} \frac{q^{\binom{n}{2}}(q - q^{-1})^n}{[n]!} ((t_i, e)(F_i^n \otimes 1 z_{w'}^\lambda))((t_i, e)(E_i^n \otimes 1 z_{w'}^\mu))$$

Now $E_i \otimes 1 z_{w'}^\mu = (E_i t_{w'}(\phi^\mu)) \otimes v_{-\mu} = (t_{w'}(t_{w'}^{-1}(E_i)(\phi^\mu))) \otimes v_{-\mu} = 0$ since $t_{w'}^{-1}(E_i)$ is a non commutative polynomial without constant term in $E_1, \ldots, E_n$ (cf. 1.3 and [20]). We deduce that

$$z_w^{\lambda+\mu} = ((t_i, e) z_{w'}^\lambda)((t_i, e) z_{w'}^\mu) = z_w^\lambda z_w^\mu$$

as desired.

1.5. Let us now specialize q to 1, i.e. consider the Hopf algebra $F_1[G] = R_q[G]/(q - 1)$. In [9] one shows that $F_1[G]$ is isomorphic as a Hopf algebra to the ring of regular functions on the algebraic group G. Furthermore the algebras $R_q[B^+]$ and $R_q[B^-]$ specialize to the ring of functions on two opposite Borel subgroups B^+ and B^- and the homomorphisms $r_\pm$ to the homomorphisms induced by the respective inclusions. The map γ specializes to the algebra homomorphism induced by the morphism $B^+ \times B^- \to G$ sending a pair (b_1, b_2) to $b_1 b_2$. If we then consider the maximal torus $T = B^+ \cap B^-$ then there is a lifting of the quotient $p : \mathscr{B} \to W = N(T)/T$ to $N(T)$ such that the $\mathscr{B} \times \mathscr{B}$ action on

$F_1[G]$ is induced, via this lifting, by the action by left and right multiplication (notice that in this case $\mathscr{B} \times \mathscr{B}$ acts by algebra automorphisms).

The fact that $F_1[G]$ is a specialization of $R_q[G]$ gives G the extra structure of a Poisson Algebraic Group (see [10]) with Poisson bracket given by

$$\{f_1, f_2\} = \frac{[\bar{f}_1, \bar{f}_2]}{q-1} \mod(q-1)$$

for any $f_1, f_2 \in F_1[G]$, with $\bar{f}_1, \bar{f}_2 \in R_q[G]$ being arbitrarily chosen representatives. This Poisson structure is well known. Here let us just recall that a symplectic leaf $\mathscr{O} \subset G$ is contained in exactly one of the sets $X_{w_1,w_2} = B^+ w_1 B^+ \cap B^- w_2 B^-$, $w_1, w_2 \in W$. If this is the case we say that $\mathscr{O}$ is associated to the pair (w_1, w_2) and we have $\dim \mathscr{O} = \ell(w_1) + \ell(w_2) + rk(w_1 - w_2)$, where the rank is taken in the reflection representation. In particular if $X_{e,e} = T$ each point is a symplectic leaf. X_{w_1,w_2} is stable under left and right multiplication by T, these transformations are compatible with the Poisson structure and they permute transitively the symplectic leaves in X_{w_1,w_2}.

1.6. We now specialize q to an arbitrary primitive ℓ-th root of unity ε, with the restriction that ℓ is odd and in case G has G_2 components ℓ is prime with 3.

Set $F_\varepsilon[G] = R_q[G]/(q - \varepsilon)$ (resp. $F_\varepsilon[B^\pm] = R_q[B^\pm]/(q - \varepsilon)$). Denote by $r_{\pm,\varepsilon}$ and γ_ε the specializations of $r_\pm$ and γ at ε. One has

Theorem. *There exist injective, $\mathscr{B} \times \mathscr{B}$ equivariant, Hopf algebra homomorphisms called Frobenius homomorphisms, $\mathscr{F} : F_1[G] \to F_\varepsilon[G]$ and $\mathscr{F}_\pm : F_1[B^\pm] \to F_\varepsilon[B^\pm]$ such that:*

(1) *Their images are central subalgebras over which $F_\varepsilon[G]$ (resp. $F_\varepsilon[B^\pm]$) are finite projective modules of rank $\ell^{\dim G}$ (resp. $\ell^{\dim B^\pm}$).*

(2) *The diagrams*

$$
\begin{array}{ccc}
F_1[G] & \xrightarrow{\mathscr{F}} & F_\varepsilon[G] \\
{\scriptstyle r_{\pm,1}}\downarrow & & \downarrow{\scriptstyle r_{\pm,\varepsilon}} \\
F_1[B^\pm] & \xrightarrow{\mathscr{F}_\pm} & F_\varepsilon[B^\pm]
\end{array}
$$

and

$$
\begin{array}{ccc}
F_1[G] & \xrightarrow{\mathscr{F}} & F_\varepsilon[G] \\
{\scriptstyle \gamma_1}\downarrow & & \downarrow{\scriptstyle \gamma_\varepsilon} \\
F_1[B^+] \otimes F_1[B^-] & \xrightarrow{\mathscr{F}_+ \otimes \mathscr{F}_-} & F_\varepsilon[B^+] \otimes F_\varepsilon[B^-]
\end{array}
$$

commute.

(3) *Let $g \in G$. Set $F_\varepsilon[G](g) = F_\varepsilon[G]/\mathscr{F}(\mathfrak{m}_g)F_\varepsilon[G]$, where $\mathfrak{m}_g$ is the maximal ideal of g in $F_1[G]$. Then if $g, h \in X_{w_1,w_2}$, for some pair $(w_1, w_2) \in W \times W$*

$$F_\varepsilon[G](g) \cong F_\varepsilon[G](h).$$

(4) $\mathcal{F}(z_w^\lambda) = (z_w^\lambda)^\ell$ (here, by abuse of notation we denote an element and its class under specialization by the same symbol).

Proof. The only statement which is not proved in [9] is the last which is shown only for $w = e$. To see the general case notice that

$$\mathcal{F}(z_w^\lambda) = \mathcal{F}((t_w, e)(z_e^\lambda)) = (t_w, e)(\mathcal{F}(z_e^\lambda)) = (t_w, e)z_e^{\lambda\ell} = (z_w^\lambda)^\ell$$

by Proposition 1.4.

1.7. In this paragraph we collect some notions and results of non commutative algebra useful for our work.

Start from the following general concept:

Definition. Given an algebra R over k, $q \in k^*$, we say that an element $a \in R$ is a q−element if there exists a basis u_i of R and integers m_i such that, for all i, we have:

$$au_i = q^{m_i}u_i a.$$

The main elementary remark is:

Proposition. If a is a q−element and a non zero divisor in R then a can be inverted in any subalgebra $A \subset R$ containing a. I.e. there is an algebra containing A whose elements are fractions r/a^k.

Proof. Set $R_i := \{r \in R | ar = q^i ra\}$, by hypothesis $R = \oplus R_i$. One has to show that a is an Ore element, i.e. that given $b \in A$ there is a $c, d \in A$ and non negative integers h, k such that $a^h b = ca$ and $ba^k = ad$. Now if $b \in A$, write $b = \sum_i b_i$, $b_i \in R_i$. If $b = b_i$ for some i we have $ab = q^i ba$. Otherwise we proceed by induction on the number of non zero b_i's. Let m be the maximum index for which $b_i \neq 0$. Write

$$ab = \sum_{i \leq m} ab_i = \sum_{i \leq m} q^i b_i a = q^m ba + \sum_{i \leq m-1}(q^i - q^m)b_i a.$$

Since $b_i a \in R_i$, and clearly the element $b' = \sum_{i \leq m-1}(q^i - q^m)b_i a \in A$, we can apply the inductive assumption to b' and find a $c' \in A$ and a non negative integer h such that $a^h b' = c'a$. We then obtain

$$a^{h+1}b = a^h(q^m ba + b') = (q^m a^h b + c')a$$

as desired. The proof of the remaining property being identical, is left to the reader.

In any algebra R denote by $(b) := RbR$ the two sided ideal generated by the element b.

Lemma. If $b \in R$ is a q−element then for any positive integer n we have that $b \in \sqrt{(b^n)}$.

Proof. Decompose $R = \oplus_i R_i$ where $R_i := \{a \in R | ab = q^i ba\}$ (if q is a primitive ℓ root of 1 we sum modulo ℓ). Then $bR_i = R_i b$, so $bR = Rb$ and $(Rb)^n \subset Rb^n$.

A basic technical problem with which one has often to deal in non commutative algebra is localization of a multiplicative set, which is only possible under special Öre conditions of which we have seen an example. We will not need very complicated localizations, but all our algebras are (left and right) noetherian rings and most of them domains which then have a quotient division algebra. So, when we invert elements in such algebras, we will always think that we do it in this division algebra.

1.8. Recall that an algebra R over a commutative ring A is called an Azumaya algebra of rank d (over A) if there is a commutative algebra B faithfully flat over A such that $R \otimes_A B \cong M_d(B)$ (where $M_d(B)$ denotes the algebra of $d \times d$ matrices over B).

We shall use the fact that, if R is an Azumaya algebra over its center Z, S a Z algebra and $\phi : R \to S$ a Z algebra homomorphism then $S = R \otimes_Z C(R)$ where $C(R)$ is the centralizer of $\phi(R)$ in S.

In particular:

(1) If both R, S are Azumaya algebras of the same degree any homomorphism $\phi : R \to S$ maps Z in the center C of S and $S = R \otimes_Z C$ (cf. [4, 3]).
(2) Any ideal I of R is extended from Z i.e. $I = R(I \cap Z)$.
(3) If V is a finite dimensional module over an Azumaya algebra R and the center Z of R acts on V as scalars, then the kernel of the action is a maximal ideal $\underline{m}$ and V is a direct sum of copies of the unique irreducible module of R/I.

A general theorem of M. Artin (cf. [2] [26]) shows that a ring R is an Azumaya algebra of rank d over its center if and only if it verifies the following two conditions:

(1) R satisfies all polynomial identities of $d \times d$ matrices.
(2) R has no homomorphism into $h \times h$ matrices over some commutative ring for $h < d$.

Condition (1) is automatically verified if R is contained in a ring of $d \times d$ matrices over a commutative ring or in a finite dimensional central simple algebra S of degree d.

If R is a finitely generated algebra over an algebraically closed field k, condition (2) is equivalent to assuming that R has no irreducible representations of dimension $< d$.

Under the same hypotheses, if we assume that R satisfies some polynomial identity and has no nilpotent ideals then the two conditions are equivalent to assuming that all irreducible representations of R have dimension d.

Let us now assume that R is a finite projective module over an algebra A contained in the center Z of R and finitely generated over an algebraically closed field k.

For any point P in the spectrum of A (which we think of as a homomorphism of A into k and for which we denote by $k(P)$ the field k as A module) denote by $R(P)$ the algebra $R \otimes_A k(P)$. We think of R as a bundle of algebras over $Spec(A)$ and of the $R(P)$ as the fibers.

Assume next that A is a domain and there is a nilpotent ideal N in R such that R/N is an order in a finite dimensional central simple algebra S of degree d, by hypothesis $A \cap N = 0$ so A is contained in the center Z of R/N and Z is a finite module over A.

Proposition. *If the algebras $R(P)$ are all isomorphic then R/N is an Azumaya algebra over its center Z and the map $Spec(Z) \twoheadrightarrow Spec(A)$ has fibers of fixed cardinality.*

Proof. The set of points in $Spec(Z)$ where R/N is an Azumaya algebra is the open set of $P \in Spec(Z)$ for which $R/N \otimes_Z k(P) \cong M_d(k)$. The set of points in $Spec(Z)$ where R/N is not an Azumaya algebra is a proper closed subvariety as also proper and closed is the subset of points where the map $\pi : Spec(Z) \rightarrow Spec(A)$ is not ètale. Thus there is a non empty open subset O of $Spec(A)$ for which R/N is Azumaya and π is ètale on $\pi^{-1}(O)$. It follows that, for $P \in O$ we have that $R/N(P) \cong M_d(k)^{\oplus n}$ where n is the degree of the map π. In particular we must have that for $P \in O$ the image of N in $R(P)$ must be its nilpotent radical and thus $R/N(P) \cong M_d(k)^{\oplus n}$ is the semisimple quotient of $R(P)$. The claim then follows, in fact by the hypothesis made the algebras $R(P)$ are all isomorphic, from Artin's theorem R/N is Azumaya of degree d.

For any $P \in Spec(A)$ the semisimple part of $R/N(P)$ equals $M_d(k)^{\oplus m}$ where m is the number of points in the fiber of π, hence m is constant.

2. QUANTUM SCHUBERT CELLS

2.1. Let us give an element $w \in W$ and a dominant weight λ and start defining:

$$v_{-w(\lambda)} := t_w(v_{-\lambda}) \qquad \phi^{w(\lambda)} := t_w(\phi^\lambda). \tag{2.1.1}$$

Notice that $v_{-w(\lambda)}$ is, up to scalars, the only vector of weight $-w(\lambda)$ in $V_{-w_0\lambda}$ (similarly for $\phi^{w(\lambda)}$).

In the classical case ($q{=}1$) this is one of the extremal vectors, its line thought of as a point in projective space lies in the flag variety and is one of the T fixed points in the center of a corresponding Schubert cell (for B^- in our conventions).

We now define the subspace $V_{w,\lambda} \subseteq V_{-w_0\lambda}$ as follows:

$$V_{w,\lambda} := \langle U_q(\mathfrak{b}^-)v_{-w(\lambda)}\rangle. \tag{2.1.2}$$

Proposition. *If $w = s_i w'$ is reduced then $v_{-w(\lambda)} = cE_i^a v_{-w'(\lambda)}$ with a a non negative integer and c a non zero scalar (resp. $\phi^{w(\lambda)} = cF_i^a \phi^{w'(\lambda)}$).*

Proof. Since $v_{-w(\lambda)} = t_i v_{-w'(\lambda)}$ it is enough to apply the representation theory of the quantum $U_q(sl_2)$ generated by $E_i, F_i, K_i^{\pm 1}$ and show that $v_{-w'(\lambda)}$ is a lowest weight vector for this algebra (Lemma 1.2). Thus we need to show that $F_i v_{-w'(\lambda)} = 0$

Now by hypothesis $x = t_{w'}^{-1}(F_i)$ is a polynomial without constant term in the F_j. Thus

$$0 = t_{w'}(x v_{-\lambda}) = F_i v_{-w'(\lambda)}. \tag{2.1.3}$$

The proof in the remaining case is identical.

2.2.

Proposition. *Let* $w = s_{i_1} s_{i_2} \ldots s_{i_k}$ *then:*

$$V_{w,\lambda} := \langle E_{i_1}^{h_1} E_{i_2}^{h_2} \ldots E_{i_k}^{h_k} v_{-\lambda} \rangle. \tag{2.2.1}$$

Proof. Work by induction.

$$V_{w,\lambda} := \langle U_q(\mathfrak{b}^-) v_{-w(\lambda)} \rangle = \langle U_q(\mathfrak{b}^-) E_{i_1}^a v_{-w'(\lambda)} \rangle. \tag{2.2.2}$$

From the basic commutation relations we have:

$$U_q(\mathfrak{b}^-) E_{i_1} \subset E_{i_1} U_q(\mathfrak{b}^-) + U_q(\mathfrak{b}^-).$$

Thus

$$\langle U_q(\mathfrak{b}^-) E_{i_1}^a v_{-w'(\lambda)} \rangle \subset \langle \sum_{b \leq a} E_{i_1}^b U_q(\mathfrak{b}^-) v_{-w'(\lambda)} \rangle.$$

Applying induction we have the inclusion:

$$V_{w,\lambda} \subset \langle E_{i_1}^{h_1} E_{i_2}^{h_2} \ldots E_{i_k}^{h_k} v_{-\lambda} \rangle.$$

Conversely by induction and the commutation relations

$$\langle E_{i_1}^{h_1} E_{i_2}^{h_2} \ldots E_{i_k}^{h_k} v_{-\lambda} \rangle \subset \langle E_{i_1}^h U_q(\mathfrak{b}^-) v_{-w'(\lambda)} \rangle \subset \langle \sum_b U_q(\mathfrak{b}^-) E_{i_1}^b v_{-w'(\lambda)} \rangle.$$

We have remarked that $v_{-w'(\lambda)}$ is a lowest weight vector for the $U_{q_i}(sl_2)$ generated by $E_i, F_i, K_i^{\pm 1}$ and $v_{-w(\lambda)}$ is the highest weight vector of the module it generates, we get:

$$\langle \sum_b U_q(\mathfrak{b}^-) E_{i_1}^b v_{-w'(\lambda)} \rangle \subset \langle \sum_b U_q(\mathfrak{b}^-) F_{i_1}^b v_{-w(\lambda)} \rangle \subset V_{w,\lambda}$$

as desired.

2.3. By the definitions and the representation theory we have

$$V_{-w_0\lambda}^* \otimes v_{-\lambda} = \{ x \in F_q[G] | (1 \otimes F_i) x = 0, \ K_\alpha x = q^{-(\alpha.\lambda)} x \}.$$

By the formula of the comultiplication in $U_q(\mathfrak{g})$ we get immediately that

$$V_{-w_0\lambda}^* \otimes v_{-\lambda} V_{-w_0\mu}^* \otimes v_{-\mu} \subset V_{-w_0(\lambda+\mu)}^* \otimes v_{-\lambda-\mu}.$$

In fact we have equality by irreducibility and this is the q−analogue of Cartan's multiplication.

Let us introduce thus the subalgebra

$$\mathcal{H} = \oplus_\lambda V_{-w_0\lambda}^* \otimes v_{-\lambda} \subset F_q[G]. \tag{2.3.1}$$

We want to study the restriction to $\mathcal{H}$ of the homomorphism $r_+ : F_q[G] \to F_q[B^+]$. We have:

Lemma. (1) *The elements* z_e^λ *are q-elements in* $\mathcal{H}$.
(2) *The restriction of* r_+ *to* $\mathcal{H}$ *extends to a isomorphism between* $\mathcal{H}[z_e^{-1}]$ *and* $F_q[B^+]$.

Proof. Part (1) follows immediately from the commutation relations 1.2.1. In particular z_e is a Öre element and it makes sense to consider part (2).

Using the Drinfeld pairing we identify $F_q[B^+]$ with $U_q^s(\mathfrak{b}^-)$. Under this identification we have that $r_+(z_e) = K_\rho$ so that it is invertible in $U_q^s(\mathfrak{b}^-)$ and hence we can extend r_+ to a homomorphism

$$\tilde{r}_+ : \mathscr{H}[z_e^{-1}] \to F_q[B^+]. \tag{2.3.2}$$

In order to show that (2.3.2) is an isomorphism we need to show that the restriction of r_+ to $\mathscr{H}$ is injective and (2.3.2) is surjective.

Notice that the decomposition (2.3.1) is the decomposition of $\mathscr{H}$ into weight spaces under the action of the algebra $1 \otimes U^0 \subset U_q(\mathfrak{g}) \otimes U_q(\mathfrak{g})$. Since this algebra also acts on $F_q[B^+]$ and r_+ is equivariant, it suffices to show injectivity for each of the spaces $V_{-w_0\lambda}^* \otimes v_{-\lambda}$. This follows immediately since $V_{-w_0\lambda} = U_q(\mathfrak{b}^+)v_{-\lambda}$.

Let us now show the surjectivity of (2.3.2). It is based on the formula:

$$r_+(z_{s_i}) = q^{2d_i}\overline{F}_i K_\rho. \tag{2.3.3}$$

From 1.2.5 and the fact that $(\alpha_i, \rho) = d_i$, we get that $z_{s_i} = -q^{d_i}F_i \otimes 1(z_e)$.

Let now $x = \prod_{i=1}^N E_{\beta_i}^{h_i} K_\alpha \in U_q(\mathfrak{b}^+)$ be a basis element as in (1.3.2). We have $z_{s_i}(x) = -q^{d_i}z_e(S(F_i)x) = 0$ unless $x = E_i K_\alpha$. If this is the case we have

$$z_{s_i}(E_i K_\alpha) = -q^{d_i}z_e(-F_i K_i E_i K_\alpha) = q^{d_i-(\alpha,\rho)+(\alpha_i,\alpha_i)-(\alpha_i,\rho)}.$$

This implies our claim by 1.3.3.

Now since the elements K_λ and F_i generate $U_q^s(\mathfrak{b}^-)$ the surjectivity of (2.3.2) follows.

Choose a reduced expression $w_0 = s_{i_1}s_{i_2}\ldots s_{i_k}s_{i_{k+1}}\ldots s_{i_N}$, let us denote by $w_k = s_{i_1}s_{i_2}\ldots s_{i_k}$, $a_k = s_{i_{k+1}}\ldots s_{i_N}$.

Definition. Set $I_k := \oplus_\lambda V_{w_k,\lambda}^\perp \otimes v_{-\lambda}$.

Set $F[B_k]$ to be the subalgebra of $F_q[B^+] \cong U_q^s(\mathfrak{b}^-)$ generated by the elements

$$F_{\beta_1}, F_{\beta_2}, \ldots, F_{\beta_k}, \ K_\lambda; \ \lambda \in P.$$

Remark. $I_k \supset I_{k+1}$, $F[B_k] \subset F[B_{k+1}]$.

Proof. The first part follows from Proposition 2.2 the second is clear.

Notice that, if we fix $w \in W$ we can always find a reduced expression for w_0 such that $w = w_k$ (for $k = \ell(w)$). I_k, $F[B_k]$ depend only on w and not on the choice of a reduced expression. If we want to stress this point of view we shall denote them by I_w, $F[B_w]$.

2.4. Consider now for each i the restriction of the function algebra $F_q[B^+]$ to the subalgebra $U_q(\mathfrak{b}_i) \subset U_q(\mathfrak{b}^+)$ generated by K_i, E_i. By duality this gives a map of $F_q[B^+] \cong U_q^s(\mathfrak{b}^-)$ to the algebra $S_i := \{K_{\omega_i}, F_i\}$. Explicitely this maps $F_j \to 0$, $j \neq i$, $F_i \to F_i$, $K_\lambda \to K_{\omega_i}^{(\lambda|\alpha_i)}$. Similarly we can deduce a mapping to the algebra $\mathscr{T}_i := \{K_{\omega_i}\}$ which can be viewed as function algebra on the subalgebra $\mathscr{T}_i^*$ generated by K_i.

Using the given reduced expression of w_0, consider the map:

$$\pi_k : F_q[B^+] \xrightarrow{\Delta^N} F_q[B^+] \otimes F_q[B^+] \ldots F_q[B^+] \to \otimes_{s=1}^k S_{i_s} \otimes \otimes_{s=k+1}^N \mathscr{T}_{i_s}. \quad (2.4.1)$$

We can identify $S_{i_1} \otimes S_{i_2} \otimes \ldots S_{i_k} \otimes \mathscr{T}_{i_{k+1}} \otimes \ldots \otimes \mathscr{T}_{i_N}$ as an algebra of linear functionals on $U_q(\mathfrak{b}_{i_1}) \otimes U_q(\mathfrak{b}_{i_2}) \otimes \ldots \otimes U_q(\mathfrak{b}_{i_k}) \otimes \mathscr{T}_{i_{k+1}}^* \otimes \ldots \otimes \mathscr{T}_{i_N}^*$ and this map by duality as a restriction of functions

$$\pi_k(f)(u_{i_1} \otimes u_{i_2} \otimes \ldots u_{i_k} \otimes m_{i_{k+1}} \otimes \ldots \otimes m_{i_N}) = f(u_{i_1} u_{i_2} \ldots u_{i_k} m_{i_{k+1}} \ldots m_{i_N}). \quad (2.4.2)$$

In particular if f is the restriction to B^+ of $c_{\phi, v_{-\lambda}}$ we have:

$$f(u_{i_1} u_{i_2} \cdots u_{i_k} m_{i_{k+1}} \cdots m_{i_N}) = \phi(u_{i_1} u_{i_2} \cdots u_{i_k} m_{i_{k+1}} \cdots m_{i_N} v_{-\lambda}),$$

so that, by Proposition 2.2, $f = c_{\phi, v_{-\lambda}} \in Ker(\pi_k)$ if and only if $\phi \in V_{w_k, \lambda}^\perp$.

From the definition of $V_{w_k, \lambda}$ we see that it has a basis of weight vectors of weight less or equal to $-w_k(\lambda)$ thus a sufficient condition for an element ϕ to be in $V_{w, \lambda}^\perp$ is that it is a weight vector of weight not greater or equal to $w_k(\lambda)$.

Definition. We set J_k to be the kernel of π_k.

Proposition. $J_k \cap \mathscr{H} = Ker(\pi_k) \cap \mathscr{H} = \oplus_\lambda V_{w_k, \lambda}^\perp \otimes v_{-\lambda} := I_k$.

Proof. We claim that, if $f \in Ker(\pi_k)$ also $(1 \otimes K_i)f \in Ker(\pi_k)$. Indeed

$$\pi_k((1 \otimes K_i)f)(u_{i_1} \otimes u_{i_2} \otimes \ldots u_{i_k} \otimes m_{i_{k+1}} \otimes \ldots \otimes m_{i_N})$$
$$= f(u_{i_1} u_{i_2} \ldots u_{i_k} m_{i_{k+1}} \ldots m_{i_N} K_i)$$

and by our definitions and the commutation relations it follows that $u_{i_1} u_{i_2} \ldots u_{i_k} m_{i_{k+1}} \ldots m_{i_N} K_i$ is again a monomial of the type considered in 2.4.2

This implies that $Ker(\pi_k) \cap \mathscr{H}$ is a direct sum of weight spaces (under the right action of U^0).

Since the decomposition $\mathscr{H} = \oplus_{\lambda \in P^+} V_{-w_0 \lambda}^* \otimes v_{-\lambda}$ is the decomposition of $\mathscr{H}$ in weight spaces under the action of $1 \otimes U^0$ the claim follows.

Corollary. (1) I_k is an ideal in $\mathscr{H}$ and $\mathscr{H}/I_k$ is an integral domain.
(2) $J_k = I_k[z_e^{-1}]$ so it depends only on w and not on the choice of a reduced expression.

Proof. (1) That I_k is an ideal is a consequence of the previous proposition. On the other hand $\mathscr{H}/I_k$ is contained in the image of the map π_k which lies in an algebra which is clearly a twisted polynomial algebra with some elements inverted and hence a domain. (2) is clear.

Remark. By the above corollary J_k can be denoted by J_w.

Example. J_e is generated by all the elements F_i. The quotient algebra $F_q[B^+]/J_e$ is the commutative (polynomial algebra) generated by K_λ, $\lambda \in P$.

For a simple reflection the ideal J_{s_i} is generated by the elements F_j, $j \neq i$. The algebra $F_q[B^+]/J_{s_i}$ is isomorphic to the subalgebra of $U_q^s(\mathfrak{b}^-)$ generated by F_i, K_λ; $\lambda \in P$.

3. THE BIRATIONAL MAP

3.1. We have chosen together with a reduced expression of w_0 the elements:

$$\beta_1, \beta_2, \ldots, \beta_N; \quad E_{\beta_1}, E_{\beta_2}, \ldots, E_{\beta_N}, \quad F_{\beta_1}, F_{\beta_2}, \ldots, F_{\beta_N}.$$

Recall that the ordering of the positive roots is *convex*, in other words if a root β_k is a linear combination with positive coefficients of other roots β_i then these roots cannot be all greater or all less than β_k.

We consider the monomials

$$K_\lambda E_{\beta_1}^{h_1} E_{\beta_2}^{h_2} \ldots E_{\beta_N}^{h_N}; \quad E_{\beta_N}^{h_N} E_{\beta_{N-1}}^{h_{N-1}} \ldots E_{\beta_1}^{h_1} K_\lambda \tag{3.1.1}$$

we call these monomials respectively the *increasing* and the *decreasing* monomials in the E. Similarly we can do this for the F. Recall that the two monomials appearing in 3.1.1 have weight $\beta := \sum_{i=1}^N h_i \beta_i$.

Definition. Let L_k^i, L_k^d be the span of all increasing (resp. decreasing) monomials in the E's with $h_i > 0$ for some $i > k$. Let $\mathcal{U}_k$ be the subalgebra generated by the K_λ's and by the elements $E_{\beta_1} \ldots, E_{\beta_k}$.

Notice that $\mathcal{U}_k$ is spanned by the increasing or decreasing monomials 3.1.1 with $h_i = 0$ for all $i > k$.

Lemma. *Any increasing (resp. decreasing) monomial of weight β_k different from $K_\alpha E_{\beta_k}$ lies in L_k^i (resp. in L_k^d).*

More generally, if M_1, M_2 are increasing monomials, and $M_1 M_2$ is of weight β_k and $M_1 \in \mathcal{U}_{k-1}$ has non zero weight then $M_2 \in L_k^i$ (similarly for decreasing).

Proof. This follows from the convexity of the order of the roots.

In what follows we shall use the following property of the comultiplication which is claimed in [17] and proved in the Appendix of [1]:

$$\Delta(E_{\beta_k}) = E_{\beta_k} \otimes 1 + K_{\beta_k} \otimes E_{\beta_k} + \sum_i a_i \otimes b_i$$

where, if we choose the a_i, b_i to be increasing monomials then $a_i \in \mathcal{U}_{k-1}$, $b_i \in L_k^i$ or, if we choose the a_i, b_i to be decreasing monomials then $a_i \in L_k^d$, $b_i \in \mathcal{U}_{k-1}$.

Lemma. (1) L_k^i *is the left ideal of $U_q(\mathfrak{b}^+)$ generated by the elements E_{β_j} with $j > k$ (resp. L_k^d is the right ideal generated by the same elements).*

(2) $S(E_{\beta_k}) = -K_{-\beta_k} E_{\beta_k} + T$ *where $T \in L_k^i \cap L_k^d$.*

(3) $S(L_k^i) = L_k^d$.

Proof. 1) It is clear that L_k^i is contained in the prescribed left ideal. Conversely consider a product $ME_{\beta_j}, j > k$ where M is an increasing monomial. It is clear from the relations L-S (cf. 1.3) that this is a linear combination of increasing monomials of the form $M'E_{\beta_t}, t \geq j$ hence the claim, (the proof for L_k^d is similar).

2) Recall that for any element a with $\Delta(a) = \sum_j u_j \otimes v_j$ one has $\epsilon(a) = \sum_j S(u_j)v_j$. Apply this to the expression for $\Delta(E_{\beta_k})$ and use the fact that $\epsilon(E_{\beta_k}) = 0$ and part 1).

3) Since $S^2(E_{\beta_j}) = q^{(\beta_j,\beta_j)}E_{\beta_j}$ so that $S^2(L_k^i) = L_k^i, S^2(L_k^d) = L_k^d$ it is enough to show that $S(E_{\beta_j}) \in L_k^d$; this follows immediately from part 2)

Remark. We can order lexicographically the multi-indices $h_N, h_{N-1}, \ldots, h_1$. If we fix one of these multi-indices $\underline{h}_0$ let us define by $L^d(\underline{h}_0)$ (resp. $L^i(\underline{h}_0)$) the span of all decreasing (resp. increasing) monomials with $\underline{h} \geq \underline{h}_0$ in the lexicographic ordering. Then in general we will have that:
$L^i(\underline{h}_0)$ is a left ideal (resp. $L^d(\underline{h}_0)$ is a right ideal).

3.2.

Definition. Set $z_k = z_{w_k}^\rho$.

Lemma. (1) z_k is annihilated under the left action of $U_q(\mathfrak{g})$ on $F_q[G]$ by L_k^i.
(2) $F[B_k]$ is the subspace orthogonal to L_k^d under the Drinfeld pairing. The image of the element z_k in $F_q[B^+] = U_q^s(\mathfrak{b}^-)$ lies in $F[B_k]$.
(3) We have

$$z_k = q^{2(\beta_k|\rho)}\overline{F}_{\beta_k}z_{k-1} + P, \ P \in F[B_{k-1}]. \tag{3.2.1}$$

Proof. (1) Set $t_{w_k} = t_k$. Clearly $t_k(\phi^\rho)$ is killed by all elements of type $t_k(M)$ where M is a positive monomial in the E's. Since by Lemma 3.1 L_k^i is the left ideal generated by the elements E_{β_j} with $j > k$ and these elements are linear combinations of elements of the previous type, 1) follows.

(2) By the very properties of the Drinfeld pairing $F[B_k]$ is the orthogonal to L_k^d since the exponents monomials defining the two spaces lie in complementary sets. From the first part $0 = (a \otimes 1)z_k$ if $a \in L_k^i$, hence $z_k(S(a)u) = 0$ for all u, in particular $(z_k, S(a)) = z_k(S(a)) = 0$ it follows from Lemma 3.1 that z_k is orthogonal to L_k^d hence $z_k \in F[B_k]$.

(3) It is similar to the proof of 2.3.3. Set $w = w_k$ and $\overline{w} = w_{k-1}$. We have

$$E_{\beta_k}(t_{\overline{w}}\phi^\rho) = t_{\overline{w}}(E_{i_k}\phi^\rho) = 0.$$

While

$$K_{\beta_k}(t_{\overline{w}}\phi^\rho) = t_{\overline{w}}(K_{i_k}\phi^\rho) = q_{\beta_k}(t_{\overline{w}}\phi^\rho). \tag{3.2.2}$$

We can apply the $U_{q_{\beta_k}}(sl_2)$ generated by $E_{\beta_k}, F_{\beta_k}, K_{\beta_k}^{\pm 1}$ (with corresponding element $t_{\beta_k} = t_{\overline{w}}t_{i_k}t_{\overline{w}}^{-1}$) for which we have seen in proposition 2.1 that $t_{\overline{w}}\phi^\rho$ is a highest weight vector. 3.2.2 implies that the representation that it generates is 2 dimensional with lowest weight vector $t_w(\phi^\rho) = t_{\beta_k}t_{\overline{w}}\phi^\rho$, getting:

$$t_w(\phi^\rho) = -q_{\beta_k}F_{\beta_k}(t_{\overline{w}}\phi^\rho).$$

We have

$$z_k = t_w(\phi^\rho) \otimes v_{-\rho} = -q_{\beta_k}(F_{\beta_k} \otimes 1)z_{k-1}, \quad z_{k-1} = -q_{\beta_k}^{-1}(E_{\beta_k} \otimes 1)z_k$$

(we are using the left action on functions). It follows from 1) that z_k is killed under left action by $L_k^i + L_{k-1}^i E_{\beta_k}$, which is the left ideal spanned by all increasing monomials for which either $h_i \geq 1$, for some $i > k$ or $h_k \geq 2$. It follows that z_k is orthogonal to $S(L_k^i + L_{k-1}^i E_{\beta_k}) = L_k^d + E_{\beta_k} L_{k-1}^d$ hence $z_k \in F_{\beta_k} F[B_{k-1}] + F[B_{k-1}]$. Write now $z_k = F_{\beta_k} u + v$, $u, v \in F[B_{k-1}]$, we want to compute the element u by applying again the Drinfeld pairing.

We have

$$(z_k | E_{\beta_k} M) = (S^{-1}(E_{\beta_k}) \otimes 1 z_k | M).$$

Recall that S^{-1} is S followed by conjugation by $K_{2\rho}$ and $S(E_{\beta_k}) = -K_{-\beta_k} E_{\beta_k} + T$ where $T \in L_k^i$ kills z_k. Thus $S^{-1}(E_{\beta_k}) \otimes 1 z_k = -q^{2(\beta_k | \rho)} z_{k-1}$. It follows that

$$(z_k | E_{\beta_k} M) = -q^{2(\beta_k | \rho)}(z_{k-1} | M).$$

Now, if M is a decreasing monomial in $\mathcal{U}_{k-1}$, we have that

$$(F_{\beta_k} u + v | E_{\beta_k} M) = (q_{\beta_k} - q_{\beta_k}^{-1})^{-1}(u | M),$$

hence $q^{2(\beta_k | \rho)}(z_{k-1} | M) = (q_{\beta_k} - q_{\beta_k}^{-1})^{-1}(u | M)$. Since we already know that $u \in F[B_{k-1}]$ we obtain $u = q^{2(\beta_k | \rho)}(q_{\beta_k} - q_{\beta_k}^{-1})z_{k-1}$ as required.

In order to continue we need to use all the integral forms of our algebras and maps.

Theorem. (1) *The induced map:*

$$\phi_w : F[B_w] \to F_q[B^+]/J_w \tag{3.2.3}$$

is injective.

(2) *The element z_w is a q−element in $F_q[B^+]/J_w$. If $\overline{w} = ws_j$, with $\ell(\overline{w}) = \ell(w) - 1$ then $z_{ws_j} \in J_{\overline{w}}$. If $w = ab$ is a reduced decomposition the elements z_a, z_w q−commute.*

(3) *After inversion (see 1.8) we have:*

$$F[B_w][z_w^{-1}] = (F_q[B^+]/J_w)[z_w^{-1}].$$

Proof. Fix a reduced expression $w = s_{i_1} \cdots s_{i_k}$ such that $i_k = j$. Thus $I_w = I_k$, $F[B_w] = F[B_k]$ and so on.

(1) We use the integral form $R_q[B_k]$, of $F[B_k]$. That is the subalgebra over R generated by the elements K_λ and $\overline{F}_{\beta_1}, \ldots, \overline{F}_{\beta_k}$. This is a free R-module with basis the increasing monomials in these generators. It suffices to show that the restriction of the map ϕ_k to $R_q[B_k]$ is injective.

J_k is by Proposition 2.4 the kernel of the map π_k defined in 2.4.1. π_k maps $R_q[B^+]$ in the corresponding tensor product of integral forms and thus it suffices to prove the injectivity after specialization at $q = 1$. At $q = 1$ the map π_k is induced on functions by a map whose image is a dense subset in $X_{e,w} \subset B^+$ (see 1.5), while the inclusion $R_q[B_k] \to R_q[B^+]$ at $q = 1$ corresponds to the projection of B^+ onto the subgroup $B_k = T \prod_{i=1}^k U_{\beta_i}$, where U_β is the root subgroup corresponding to the root β given by the decomposition $B^+ = B_k \times \prod_{i=k+1}^{N} U_{\beta_i}$. It is shown in [9], [10], that the composed map from $X_{e,w}$ to B_k is birational hence the injectivity follows.

(2) The fact that z_k is a q−element in $F_q[B^+]/J_k$ is proved in [17] and follows from the commutation relations 1.2.1; in fact if $M(E)$ is a non constant monomial in the E_i the element $M(E)t_{w_k}\phi^\rho$ has weight strictly greater than $w_k(\rho)$ and so it is in $V_{w_k,\lambda}^\perp$.

For the same reason $t_{w_k}\phi^\rho \in V_{w_{k-1},\lambda}^\perp$; thus by Proposition 2.4 $z_k \in J_{k-1}$.

For the last part we are claiming that z_i, z_k q−commute $(i < k)$. By part 1) it is enough to show that they q−commute modulo J_k and this follows again by 1.2.1.

(3) Let us prove by induction that, for every $i > 0$, we have $F_{\beta_{k+i}} z_k^M \in F[B_k] + J_k$ for some M. This is formula 3.2.1 for $i = 1$ in view of 2.

Work by decreasing induction assuming that, if $i \geq 2$ we have

$$F_{\beta_{k+i}} z_{k+1}^M = u, \text{ mod } J_{k+1} \tag{3.2.4}$$

where $u = \sum_{j=0}^{m} u_j F_{\beta_{k+1}}^j$ and $u_j \in F[B_k]$.

We now claim that, for each j there is a sufficiently high integer t depending on j such that:

$$F_{\beta_{k+1}}^j z_k^t = \sum_{s=0}^{h} x_s z_{k+1}^s, \text{ mod } J_{k+1}$$

with $x_s \in F[B_k]$. If $j = 1$ this is just relation 3.2.1, so we can proceed by induction on j. If we multiply $F_{\beta_{k+1}}^j$ by z_k we obtain, using 3.2.1,

$$F_{\beta_{k+1}}^j z_k = a F_{\beta_{k+1}}^{j-1} z_{k+1} + F_{\beta_{k+1}}^{j-1} b$$

with $a, b \in F[B_k]$. Now using the commutation relations between the F_β's one easily gets that $F_{\beta_{k+1}}^{j-1} b = \sum_{r<j} b_r F_{\beta_{k+1}}^r$. Thus using the inductive hypothesis and the fact that z_k and z_{k+1} q−commute mod J_{k+1}, we get the desired conclusion.

From what we have seen it now follows that if we multiply both sides of 3.2.4 by a sufficiently high power of z_k, we obtain

$$F_{\beta_{k+i}} z_{k+1}^M z_k^P = \sum_{j=0}^{h} v_j z_{k+1}^j, \text{ mod } J_{k+1}$$

with $v_j \in F[B_k]$. If $M = 0$ we are done, otherwise since reducing modulo J_{k+1} we have a domain, since z_k, z_{k+1} q−commute we may divide by the highest power of z_{k+1} and assume that $v_0 \neq 0$. Now reduce modulo z_{k+1} which lies in J_k getting $F_{\beta_{k+i}} z_{k+1}^M z_k^P = v_0$, mod J_k. Since $M > 0$ and $z_{k+1} \in J_k$, we get $v_0 = 0$ mod J_k. This contradicts step 1.

Remark. From 3.2.1 and 1.3 it follows easily that the ordered monomials in the elements $K_{\omega_1}, \ldots, K_{\omega_n}, z_1, \ldots, z_k$ are linearly independent.

We work now in the division algebra D of fractions of $F[B_k]$ and use the notation $A < a_1, \ldots, a_k >$ for the subalgebra of D generated by a subalgebra A and elements a_i.

Corollary. (1) *The algebra $F[B_k]$ is contained in the twisted Laurent polynomial ring generated by $K_{\omega_1}, \ldots, K_{\omega_n}, z_1, \ldots, z_k$ and their inverses.*

(2) *Let $a \in F_q[B^+]$, there exists a non negative integer p such that $az_k^p \in F[B_k]$ mod $J_{k+1} + (z_{k+1})$. In particular if $a \in J_k$, $az_k^p \in J_{k+1} + (z_{k+1})$.*

Proof. (1) Since $F[B_{k-1}] < z_{k-1}^{-1}, z_k >$ contains $F[B_{k-1}]$, to show that it contains $F[B_k]$ we compute $\overline{F}_{\beta_k}$. By 3.2.1 this is $-q^{-2(\beta_k|\rho)}(z_k - P)z_{k-1}^{-1}$, where $P \in F[B_{k-1}]$ is as in 3.2.1. The rest of our statement follows easily by induction since for $k = 0$ the algebra $F[B_k] = F[T]$ is exactly the Laurent polynomial ring generated by $K_{\omega_1}, \ldots, K_{\omega_n}$.

(2) From the proof of step (3) in the Theorem we get that our statement holds for the F_β's. Since z_k is an Öre element in $F[B_k]$ the claim follows from this simple fact that we leave to the reader:

If $A \subset B$ are algebras, J an ideal of B, a an Öre element in A then the set

$$C := \{x \in B | \exists p \geq 0, \ xa^p \in A + J\}$$

is a subalgebra.

For the second part use the fact that by step 1 in the Theorem $J_k \cap F[B_k] = 0$.

Remark. For $w = s_i$ a simple reflection we have that the map ϕ_{s_i} is already an isomorphism (cf. last Remark of 2.4).

4. ROOTS OF 1

4.1. In this section ε is a primitive ℓ^{th} root of 1, with ℓ as in 1.6.

We have already remarked that the map π_k defined in 2.4.1 induces a map of integral forms:

$$\pi_k : R_q[B^+] \xrightarrow{\Delta^N} R_q[B^+]^{\otimes N} \to \otimes_{s=1}^k R_q[B_{i_s}] \otimes_{s=k+1}^N R_q[T_{i_s}]. \quad (4.1.1)$$

This map clearly specializes at $q = \varepsilon$ and let us indicate by $\pi_{\varepsilon,k}$ the induced map and by $J_{\varepsilon,k}$ (or $J_{\varepsilon,w}$ if we wish to stress its dependence only on $w = s_{i_1}s_{i_2}\cdots s_{i_k}$) its kernel.

Note that the map $\pi_{\varepsilon,k}$ takes values again in a twisted polynomial algebra which is a domain and so $J_{\varepsilon,k}$ is a (completely) prime ideal.

We are now ready to study the irreducible representations of the algebras $F_\varepsilon[G], F_\varepsilon[B^+]$ introduced in 1.6.

In both cases we have the Frobenius map (see 1.6), i.e. they contain the coordinate ring $\mathbb{C}[G], \mathbb{C}[B^+]$ respectively as Hopf subalgebras in the center and are projective modules of rank $\ell^{dim(G)}$ and $\ell^{dim(B^+)}$ respectively over them.

Let us denote by R, Z_0 one of these two algebras and the corresponding commutative Hopf subalgebra. The spectrum of Z_0 is either the group G or B^+ and we stratify it respectively by strata X_{w_1,w_2} and $X_{e,w} = X_w$, each of these is a union of symplectic leaves permuted by T transitively. It follows that up to the isomorphisms given by T and the ones given by the Hamiltonian flows, the algebras $R(p)$ are constant on these strata (cf. 1.6 3)). Our final task is to study these algebras $R(p)$ for each stratum and prove the results announced in the introduction.

4.2. We start with $F_\varepsilon[B^+]$. In this case the corresponding algebra Z_0 is the coordinate ring of B^+ and its spectrum B^+ is stratified by the locally closed sets X_w.

We shall use the notations of §2. Let $w \in W$. Define Y_w to be the ideal of $Z_0 = \mathbb{C}[B^+]$ formed by the elements vanishing on X_w. Let $Y_{\varepsilon,w}$ be the ideal of the algebra $F_\varepsilon[B^+]$ generated by Y_w.

Lemma. $Y_w = Z_0 \cap J_{\varepsilon,w}$.

Proof. We can assume, as usual that $w = w_k$. We have to understand the restriction of the map $\pi_{\varepsilon,k}$ to $Z_0 = \mathbb{C}[B^+]$. For this remark that Z_0 is a sub Hopf algebra and so the comultiplication is the one induced by the group B^+. The projections to the algebras S_i and $\mathscr{T}_i$ restricted to Z_0 correspond to the inclusions of subgroups B_i^+, T_i which are respectively the positive Borel and the maximal torus of the canonical Sl_2 associated to the node of the Dynkin diagram of index i. The claim follows from the fact that the corresponding product of subgroups is dense in X_w [11].

4.3. We come now to the main lemma.

Lemma. $\sqrt{Y_{\varepsilon,w}} = J_{\varepsilon,w}$.

Proof. We can assume as usual that $w = w_k$. From the previous lemma we have that $Y_{\varepsilon,k} \subset J_{\varepsilon,k}$. Since $J_{\varepsilon,k}$ is a prime ideal it is sufficient to show that $J_{\varepsilon,k}$ is in the nilpotent radical of $Y_{\varepsilon,k}$. By induction J_{k+1} is in the nilpotent radical of $Y_{\varepsilon,k+1}$. Now since $F_\varepsilon[B^+]$ is a projective module over Z_0 we have that $F_\varepsilon[B^+]/Y_{\varepsilon,k} = F_\varepsilon[B^+] \otimes_{Z_0} Z_0/Y_k$, which is projective over Z_0/Y_k. Since $z_k^\ell \in Z_0$ (see 1.6) and $z_k \notin J_k$ the element z_k^ℓ will be a non zero divisor in $F_\varepsilon[B^+]/Y_{\varepsilon,k}$. By the Lemma in 1.7 and the fact that z_{k+1} is a q-element mod J_{k+1}, we have that $J_{k+1} + (z_{k+1})$ is in the nilpotent radical of $J_{k+1} + (z_{k+1}^\ell)$. By Corollary 3.2 for any element $c \in J_{\varepsilon,k}$ there is a product $cz_k^{\ell r} \in J_{k+1} + (z_{k+1}^\ell)$. Thus there is a power $c^m z_k^{\ell rm} \in J_{k+1} + (z_{k+1}^\ell)$. Since $z_{k+1}^\ell \in Y_k$ we have that $c^m z_k^{\ell rm} \in \sqrt{Y_{\varepsilon,k}}$ hence $c^p z_k^{\ell rp} \in Y_{\varepsilon,k}$ for some p. Now z_k^ℓ is a non zero divisor modulo $Y_{\varepsilon,k}$ hence it follows that $c^p \in Y_{\varepsilon,k}$ for some p, hence $J_{\varepsilon,k}$ is in the radical of $Y_{\varepsilon,k}$.

4.4. We now consider the specializations at $q = 1$ of z_w and ζ_w as functions on the group G. We have

Proposition. *Let D_w (resp. C_w) be the divisor of z_w (resp. ζ_w). Then $D_w \cap \overline{B^- w B^-} = \overline{B^- w B^-} - B^- w B^-$ and $C_w \cap \overline{B^+ w B^+} = \overline{B^+ w B^+} - B^+ w B^+$.*

Proof. We prove the proposition for D_w the proof for C_w being the same. Since D_e is the complement of the big cell $B^+ B^-$ and $D_w = w D_e$ by definition, we clearly have that $G - D_w = w B^+ B^-$. Thus we have to show that $B^- w B^- = w B^+ B^- \cap \overline{B^- w B^-}$. The fact that $B^- w B^- \subset w B^+ B^-$ is clear since setting $U_w^- = w U^+ w^{-1} \cap U^-$, one has $B^- w B^- = U_w^- w B^- = w w^{-1} U_w^- w B^- \subset w B^+ B^-$.

To complete the proof it suffices to show that B^-wB^- is relatively closed in wB^+B^-. Now the product map gives an isomorphism of algebraic varieties between wU^+w^{-1} and $U_w^+ \times U_w^-$ where $U_w^+ = wU^+w^{-1} \cap U^+$. Decompose now, again using the product map, wB^+B^- as the product $wU^+w^{-1} \times wB^-$. Substituting we obtain an isomorphism between wB^+B^- and $U_w^+ \times U_w^- \times wB^-$. Under this isomorphism $\{1\} \times U_w^- \times wB^-$ maps isomorphically onto B^-wB^-, so we obtain an isomorphism of wB^+B^- with $U_w^+ \times B^-wB^-$ proving that B^-wB^- is relatively closed in wB^+B^-.

Let us consider now the algebra $F_\varepsilon[B^+]/J_{\varepsilon,w}$. We are now ready to prove:

Theorem. (1) *The algebra* $F_\varepsilon[B^+]/J_{\varepsilon,w}[z_w^{-1}]$ *is an Azumaya algebra.*

(2) *The map* $Spec(F_\varepsilon[B^+]/J_{\varepsilon,w}[z_w^{-1}]) \to Spec(F_\varepsilon[B^+])$ *is bijective onto* $Spec(F_\varepsilon[B_w^+])$.

(3) *If* ℓ *is prime with the bad primes (cf. [5]) for the root system, the degree of* $F_\varepsilon[B^+]/J_{\varepsilon,w}[z_w^{-1}]$ *equals* $\ell^{1/2(\ell(w)+\mathrm{rank}(e-w))}$.

Proof. (1) From Lemma 4.3 it follows that, given an irreducible $F_\varepsilon[B^+]$ module, it defines a $F_\varepsilon[B^+]/J_{\varepsilon,w}$ module if and only if its central character in $Spec(Z_0)$ lies in $\overline{X}_w$ and among these the $F_\varepsilon[B^+]/J_{\varepsilon,w}[z_w^{-1}]$ modules are the ones with central character in X_w since $z_w^\ell \in Z_0$ and its divisor in $\overline{X}_w$ is the complement of X_w (by 1.6 and the previous Proposition). On the other hand we know that the algebras $F_\varepsilon[B^+](g)$, $g \in X_w$ are all isomorphic (Theorem 1.6), hence from Proposition 1.7 (1) and (2) follow. From Theorem 3.2 we know that the degree of this algebra equals that of the algebra $F_\varepsilon[B_w]$ which has been computed in [6] proving (3).

4.5. From now on we shall assume that ℓ is prime with the bad primes for the root system.

We want now to determine the spectrum of $F_\varepsilon[B^+]/J_{\varepsilon,w}[z_w^{-1}]$.

For this we start noticing that the algebra $F_\varepsilon[B^+]/J_{\varepsilon,e}$ can be identified with the coordinate ring of the torus T (with character group P). Using this we get a homomorphism $ev_t : F_\varepsilon[B^+] \to k$, $\forall t \in T$ and, since $J_{\varepsilon,e}$ is a Hopf ideal, an action of T on $F_\varepsilon[B^+]$ by algebra automorphisms, defined by $t(f) = ev_t \otimes id \cdot \Delta(f)$ for all $t \in T$, $f \in F_\varepsilon[B^+]$.

The reader can easily verify that, given a matrix coefficient $c_{\phi,v}$ with ϕ a weight vector of weight μ one has $t(c_{\phi,v}) = t^\mu c_{\phi,v}$, so that in particular, $t(z_w^\lambda) = t^{w\lambda} z_w^\lambda$. This clearly implies that the ideal $J_{\varepsilon,w}$ is stable under the action of T so that we get a T-action on $F_\varepsilon[B^+]/J_{\varepsilon,w}$ and, since z_w is a weight vector for T, also a T-action on $F_\varepsilon[B^+]/J_{\varepsilon,w}[z_w^{-1}]$.

Remark also that by the expression $\Delta(F_i) = F_i \otimes K_i^{-1} + 1 \otimes F_i$ and the fact that the elements $\overline{F}_\beta \in J_{\varepsilon,e}$ it follows immediately that the $\overline{F}_\beta$'s are fixed by T.

Having made this considerations we are ready to prove

Theorem. *The spectrum of* $F_\varepsilon[B^+]/J_{\varepsilon,w}[z_w^{-1}]$ *is a Galois covering of* X_w *with Galois group the group* $T_{w,\ell}$, *of elements of* ℓ *torsion in* T/S_w *where* $S_w := \{t \in T \mid t^\lambda = 1 \ \forall \lambda \in P \ such \ that \ w\lambda = \lambda\}$.

Proof. Since $F_\varepsilon[B^+]/J_{\varepsilon,w}[z_w^{-1}]$ is an Azumaya algebra we know that its spectrum equals the spectrum of its center Z_w, so we shall determine Z_w. We start by exhibiting some elements in Z_w.

We have from the q-commutation relations 1.2.1 the fact that the z_w^λ are ε-elements in $F_\varepsilon[B^+]/J_{\varepsilon,w}$. Explicitly modulo $J_{\varepsilon,w}$ we have

$$z_{w'}^\mu z_w^\lambda = q^{(w'\mu|w\lambda)-(\mu|\lambda)} z_w^\lambda z_{w'}^\mu$$

and

$$z_e^\lambda z_{w'}^\mu = q^{-(\mu-w'\mu|\lambda)} z_{w'}^\mu z_e^\lambda$$

for all $w' \in W$. Notice now that if we work in $F_\varepsilon[B^+]/J_{\varepsilon,w}[z_w^{-1}]$ we get, using Proposition 1.4, that, for $\lambda \in P^+$, each z_w^λ is invertible. Take $\lambda \in P$, write $\lambda = \lambda_+ - \lambda_-$ with $\lambda_\pm \in P^+$ and set $z_w^\lambda = z_w^{\lambda+}(z_w^{\lambda-})^{-1}$. It is then clear from the definition that the above relations continue to hold even if λ is not dominant.

Let now $P^w \subset P$ denote the sublattice of weights fixed by w. If $\lambda \in P^w$ we set

$$d_w^\lambda = z_w^\lambda z_e^\lambda.$$

It is then clear that the elements d_w^λ are central in $F_\varepsilon[B^+]/J_{\varepsilon,w}[z_w^{-1}]$ and that the inverse of d_w^λ is $d_w^{-\lambda}$. Furthermore if we set $Z_{0,w} = Z_0/Y_w[z_w^{-\ell}]$, $d_w^{\ell\lambda} \in Z_{0,w}$. Choose a basis $\lambda_1, \dots, \lambda_s$ for P^w. We claim that

$$Z_w = Z_{0,w}[d_w^{\lambda_1}, \dots, d_w^{\lambda_s}] \tag{4.5.1}$$

(of course if $P^w = \{0\}$ this just means that $Z_w = Z_{0,w}$). The elements d_w^λ with $\lambda \in P_\ell^w$ where $P_\ell^w = \{\lambda | \lambda = \sum_{i=1}^s m_i \lambda_i$ with $0 \le m_i \le \ell - 1\}$ clearly span $Z_{0,w}[d_w^{\lambda_1}, \dots, d_w^{\lambda_s}]$. We claim that they are linearly independent on $Z_{0,w}$. Indeed T acts on $F_\varepsilon[B^+]/J_{\varepsilon,w}[z_w^{-1}]$. Let us restrict our action to the subgroup T_ℓ of ℓ-torsion points in T. It is clear from the above considerations and from the fact that Z_0 is generated by the elements $\overline{F}_\beta$ and $K_{\ell\mu}$ that T_ℓ acts trivially on $Z_{0,w}$. On the other hand if $\lambda \in P_\ell^w$ and $t \in T$ we have $t(d_w^\lambda) = t^{2\lambda} d_w^\lambda$. Since P^w is a split direct summand in P and ℓ is odd, the action of T_ℓ factors through an action of $T_{w,\ell}$ and, as λ runs through P_ℓ^w, the characters $t^{2\lambda}$ run through a complete set of distinct characters for $T_{w,\ell}$. This clearly implies our claim and proves that the natural map

$$Spec(Z_{0,w}[d_w^{\lambda_1}, \dots, d_w^{\lambda_s}]) \longrightarrow X_w = Spec(Z_{0,w}),$$

is an unramified Galois covering with structure group $T_{w,\ell}$ (remember that the elements d_w^λ are invertible). On the other hand we know that the algebra $F_\varepsilon[B^+]/J_{\varepsilon,w}[z_w^{-1}]$ is a free module over $Z_{0,w}$ of rank $\ell^{\ell(w)+n}$ and has degree $\ell^{\ell(w)+rk(e-w)} = \ell^{\ell(w)+n-|T_{w,\ell}|}$. It follows immediately that the quotient fields of Z_w and of $Z_{0,w}[d_w^{\lambda_1}, \dots, d_w^{\lambda_s}]$ coincide, so since both algebras are finite as $Z_{0,w}$-modules and $Z_{0,w}[d_w^{\lambda_1}, \dots, d_w^{\lambda_s}]$ is integrally closed, (4.5.1) and the theorem follows.

We finish this section remarking that all we have stated for the algebras $F_q[B^+]$, $F_\varepsilon[B^+]$ has an analogue for the algebras $F_q[B^-]$, $F_\varepsilon[B^-]$. In particular the coordinate ring $\mathbb{C}[B^-]$ embeds as a central Hopf subalgebra in $F_\varepsilon[B^-]$. We can define, for any $w \in W$ the ideals $J_w^- \in F_q[B^-]$, $J_{\varepsilon,w}^- \subset F_\varepsilon[B^-]$. We can show that the intersection of this second ideal with $\mathbb{C}[B^-]$ is the ideal Y_w^- of functions vanishing on the set $X_w^- = B^+wB^+ \cap B^-$, while if we consider the ideal $Y_{\varepsilon,w}^- \subset F_\varepsilon[B^-]$ generated by Y_w^- we have that $\sqrt{Y_{\varepsilon,w}^-} = J_{\varepsilon,w}^-$. In this situation the role of the elements z_w is taken by the elements ζ_w so that in particular ζ_w is a ε-element in $F_\varepsilon[B^-]/J_{\varepsilon,w}^-$ and $F_\varepsilon[B^-]/J_{\varepsilon,w}^-[\zeta_w^{-1}]$ is an Azumaya algebra of degree $\ell^{\frac{1}{2}(\ell(w)+rk(e-w))}$.

4.6. We now pass to the study of the irreducible representations of $F_\varepsilon[G]$.

In this situation we stratify G by the subsets $X_{w_1,w_2} = B^-w_1B^- \cap B^+w_2B^+$ (notice that $X_w = X_{w,e}$ while $X_w^- = X_{e,w}$).

Consider the homomorphism

$$\gamma_\varepsilon : F_\varepsilon[G] \to F_\varepsilon[B^+] \otimes F_\varepsilon[B^-]$$

defined in (1.6). Recall that γ_ε is injective and can be extended to an isomorphism of $F_\varepsilon[G][z_\varepsilon^{-1}]$ onto the subalgebra $\mathscr{A} \subset F_\varepsilon[B^+] \otimes F_\varepsilon[B^+]$ generated by the elements $\overline{F}_\beta \otimes 1$, $1 \otimes \overline{E}_\beta$, $K_\lambda \otimes K_{-\lambda}$ (here we are using the isomorphisms given by Drinfeld duality). In particular $F_\varepsilon[B^+] \otimes F_\varepsilon[B^-]$ is the twisted Laurent polynomial algebra $\mathscr{A}[u_1^{\pm 1}, \ldots, u_n^{\pm 1}]$ with $u_i = 1 \otimes z_\varepsilon^{\omega_i}$.

Definition. For any pair $(w_1, w_2) \in W_1 \times W_2$ we define $J_{\varepsilon,w_1,w_2} \subset F_\varepsilon[G]$ to be the ideal $\gamma_\varepsilon^{-1}(J_{\varepsilon,w_1} \otimes F_\varepsilon[B^-] + F_\varepsilon[B^+] \otimes J_{\varepsilon,w_2}^-)$

Remark. If as usual w_0 denotes the longest element in W, the ideal

$$J_{\varepsilon,w_0,e} = \gamma_\varepsilon^{-1}(F_\varepsilon[B^+] \otimes J_{\varepsilon,e}^-)$$

equals the kernel of r_ε^+ since the mapping

$$F_\varepsilon[B^+] \to F_\varepsilon[B^+] \otimes F_\varepsilon[B^-]/J_{\varepsilon,e}^- = F_\varepsilon[B^+] \otimes U^0$$

maps $\overline{F}_\beta \to \overline{F}_\beta \otimes 1$, $K_\lambda \to K_\lambda \otimes K_{-\lambda}$ and is clearly injective.

The algebra $F_\varepsilon[G][z_\varepsilon^{-1}]/J_{\varepsilon,w_1,e}[z_\varepsilon^{-1}] \cong F_\varepsilon[B^+]/J_{\varepsilon,w_1}$.

We take now as $Z_0 \subset F_\varepsilon[G]$ the image of $\mathbb{C}[G]$ under the Frobenius homomorphism. We set $Y_{w_1,w_2} \subset Z_0$ equal to the ideal of functions vanishing on X_{w_1,w_2} and Y_{ε,w_1,w_2} equal to the ideal in $F_\varepsilon[G]$ generated by Y_{w_1,w_2}. As before we have:

Lemma. $J_{\varepsilon,w_1,w_2} \cap Z_0 = Y_{w_1,w_2}$.

Proof. Using Lemma 4.2, the commutative diagram (see 1.6)

$$
\begin{array}{ccc}
F_1[G] \cong Z_0 & \xrightarrow{\mathscr{F}} & F_\varepsilon[G] \\
\gamma_1 \downarrow & & \gamma_\varepsilon \downarrow \\
F_1[B^+] \otimes F_1[B^-] & \xrightarrow{\mathscr{F}+\otimes\mathscr{F}-} & F_\varepsilon[B^+] \otimes F_\varepsilon[B^-]
\end{array}
$$

and recalling that the map γ_1 is induced by the product map

$$\gamma_1^* : B^+ \times B^- \to G$$

sending a pair (b_+, b_-) to $b_+ b_-$, we reduce to proving that $X_{w_1,e} X_{e,w_2}$ is a dense subset in X_{w_1,w_2}. The fact that it is contained in X_{w_1,w_2} is clear from the definitions. To show that it is dense it suffices to see, since X_{w_1,w_2} is irreducible, that $dim(X_{w_1,e} X_{e,w_2}) \geq dim(X_{w_1,w_2})$. Now $dim(X_{w_1,w_2}) = \ell(w_1) + \ell(w_2) + n$ while γ_1^* is a principal T bundle onto its image; it follows that $dim(X_{w_1,e} X_{e,w_2}) \geq \ell(w_1) + \ell(w_2) + n$, proving the lemma.

4.7.

Lemma. *The algebra* $F_\varepsilon[B^+]/J_{\varepsilon,w_1} \otimes F_\varepsilon[B^-]/J_{\varepsilon,w_2}^-$ *is a domain. Furthermore*

$$\sqrt{Y_{\varepsilon,w_1} \otimes F_\varepsilon[B^-] + F_\varepsilon[B^+] \otimes Y_{\varepsilon,w_2}^-} = J_{\varepsilon,w_1} \otimes F_\varepsilon[B^-] + F_\varepsilon[B^+] \otimes J_{\varepsilon,w_2}^-$$

Proof. We have by Corollary 3.2 that the algebra $F_\varepsilon[B^+]/J_{\varepsilon,w_1}$ is a subalgebra in a twisted Laurent polynomial algebra and similarly for $F_\varepsilon[B^-]/J_{\varepsilon,w_2}^-$. Hence their tensor product is also a subalgebra of a twisted Laurent polynomial algebra and hence has no zero divisors. To see the second part of the Lemma notice that by Lemma 4.6 we have that

$$Y_{\varepsilon,w_1} \otimes F_\varepsilon[B^-] + F_\varepsilon[B^+] \otimes Y_{\varepsilon,w_2}^- \subset J_{\varepsilon,w_1} \otimes F_\varepsilon[B^-] + F_\varepsilon[B^+] \otimes J_{\varepsilon,w_2}^-,$$

while Lemma 4.3 implies that

$$\sqrt{Y_{\varepsilon,w_1} \otimes F_\varepsilon[B^-] + F_\varepsilon[B^+] \otimes Y_{\varepsilon,w_2}^-} \supset J_{\varepsilon,w_1} \otimes F_\varepsilon[B^-] + F_\varepsilon[B^+] \otimes J_{\varepsilon,w_2}^-.$$

The other inclusion follows immediately from the first part.

4.8.   Let us now consider in $\mathscr{A}$ the ideals $J_{\varepsilon,w_1,w_2}[z_e^{-1}]$ and $Y_{\varepsilon,w_1,w_2}[z_e^{-1}]$. We have

Proposition. $\sqrt{Y_{\varepsilon,w_1,w_2}[z_e^{-1}]} = J_{\varepsilon,w_1,w_2}[z_e^{-1}].$

Proof. By Lemma 4.7 and the definitions our proposition will follow from the following

Lemma. *Let* S *be an algebra with an automorphism* ϕ, *let* $S_\phi[x^{\pm 1}]$ *be the associated twisted polynomial ring. Let* I *be an ideal in* $S_\phi[x^{\pm 1}]$ *and* $\sqrt{I}$ *its radical. Then:*

$$S \cap \sqrt{I} = \sqrt{S \cap I}.$$

Proof. The fact that $S \cap \sqrt{I} \subset \sqrt{S \cap I}$ is clear. As for the reverse inclusion, remark that $I \cap S$ is ϕ stable and thus we may reduce modulo $I[x^{\pm 1}]$ and assume that $S \cap I = 0$. Thus let N be the nilpotent radical of S (clearly ϕ stable), we need to show that $N[x^{\pm 1}]$ is nilpotent, but this is clear by the commutation relations between N and x.

4.9.   Let us denote by $\tilde{J}_{\varepsilon,w_1,w_2}$ the ideal $J_{\varepsilon,w_1,w_2}[z_\varepsilon^{-1}]$ of the algebra $\mathscr{A}$. Also denote by D the subalgebra of $\mathscr{A}$ generated by the elements $\overline{F}_{\beta_i} \otimes 1$, $1 \otimes \overline{E}_{\delta_j}$, $K_\lambda \otimes K_{-\lambda}$ where $\beta \in \Phi_{w_1}$, $\delta \in \Phi_{w_2}$ (in fact this algebra is independent of the choice of the reduced expressions). Notice that the algebra $F_\varepsilon[B_{w_1}^+] \otimes F_\varepsilon[B_{w_2}^-]$ is just the twisted Laurent polynomial algebra $D[u_1^{\pm 1}, \ldots, u_n^{\pm 1}]$. We have

Proposition. *The algebra D contains the elements $\gamma_\varepsilon(z_{w_1})$ and $\gamma_\varepsilon(\zeta_{w_2})$.*
If we restrict the quotient homomorphism

$$\pi : \mathscr{A} \to \mathscr{A}/\tilde{J}_{\varepsilon,w_1,w_2},$$

to D, we can extend π to an isomorphism between the algebras $D[\gamma_\varepsilon(z_{w_1}\zeta_{w_2})^{-1}]$ and $\mathscr{A}/\tilde{J}_{\varepsilon,w_1,w_2}[\gamma_\varepsilon(z_{w_1}\zeta_{w_2})^{-1}]$

Proof. We start with a general remark. Consider matrix coefficients of the form $c_{\phi,v_{-\lambda}} \in V_{-w_0\lambda}^* \otimes v_{-\lambda}$ (resp. $c_{v^\lambda,\phi} \in v^\lambda \otimes V_{-w_0\lambda}$). For such matrix coefficients one has

$$\gamma_\varepsilon(c_{\phi,v_{-\lambda}}) = c_{\phi,v_{-\lambda}} \otimes z_\varepsilon^\lambda \qquad \gamma_\varepsilon(c_{\phi^\lambda,v}) = z_\varepsilon^\lambda \otimes c_{\phi^\lambda,v}.$$

In particular

$$\gamma_\varepsilon(z_{w_1}) = z_{w_1} \otimes z_\varepsilon^\lambda \qquad \gamma_\varepsilon(\zeta_{w_2}) = z_\varepsilon^\lambda \otimes \zeta_{w_2}.$$

This clearly implies, by Lemma 3.2 that the elements $\gamma_\varepsilon(z_{w_1})$ and $\gamma_\varepsilon(\zeta_{w_2})$ lie in the algebra $D[u_1^{\pm 1}, \ldots, u_n^{\pm 1}]$. Since they also lie in $\mathscr{A}$ they must lie in D proving our first claim.

Let us now show the second part. It follows by the above remark that the algebras $F_\varepsilon[B^+]/J_{\varepsilon,w_1} \otimes F_\varepsilon[B^-]/J_{\varepsilon,w_2}^-[\gamma_\varepsilon(z_{w_1}\zeta_{w_2})^{-1}]$ and $F_\varepsilon[B_{w_1}^+] \otimes F_\varepsilon[B_{w_2}^-][\gamma_\varepsilon(z_{w_1}\zeta_{w_2})^{-1}]$ are just the algebras $F_\varepsilon[B^+]/J_{\varepsilon,w_1}[z_{w_1}^{-1}] \otimes F_\varepsilon[B^-]/J_{\varepsilon,w_2}^-[\zeta_{w_2}^{-1}]$ and $F_\varepsilon[B_{w_1}^+][z_{w_1}^{-1}] \otimes F_\varepsilon[B_{w_2}^-][\zeta_{w_2}^{-1}]$ respectively, which are isomorphic. It also follows, since the ideals J_{ε,w_1} and J_{ε,w_2}^- are generated by matrix coefficients of the form considered above that the ideal $J_{\varepsilon,w_1} \otimes F_\varepsilon[B^-] + F_\varepsilon[B^+] \otimes J_{\varepsilon,w_2}^-$ is generated by J_{ε,w_1,w_2} so that

$$F_\varepsilon[B^+]/J_{\varepsilon,w_1} \otimes F_\varepsilon[B^-]/J_{\varepsilon,w_2}^- \cong \mathscr{A}/\tilde{J}_{\varepsilon,w_1,w_2}[u_1^{\pm 1}, \ldots, u_n^{\pm 1}].$$

Putting this together we obtain an isomorphism of twisted Laurent polynomial algebras between $D[\gamma_\varepsilon(z_{w_1}\zeta_{w_2})^{-1}][u_1^{\pm 1}, \ldots, u_n^{\pm 1}]$ and $\mathscr{A}/\tilde{J}_{\varepsilon,w_1,w_2}[\gamma_\varepsilon(z_{w_1}\zeta_{w_2})^{-1}][u_1^{\pm 1}, \ldots, u_n^{\pm 1}]$ which coincides with π on D. This clearly implies our proposition.

4.10.

Proposition. *The algebra $F_\varepsilon[G]/\sqrt{Y_{\varepsilon,w_1,w_2}}[z_{w_1}^{-1}, \zeta_{w_2}^{-1}]$ is an Azumaya algebra of degree $\ell^{1/2(\ell(w_1)+\ell(w_2)+rk(w_1-w_2))}$ over its center.*

Proof. By Proposition 4.4 our algebra is an algebra over the coordinate ring of X_{w_1,w_2}. Moreover by Theorem 1.6 the semisimple parts of the algebras obtained by evaluation of C at the points of X_{w_1,w_2} are all isomorphic.

On the other hand if we invert z_ε we obtain, by 4.8 the algebra $\mathscr{A}/\tilde{J}_{\varepsilon,w_1,w_2}$ which by 4.7 is a domain. Since a domain is generically an Azumaya algebra (of degree equal to the degree of the domain), it follows that all irreducible representations of C have the same dimension and hence, since by construction C is semiprime it follows by Artin's theorem and the Nullstellensatz (cf. [22]) that C is Azumaya (cf 1.8).

It remains to determine its degree. Let $\mathbb{Z}' := \mathbb{Z}[\frac{1}{2d}]$ where d is the product of the bad primes for our root system. For $w \in W$ let $\Phi_w = \{\beta_1, \beta_2, \ldots, \beta_k\}$ be ordered by the convex ordering associated to a reduced expression $w = s_{i_1} s_{i_2} \cdots s_{i_k}$. Consider the free $\mathbb{Z}'$ module V_w with basis $u_1, u_2, \ldots u_k$ and define an antisymmetric bilinear form on V_w by setting

$$a_{i,j} = \langle u_i . u_j \rangle = (\beta_i | \beta_j) \text{ if } i < j.$$

We identify V_w with its dual V_w^* using the given basis. Then the matrix $A_w = (a_{i,j})$ can be thought, as a linear operator from V_w to itself. Also we can consider the matrix $C_w = ((\omega_i | \beta_j))$ as the matrix of a linear map from the module V_w with the basis $u_1, \ldots, u_k$ to the module $Q^{\vee'} = Q^\vee \otimes_\mathbb{Z} \mathbb{Z}'$ with the basis $\alpha_1^\vee, \ldots, \alpha_n^\vee$.

Take now the following antisymmetric matrix

$$S = \begin{pmatrix} A_{w_1} & 0 & -{}^t C_{w_1} \\ 0 & -A_{w_2} & {}^t C_{w_2} \\ C_{w_1} & -C_{w_2} & 0 \end{pmatrix},$$

We shall consider S as a linear map $V_{w_1} \oplus V_{w_2} \oplus P' \to V_{w_1} \oplus V_{w_2} \oplus Q'$ where $P' = P \otimes_\mathbb{Z} \mathbb{Z}'$.

By 4.9 the degree of $F_\epsilon[G]/\sqrt{Y_{\epsilon,w_1,w_2}}[z_{w_1}^{-1}, \zeta_{w_2}^{-1}]$ equals the degree of the algebra D considered in 4.9 which by [6] (see also [9]) equals the square root of the cardinality of the image of S composed with reduction mod ℓ. Thus in order to show our claim it suffices to see that $rk\, S = \ell(w_1) + \ell(w_2) + rk(w_1 - w_2)$ and that the image of S is a direct summand.

To see this let us recall that in [7] it is proven that the matrix $\begin{pmatrix} A_w & -{}^t C_w \end{pmatrix}$ gives a surjective linear map whose kernel is the $\mathbb{Z}'$ module spanned by the elements $(v_{\omega_i}, (1+w)\omega_i)$, with $v_{\omega_i} = \sum_{r|i_r=i} u_r$ which can be identified with P'. Furthermore under this identification the restriction of C_w to the above kernel is just $1 - w$ whose image is a direct summand.

It follows that we can identify the kernel of $\begin{pmatrix} A_{w_1} & 0 & -{}^t C_{w_1} \end{pmatrix}$ with $V_{w_2} \oplus P'$. The restriction of the linear map associated to the matrix

$$\begin{pmatrix} 0 & -A_{w_2} & {}^t C_{w_2} \\ C_{w_1} & -C_{w_2} & 0 \end{pmatrix}$$

to the kernel of $\begin{pmatrix} A_{w_1} & 0 & -{}^t C_{w_1} \end{pmatrix}$ under the previous identification is a linear map $V_{w_2} \oplus P' \to V_{w_2} \oplus Q^{\vee'}$ given by the matrix.

$$T = \begin{pmatrix} -A_{w_2} & {}^t C_{w_2}(1 + w_1) \\ -C_{w_2} & 1 - w_1 \end{pmatrix}$$

We need to show that the image of T is a direct summand and its rank equals $l(w_2) + rk(w_1 - w_2)$. This can as well be proven for

$$^tT = \begin{pmatrix} A_{w_2} & -^tC_{w_2} \\ (1 + w_1^{-1})C_{w_2} & 1 - w_1^{-1} \end{pmatrix}.$$

Now we know, again by the above result applied this time for w_2, that $\left(A_{w_2} \ -^tC_{w_2} \right)$ is surjective, its kernel can be identified with P' and the restriction of the linear map given by $\left((1 + w_1^{-1})C_{w_2} \ 1 - w_1^{-1} \right)$ to this kernel is given by $(1 + w_1^{-1})(1 - w_2) + (1 - w_1^{-1})(1 + w_2) = 2(1 - w_1^{-1}w_2)$. Since 2 is invertible in $\mathbb{Z}'$ we deduce that its image is a direct summand whose rank equals $rk(w_1 - w_2)$ as desired.

Remark. The computation of the degree given above is contained in [7] in the case in which $w_2 = id$ and in fact we use that proof heavily in the above argument. The general case is claimed in [9] but the proof given there is incomplete giving only the computation of the rank of the matrix S. We are grateful to M. Costantini for pointing out the gap in [9] and providing us with the above argument.

We are now ready to prove the main result of our paper.

Theorem. (1) *The algebras* $F_\varepsilon[G]/\sqrt{Y_{\varepsilon,w_1,w_2}}[z_{w_1}^{-1}, \zeta_{w_2}^{-1}]$ *and* $F_\varepsilon[G]/J_{\varepsilon,w_1,w_2}[z_{w_1}^{-1}, \zeta_{w_2}^{-1}]$ *are isomorphic.*
(2) *The spectrum of* $F_\varepsilon[G]/J_{\varepsilon,w_1,w_2}[z_{w_1}^{-1}, \zeta_{w_2}^{-1}]$ *is a Galois covering of* X_{w_1,w_2} *with Galois group the group* $T_{w_2^{-1}w_1,\ell}$ *(see 4.5 for the definition).*

Proof. Since by Lemma 4.7 the algebra $F_\varepsilon[G]/J_{\varepsilon,w_1,w_2}[z_{w_1}^{-1}, \zeta_{w_2}^{-1}]$ is a domain, it is a quotient of $F_\varepsilon[G]/\sqrt{Y_{\varepsilon,w_1,w_2}}[z_{w_1}^{-1}, \zeta_{w_2}^{-1}]$ so in particular it is an Azumaya algebra. Thus the spectrum of these algebras equals the spectrum of their centers and to show their isomorphism it suffices to show that the spectra of their centers are isomorphic as algebraic varieties. Call these spectra S_1 and S_2 respectively. Both S_1 and S_2 project to X_{w_1,w_2} and we have a commutative diagram

$$\begin{array}{ccc} S_2 & \xrightarrow{\ i\ } & S_1 \\ {\scriptstyle p}\downarrow & & \downarrow{\scriptstyle r} \\ X_{w_1,w_2} & \xrightarrow{\ id\ } & X_{w_1,w_2}. \end{array}$$

i injects S_2 as an irreducible Zariski closed subvariety of S_1. The maps p and r are finite, and the cardinality of the fibers of r is constant.

Furthermore, since after inverting z_e our two algebras become isomorphic, i is an isomorphism outside the divisor defined by the element Z_e^ℓ. It follows from these considerations that, if we prove that also the cardinality of the fibers of p is constant we shall get the isomorphism of S_2 and S_1.

Notice now that using 4.9 we have that the rank of $F_\varepsilon[G]/J_{\varepsilon,w_1,w_2}[z_{w_1}^{-1}, \zeta_{w_2}^{-1}]$ as a module over $\mathbb{C}[X_{w_1,w_2}] = Z_0/Y_{w_1,w_2}[z_{w_1}^{-\ell}, \zeta_{w_2}^{-\ell}]$ equals the rank of the algebra D as a module over the subalgebra generated by the elements $\overline{F}_{\beta_i}^\ell \otimes 1$, $1 \otimes \overline{E}_{\delta_j}^\ell$, $K_\lambda^\ell \otimes K_{-\lambda}^\ell$

where $\beta \in \Phi_{w_1}$ (resp $\delta \in \Phi_{w_2}$) so it equals $\ell^{\ell(w_1)+\ell(w_2)+n}$. It follows that the map p has degree $\ell^{n-rk(w_1-w_2)} = \ell^{|T_{w_2^{-1}w_1}\ell|}$.

Now, for $\lambda \in P^{w_1^{-1}w_2}$ consider the elements

$$d_\lambda := d_{\lambda,w_1,w_2} := z_{w_1}^\lambda \zeta_{w_2}^{w_1(\lambda)}$$

One checks exactly as in 4.5 that these elements are central in $F_\varepsilon[G]/J_{\varepsilon,w_1,w_2}[z_{w_1}^{-1}, \zeta_{w_2}^{-1}]$. Also, we can repeat word by word the argument in 4.5 and show that the center of $F_\varepsilon[G]/J_{\varepsilon,w_1,w_2}[z_{w_1}^{-1}, \zeta_{w_2}^{-1}]$ is generated by $\mathbb{C}[X_{w_1,w_2}]$ and these elements and that S_2 is a Galois covering of X_{w_1,w_2} with Galois group $T_{w_2^{-1}w_1,\ell}$. This in particular implies that the cardinality of the fibers of p is constant and completely proves the theorem.

Definition. We shall denote by $F_\varepsilon[G]_{w_1,w_2}$ the algebra

$$F_\varepsilon[G]/\sqrt{Y_{\varepsilon,w_1,w_2}}[z_{w_1}^{-1}, \zeta_{w_2}^{-1}] \cong F_\varepsilon[G]/J_{\varepsilon,w_1,w_2}[z_{w_1}^{-1}, \zeta_{w_2}^{-1}].$$

From the previous proof it follows that, if we denote by Z_{1,w_1,w_2} the algebra generated by the elements d_λ, $\lambda \in P^{w_1^{-1}w_2}$, the intersection $\mathbb{C}[X_{w_1,w_2}] \cap Z_{1,w_1.w_2}$ equals the algebra generated by the elements $d_{\ell\lambda}$, $\lambda \in P^{w_1^{-1}w_2}$. The center of $F_\varepsilon[G]_{w_1,w_2}$ is

$$\mathbb{C}[X_{w_1,w_2}] \otimes_{\mathbb{C}[X_{w_1,w_2}] \cap Z_{1,w_1,w_2}} Z_{1,w_1,w_2}.$$

Let us denote by S_{w_1,w_2} its spectrum. By construction we have a cartesian square

$$\begin{array}{ccc} S_{w_1,w_2} & \xrightarrow{\kappa} & T_{w_2^{-1}w_1} \\ {\scriptstyle p}\big\downarrow & & \big\downarrow{\scriptstyle \ell\text{-th power}} \\ X_{w_1,w_2} & \xrightarrow{\tilde\kappa} & T_{w_2^{-1}w_1} \end{array} \qquad (4.10.1)$$

where all the maps in the diagram are induced by inclusion.

Remark. The torus T acts on each of the varieties in the above commutative diagram. We have $\kappa(tx) = t^2\kappa(x)$ and similarly for $\tilde\kappa$. The reader may wish to substitute $T_{w_2^{-1}w_1}$ with its quotient module elements of 2-torsion to get that they are equivariant maps.

4.11. We want to summarize the results of the previous section with more appealing notations. Let us denote the spectrum of $F_\varepsilon[G]$, i.e. the set of equivalence classes of irreducible representations by G_ε and by $K_\varepsilon(G)$ the Grothendieck ring of finite dimensional representations of $F_\varepsilon[G]$, which has as basis the set G_ε. The Frobenius map induces a projection map (central character) and a ring homomorphism:

$$\mathscr{F}^* : G_\varepsilon \to G; \ K_\varepsilon(G) \to K(G).$$

We have then compatible decompositions $G = \cup_{w_1,w_2} X_{w_1,w_2}$, $G_\varepsilon = \cup S_{w_1,w_2}$ so that the diagram:

$$
\begin{array}{ccc}
S_{w_1,w_2} & \xrightarrow{\;i\;} & G_\varepsilon \\
{\scriptstyle p}\downarrow & & \downarrow{\scriptstyle \mathscr{F}} \bullet \\
X_{w_1,w_2} & \xrightarrow{\;i\;} & G
\end{array}
\tag{4.11.1}
$$

commutes (i denotes inclusion).

5. A SEMISIMPLICITY PROPERTY AND THE CONSTRUCTION OF IRREDUCIBLE MODULES.

5.1. In this last section we will discuss the problem of construction of irreducible representations from basic ones. So first of all let us discuss the part $S_{s_i,e}$ of the spectrum with s_i a simple reflection (a similar discussion applies to S_{e,s_i}).

We have seen in 2.4 that for a simple reflection the ideal J_{s_i} is generated by the elements F_j, $j \neq i$. The algebra $F_q[B^+]/J_{s_i}$ is isomorphic to the subalgebra of $U_q^s(\mathfrak{b}^-)$ generated by F_i, K_λ; $\lambda \in P$. On the other hand the computation of 2.3 and the analogue of formula 3.2.1 imply that the image z_i of z_{s_i} in $F_\varepsilon[B^+]$ is $-q^{2d_i}\overline{F}_i K_\varrho$ it follows that the algebra $F_\varepsilon[G]_{s_i,e}$ is canonically isomorphic to the twisted Laurent polynomials $\mathbb{C}[P][F_i, F_i^{-1}]$ where $\mathbb{C}[P]$ is the group ring of the weight lattice and the automorphism is of course induced by the commutation relations $F_i K_\lambda F_i^{-1} = \varepsilon^{(\alpha_i,\lambda)} K_\lambda$. In other words this is the ordinary Laurent polynomial ring in the variables K_{ω_j}, $j \neq i$ over the quantum torus $\mathbb{C}[F_i^{\pm 1}, K_{\omega_i}^{\pm 1}]$ with commutation relation $F_i K_{\omega_i} = \varepsilon^{d_i} K_{\omega_i} F_i$ and $\eta = \varepsilon^{d_i}$ is also a primitive ℓ root of 1 since d_i is prime with ℓ by assumption.

Now the irreducible representations of a quantum torus $\mathbb{C}[x^{\pm 1}, y^{\pm 1}]$, $xy = \eta yx$, which is an Azumaya algebra of rank ℓ, are well known and easy to describe. One fixes two non zero numbers α, β and a vector space with basis vectors $v_1, \dots, v_\ell$ and sets $xv_i = \eta^i \alpha v_i$, $yv_i = \beta v_{i+1}$ ($\ell + 1 = 1$). Two pairs α, β, α', β' give rise to isomorphic representations if and only if $\alpha = \varepsilon_1 \alpha'$, $\beta = \varepsilon_2 \beta'$ with $\varepsilon_1, \varepsilon_2$ both ℓ roots of 1. The spectrum of this quantum torus is thus the torus of coordinates α^ℓ, β^ℓ which are values of the central character on the two generators x^ℓ, y^ℓ of the center of $\mathbb{C}[x^{\pm 1}, y^{\pm 1}]$.

5.2. We start working at generic q. We fix two elements $w_1, w_2 \in W$ and we define $J_{w_1,w_2} \subset F_q[G]$ to be the ideal $\gamma^{-1}(J_{w_1} \otimes F_q[B^-] + F_q[B^-] \otimes J_{w_2})$. Take now a simple reflection s_i such that $\ell(s_i w_1) > \ell(w_1)$. Consider the homomorphism

$$
\psi : F_q[G] \to F_q[G]/J_{s_i,e} \otimes F_q[G]/J_{w_1,w_2}
$$

defined as the composition $(f \otimes g) \cdot \Delta$ where $f : F_q[G] \to F_q[G]/J_{s_i,e}$ (resp. $g : F_q[G] \to F_q[G]/J_{w_1,w_2}$) is the quotient homomorphism.

Lemma. *Let $\lambda \in P^+$. Then:*

(1) $\psi(z_{s_i w_1}^\lambda) = z_{s_i}^{w_1(\lambda)} \otimes z_{w_1}^\lambda$.

(2) $\psi(\zeta_{w_2}^\lambda) = \zeta_e^\lambda \otimes \zeta_{w_2}^\lambda$.

Proof. (1) Consider in $U_q(\mathfrak{g})$ the subalgebra $\mathscr{L}_i$ generated by E_1, F_i and by the $K_j^{\pm 1}$'s with $j = 1, \ldots, n$. It follows from Proposition 2.4 and the definitions that the ideal J_{s_i,s_i} is the kernel of the restriction of elements of $U_q(\mathfrak{g})$ to $\mathscr{L}_i$. Since $J_{s_i,e} \supset J_{s_i,s_i}$, it suffices to show that $\psi'(z_{s_i w_1}^\lambda) = z_{s_i}^{w_1(\lambda)} \otimes z_{w_1}^\lambda$ where

$$\psi' : F_q[G] \to F_q[G]/J_{s_i,s_i} \otimes F_q[G]/J_{w_1,w_2}$$

defined as the composition $(f' \otimes g) \cdot \Delta$, $f' : F_q[G] \to F_q[G]/J_{s_i,s_i}$ being the quotient homomorphism.

For this restrict the representation $V_{-w_0(\lambda)}$ to the subalgebra $\mathscr{L}_i$. It is then easy to see that $E_i v_{-s_i w_1(\lambda)} = 0$ and $K_j v_{-s_i w_1(\lambda)} = q^{-(\alpha_j, s_i w_1(\lambda))} v_{-s_i w_1(\lambda)}$. Set $m = -\frac{(\alpha_i, s_i w_1(\lambda))}{d_i} = \frac{(\alpha_i, s_i w_1(\lambda))}{d_i} \geq 0$. We get that $v_{-s_i w_1(\lambda)}$ is a highest weight vector for a $m+1$ dimensional irreducible $\mathscr{L}_i$-module for which $v_{-w_1(\lambda)}$ is a lowest weight vector. Take the basis $v_0, \ldots, v_m$ with $v_j = F^j v_{-s_i w_1(\lambda)}$ for this module and complete it to a basis $v_0, \ldots, v_M$ for $V_{-w_0(\lambda)}$, in such a way that the remaining vectors also span a $\mathscr{L}_i$ submodule, we clearly have that the matrix coefficient $c_{\phi^{s_i w_1(\lambda)}, v_j} \in J_{s_i,e}$ unless $j \leq m$. Consider now the dual basis $\phi_0, \ldots, \phi_M$ to $v_0, \ldots, v_M$ and notice that, up to scalars ϕ_m, v_m are multiples of $\phi^{w_1(\lambda)}, v_{-w_1(\lambda)}$ so we renormalize them and assume $\phi_m = \phi^{w_1(\lambda)}$. Also if $j < m$, then $c_{\phi_j, v_{-\lambda}} \in J_{w_1,w_2}$. Since $\Delta(c_{\phi^{s_i w_1(\lambda)}, v_{-\lambda}}) = \sum_{j=1}^M c_{\phi^{s_i w_1(\lambda)}, v_j} \otimes c_{\phi_j, v_{-\lambda}}$ it follows that

$$\psi'(z_{s_i w_1}^\lambda) = \psi'(c_{\phi^{s_i w_1(\lambda)}, v_m} \otimes c_{\phi_m, v_{-\lambda}}) = c_{\phi^{s_i w_1(\lambda)}, v_m} \otimes z_{w_1}^\lambda$$

It follows that in order to show our claim we need to show that $f'(c_{\phi^{s_i w_1(\lambda)}, v_m}) = z_{s_i}^{w_1(\lambda)}$ (notice that this makes sense since if $\mu \in P^+$ is orthogonal to α_i, $z_{s_i}^\mu = z_e^\mu$ is invertible in $F_q[G]/J_{s_i,s_i}$). It is clear from weight considerations that $f'(c_{\phi^{s_i w_1(\lambda)}, v_m}) = a z_{s_i}^{w_1(\lambda)}$ for some non zero scalar a, we only need to show that this scalar is 1. Now f' is t_i equivariant so that we get $f'(c_{\phi^{s_i w_1(\lambda)}, v_m}) = (t_i, e)(f'(c_{\phi^{w_1(\lambda)}, v_m})$. Since $\phi^{s_i w_1(\lambda)}(v_m) = 1$, we obtain that $f'(c_{\phi^{w_1(\lambda)}, v_m}) = z_e^{w_1(\lambda)}$ (they have the same value on the identity element). This clearly implies our claim.

The easy part (2) is left to the reader.

We now specialize to ε, our ℓ root of 1. Notice that, since ψ preserves the integral forms, it induces a map

$$\psi_\varepsilon : F_\varepsilon[G] \to F_\varepsilon[G]/J_{\varepsilon,s_i,e} \otimes F_\varepsilon[G]/J_{\varepsilon,w_1,w_2}$$

and the conclusion of our lemma holds after specialization.

For $q = 1$ we get that this map is induced by the multiplication in G. By the Bruhat decomposition the product gives us a morphism:

$$X_{s_i,e} \times X_{w_1,w_2} \to X_{s_i w_1, w_2} \tag{5.2.1}$$

and $X_{s_i,e} X_{w_1,w_2}$ is a dense subset in $X_{s_i w_1, w_2}$. It follows that the map ψ_1 factors through $\mathbb{C}[G]/Y_{s_i w_1, w_2}$ and localizes to give a map

$$\mathbb{C}[G]/Y_{s_i w_1, w_2}[z_{s_i w_1}^{-1}, \zeta_{w_2}^{-1}] \to \mathbb{C}[G]/Y_{s_i,e}[z_{s_i}^{-1}] \otimes \mathbb{C}[G]/Y_{w_1, w_2}[z_{w_1}^{-1}, \zeta_{w_2}^{-1}].$$

Using the compatibility of the Frobenius homomorphism (cf 1.6) and Proposition 1.4, we have that the map ψ_ε factors first through the ideal $Y_{\varepsilon, s_i w_1. w_2}$, and also through its radical since the algebra $F_\varepsilon[G]/J_{\varepsilon,s_i,e} \otimes F_\varepsilon[G]/J_{\varepsilon,w_1,w_2}$ is contained in the tensor product of twisted Laurent polynomial rings and so it is a domain.

By Theorem 4.10, we can then localize getting a map

$$F_\varepsilon[G]/J_{\varepsilon,s_iw_1,w_2}[z_{s_iw_1}^{-1},\zeta_{w_2}^{-1}] \to F_\varepsilon[G]/J_{\varepsilon,s_i,e}[z_{s_i}^{-1}] \otimes F_\varepsilon[G]/J_{\varepsilon,w_1,w_2}[z_{w_1}^{-1},\zeta_{w_2}^{-1}]$$

which by the way it has been constructed is injective.

5.3. We will need a simple geometrical fact. If v is a non zero vector in a Euclidean space V let us denote by r_v the orthogonal reflection relative to the hyperplane orthogonal to v. The map $1 - r_v$ has image $\mathbb{R}v$. Let $v_1, v_2, \ldots, v_k \in V$ be non zero vectors.

Lemma. *The image of $1 - r_{v_1}r_{v_2}\ldots r_{v_k}$ is contained in the span of $v_1, \ldots, v_k$ its kernel contains the orthogonal to the span of $v_1, \ldots, v_k$.*

Its rank is k iff the vectors are independent, in this case the kernel equals the orthogonal of the span of $v_1, \ldots, v_k$.

Proof. It is a simple induction from the formula $1 - r_{v_1}r_{v_2}\ldots r_{v_k} = 1 - r_{v_2}\ldots r_{v_k} + (1 - r_{v_1})r_{v_2}\ldots r_{v_k}$.

Let $A = \mathbb{Z}[d^{-1}]$ where d is the product of the bad primes, then under the canonical scalar product P_A, Q_A are in perfect duality.

We need two facts from the theory of root systems. First, if w is an element of the Weyl group we can write $w = s_{\beta_1}s_{\beta_2}\ldots s_{\beta_k}$ for independent roots β_i (so that from the previous lemma P^w is the orthogonal to these roots).

The second fact is that any maximal independent sets of roots spans a lattice in Q which intersects the root system in a new root system. Choose a basis for this new root system, then this basis can be obtained from the basis Δ of Q by a sequence of steps of the following type. Replace in Δ one of the simple roots with the negative of the longest root take the root system with this set as basis and continue. From the definition of bad primes and this fact it follows that any maximal set of linearly independent roots is a basis over A of Q_A

We need some degree computations. Notice first that $w \in W$ and s a simple reflection, $rk(sw - e) = rk((s - e)w + w - e)$ so

$$rk(w - e) - 1 \le rk(sw - e) \le rk(w - e) + 1.$$

Since $\ell(sw) + rk(sw - e)$, $\ell(w) + rk(w - e)$ are both even it follows that $rk(sw - e) = rk(w - e) \pm 1$.

Changing w with sw if necessary we are reduced to discuss the following case.

Proposition. *Let α be the simple root associated to s. If $rk(sw - e) = rk(w - e) - 1$ then:*

$$P^w = \{x \in P^{sw}|(\alpha, x) = 0\}, \quad \alpha(P^{sw}) + \ell\mathbb{Z} = \mathbb{Z}.$$

Proof. Write $s_iw = s_{\beta_1}s_{\beta_2}\ldots s_{\beta_k}$ for independent roots β_i. From the previous lemma applied to $w = s_is_iw$ we have that α_i is independent from the β_j, P^w is the orthogonal to the roots $\alpha_i, \beta_1, \beta_2, \ldots, \beta_k$ while P^{sw} is the orthogonal to the roots $\beta_1, \beta_2, \ldots, \beta_k$. The elements $\alpha_i, \beta_1, \beta_2, \ldots, \beta_k$ are part of an A basis so the functions they induce on P_A by scalar product are part of a system of linear coordinates and this clearly proves the proposition.

Take w_1, w_2, s_i with $\ell(s_i w_1) = \ell(w_1) + 1$. We have two cases (notice that $P^{w_1^{-1} w_2} = w_2^{-1} P^{w_1 w_2^{-1}}$).

Case 1) $rk(s_i w_1 - w_2) = rk(w_1 - w_2) + 1$. In this case $P^{w_1^{-1} w_2} \cap w_2^{-1} P^{s_i} = P^{(s_i w_1)^{-1} w_2}$. This implies that we have a surjective homomorphism:

$$T_{w_1^{-1} w_2} \times T_{s_i} \to T_{(s_i w_1)^{-1} w_2}.$$

Using the cartesian square 4.10.1 and the map 5.2.1 we obtain a morphism:

$$j : S_{s_i, e} \times S_{w_1, w_2} \to S_{s_i w_1, w_2}. \tag{5.3.1}$$

Take $(p, q) \in S_{s_i, e} \times S_{w_1, w_2}$ and V_p, V_q the corresponding irreducible representations of $F_\varepsilon[G]$. By Lemma 5.2 $V_p \otimes V_q$ has central character $p \cdot q := j(p, q)$. This point corresponds to an irreducible representation of the same dimension as $V_p \otimes V_q$ hence $V_p \otimes V_q = V_{p \cdot q}$.

Case 2) $rk(s_i w_1 - w_2) = rk(w_1 - w_2) - 1$. In this case $P^{(s_i w_1)^{-1} w_2} \cap w_2^{-1} P^{s_i} = P^{w_1^{-1} w_2}$. Consider the lattice $P^{w_1^{-1} w_2} + \ell P^{(s_i w_1)^{-1} w_2}$ which is the kernel of the surjective map $\alpha_i : P^{(s_i w_1)^{-1} w_2} \to \mathbb{Z}/(\ell)$. To this map corresponds a covering $\nu : S_{s_i w_1, w_2} \to \tilde{S}_{s_i w_1, w_2}$.

Reasoning as above we obtain a morphism:

$$j : S_{s_i, e} \times S_{w_1, w_2} \to \tilde{S}_{s_i w_1, w_2}. \tag{5.3.2}$$

Take $(p, q) \in S_{s_i, e} \times S_{w_1, w_2}$ and V_p, V_q the corresponding irreducible representations of $F_\varepsilon[G]$. By Lemma 5.2 the coordinate ring of $\tilde{S}_{s_i w_1, w_2}$ acts on $V_p \otimes V_q$ by the central character $p \cdot q := j(p, q)$. Now this point corresponds to ℓ irreducible representations of dimension $dim(V_p \otimes V_q)/\ell$ in the fiber of the covering map ν.

Claim. $V_p \otimes V_q = V_{p \cdot q}$ decomposes as the direct sum of all these representations.

In order to prove it we choose a character $\lambda \in P^{(s_i w_1)^{-1} w_2}$ with $(\alpha, \lambda) = 1 \bmod(\ell)$. The function $d_\lambda = d_{\lambda, s_i w_1, w_2}$ takes ℓ different values on the fiber $\nu^{-1}(u)$ of a point $u \in \tilde{S}_{s_i w_1, w_2}$. Suppose we prove that d_λ has at least ℓ distinct eigenvalues on $V_p \otimes V_q$ then our claim is proved, in fact, given such an eigenvalue of d_λ, its eigenspace is an $F_\varepsilon[G]$ subrepresentation of $V_p \otimes V_q$ which factors as an $F_\varepsilon[G]_{s_i w_1, w_2}$ module with central character the point $r \in \nu^{-1}(p \cdot q)$ determined by the chosen eigenvalue for d_λ. From 1.8 we know that this eigenspace decomposes as a sum of irreducible representations isomorphic to V_r. Since $\ell \, dim(V_r) = dim(V_p \otimes V_q)$ the claim follows.

It remains to perform the computation of the eigenvalues. It is clearly enough to prove that on V_p the operator $z_{s_i}^{w_1(\lambda)} \zeta^\lambda{}_e$ has ℓ distinct eivenvalues. Recall that on V_p this operator acts (up to a non zero constant) as $F_i K_\lambda^2$ hence the claim follows from the remarks in 5.1 ($F_i K_\lambda^2, K_{\omega_i}$ can be taken as the two generators x, y of a quantum torus).

Summarizing take w_1, w_2, s_i with $\ell(s_i w_1) = \ell(w_1) + 1$. We have proved the following theorem.

Theorem. *We have two cases:*

(1) $rk(s_i w_1 - w_2) = rk(w_1 - w_2) + 1$. *We obtain a morphism* $j : S_{s_i,e} \times S_{w_1,w_2} \to S_{s_i w_1, w_2}$. *If* $(p,q) \in S_{s_i,e} \times S_{w_1,w_2}$ *and* V_p, V_q *the corresponding irreducible representations of* $F_e[G]$. $V_p \otimes V_q$ *is isomorphic to the irreducible representation* $V_{j(p,q)}$.

(2) $rk(s_i w_1 - w_2) = rk(w_1 - w_2) - 1$. *We obtain morphisms* $j : S_{s_i,e} \times S_{w_1,w_2} \to \tilde{S}_{s_i w_1, w_2}$; $\nu : S_{s_i w_1, w_2} \to \tilde{S}_{s_i w_1, w_2}$. $V_p \otimes V_q = V_{p \cdot q}$ *decomposes as the direct sum of all the irreducible representations in the fiber* $\nu^{-1}(j(p,q))$.

Similarly for pairs (w_1, w_2), $(w_1, w_2 s_i)$ *with* s_i *a simple reflection such that* $\ell(w_2 s_i) = \ell(w_2) + 1$.

Remarks. (a) Under the previous hypotheses one can easily check that the multiplication map (5.2.1) is not in general surjective. The complement of its image appears to be a remarkable set which should be analyzed.

(b) Set for a pair (w_1, w_2), $X^0_{w_1, w_2}$ equal to the subset in X_{w_1, w_2} consisting of elements which lie in the image of the iterated multiplication map

$$\prod_{t=1}^{h} X_{s_{i_t}, e} \times \prod_{r=1}^{k} X_{e, s_{j_r}} \to X_{w_1, w_2},$$

for some reduced expressions $w_1 = s_{i_1} \cdots s_{i_h}$, $w_2 = s_{j_1} \cdots s_{j_k}$. Then an easy induction based on the previous theorem shows that, if we take two pairs (w_1, w_2) and (w'_1, w'_2) such that $\ell(w_1 w'_1) = \ell(w_1) + \ell(w'_1)$ and $\ell(w_2 w'_2) = \ell(w_2) + \ell(w'_2)$ and two elemnts $g \in X^0_{w_1, w_2}$ and $g' \in X^0_{w'_1, w'_2}$, the tensor product of two irreducible $F_e[G]$-modules, one lying above g and the other above g' is always semisimple.

REFERENCES

1. H. H. Andersen, J. C. Jantzen, and W. Soergel, *Representations of Quantum groups at a p-th root of unity and of semisimple groups in characteristic p: independence of p*, preprint, 1992.
2. M. Artin, *On Azumaya algebras and finite–dimensional representations of rings*, J. Algebra **11** (1969), 532–563.
3. M. Auslander and O. Goldman, *The Brauer Group of a Commutative Ring*, Trans. A.M.S. **97** (1960), 367–409.
4. ———, *Maximal Orders*, Trans. A.M.S. **97** (1960), 1–24.
5. N. Bourbaki, Groupes et Algebre de Lie, ch. 4,5,6, Hermann, Paris, 1980.
6. C. De Concini, V. G. Kac, and C. Procesi, *Quantum coadjoint action*, Journal of AMS **5** (1992), 151–190.
7. ———, *Some quantum analogues of solvable groups*, preprint, 1992.
8. ———, *Some remarkable degenerations of quantum groups*, Comm. Math. Phys. (1993).
9. C. De Concini and V. Lyubashenko, *Quantum coordinate ring at roots of unity*, Adv. in Math., (to appear).
10. C. De Concini and C. Procesi, *Quantum groups*, Springer LNM **1565**, 31–140.
11. M. Demazure, *Désingularisation des variétés de Schubert généralisées*, Ann. scient. Éc. Norm. Sup., 4e série, t. 7 (1974), 53–88.
12. V. G. Drinfeld, *Quantum groups*, Proc. ICM Berkeley 1 (1986), 789–820.
13. T. Hodges and T. Levasseur, *Primitive Ideals of* $\mathbb{C}_q[G]$, preprint, 1993.
14. M. Jimbo, *A q-difference analogue of* $U(q)$ *and the Yang-Baxter equation*, Lett. Math. Phys. **10** (1985), 63–69.

15. A. Joseph, *On the Prime and Primitive spectra of the Algebra of Functions on a Quantum Group*, preprint, 1992.
16. V. Lakshmibai and N. Reshetikhin, *Quantum deformations of Flag and Schubert schemes.*
17. S. Z. Levendorskii and Ya. S. Soibelman, *Algebras of functions on compact quantum groups, Schubert cells and quantum tori*, Comm. Math. Physics **139** (1991), 141–170.
18. _____, *Quantum Weyl group and multiplicative formula for the R-matrix of a simple Lie algebra*, Funct. Analysis and its Appl. **25** (1991), no. 2, 143–145.
19. G. Lusztig, *Quantum deformations of certain simple modules over enveloping algebras*, Adv. in Math. **70** (1988), 237–249.
20. _____, *Quantum groups at roots of 1*, Geom. Ded. **35** (1990), 89–114.
21. M. Matsumoto, *Generateurs et relations de groupes de Weyl generalises*, C.R. Acad. Sci. Paris **258** (1964), 3419–3422.
22. C. Procesi, *Rings with polynomial identities*, Pure and Applied Mathematics, no. 17, M. Dekker, 1973.
23. Reshetikhin, Takhtadzyan, and Faddeev, *Quantization of Lie groups and Lie algebras*, Leningrad Math. J. **1** (1990), 193–323.
24. C. Ringel, *Hall algebras and quantum groups*, Inv. Math. **101** (1990), 583–592.
25. M. Rosso, *Finite dimensional representations of the quantum analogue of the enveloping algebra of a semisimple Lie algebra*, Comm. Math. Phys. **117** (1988), 581–593.
26. W. Schelter, *Azumaya algebras and Artin's theorem*, Jour. of Algebra **46** (1977), 303–304.
27. Ya. S. Soibelman, *The algebra of functions on a compact quantum group, and its representations*, Leningrad Math. J. **2** (1991), 161–178.
28. _____, *Quantum Weyl group and some of its applications*, Rend. Circolo Mat. Pal. s. II **26** (1991), 232–235.
29. Ya. S. Soibelman and L. Vaksman, *Algebra of functions on quantum group SU(n + 1) and odd dimensional quantum spheres*, Algebra i analiz **2** (1990), no. 5.
30. T. Tanisaki, *Killing forms, Harish-Chandra isomorphisms, and universal r-matrices for quantum algebras*, Infinite Analysis Part A, Adv. Series in Math Phys. **16** (1992), 942–962.
31. J. Tits, *Sur les constants de structure et le théorème d'existence des algebres de Lie semi-simple*, Publ. Math. IHES **31** (1966), 21–58.

SCUOLA NORMALE SUPERIORE.
E-mail address: deconcin@ux1sns.sns.it

UNIVERSITÀ DI ROMA "LA SAPIENZA".
E-mail address: Procesi@sci.uniroma1.it

PURITY AND EQUIVARIANT WEIGHT POLYNOMIALS

A. DIMCA AND G. I. LEHRER

Dedicated to the memory of Roger W. Richardson

INTRODUCTION

Suppose Γ is a finite group of homeomorphisms of a topological space X which in this work will usually be a complex algebraic variety. The fundamental problem which motivates this work is the determination of the linear representations of Γ on the cohomology spaces $H_c^i(X, \mathbb{C})$ (and $H^i(X, \mathbb{C})$), which are assumed to be finite dimensional. The term "determine" above means that we seek expressions for the character of the cohomology representations in terms of the geometry of X. Examples may be found in [23], [20], [19].

Our approach is as follows. Clearly the problem is equivalent to the determination of the polynomials $P_c^{\Gamma}(g, X, t) := \sum_i \text{trace}(g, H_c^i(X, \mathbb{C}))t^i$ $(g \in \Gamma,$ t an indeterminate), or equivalently of the element $P_c^{\Gamma}(X, t) := \sum_i H_c^i(X, \mathbb{C})t^i$ of $R(\Gamma)[t]$, where $R(\Gamma)$ is the Grothendieck ring of Γ. In order to achieve this, we replace $P_c^{\Gamma}(X, t)$ by a *weight polynomial* $W_c^{\Gamma}(X, t)$ (see (1.5)(ii) below) which is *additive* in the sense that if U is a Γ–invariant open subset of X then $W_c^{\Gamma}(X, t) = W_c^{\Gamma}(U, t) + W_c^{\Gamma}(X \backslash U, t)$. The determination of the weight polynomials is approached by decomposing X appropriately.

Now the weight polynomial $W_c^{\Gamma}(X, t)$ determines the Poincaré polynomial $P_c^{\Gamma}(X, t)$ in case $H_c^*(X, \mathbb{C})$ satisfies a condition which we call "*separable purity*" (see (3.1) below). Thus in §3 we determine some sufficient conditions for X to be separably pure and we define and study a stronger property called minimal purity. In §4 we apply these results to "toral arrangements" i.e. to the case when X is the complement of a finite union of kernels of characters of an algebraic torus. A particular result here is an explicit formula (4.3) for the action of μ_n on the cohomology of the variety of regular elements in a maximal torus of $SL(n, \mathbb{C})$.

There are close connections between this work and corresponding results for varieties over $\mathbb{F}_q$, in which the Hodge filtration is replaced by Deligne's filtration which arises from the action of a Frobenius endomorphism on ℓ–adic cohomology. This makes it possible in some cases to determine the polynomials $W_c^{\Gamma}(X, t)$ by counting $\mathbb{F}_q$–points. We treat this aspect in §5 (cf. also (3.12), (3.13) below and [18], [21]).

In §6 we prove a multiplicative formula for the weight polynomial which applies to Γ–equivariant fibrations. This is applied in the context of Lie groups and their homogeneous spaces.

If ρ, σ are complex representations of the finite group Γ, we denote by $(\rho, \sigma)_{\Gamma}$ their intertwining number, or the inner product of their characters. The same notation is used if ρ and σ are any ring valued class functions on Γ.

1. EULER CHARACTERISTICS AND WEIGHT POLYNOMIALS

Let X be a topological space with finite dimensional compactly supported cohomology groups $H_c^j(X, \mathbb{C})$ for any $j \in \mathbb{N}$ and assume that only finitely many of them non-zero.

If Γ is a finite group acting continuously on X, then each of the above cohomology groups is a Γ-module. Let $R(\Gamma)$ denote the complex representation ring of the group Γ. Then $R(\Gamma)$ may be canonically identified with the character ring G, as in Serre [27].

Definition 1.1. The *equivariant Euler characteristic* of the Γ-space X is the element of $R(\Gamma)$ given by

$$E_c^\Gamma(X) = \sum_j (-1)^j H_c^j(X, \mathbb{C}).$$

When Γ is trivial, we use the notation $E_c(X)$ and identify $R(\Gamma)$ with $\mathbb{Z}$.

Under suitable finiteness conditions, one may similarly define an equivariant Euler characteristic $E^\Gamma(X)$ using the usual cohomology of X. When X is a connected orientable (smooth) n-dimensional manifold, Poincaré duality implies that

$$E^\Gamma(X) = (-1)^n E_c^\Gamma(X) \otimes \varepsilon^\Gamma \tag{1.2}$$

where $\varepsilon^\Gamma(g) = deg(g) \in \{-1, +1\}$, with deg being the degree of the proper mapping $g : X \to X$, $x \mapsto gx$. This formula is also valid for manifolds if we assume either (i) g preserves the given orientation (and in this case $\varepsilon^\Gamma(g) = 1$, see for instance [13, p.195]) or (ii) g preserves the connected components of X and reverses the orientation (an important special case of this is given by complex conjugation in certain complex algebraic varieties).

The information contained in these equivariant Euler characteristics is exactly the same as that given by the knowledge of the Lefschetz numbers $\Lambda(g)$, resp. $\Lambda_c(g)$ for all $g \in \Gamma$. Indeed, we have by definition

$$E^\Gamma(X)(g) = \Lambda(g)$$

and a similar relation holds in the case of compact supports.

In case X is compact, one can use the Lefschetz formula

$$\Lambda(g) = E(X^g) \text{ where } X^g = \{x \in X | gx = x\}$$

(see for instance Wall [30]), to compute these invariants explicitly.

Example 1.3. (Fermat curves and S_3 actions)

Let $X_d : x^d + y^d + z^d = 0$ be the Fermat curve of degree d in the complex projective plane $\mathbb{P}^2(\mathbb{C})$. The symmetric group S_3 acts on X_d by permuting coordinates. Let 1 denote the trivial 1-dimensional representation of S_3, ε denote the sign representation and θ denote the unique 2-dimensional irreducible S_3-representation.

Define two functions $\alpha, \beta : \mathbb{N} \to \mathbb{N}$ by

$$\alpha(d) = 0 \text{ for } d \text{ even}; \quad \alpha(d) = 1 \text{ for } d \text{ odd}$$
$$\beta(d) = 0 \text{ if } 3|d; \quad \beta(d) = 1 \text{ otherwise}.$$

With this notation, it is easy to show that

$$E^{S_3}(X_d) = -a\varepsilon - b1 - c\theta$$

where

$$a = (d^2 - 6d - 3\alpha(d) - 4\beta(d))/6, b = (d^2 + 3\alpha(d) - 4\beta(d))/6 \quad \text{and}$$
$$c = (d^2 - 3d + 2\beta(d))/3.$$

In particular, it follows that X_d/S_3 is a smooth rational curve precisely when $d \in \{1, 2, 3, 5\}$.

Remark 1.4. There is a useful version of the Lefschetz fixed point theorem for closed subvarieties X of $\mathbb{C}^N$. For any such variety X there is a number $r_0 > 0$ such that for any $r \geq r_0$ the inclusion of $X_r = \{x \in X | |x| \leq r\}$ in X is a homotopy equivalence, (see for instance [7, p.26]). If we have an algebraic morphism $f : X \to X$ such that $f(X_r) \subset X_r$ for r large enough, then we have

$$\Lambda(f) = E(X^f)$$

Indeed, in this case the fixed point set $Y = X^f$ is itself an affine variety, and we can choose r good in the above sense for both X and Y and apply the classical Lefschetz theorem to the compact space X_r and the restricted map $f : X_r \to X_r$.

A special case of this occurs when the group Γ acts not only on X but also on the ambient space $\mathbb{C}^N \cong \mathbb{R}^{2N}$ in such a way that g is an $\mathbb{R}$-linear transformation for any $g \in \Gamma$. The result follows in this case by replacing the usual norm $|x|$ on $\mathbb{R}^{2N}$ by a Γ-invariant norm.

Another important case is when the variety X is defined over $\mathbb{R}$ and $\Gamma = \{1, \iota\}$, ι being complex conjugation. It is well-known that in this case

$$\varepsilon^\Gamma(\iota) = (-1)^n$$

for a smooth connected n-dimensional variety X.

Assume from now on that X is an n-dimensional complex algebraic variety. Each cohomology group $H_c^j(X, \mathbb{C})$ then has a *weight* filtration

$$0 \subset W_0 H_c^j(X, \mathbb{C}) \subset \cdots \subset W_{2n} H_c^j(X, \mathbb{C}) = H_c^j(X, \mathbb{C})$$

as part of the mixed Hodge structure constructed by Deligne [4]. When Γ acts algebraically on X, it respects this filtration and hence the graded pieces

$$Gr_m^W H_c^j(X, \mathbb{C}) = W_m H_c^j(X, \mathbb{C})/W_{m-1} H_c^j(X, \mathbb{C})$$

are all Γ-modules.

Definition 1.5.   (i) The *weight m equivariant Euler characteristic* of the Γ-variety X is the element of $R(\Gamma)$ given by

$$E_c^{\Gamma,m}(X) = \sum_j (-1)^j Gr_m^W H_c^j(X,\mathbb{C})$$

(ii) The *weight equivariant Euler polynomial* (*weight polynomial* for short) of the Γ-variety X is the element of $R(\Gamma)[t]$ given by

$$W_c^{\Gamma}(X,t) = \sum_m E_c^{\Gamma,m}(X)t^m$$

Such polynomials were first considered by Serre (unpublished) in the non equivariant case and later by Danilov-Khovanski [3] and by several other authors who took into account the Hodge decomposition, not only the weights; see for instance Batyrev-Dais [1], Fulton [10], Fulton-MacPherson [11],Lehrer [19], Looijenga [22]; see also Durfee [8] for the weight m Euler characteristics.

As remarked above for equivariant Euler characteristics, one may similarly use the usual cohomology of X to define a weight polynomial $W^\Gamma(X)[t]$. When X is smooth and connected, the Poincaré duality isomorphism takes classes of type (p,q) in $H_c^j(X,\mathbb{C})$ to classes of type $(n-p,n-q)$ in $H^{2n-j}(X,\mathbb{C})$, see Fujiki [9]. This remark, together with the fact that for algebraic actions one has $\varepsilon^\Gamma = 1$ imply the following relation when Γ is a group of algebraic automorphisms of X

$$W^\Gamma(X,t) = t^{2n}W_c^\Gamma(X,t^{-1}) \tag{1.6}$$

When $\Gamma = \{1,\iota\}$ is the conjugation group, then the weight filtration is again Γ-invariant, so that one can define the corresponding polynomials $W^\Gamma(X,t)$ and $W_c^\Gamma(X,t)$. Note however that in this case the relation (1.6) becomes

$$W^\Gamma(X,t) = t^{2n}W_c^\Gamma(X,t^{-1}) \otimes \varepsilon^\Gamma \tag{1.6}'$$

If X is a smooth projective variety then it is known that $Gr_m^W H_c^j(X,\mathbb{C}) = 0$ unless $j = m$.

Example 1.7. Consider again the S_3-action on the Fermat curve X_d. In the notation of (1.3) one sees easily that

$$W^{S_3}(X_d,t) = \varepsilon - [(a+2)\varepsilon + b1 + c\theta]t + \varepsilon t^2$$

Note that there is no "weighted Lefschetz formula" in general; i.e. for a compact algebraic variety X the numbers $E^{\Gamma,m}(X,g)$ and $E^m(X^g)$ are generally different. As an example take $m = 2$ and $g(x:y:z) = (y:x:z)$ above.

2. EULERIAN COLLECTIONS AND ADDITIVITY

Let B be a finite subalgebra of the Boolean algebra of constructible subsets of the algebraic variety X. Recall that this means that B contains X and is closed under the taking of intersections and complements.

Definition 2.1.   (i)   Let A be an abelian group. We say that the function $\chi : B \to A$ is additive if it satisfies

$$\chi(Z) + \chi(Y \setminus Z) = \chi(Y)$$

for any pair of subsets $Y, Z \in B$ such that $Z \subset Y$.

(ii)   A finite collection L of (Zariski) closed subsets of X is called an *Eulerian collection* if it satisfies (a) $X \in L$ and (b) if $Y, Z \in L$ then $Y \cap Z \in L$.

For any Eulerian collection L, we denote by $B(L)$ the Boolean algebra of subsets of X generated by L. For $Y \in L$ we define $Y^* = Y \setminus \cup Z \in B(L)$, where the union is over all $Z \in L, Z \subset Y, Z \neq Y$. Note that Y^* may be empty.

Proposition 2.2. *For any Eulerian collection L in X and any additive function $\chi : B(L) \to A$ one has the following equality.*

$$\chi(X) = \sum_{Y \in L} \chi(Y^*)$$

Proof. This proof is by induction on $|L|$, and is elementary.

Corollary 2.3. *For any set $Y \in L$ one has*

$$\chi(Y^*) = \sum_{Z \subset Y} \mu(Z, Y) \chi(Z)$$

where μ is the Möbius function of L regarded as a poset under inclusion.

Proof. For any poset L and abelian group A, if $f, g : L \to A$ are functions, then it is standard that

$$f(Y) = \sum_{Z \leq Y} g(Z)$$

if and only if

$$g(Y) = \sum_{Z \leq Y} \mu(Z, Y) f(Z)$$

(see for instance Rota [25]). If we take $f(Y) = \chi(Y)$ and $g(Y) = \chi(Y^*)$ the result follows immediately from (2.2). (Compare also with Proposition (3.6) in [1]).

Assume now that there is an algebraic Γ-action on the variety X as in the first section. Let $Y \subset X$ be a Zariski closed Γ-invariant subset. Then there is a long exact sequence of compactly supported cohomology spaces

$$\cdots \to H^j_c(U, \mathbb{C}) \to H^j_c(X, \mathbb{C}) \to H^j_c(Y, \mathbb{C}) \to H^{j+1}_c(U, \mathbb{C}) \to \ldots \qquad (2.4)$$

where $U = X \setminus Y$. This sequence respects both the mixed Hodge and Γ-module structures on the cohomology. It follows (as in Durfee [8]) that the equivariant Euler characteristic and the weight polynomial are additive functions, i.e. that

$$E^\Gamma_c(X) = E^\Gamma_c(Y) + E^\Gamma_c(U) \qquad (2.5)$$

and

$$W^\Gamma_c(X) = W^\Gamma_c(Y) + W^\Gamma_c(U) \qquad (2.6)$$

Proposition (2.2) indicates that these invariants might be computed by finding a partition with known constructible pieces. We call such a partition *constructible*.

Remark 2.7. For any Γ- variety X (where Γ is a group of algebraic automorphisms of X) one has

$$E^\Gamma(X) = E_c^\Gamma(X).$$

Proof. The case Γ trivial is treated in Fulton [10], pp. 141-142. The general case follows by the same arguments, using Γ-invariant resolutions and tubular neighbourhoods.

Example 2.8. (Toric varieties)

Let $X = X(\Delta)$ be the toric variety associated with a fan Δ (see for instance Fulton [10] for details concerning the basic properties of toric varieties).

If $n = dim(X)$, then the n-dimensional algebraic torus T^n acts on X in such a way that for each cone σ in Δ there is exactly one T^n-orbit O_σ, and this orbit is isomorphic to $(\mathbb{C}^*)^{n-k}$ with $k = dim(\sigma)$. In this way we obtain a constructible partition of X.

It is evident that $W_c(O_\sigma, t) = (t^2 - 1)^{n-k}$. Using the additivity of the weight polynomial we get

$$W_c(X, t) = \sum_{k=0}^{n} d_k (t^2 - 1)^{n-k}$$

where d_k is the number of k-dimensional cones in the fan Δ.

In particular $E(X) = E_c(X) = d_n$.

3. PURITY

Let X be a complex algebraic variety of dimension n. It is well known that if X is smooth, any weight w occurring in $H_c^j(X)$ satisfies $w \geq 2j - 2n$; moreover the weight space of weight $2j - 2n$ has Hodge type $(j - n, j - n)$. This motivates the following

Definition 3.1.   (i)   The compactly supported cohomology $H_c^j(X)$ ($\neq 0$) is a *pure Hodge structure of weight w* if $Gr_k^W H_c^j(X) = 0$ unless $k = w$.

  (ii)   The variety X is *separably pure* (sp for short) if each cohomology group $H_c^j(X)$ is either 0 or a pure Hodge structure of weight w_j such that $w_i \neq w_j$ for $i \neq j$.

 (iii)   An irreducible variety X is *minimally pure* (mp for short) if X is separably pure and $w_j = 2j - 2dimX$ for any j with $H_c^j(X) \neq 0$ (so $\emptyset$ is mp).

 (iv)   A pure dimensional variety X (i.e. one all of whose irreducible components have the same dimension) is minimally pure if for any collection $\{X_1, X_2, \ldots, X_r\}$ of irreducible components of X, the irreducible variety $X_1 \setminus (X_2 \cup \cdots \cup X_r)$ is mp. In particular, the case $r = 1$ implies that all irreducible components of X are mp.

Note that the condition that a variety X be sp is equivalent to stipulating that the weight polynomial $W_c^\Gamma(X)$ determine the individual Γ-modules $H_c^j(X)$.

The condition mp is a special case of sp, (see Lemma(3.2) below), and occurs naturally in situations like hyperplane arrangements, (cf. Example (3.3) below). Note in particular that for an mp variety X, one has $H_c^j(X) = 0$ for $j < \dim X$; e.g. if X is compact and mp then $\dim X = 0$.

Lemma 3.2.   (i) *If the variety X is mp, then any difference $X_S \setminus X_T$ of unions $X_S = \cup_{s \in S} X_s$ and $X_T = \cup_{t \in T} X_t$ of irreducible components of X is mp.*

(ii) *If the variety X is mp, then $H_c^j(X)$ is either 0 or is a pure Hodge structure of weight $w_j = 2j - 2\dim X$.*

(iii) *Let X be an irreducible mp variety and $D \subset X$ a divisor which is also mp. Then the complement $X \setminus D$ is mp.*

Proof. (i) is obvious from Definition (3.1).

(ii) Let $X_1, \ldots, X_r$ be the set of irreducible components of X. For $r = 1$ there is nothing to prove. The general case ($r > 1$) follows by induction on r: in the exact sequence (2.4) take $Y = X_2 \cup \cdots \cup X_r$ and note that $X \setminus Y = X_1 \setminus Y$ is irreducible, since it is open in X_1.

(iii) This follows again from the exact sequence (2.4) by taking $Y = D$ and using (ii) above. Note however that if U and D are mp it does not follow that X itself is mp, e.g. take $X = \mathbb{P}^1$ and D a point.

Examples 3.3. (i) Toric varieties

If the fan Δ is simplicial and complete, then the toric variety $X = X(\Delta)$ is a compact $\mathbb{Q} - manifold$ and hence it has a pure Hodge structure, i.e. $H^j(X)$ is pure of weight j if it is non-zero. Thus the formula in Example (2.6) gives the individual Betti numbers of X: $b_j(X) = 0$ for j odd and

$$b_{2j}(X) = \sum_{k=0}^{n-j}(-1)^{n-k-j}d_k \binom{n-k}{j}$$

for $0 \le j \le n$, (compare with Fulton [10], p.104).

(ii) Hyperplane arrangements

Let $(H_i)_{i \in I}$ be a hyperplane arrangement in $\mathbb{C}^n$. Then the union $D = \cup_i H_i$ and the complement $U = \mathbb{C}^n \setminus D$ are mp varieties.

Since $H_c^j(\mathbb{C}^n)$ is zero for $j \ne 2n$ and has Hodge type (n, n) (and hence weight $2n$) it follows that $\mathbb{C}^n$ is mp. For $n = 1$, D is just a finite collection of points, hence is mp. Using (2.2) (iii) it follows that U is also mp in this case.

The general case follows by induction on n: indeed, to check that D is mp one considers differences $U' = H_{i_1} \setminus (H_{i_2} \cup \cdots \cup H_{i_r})$. But this U' is the complement of a hyperplane arrangement in $\mathbb{C}^{n-1}$, whence we are done by induction.

Hyperplane arrangements can be generalized in several ways. Here is one such. We say that a linear arrangement $(E_i)_{i \in I}$ in $\mathbb{C}^n$ has codimension r if for any collection of linear spaces $\{E_1, \ldots, E_p\}$ in the arrangement we have either

$$E_1 \cap \cdots \cap E_{p-1} = E_1 \cap \cdots \cap E_p \quad \text{or} \quad \text{codim}_{E_1 \cap \cdots \cap E_{p-1}} E_1 \cap \cdots \cap E_p = r.$$

In particular, this implies $codim_{\mathbb{C}^n} E_i = r$ for any $i \in I$. It is clear that a hyperplane arrangement has codimension 1.

Proposition 3.4. *If* $(E_i)_{i \in I}$ *is a nonempty codimension* r *linear arrangement in* $\mathbb{C}^n$, *then the complement* $U = \mathbb{C}^n \setminus \cup_i E_i$ *is a sp variety. More precisely,* $H_c^j(U)$ *is zero or has pure weight* $w_j = 2n - 2r(2n - j)/(2r - 1)$ *for* $2n - j \equiv 0 \bmod(2r - 1)$, *and is zero otherwise.*

Proof. By Poincaré duality the statement is equivalent to

the cohomology spaces $H^j(U)$ are zero unless $j \equiv 0 \bmod(2r - 1)$

and then $H^j(U)$ has pure weight $w'_j = jr/(2r - 1)$ if it is non-zero.

We use induction on n starting with $n = r$. Then $A = \cup_i E_i$ consists of finitely many points and there is a Gysin sequence

$$\cdots \to H^j(\mathbb{C}^r) \to H^j(U) \to H^{j-2r+1}(A) \to H^{j+1}(\mathbb{C}^r) \to \cdots \qquad (3.5)$$

where the middle morphism is the Leray residue morphism, and has Hodge type $(-r, -r)$. It follows that the only nonzero cohomology groups $H^j(U)$ are for $j = 0$ (type $(0,0)$) and for $j = 2r - 1$ (type (r, r)).

Assume now that (3.4) is true for arrangements in $\mathbb{C}^m$ for $m < n$. If our arrangement in $\mathbb{C}^n$ consists of just one linear subspace, then $U = (\mathbb{C}^* \setminus 0) \times \mathbb{C}^{n-r}$ and the result follows from the previous case $n = r$.

When $|I| > 1$ define $U_d = \mathbb{C}^n \setminus \cup_{i \neq i_1} E_i$ (the deleted arrangement) and $U_r = E_{i_1} \setminus \cup_{i \neq i_1} E_i$ (the restricted arrangement) for some $i_1 \in I$. Since U_r is a closed smooth algebraic subvariety in U_d, there is a Gysin sequence similar to (3.5) relating the cohomology of U, U_d and U_r. By double induction (on n and $|I|$), the result is true for U_d and U_r. Then the exact sequence shows that (3.4) is true for U as well.

Another generalization of hyperplane arrangements is the following non-linear one.

Definition 3.6. Let X be a mp variety and $\mathscr{A} = (A_i)_{i \in I}$ a finite collection of closed subsets of X. Let L be the lattice of intersections of elements of $\mathscr{A}$. We say that $\mathscr{A}$ is a *mp arrangement* if the following conditions hold.

(α) For each $i \in I$, either A_i is a union of irreducible components of X or else $\mathrm{codim}_{X_j} A_i \cap X_j = 1$ for all irreducible components X_j of X with $A_i \cap X_j \neq \emptyset$.

(β) For each $Y \in L$ we have

(i) Y is a mp variety;

(ii) $\dim(Y \cap X_j) = \dim(Y)$ for any irreducible component X_j of X with $Y \cap X_j \neq \emptyset$.

(iii) The family of subsets $\mathscr{A}_Y = \{A_i \cap Y \mid i \in I, A_i \cap Y \neq Y\}$ of Y satisfies the conditions $\alpha, \beta(i)$ *and* $\beta(ii)$ above with X replaced by Y.

Theorem 3.7. *Let* $\mathscr{A} = (A_i)_{i \in I}$ *be a mp arrangement in the mp variety* X. *Then the complement* $X^* = X \setminus \cup_{i \in I} A_i$ *is mp.*

Proof. First observe that it suffices to prove the result for X irreducible. Indeed, if X_j is an irreducible component of X, then $X_j^* = X_j \cap X^*$ is either empty, or an irreducible component of X^*. Moreover, any irreducible component of X^* has this form. The key fact is that the restricted family $\mathscr{A}_j = (A_i \cap X_j)_{i \in I}$ is a mp arrangement in X_j. For instance, to prove $\beta(i)$ for $\mathscr{A}_j$ note that $\beta(ii)$ for $\mathscr{A}$ implies that $Y \cap X_j$ is a union of irreducible components of Y for each $Y \in L$.

More generally, to check that a difference of irreducible components of X^* is mp we consider the restriction of the arrangement $\mathscr{A}$ to an irreducible variety of type $Z = X_1 \setminus \cup_{j>1} X_j$. Again the intersections $Y \cap Z$ are mp by Lemma (3.2.i).

We now prove the Theorem for irreducible X by double induction on the dimension n of X and the cardinality $|I|$. The statement is trivial for $|I| = 0$. If all subsets A_i also have dimension n, then X^* is a difference of unions of components of X and the result follows by Lemma (3.2.i).

Assume that there is an element $A_1 \in \mathscr{A}$ such that $dim(A_1) = n - 1$. Let

$$X' = X \setminus \cup_{i \neq 1} A_i, \qquad X'' = A_1 \setminus \cup_{i \neq 1}(A_1 \cap A_i).$$

Now X'' is closed in X', $X^* = X' \setminus X''$ and we may consider the corresponding exact sequence (2.4). By induction on dimension, X'' is mp of dimension $n - 1$, and it follows by induction on $|I|$, that X' is mp of dimension n. Hence the long exact sequence splits into exact sequences of type

$$0 \to H_c^{k-1}(X'') \to H_c^k(X^*) \to H_c^k(X') \to 0 \qquad (3.8)$$

Since this sequence is weight preserving $H_c^k(X^*)$ is a pure Hodge structure of weight $k - 2n$. This completes the proof of (3.7).

Note that in case $dim(A_1) = n$ we also have exact sequences relating the cohomology of X^*, X'' and X', namely

$$0 \to H_c^k(X^*) \to H_c^k(X') \to H_c^k(X'') \to 0 \qquad (3.8)'$$

Definition 3.9. Let $\mathscr{A}$ be a mp arrangement in the n-dimensional mp variety X. Let

$$\Psi_c(\mathscr{A}, t) = \sum_{i \geq 0} \dim H_c^i(X^*) t^i$$

be the compactly supported Poincaré polynomial of the complement X^* of the arrangement $\mathscr{A}$ in X.

Remarks 3.10. (i) With the notation from the proof of (3.7) we have

$$\Psi_c(\mathscr{A}, t) = \Psi_c(\mathscr{A}', t) - (-t)^{\mathrm{codim}(A_1)} \Psi_c(\mathscr{A}'', t)$$

(ii) The weight polynomial and the Poincaré polynomial satisfy the relation:

$$W_c(X^*, t) = t^{-2n} \Psi(\mathscr{A}, -t^2)$$

with $n = \dim X$. There is also an obvious equivariant version of this result.

Using (2.3) and (3.7) we obtain the following result.

Corollary 3.11. *With the notation of (3.7), assume that the finite group Γ acts algebraically on the mp variety X so that each subset $Y \in L$ is setwise invariant. Then we have the following equality in $R(\Gamma)$*

$$H_c^j(X^*) = \sum_{Y \in L} (-1)^{c(Y)} \mu(Y, X) H_c^{j-c(Y)}(Y)$$

where $c(Y) = codim(Y)$.

3.12. ℓ-adic theory. Let X be a scheme, separated and of finite type over $\mathbb{Z}$. Write $X(K)$ for its set of K-points (K a field) and let F be the Frobenius endomorphism of $X(\bar{\mathbb{F}}_q)$ corresponding to its $\mathbb{F}_q$-structure.

The action of F on the ℓ-adic cohomology $H_c^*(X(\bar{\mathbb{F}}_q), \bar{\mathbb{Q}}_\ell)$ determines a weight filtration and one may consider the corresponding weight polynomial. When $X(\bar{\mathbb{F}}_q)$ is a "good reduction" of $X(\mathbb{C})$, then the ℓ-adic and Hodge–theoretic weight polynomials of X coincide. In case X is a hyperplane complement, these were determined by Lehrer [18], who showed in particular that these polynomials could be determined by computing $|X(\bar{\mathbb{F}}_q)^F|$, the cardinality of the set of $\mathbb{F}_q$-points of X, as a polynomial on q [see §5 below].

Remark 3.13. Say that X above is *polynomial* if it is smooth, projective and there is a polynomial $f(X, t) \in \mathbb{Q}[t]$ such that $f(X, q^n) = |X(\mathbb{F}_{q^n})| \, (= |X(\bar{\mathbb{F}}_q)^{F^n}|)$ for all n.

Göttsche [12, p.6] has remarked that if X is polynomial and X_q is a good reduction of $X(\mathbb{C})$, then $W(X, t) = f(Y, t^2)$.

This relation may be extended to a larger class of varieties (including those in [18]) as follows. Choose a compactification (Z, D) of X. This means that Z is smooth and projective, D is a divisor with normal crossings and smooth irreducible components D_j and $X = Z \backslash D$. Assume that $Z(\mathbb{C})$ and $D_j(\mathbb{C})$ have good reductions mod q. If all of Z and $D_{j_1} \cap \cdots \cap D_{j_k}$ are polynomial, then

$$W_c(X, t) = f(Y, t^2).$$

This follows from Göttsche's observation, since both W_c and f are additive, so that (2.3) may be applied.

Remark 3.14. The minimal purity of hyperplane complements furnishes a direct and easy proof of "Brieskorn's Lemma" (cf. [24]) for them (see[18] for the $\ell - adic$ case). This reduces the computation of their cohomology to the study of the top cohomology of various related arrangements. Sequences such as those of (3.10) may be used to prove "generalised Brieskorn lemmas" which may also serve to link the complex and $\ell - adic$ theories (see §5 below).

4. TORAL ARRANGEMENTS

In this section we discuss toral arrangements. To define them, let T be an algebraic torus, i.e. a diagonalisable algebraic group with connected component of the identity isomorphic to the standard torus $(\mathbb{C}^*)^n$. Let $(\chi_i)_{i \in I}$ be a finite family of characters of the torus T. A toral arrangement is a family $\mathscr{A} = (A_i)_{i \in I}$ with $A_i = ker\chi_i$.

The referee has kindly pointed out that the following result and its corollary were proved by Looijenga [22, (2.4.3)]

Theorem 4.1.   (i) *Any algebraic torus T is a mp variety.*
(ii) *Any toral arrangement is a rp arrangement.*

Proof. (i) If T is connected, then T is isomorphic to $(\mathbb{C}^*)^n$ and the result follows from (3.3.2) since $(\mathbb{C}^*)^n$ is a hyperplane complement. In general T is a finite union of disjoint connected components, each isomorphic to $(\mathbb{C}^*)^n$, hence Definition (3.1.iii) is obviously satisfied.

(ii) A careful check of the conditions of Definition (3.6) yields the result.

Corollary 4.2. *Let T be an algebraic torus. The complement of a toral arrangement in T is mp.*

We consider now the case of the variety of regular elements in a maximal torus of $SL(n, \mathbb{C})$. We realise this variety as T^* for a mp arrangement $\mathscr{A}$ in T as follows.

$$T = \{z = (z_1, \ldots, z_n) \in \mathbb{C}^n | z_1 \cdot \ldots \cdot z_n = 1\}$$

$$\mathscr{A} = (A_{ij})_{1 \le i < j \le n}, \quad A_{ij} = \{z \in T | z_i = z_j\}$$

Let μ_n be the cyclic group of the n-th roots of unity. There is an obvious algebraic μ_n action on T which leaves each A_{ij} invariant.

To compute the equivariant weight polynomial $W_c^{\mu_n}(T^*)$ we use (3.11). In the situation at hand, the lattice L is isomorphic to the lattice of partitions of the set $\{1, 2, \ldots, n\}$. For each $Y \in L$ we let $\lambda(Y)$ be the partition of n corresponding to Y; write $\lambda(Y) := \lambda_1 \ge \cdots \ge \lambda_k > 0$, $\sum \lambda_j = n$. Let $d(Y) = gcd(\lambda_1, \ldots, \lambda_k)$, $k(Y) = k$, $\mu(Y) = \mu(Y, T)$.

Theorem 4.3.   (i) *The complement $T^* = T \setminus \cup A_{ij}$ is a mp variety.*
(ii) *The equivariant weight polynomial of the toral arrangement T^* is given by*

$$W_c^{\mu_n}(T^*, t) = \sum_{d|n} \sum_{Y \in L, d(Y) = d} \mu(Y)(t^2 - 1)^{k(Y) - 1} Ind_{\mu_n^d}^{\mu_n}(1)$$

where μ_n^d is the unique subgroup in μ_n of index d. In particular, the 1_{μ_n}-isotypic part of the weight polynomial above is given by

$$W_c^{\mu_n}(T^*)_{1_{\mu_n}} = (W_c^{\mu_n}(T^*, t), 1_{\mu_n})$$
$$= \sum_{Y \in L} \mu(Y)(t^2 - 1)^{k(Y) - 1}$$
$$= (t^2 - 1)(t^2 - 2) \cdot \ldots \cdot (t^2 - n).$$

(iii) *Any faithful character ζ of μ_n occurs only in $H_c^{n-1}(T^*)$, and*

$$(H_c^{n-1}(T^*), \zeta)_{\mu_n} = (n - 1)!.$$

Proof. (i) This follows from (4.1.ii).

(ii) All statements follow from (3.11) except the last claim. To justify this, note that the quotient space $\bar{T}^* = T^*/\mu_n$ is isomorphic to the hyperplane arrangement complement corresponding to the arrangement in $\mathbb{C}^{n-1}$ defined by the equations $z_i \neq 0$, $z_i \neq z_j$ for $i \neq j$. The Poincaré polynomial of this arrangement (homotopy equivalent to the braid arrangement) is well-known, see [24], p. 45, or [21] and the result follows via (3.10.ii).

(iii) By (ii) the faithful characters may occur only for $d = n$ and then the corresponding induced representation is the regular representation of μ_n. Its coefficient in the first sum of (ii) is $(-1)^{n-1}(n-1)!$. This corresponds to the action on $H_c^{n-1}(T^*)$ as this cohomology group is the only one of weight 0.

In particular, note that for n prime (ii) and (iii) above determine the polynomial $W_c^{\mu_n}(T^*)$ since we have (writing M_ζ for the ζ-isomorphic component of any μ_n-module M)

$$W_c^{\mu_n}(T^*) = W_c^{\mu_n}(T^*)_{1_{\mu_n}} + (-1)^{n-1} \sum_{\zeta \neq 1} H_c^{n-1}(T^*)_\zeta. \qquad (4.3)$$

4.4. The action of Sym(n).

In addition to its μ_n-action, the variety T^* has an obvious $Sym(n)$ action, where $Sym(n)$ is the symmetric group of degree n. The weight polynomial $W_c(T^*, t)$ may therefore be considered as $W_c^{\mu_n \times Sym(n)}(T^*, t)$ and by purity, we still have

$$W_c^{\mu_n \times Sym(n)}(T^*, t) = t^{-2(n-1)} \sum_{j=0}^{2(n-1)} (-t^2)^j H_c^j(T^*) \qquad (4.4.1)$$

where the cohomology spaces on the right are regarded as $\mu_n \times Sym(n)$ modules. For $x = (\alpha, g) \in \mu_n \times Sym(n)$, write

$$W_c^{\mu_n \times Sym(n)}(T^*, t)(x) = (-t)^{-2(n-1)} \sum_{j=0}^{2(n-1)} (-t^2)^j \operatorname{trace}(x, H_c^j(T^*)). \qquad (4.4.2)$$

Then the Lefschetz fixed point theorem of (1.4) implies that

$$W_c^{\mu_n \times Sym(n)}(T^*, 1)(x) = E_c(T^{*x}) \qquad (4.4.3)$$

where E_c denotes the Euler characteristic (with compact support).

We now wish to study $H_c^*(T^*)$ as $Sym(n)$ module. For this purpose, observe that $H_c^*(T^*)$ is a direct sum of its μ_n-isotypic components and that these are $Sym(n)$-modules. For any character $\zeta \in \hat{\mu}_n$ and μ_n-module M, write M_ζ for the ζ-isotypic component of M. Then write

$$W_c^{Sym(n)}(T^*, t)_\zeta = t^{-2(n-1)} \sum_{j=0}^{(n-1)} H_c^j(T^*)_\zeta (-t^2)^j. \qquad (4.4.4)$$

Clearly we have

$$W_c^{Sym(n)}(T^*, t) = \sum_{\zeta \in \hat{\mu}_n} W_c^{Sym(n)}(T^*, t)_\zeta. \qquad (4.4.5)$$

Now $W_c^{Sym(n)}(T^*, t)_{1_{\mu_n}} = W_c^{Sym(n)}(T^*/\mu_n, t)$ is known from [21, Theorem C]. In the case when n is prime, this, together with (4.4.3) may be used to determine $W_c^{Sym(n)}(T^*, t)$ as follows. From (4.3) it follows that

$$W_c^{Sym(n)}(T^*, t) = W_c^{Sym(n)}(T^*, t)_{1_{\mu_n}} + (-1)^{n-1} \sum_{\zeta \neq 1} H_c^{n-1}(T^*)_\zeta. \quad (4.4.6)$$

Taking the trace of $g \in Sym(n)$ on both sides and setting $t = 1$, we have from (4.4.3)

$$(-1)^{n-1} \sum_{\zeta \pm 1} \text{trace}(g, H_c^{n-1}(T^*)_\zeta) = E_c(T^{*g}) - E_c(\bar{T}^{*g}) \quad (4.4.7)$$

where $\bar{T}^* = T^*/\mu_n$. If $g \neq 1$, $T^{*g} = \emptyset$, whence $E_c(T^{*g}) = 0$. Moreover, a short computation shows that for n prime and $g \neq 1$,

$$E_c(\bar{T}^{*g}) = \begin{cases} 0 & \text{unless } g \text{ is an } n\text{-cycle} \\ n - 1 & \text{if } g \text{ is an } n\text{-cycle.} \end{cases} \quad (4.4.8)$$

It follows from (4.4.7) that if $g \neq 1$,

$$\sum_{\zeta \neq 1} trace(g, H_c^{n-1}(T^*)_\zeta) = \begin{cases} (-1)^n(n-1) & \text{if } g \text{ is } n\text{-cyclic} \\ 0 & \text{otherwise.} \end{cases} \quad (4.4.9)$$

To treat the case $g = 1$, observe that $E_c(\bar{T}^{*g}) = (-1)^{n-1}(n-1)!$ by [21, loc. cit.] and that $E_c(T^*) = (-1)^{n-1}n!$ by (4.2)(ii). Hence for $g = 1$ the left side of (4.4.9) is $(-1)^{n-1}(n-1)(n-1)!$. Putting together (4.4.4) to (4.4.9) above, we obtain

Proposition 4.5. *We have, for n prime*

$$W_c^{Sym(n)}(T^*, t) = W_c^{Sym(n)}(\bar{T}^*, t) + (-1)^{n-1} Ind_{\langle c_n \rangle}^{Sym(n)}(\xi)$$

where c_n is an n-cycle, ξ is a faithful character of $\langle c_n \rangle$, $\bar{T}^* = T^*/\mu_n$ and $W_c^{Sym(n)}(\bar{T}^*)$ is given by [21, Theorem C], i.e.

$$W_c^{Sym(n)}(\bar{T}^*, t) = t^{2(n-1)} P_{\bar{T}^*}(-t^{-2})$$

where $P_{\bar{T}^*}(t) = \sum H^i(\bar{T}_*, \mathbb{C})t^i$ is the polynomial $P_{\bar{M}_n}(t)$ of [21, Theorem C(i)].

5. PURITY AND FROBENIUS

The results of §§3, 4 may also be discussed in terms of the ℓ-adic cohomology of varieties over fields of positive characteristic. We give a brief description of the statements and one or two consequences here. See also [5].

Let X be a variety over $\bar{\mathbb{F}}_q$ which is defined over $\mathbb{F}_q$ and let $F : X \to X$ be the Frobenius morphism corresponding to the $\mathbb{F}_q$-structure on X. Let ℓ be a prime with $(\ell, q) = 1$ and write $H_c^*(X, \bar{Q}_\ell)$ for ℓ-adic cohomology with compact support.

Definition 5.1. (cf. (3.1))

(i)   Say that X is separably pure (sp) if all the eigenvalues of F on $H^j_c(X, \bar{\mathbb{Q}}_\ell)$ are equal to q^{m_j} and $j \mapsto m_j$ is an injective map.

(ii) An irreducible $\bar{\mathbb{F}}_q$-variety X is minimally pure (mp) if F acts on $H^j_c(X, \bar{\mathbb{Q}}_\ell)$ with all eigenvalues equal to $q^{j - \dim X}$.

(iii) A pure–dimensional $\mathbb{F}_q$-variety X is mp if the property analogous to (3.1)(iv) holds in this context.

The notion of *mp arrangement* is defined for varieties over $\mathbb{F}_q$ exactly as in (3.6), with the term "mp" (for varieties) interpreted according to (5.1). With this understanding, we have

Theorem 5.2. *Let $\mathscr{A} = (A_i)_{i \in I}$ be a mp arrangement in the mp $\mathbb{F}_q$-variety X. Then $X^* = X \backslash \cup_{i \in I} A_i$ is mp.*

The proof is exactly the same as that of (3.7).

5.3. Rational points. If X is a mp $\mathbb{F}_q$-variety, its set of $\mathbb{F}_q$-points X^F may be counted as follows. By Grothendieck's fixed point theorem [6] we have

$$|X^F| = \sum_j (-1)^j \operatorname{trace}(F, H^j_c(X, \bar{\mathbb{Q}}_\ell)). \qquad (5.3.1)$$

But using the mp property and the ℓ-adic analogue of (3.2)(ii), one sees easily that

$$|X^F| = q^{-\dim X} P_c(X, -q) \qquad (5.3.2)$$

where $P_c(X, t) = \sum_j \dim H^j_c(X, \bar{\mathbb{Q}}_\ell) t^j$.

This formula is of particular interest when X is a scheme to which a comparison theorem applies, i.e. $\dim_{\mathbb{C}} H^j_c(X(\mathbb{C}), \mathbb{C}) = \dim_{\bar{\mathbb{Q}}_\ell} H^j_c(X(\bar{\mathbb{F}}_q), \bar{\mathbb{Q}}_\ell)$. In such cases we have (cf. (4.4.1))

$$|X^F| = W_c(X, q^{\frac{1}{2}}) \qquad (5.3.3)$$

where $X_c(X, t)$ is the weight polynomial of (1.5).

The following result has exactly the same proof as (4.1).

Theorem 5.4. *Let T be an algebraic torus and $\{\chi_i\}_{i \in I}$ a finite set of characters of T, all defined over $\mathbb{F}_q$. Then $\{\ker \chi_i\}_{i \in I}$ is a mp arrangement in the sense of (5.2).*

Let X be a scheme, separated and of finite type over $\mathbb{Z}$. Let Γ be a finite group of automorphisms of X. As in (3.12) write $X(K)$ for the variety of K-points of X and for a prime power q let F be the Frobenius endomorphism of $X(\bar{\mathbb{F}}_q)$ corresponding to its $\mathbb{F}_q$-structure. Consider the following two properties of X.

(CT) $\dim_{\mathbb{C}} H^i_c(X(\mathbb{C}), \mathbb{C}) = \dim_{\bar{\mathbb{Q}}_\ell} H^i_c(X(\bar{\mathbb{F}}_q), \bar{\mathbb{Q}}_\ell) \forall i$ (where ℓ is a prime not dividing q).

(MP) $X(\mathbb{C})$ and $X(\bar{\mathbb{F}}_q)$ are mp.

Clearly if X satisfies (MP), so does X/Γ since the respective cohomology spaces are the 1_Γ–isotypic subspaces of those of X (by the transfer theorem) and therefore have the correct weights.

Proposition 5.5. *Suppose X is a scheme as above and assume the following:*

(i) *X satisfies (MP) and*

(ii) *X and X/Γ satisfy (CT).*

Then

$$|(X/\Gamma)^F| = |\Gamma|^{-1} \sum_{w \in \Gamma} |X(\mathbb{F}_q)^{wF}|$$

$$= (W_c^\Gamma(X(\mathbb{C}), q^{1/2}), 1)_\Gamma$$

Proof. By (i) and the remark immediately preceding (5.5), we have

$$W_c^\Gamma(X, t) = t^{-2\dim X} \sum_{j=0}^{2\dim X} H_c^j(X)(-t^2)^j \text{ and} \tag{5.5.1}$$

$$W_c(X/\Gamma, t) = t^{-2\dim X} \sum_{j=0}^{2\dim X} (H_c^j(X), 1)_\Gamma(-t^2)^j \tag{5.5.2}$$

where X is interpreted as either $X(\mathbb{C})$ or $X(\bar{\mathbb{F}}_q)$ and H_c^j denotes the appropriate cohomology.

Now $H_c^j(X/\Gamma(\bar{\mathbb{F}}_q), \bar{\mathbb{Q}}_\ell) = H_c^j(X(\bar{\mathbb{F}}_q), \bar{\mathbb{Q}}_\ell)_{1_\Gamma}$ and by (i), F acts on this space with all eigenvalues equal to $q^{j-\dim X}$. It follows from the Grothendieck–Lefschetz fixed point theorem [6] that

$$|(X/\Gamma)^F| = |\Gamma|^{-1} \sum_{w \in \Gamma} (X(\bar{\mathbb{F}}_q)^{wF}| \tag{5.5.3}$$

$$= q^{-\dim X} \sum_{j=0}^{2\dim X} (-q)^j (1, H_c^j(X(\bar{\mathbb{F}}_q), \bar{\mathbb{Q}}_\ell))_\Gamma.$$

But (ii) implies that $(1, H_c^j(X(\bar{\mathbb{F}}_q), \bar{\mathbb{Q}}_\ell))_\Gamma = (1, H_c^j(X(\mathbb{C}), \mathbb{C}))_\Gamma$, whence the result.

Note that there are examples where X satisfies (i) and (ii) but the Γ–modules $H_c^j(X, \mathbb{C}))$ and $H_c^j(X(\bar{\mathbb{F}}_q), \bar{\mathbb{Q}}_\ell)$ have different characters. However it is possible to use variations of (3.11) and (5.3.3) to prove equivariant versions of (CT) for varieties of the form X^*, given (CT) for all the elements Y of $L(\mathscr{A})$ for an mp arrangement $\mathscr{A}$ (cf. [18]).

Example 5.6. Let $X = T$ be the maximal torus of a simply connected semisimple algebraic group scheme G of rank r over $\mathbb{Z}$ and take $\Gamma = W$ be the Weyl group of G with respect to T. Then the character map (cf. Steinberg [29, (3.8)]) identifies T/W with $\mathbb{A}^r$, whence we deduce from (5.5) that $|(T/W(\bar{\mathbb{F}}_q))^F| = q^r$. This gives a proof of Steinberg's result (cf. [19]) that $G(\bar{\mathbb{F}}_q)$ has q^r F–stable semisimple conjugacy classes and that if χ is the class function on W given by $\chi(w) = |T(\bar{\mathbb{F}}_q)^{wF}|(w \in W)$, then $(\chi, 1)_W = q^r$. This latter result is of course also well known (see e.g. [19]).

Remark 5.7. The property (CT) is discussed in [14]. A sufficient condition for (CT) is as follows. Suppose we have $X \to Spec(V)$, V a valuation ring, where X is smooth and that there exists $\bar{X}$ proper and smooth over $Spec(V)$ such that

$$
\begin{array}{ccc}
X & \hookrightarrow & \bar{X} \\
& \searrow \quad \swarrow & \\
& Spec(V) &
\end{array}
$$

commutes. If $\bar{X} \backslash X$ is a relative divisor with normal crossings, then (CT) applies. The authors thank P. Deligne for pointing this out.

6. FIBRATIONS, CLASSICAL GROUPS AND HOMOGENEOUS SPACES

Let E, B and F be complex algebraic varieties. Assume that the finite group Γ acts algebraically on each of them, the action on B being trivial. Let $p : E \to B$ be an algebraic morphism which is Γ-invariant, i.e. $p(gx) = p(x)$ for all $g \in \Gamma$ and $x \in E$. Assume moreover that each fibre $F_b = p^{-1}(b)$ is equivariantly isomorphic to F.

We then have the following result.

Theorem 6.1. *Assume that either*

(i) *p is locally trivial in the Zariski topology of B, or*

(ii) *E, B and F are smooth, p is locally trivial in the strong complex topology of B and the local systems $R^j p_* \mathbb{C}_E$ are constant for each j.*

Then

$$
W_c^\Gamma(E) = W_c^\Gamma(B) W_c^\Gamma(F).
$$

Before giving the proof of (6.1), we make some remarks.

Remarks 6.2. (i) For any topologically locally trivial fibration $F \to E \to B$, the Euler numbers behave multiplicatively, i.e. $E(E) = E(B)E(F)$, see for instance [2], p.182. On the other hand, one may consider the Poincaré polynomial $\Psi(X)$ of an algebraic variety X (defined similarly to (3.9)). Even under the assumption (6.1.ii), the relation $\Psi(E) = \Psi(B)\Psi(F)$ holds iff the Leray-Serre spectral sequence of the fibration p degenerates at the E_2-term, i.e. iff

$$
H^*(E) = H^*(B) \otimes H^*(F).
$$

However, even for basic fibrations in algebraic geometry (e.g. the Hopf fibration defining $\mathbb{P}^n$) this may fail.

(ii) Consider the fibration p, where $E = \mathbb{C}^2 \setminus C$ with

$$
C = \{(x, y) \in \mathbb{C}^2 \mid x^3 - y^2 = 0 \}
$$

(a cuspidal cubic), $B = \mathbb{C}^*$, $p(x, y) = x^3 - y^2$ and Γ is trivial.

Then F is the Milnor fibre of the A_2-singularity, and its mixed Hodge structure is known by Steenbrink [28], see also [7], pp. 243-244. The homeomorphism $\mathbb{C} \to C$ implies that $W_c(C, t) = t^2$. Hence we get

$$
W_c(E, t) = t^4 - t^2, \quad W_c(B, t) = t^2 - 1, \quad W_c(F, t) = t^2 - 2t
$$

Thus there are algebraic fibrations for which the multiplicativity property of the weight polynomials does not hold. In this case neither (6.1.i) nor (6.1.ii) holds.

(iii) If we drop the smoothness assumption in (6.1.ii) we still have the multiplicativity property for weight polynomials, but not the compactly supported ones, (see (6.4) below).

Proof of Theorem 6.1. We give the proof only in the case Γ trivial, the general case involving no other ideas.

(i) This is essentially known, see for instance [10] and [1]. Note also that when all spaces involved are smooth, then the condition (6.1.i) implies the condition (6.1.ii).

(ii) In general, the Leray-Serre spectral sequence

$$E_2^{a,b} = H^a(B, R^b p_* \mathbb{C}_E) \quad \Longrightarrow \quad H^{a+b}(E)$$

in the category of algebraic varieties is a spectral sequence of mixed Hodge structures. This follows from the existence of the t-structure on $D^b MHM(B)$, the derived category of mixed Hodge Modules on B whose underlying t-structure on $D_c^b(\mathbb{C}_B)$ is the classical one, see M. Saito [26, Remark 2 in 4.6].

In addition, note that if the local system $R^b p_* \mathbb{C}_E$ is constant, then the corresponding variations of MHS are trivial since the natural morphism $H^0(B, R^b p_* \mathbb{C}_E)$ $\to (R^b p_* \mathbb{C}_E)_y$ is a morphism of MHS for any $y \in B$. We also have $H^a(B, R^b p_* \mathbb{C}_E)$ $= H^a(B) \otimes (R^b p_* \mathbb{C}_E)_y$ as MHS.

Therefore there are MHS isomorphisms

$$E_2^{a,b} \cong H^a(B) \otimes H^b(F)$$

In particular, we have

$$Gr_m^W(E_2^{a,b}) = \sum_{c+d=m} Gr_c^W H^a(B) \otimes Gr_d^W H^b(F). \tag{6.3}$$

Since all the differentials in this spectral sequence are MHS morphisms, we get a spectral sequence for each weight m, namely

$$E(m)_2^{a,b} = Gr_m^W E_2^{a,b} \quad \Longrightarrow \quad Gr_m^W H^{a+b}(E).$$

We set

$$P_m(t) = \sum_{s \geq 0} dim(\oplus_{a+b=s} E(m)_2^{a,b}) t^s.$$

Then $P_m(-1) = E^m(E)$, the weight m Euler characteristic of the variety E, defined in a similar way to (1.5.i).

On the other hand, using (6.3), it follows that

$$P_m(-1) = \sum_{a+b=m} E^a(B) E^b(F).$$

Thus we have shown

$$W(E) = W(B)W(F). \tag{6.4}$$

Finally we use the smoothness assumption to pass from weight polynomials in (6.4) to compactly supported weight polynomials in (6.1) via (1.6).

The equality (6.4) can be used to compute the weight polynomials of complex algebraic groups G and of their homogeneous spaces G/H regarded as algebraic varieties in the usual way, see Humphreys [16], Chap. IV. Indeed, let G be a connected linear algebraic group and H a closed subgroup. If H itself is connected, it follows that the canonical projection

$$p : G \to G/H$$

satisfies the assumption (6.1.ii).

If H is not connected, let H_0 denote the connected component of the identity of H. Then $W = H/H_0$ is a finite group acting on G/H_0 with quotient G/H. Thus $H^*(G/H) = H^*(G/H_0)^W$; but W may act non-trivially on $H^*(G/H_0)$, a classical example being when H is the normalizer of a maximal torus of a semisimple Lie group G, and W is the Weyl group. In this case $H^*(G/H_0)$ is the coinvariant algebra of W. However, when H is finite its action on $H^*(G)$ is trivial, so that $H^*(G/H) \cong H^*(G)$.

Theorem (9.1.5) in Deligne [4], Hodge III, determines the mixed Hodge structure on the cohomology $H^*(G)$. As an easy consequence we have

6.5. *Let G be a connected reductive linear algebraic group with Poincaré polynomial*

$$P(G, t) = (1 + t^{2k_1 - 1}) \cdot \ldots \cdot (1 + t^{2k_m - 1}).$$

Then the weight polynomial of the group G is given by

$$W(G, t) = (1 - t^{2k_1}) \cdot \ldots \cdot (1 - t^{2k_m})$$

The integers k_j are called the degrees of the group G and are known for the classical groups. Thus we obtain

6.6. (i) $W(GL(n, \mathbb{C}), t) = (1 - t^2) \cdot \ldots \cdot (1 - t^{2n}), \qquad n \geq 1;$

  (ii) $W(SO(n, \mathbb{C}), t) = (1 - t^4) \cdot \ldots \cdot (1 - t^{2n-4})(1 - t^n), \qquad$ for n even, $n \geq 2;$

  $W(SO(n, \mathbb{C}), t) = (1 - t^4) \cdot \ldots \cdot (1 - t^{2n-2}), \qquad$ for n odd, $n \geq 1;$

  (iii) $W(Sp(2n, \mathbb{C}), t) = (1 - t^4) \cdot \ldots \cdot (1 - t^{4n}), \qquad$ for $n \geq 1.$

A direct, more elementary approach to the formulas in (6.6) as well as interesting applications can be found in Jones [17].

Many applications of (6.1) stem from the fact that when an algebraic group G acts transitively on an algebraic variety B, and H is the isotropy subgroup of a point $b \in B$, then (Chevalley) there is an isomorphism $B \cong G/H$. It is known that the homogeneous space G/H is always smooth and it is projective if and only if the subgroup H is parabolic (see [16]). In the latter situation the weight polynomial of G/H determines all the Betti numbers $b_j(G/H)$ by an argument similar to (3.3.i).

6.7. Applications. (i) Complex Grassmannians (see also [2], p. 292)

Let $G(k, n) = GL(n, \mathbb{C})/H$ be the complex grassmannian of k-dimensional linear subspaces on $\mathbb{C}^n$. It is well known that $G(k, n)$ is smooth and projective, hence sp. The parabolic subgroup H fibres as follows (Levi decomposition)

$$\mathbb{C}^{k(n-k)} \to H \to GL(k, \mathbb{C}) \times GL(n - k, \mathbb{C})$$

Using this fibration, (6.1) and (6.6.i) we can compute the weight polynomial of the grassmannian variety. The result is the following.

$$\begin{aligned} W(G(k, n), t) &= \frac{W(GL(n, \mathbb{C}), t)}{W(GL(k, \mathbb{C}))(t)W(GL(n - k, \mathbb{C}), t)} \\ &= \frac{(1 - t^{2(n-k)+2}) \cdot \ldots \cdot (1 - t^{2n})}{(1 - t^2) \cdot \ldots \cdot (1 - t^{2k})} \end{aligned}$$

(ii) Complex Lagrangian Grassmannians

Let $\Lambda(\mathbb{C}^{2n})$ be the complex lagrangian grassmannian of n-dimensional lagrangian subspaces of $\mathbb{C}^{2n}$ endowed with the canonical symplectic structure. This is a homogeneous space $Sp(2n, \mathbb{C})/H$, where the parabolic subgroup H fibres as follows (again the Levi decomposition)

$$\mathbb{C}^{n(n+1)/2} \to H \to GL(n, \mathbb{C})$$

Using this fibration, (6.1) and (6.6.iii) it follows that

$$W(\Lambda(\mathbb{C}^{2n}), t) = W(Sp(2n, \mathbb{C}), t)/W(GL(n, \mathbb{C}), t) = (1 + t^2) \cdot \ldots \cdot (1 + t^{2n}).$$

Remark 6.8. Several classes of algebraic varieties considered in this paper are Tate varieties in the sense that their cohomology groups contain only classes of Hodge type (p, p) for various p. This is true for toric varieties considered in (2.8), the hyperplane arrangement complements in (3.3.ii), for complements of linear arrangements in (3.4), for toral arrangements in (4.2) and for linear algebraic groups and theirs homogeneous spaces by Deligne result referred to in the proof of (6.5).

If such a Tate variety X is also smooth, projective and of even dimension, then the usual relation between the signature $\sigma(X)$ of X and the Hodge numbers of X, see for instance [15], implies the following.

$$\sigma(X) = W(X, i), \quad \text{where} \quad i = \sqrt{-1}$$

Applying this formula we get:

(i) $\sigma(X) = \sum_{k=0,n}(-1)^k 2^{n-k} d_k$ for a smooth (or even simplicial) projective toric variety X as in (3.3.i) with n even;

(ii) $\sigma(G(k, n)) = \frac{n?}{k?(n-k)?}$ for a complex Grassmann manifold of even dimension $d = k(n - k)$, where $m?$ means the product of all even numbers between (and including) 2 and m;

(iii) $\sigma(\Lambda(\mathbb{C}^{2n})) = 0$ for a complex Lagrangian Grassmannian of even dimension $d = n(n+1)/2$.

REFERENCES

1. V. V. Batyrev and D. I. Dais, *Strong McKay correspondence, string-theoretic Hodge numbers and mirror symmetry*, preprint MPI-Bonn, 1994.
2. R. Bott and L. W. Tu, *Differential Forms in Algebraic Topology*, Grad. Texts in Maths. 82, Springer-Verlag, 1982.
3. V. I. Danilov and Khovanski, *Newton polyhedra and an algorithm for computing Hodge-Deligne numbers*, Math. USSR Izvestiya **29** (1987), 279–298.
4. P. Deligne, *Theorie de Hodge, II and III*, Publ. Math. IHES **40** (1971), 5–58 and **44** (1974), 5–77.
5. _____, *Poids dans la cohomologie des variétés algébriques*, Actes I.C.M. Vancouver (1974), 79–85.
6. _____, *SGA 4½, Cohomologie étale*, Lecture Notes in Math. 469, Springer, Berlin, 1977.
7. A. Dimca, *Singularities and Topology of Hypersurfaces*, Springer-Verlag, New York, 1992, Universitext.
8. A. Durfee, *Algebraic varieties which are disjoint unions of subvarieties*, Geometry and Topology: Manifolds, varieties, and knots (New York), Marcel Dekker, New York, 1987, pp. 99–102.
9. A. Fujiki, *Duality of mixed Hodge structures of algebraic varieties*, Publ. RIMS Kyoto Univ. **16** (1980), 635–667.
10. W. Fulton, *Introduction to Toric Varieties*, Annals of Math. Studies 131, Princeton Univ. Press, 1993.
11. W. Fulton and R. MacPherson, *A compactification of configuration spaces*, Ann. Math. **139** (1994), 183–225.
12. L. Göttsche, *Hilbert Schemes of Zero-dimensional Subschemes of Smooth Varieties*, Lecture Notes in Math. 1572, Springer-Verlag, 1994.
13. W. Greub, S. Halperin, and R. Vanstone, *Connections, Curvature and Cohomology*, vol. I, Academic Press, New York, 1972.
14. A. Grothendieck et al, *Séminaire de géometrie algébrique du bois-marie 1965-66*, Lecture Notes in Math 589, Springer-Verlag, Berlin, 1977.
15. F. Hirzebruch, *Topological Methods in Algebraic Geometry*, 3rd ed., Springer-Verlag, Berlin, 1966.
16. J. E. Humphreys, *Linear Algebraic Groups*, Grad. Texts in Maths. 21, Springer, 1975,
17. O. Jones, *The topology of symmetric and skew-symmetric matrices*, preprint Sydney Univ., 1994.
18. G. I. Lehrer, *The l-adic cohomology of hyperplane complements*, Bull. London Math. Soc. **24** (1992), 76–82.
19. _____, *Rational tori, semisimple orbits and the topology of hyperplane complements*, Comment Math. Helvetici **67** (1992), 226–251.
20. _____, *Poincaré polynomials for unitary reflection groups*, Inventiones Math. **120** (1995), 411–425.
21. _____, *A toral configuration space and regular semisimple conjugacy classes*, Math. Proc. Camb. Phil. Soc. **118** (1995), 105–113.
22. E. Looijenga, *Cohomology of M_3 and M_3^1*, in Mapping class groups and moduli of Riemann surfaces, Contemporary Mathematics (C.-F. Boedigheimer and R. Hain, eds.), vol. 150, 1993, pp. 205–228.
23. P. Orlik and L. Solomon, *Combinatorics and the topology of hyperpalne complements*, Invent. Math. **56** (1980), 167–189.
24. P. Orlik and H. Terao, *Arrangements of Hyperplanes*, Grund. math. Wissen. 300, Springer-Verlag, 1992.

25. G.-C. Rota, *On the foundations of combinatorial theory I; Möbius functions*, Zeit. für Wahrsch. Verw. Geb. **2** (1964), 340–368.
26. M. Saito, *Mixed Hodge modules*, Publ. RIMS Kyoto Univ. **26** (1990), 221–333.
27. J. P. Serre, *Linear Representations of Finite Groups*, Grad. Texts in Maths. 42, Springer-Verlag, 1977.
28. J. Steenbrink, *Intersection forms for quasihomogeneous singularities*, Compositio Math. **34** (1977), 211–223.
29. R. Steinberg, *Conjugacy classes in algebraic groups*, Lecture Notes in Math. 366, Springer-Verlag, Berlin, 1974.
30. C. T. C. Wall, *A note on symmetry of singularities*, Bull. London Math. Soc. **12** (1980), 169–175.

THE RESTRICTION OF THE REGULAR MODULE FOR A QUANTUM GROUP

STEPHEN DONKIN

Dedicated to the memory of Roger W. Richardson

INTRODUCTION

If G is a quantum group over a field k with coordinate algebra $k[G]$ and H is a (quantum) subroup it is extremely useful, in comparing the representation theory G and that of H to know that the regular module $k[G]$ is injective as an H-module. We prove here that in fact $k[G]|_H$ is a direct sum of copies of $k[H]$ when H is finite, under the hypothesis that the rate of growth of each finitely generated subalgebra of $k[G]$ is polynomial. The result is used in our related paper, [3]. The proof we shall give has elements in common with that of the freeness result of Nichols and Zoeller, [8].

TERMINOLOGY

We fix a field k. We adopt the philosophy of Parshall-Wang, [9], in regarding the category of quantum groups as the dual of the category of k-Hopf algebras and identifying the category of modules for a quantum group with the category of comodules for the corresponding Hopf algebra. Thus we shall use the expression "let G be a quantum k-group" (or "let G be a quantum group over k", or simply "let G be a quantum group") to indicate that we have in mind a Hopf algebra over k, called the coordinate algebra of G and denoted $k[G]$. If G is a quantum group we denote by $\delta_G : k[G] \to k[G] \otimes k[G]$, $\epsilon_G : k[G] \to k$, $\sigma_G : k[G] \to k[G]$ the comultiplication, augmentation and antipode maps of the Hopf algebra $k[G]$. We say that a quantum group G is finite if its coordinate algebra $k[G]$ is a finite dimensional k-space. We shall use the expression "$\theta : G \to H$ is a morphism of quantum groups" to indicate that G and H are quantum groups and we have in mind a morphism of Hopf algebras $\hat{\theta} : k[H] \to k[G]$, called the comorphism of θ. We shall use the expression "H is a quantum subgroup of G", or simply "H is a subgroup of G" to indicate that we have in mind a Hopf ideal I_H of $k[G]$ and that $k[H] = k[G]/I_H$. Thus if H is a subgroup of G we have the morphism $\theta : H \to G$ whose comorphism $\hat{\theta} : k[G] \to k[H]$ is the natural map. We call θ the inclusion map and call $\hat{\theta}$ the restriction map.

For a set X we write id_X for the identity map $X \to X$. Let G be a quantum group over k. By a (left) G-module we mean a right $k[G]$-comodule and by a morphism of (left) G-modules we mean a morphism of right $k[G]$-comodules. We write $\mathrm{Mod}(G)$ for the category of G-modules and write $V \in \mathrm{Mod}(G)$ to indicate that V is a G-module. For $V, W \in \mathrm{Mod}(G)$ we write $\mathrm{Hom}_G(V, W)$ for the space of G-module homomorphisms from V to W. For $V \in \mathrm{Mod}(G)$ we write τ_V for the structure map $V \to V \otimes k[G]$. For $V, W \in \mathrm{Mod}(G)$ the k-space $V \otimes W$ is naturally a G-module with structure map given by $\tau_{V \otimes W}(v \otimes w) = \sum_{i,j} v_i \otimes w_j \otimes f_i g_j$, for $v \in V$, $w \in W$ with $\tau_V(v) = \sum_i v_i \otimes f_i$ and $\tau_W(w) = \sum_j w_j \otimes g_j$. For $V \in \mathrm{Mod}(G)$ with k-basis $\{v_i : i \in I\}$ and $\tau_V(v_i) = \sum_j v_j \otimes f_{ji}$, for $i \in I$, the coefficient space $\mathrm{cf}(V) \leq k[G]$ is the k-sp an of $\{f_{ij} : i, j \in I\}$. The coefficient space is independent of the choice of basis and, for $V, W \in \mathrm{Mod}(G)$, we have $\mathrm{cf}(V \otimes W) = \mathrm{cf}(V) \cdot \mathrm{cf}(W)$, the k-span of all products fg, with $f \in \mathrm{cf}(V)$, $g \in \mathrm{cf}(W)$. If V is a G-module or k-space and $W \in \mathrm{Mod}(G)$ we write $|V| \otimes W$ for the k-space $V \otimes W$ regarded as a G-module via the structure map $\tau_{|V| \otimes W} = \mathrm{id}_V \otimes \tau_W : V \otimes W \to V \otimes W \otimes k[G]$. If V is a finite dimensional G-module we denote by V^* the dual module. If V has basis $v_1, \ldots, v_n$ and $\alpha_1, \ldots, \alpha_n$ is the dual basis of $V^* = \mathrm{Hom}_k(V, k)$ and $\tau_V(v_i) = \sum_j v_j \otimes f_{ji}$ then $\tau_{V^*}(\alpha_i) = \sum_j \alpha_j \otimes \sigma_G(f_{ij})$, for $1 \leq i, j \leq n$. The coordinate algebra $k[G]$, viewed as a G-module via $\delta_G : k[G] \to k[G] \otimes k[G]$, is called the regular G-module. For any $V \in \mathrm{Mod}(G)$ the structure map $\tau_V : V \to |V| \otimes k[G]$ is an injective G-module map. The regular G-module $k[G]$ is injective and $\mathrm{Mod}(G)$ has enough injectives. For $V, W \in \mathrm{Mod}(G)$ and $i \geq 0$ we write $\mathrm{Ext}_G^i(V, W)$ for the ith derived functor of $\mathrm{Hom}_G(V, -)$ evaluated at W. Let H be a subgroup of G and let $\theta : H \to G$ be inclusion. If V is a G-module we write $V|_H$ for the k-space V viewed as a H-module via the structure map $(\mathrm{id}_V \otimes \hat{\theta}) \circ \tau_V$ and call $V|_H$ the restriction of V to H.

THE THEOREM

For $V \in \mathrm{Mod}(G)$ we have a G-module isomorphism $\phi_V : V \otimes k[G] \to |V| \otimes k[G]$, defined by $\phi_V(v \otimes a) = \sum_i v_i \otimes f_i a$ for $v \in V$, $a \in k[G]$ with $\tau_V(v) = \sum_i v_i \otimes f_i$. (The inverse map $\psi_V : |V| \otimes k[G] \to V \otimes k[G]$ is given by $\psi_V(v \otimes a) = \sum_i v_i \otimes \sigma_G(f_i)a$ for $v \in V$, $a \in k[G]$ with $\tau_V(v) = \sum_i v_i \otimes f_i$.) A G-module I is injective if and only if it is a direct summand of a direct sum of copies of $k[G]$ and hence if I is injective then $V \otimes I$ is injective for every $V \in \mathrm{Mod}(G)$. We shall use ϕ_V and this remark in the proof of the Lemma. Also we shall use many times in this section the fact that a finite dimensional Hopf algebra is self injective, [6], hence in the category of modules for a finite quantum group H, the regular module $k[H]$ is both injective and projective. Furthermore we have $I \otimes X$ injective for $I, X \in \mathrm{Mod}(H)$ with I injective (and H finite). (We have that I is injective and hence projective for the dual Hopf algebra $k[H]^*$ (by self injectivity of finite dimensional Hopf algebras). Thus $I \otimes X$ is projective (e.g. by [4], Appendix T) and hence injective as a $k[H]^*$-module and H-module. See also [9], (2.8.2) Proposition,(2).)

Lemma. *Let G be a quantum group and H a finite subgroup.*

(i) *If $k[G]|_H$ is injective then it is a direct sum of copies of $k[H]$.*

(ii) *Suppose that there exists a G-module V which has an H-module decomposition $V = I \oplus Z$, where I is an injective H-module and the G-submodule W, say, generated by Z is not equal to V. Then $k[G]|_H$ is a direct sum of copies of $k[H]$.*

Proof. (i) Let $E_1, \ldots, E_m$ be a complete set of inequivalent injective indecomposable H-modules. The restriction map $k[G] \to k[H]$ is a split epimorphism of H-modules ($k[H]$ is an injective and hence projective H-module). Thus each E_i occurs as a summand of $k[G]|_H$.

Suppose first that G is finite. It follows from the Krull-Schmidt theorem that for some $r \geq 1$ we have an H-module decomposition $k[G]^{(r)} = F \oplus Y$, where F is a non-zero direct sum of copies of $k[H]$ and, for some $1 \leq j \leq m$, the injective module E_j does not occur as a summand of Y (cf [8], p. 383, paragraph 6). Now we have $V \otimes k[G] \cong |V| \otimes k[G] \cong k[G]^{(\dim V)}$ as G-modules, for any $V \in \mathrm{Mod}(G)$, and hence $X \otimes X \cong X^{(n)}$, where $X = k[G]^{(r)}$ and $n = \dim X$. We get $X^{(n)} \cong X \otimes F \oplus F \otimes Y \oplus Y \otimes Y \cong F^{(n)} \oplus F^{(\dim Y)} \oplus Y \otimes Y$, as H-modules. If s is the multiplicity of E_j as a summand of F, we get $ns = ns + (\dim Y)s + t$, where t is the multiplicity of E_j as a summand of $Y \otimes Y$. Thus we have $Y = 0$ so that $X = k[G]^{(r)}$ is a direct sum of copies of $k[H]$ and hence, by the Krull-Schmidt theorem, so is $k[G]$.

Now suppose that G is infinite. Since $k[G]|_H$ is injective it is a direct sum of copies of the modules $E_1, \ldots E_m$ and the assertion that $k[G]|_H$ is a direct sum of copies of $k[H]$ is equivalent to the assertion that each each E_i occurs with multiplicity the dimension of $k[G]$. We have an H-module decomposition $k[G] = D \oplus D'$ with $D \cong k[H]$, because $k[G] \to k[H]$ is a split surjection. Let V be a finite dimensional G-submodule of $k[G]$ containing D. We have $V = D \oplus Y$ for some H-submodule Y of V, by injectivity. Let $d = \dim V$. We have $k[G]^{(d)} \cong V \otimes k[G] = D \otimes k[G] \oplus Y \otimes k[G]$. But $D \cong k[H]$ so that, as an H-module, $D \otimes k[G]$ is isomorphic to the direct sum of $\dim k[G]$ copies of $k[H]$. It follows that, for $1 \leq i \leq m$, the module E_i occurs with multiplicity $\dim k[G]$ as a direct summand of $k[G]$. This finishes the proof.

(ii) By part (i) it suffices to show that $k[G]|_H$ is injective. Let $X, Y \in \mathrm{Mod}(H)$. Then $I \otimes Y$ is an injective H-module and hence inclusion $Z \otimes Y \to V \otimes Y$ induces an isomorphism $\mathrm{Ext}^1_H(X, Z \otimes Y) \to \mathrm{Ext}^1_H(X, V \otimes Y)$ and, since inclusion factors through $W \otimes Y$, the map $\mathrm{Ext}^1_H(X, W \otimes Y) \to \mathrm{Ext}^1_H(X, V \otimes Y)$ is surjective. We now take $Y = k[G]$. We have natural isomorphisms $\phi_W : W \otimes k[G] \to |W| \otimes k[G]$ and $\phi_V : V \otimes k[G] \to |V| \otimes k[G]$ and ϕ_W is the restriction of ϕ_V. Thus we have a commutative diagram

$$\begin{CD} \mathrm{Ext}^1_H(X, W \otimes k[G]) @>>> \mathrm{Ext}^1_H(X, V \otimes k[G]) \\ @VVV @VVV \\ \mathrm{Ext}^1_H(X, |W| \otimes k[G]) @>>> \mathrm{Ext}^1_H(X, |V| \otimes k[G]) \end{CD}$$

and the vertical maps are isomorphisms. Thus we get that the lower map is surjective and since $|W| \otimes k[G] \to |V| \otimes k[G]$ is a split injection we must have $\mathrm{Ext}^1_H(X, |V/W| \otimes k[G]) = 0$. But $W \neq V$ so we get $\mathrm{Ext}^1_H(X, k[G]) = 0$ for every $X \in \mathrm{Mod}(H)$ and $k[G]|_H$ is injective.

Theorem. *Let G be a quantum group over k and suppose that every finitely generated subalgebra of $k[G]$ has polynomial growth. If H is any finite subgroup of G then the restriction $k[G]|_H$ is a direct sum of copies of $k[H]$.*

Proof. Since H is finite the regular H-module $k[H]$ is projective and hence the restriction map $k[G] \to k[H]$ is a split H-module epimorphism. Thus $k[G]$ contains an H-submodule A, say, isomorphic to $k[H]$. Let V be the G-submodule of $k[G]$ generated by A. Then we have an H-module decomposition $V = A \oplus Y$. Let r be a positive integer. For any H-module X the tensor product $X \otimes k[H]$ is isomorphic to a direct sum of copies of $k[H]$. Hence we have an H-module decomposition $V^{\otimes r} = I_r \oplus Y_r$, where $Y_r = Y^{\otimes r}$ and I is injective. By the Lemma, it suffices to prove that, for r sufficiently large, Y_r generates a proper submodule of $V_r = V^{\otimes r}$. Let $C = \mathrm{cf}(V)$. Then $C_r = \mathrm{cf}(V^{\otimes r}) = C.C \ldots C$ (r times). Now any C_r-comodule may be naturally regarded as a module for the dual algebra $S_r = C_r^*$ and, for a C_r-comodule T a subspace S of T is a C_r-subcomodule if and only if it is stable under the action of S_r. The G-submodule W_r, say, of V_r generated by Y_r is the C_r-subcomodule generated by Y_r, i.e. the S_r-submodule generated by Y_r. We therefore have $\dim W_r \leq (\dim S_r).(\dim Y_r) = (\dim C_r).(\dim Y)^r$. By the hypothesis, there is a positive integer α such that $\dim C_r \leq r^\alpha$ for all $r \gg 0$. Thus we have

$$\dim W_r / \dim V_r = \dim W_r / (\dim V)^r \leq (\dim C_r)(\dim Y / \dim V)^r \leq \lambda^r r^\alpha$$

for $r \gg 0$, where $\lambda = \dim Y / \dim V$. Since $\lambda < 1$ we have $\lim\limits_{r \to \infty} \dim W_r / \dim V_r = 0$ and in particular $W_r \neq V_r$ for some r. Hence by the Lemma (with V_r in place of V and Y_r in place of Z) $k[G]|_H$ is a direct sum of copies of $k[H]$.

COMPLEMENTS

(1) It is easy to check that hypothesis of the Theorem is satisfied by the quantum version of GL_n introduced by R. Dipper and the author, [2], by the Manin quantization studied by Parshall-Wang,[9]and indeed by the two parameter quantization due to Takeuchi, [12]. The hypothesis is also satisfied by the Drinfeld-Jimbo quantization of an enveloping algebra studied by Lusztig, [7].

(2) If G is a linear algebraic group or more generally an affine group scheme over k then $k[G]$ is commutative so the hypothesis of the Theorem is satisfied and therefore the restriction of $k[G]$ to any (not necessarily reduced) finite sugbroup scheme H is a direct sum of copies of $k[H]$. Injectivity in this situation also follows from [5], I, 5.13 Corollary, (b) (which depends on Demazure-Gabriel, [1], III, §2, n°4 and §1,2.10) together with 4.12 Proposition.

Note also that the hypothesis is satisfied if G is finite.

(3) Let A be a Hopf algebra and let B be a finite dimensional subHopf algebra. For an A-module X we write $X|_B$ for the B-module obtained from X by restricting the action. The argument of the Lemma gives the following.

   (i) If B embeds in a finite dimensional A-module and $A|_B$ is projective then $A|_B$ is free.

   (ii) If there is an A-module V which has a B-module decomposition $V = P \oplus Z$ such that P is a projective B-module and Z does not generate V as an A-module then $A|_B$ is projective.

The argument of the Theorem gives the following.

(iii) If V is a finite dimensional A-module which contains a non-zero projective B-summand and the rate of growth of dim $A/\operatorname{Ann}_A(V^{\otimes r})$ is polynomial in r (where $\operatorname{Ann}_A(V^{\otimes r})$ is the annihilator in A of $V^{\otimes r}$) then $A|_B$ is projective.

If A is finite dimensional then we can take $V = A$ so that $A|_B$ is free and we obtain the result of Nichols and Zoeller, [8]. (This can also be obtained from the Theorem, 2) above, and duality for finite dimensional Hopf algebras.) Now assume A is commutative. Radford has shown (see [10], Theorem 1) that $A|_B$ is free in this case and we indicate how this may be obtained in the present context. The Hopf algebra A is the ascending union of finitely generated subHopf algebras containing B. If each of these is a B-module direct sum of copies of B then so is A (by the injectivity of B). Thus we may assume that A is finitely generated. If k has characteristic 0 then B is reduced, [11], Theorem 13.1.2, hence semisimple so that $A|_B$ is projective and hence free by (i). Now assume that k has characteristic $p > 0$. If X is a B-module and K is a field extension of k then $X_K = K \otimes_k X$ is a free $B_K = K \otimes_k B$-module if and only if X is a free B-module. Thus we may (and do) assume that k is algebraically closed. Now A is residually finite dimensional so we can choose a cofinite ideal N, say, such that $B \cap N = 0$. The B-module $T = A/N$ has a free B-module component. Let S be an indecomposable A-module component of T which has a non-zero projective B-module summand. Then (since A is commutative and k is algebraically closed) T has an A-module composition series with all factors isomorphic to some one dimensional A-module L, say. Now $V = T \otimes L^*$ has a non-zero projective B-module summand and the trivial module k is the only A-module composition factor of V. Thus V is annihilated by M^d, where $d = \dim V$ and M is the augmentation ideal of A. But M^d contains the ideal I generated by f^{p^d}, $f \in M$ and I is a Hopf ideal of finite codimension. So I annihilates $V^{\otimes r}$ we have dim $A/(\operatorname{Ann}_A(V^{\otimes r}) \le \dim A/I$ for all r and $A|_B$ is free, by (iii) and (i).

REFERENCES

1. M. Demazure and P. Gabriel, *Groupes Algébriques, Tome I*, North Holland, 1970.
2. R. Dipper and S. Donkin, *Quantum GL$_n$*, Proc. Lond. Math. Soc. (3) **63** (1991), 165–211.
3. S. Donkin, *Standard homological properties for quantum GL$_n$*, J. Algebra, to appear.
4. J. E. Humphreys, *Ordinary and Modular Representations of Chevalley Groups*, Lecture Notes in Mathematics 528, Springer, Berlin/Heidelberg/New York, 1976.
5. J. C. Jantzen, *Representations of Algebraic Groups*, Pure and Applied Mathematics 131, Academic Press, 1987.
6. R. G. Larson and M. Sweedler, *An associative orthogonal bilinear form for Hopf algebras*, Amer. J. Math. **92** (1969), 75–93.
7. G. Lusztig, *Introduction to Quantum Groups*, Progress in Mathematics 110, Birkhäuser, 1993.
8. W. D. Nichols and M. B. Zoeller, *A Hopf algebra freeness theorem*, Amer. J. Math. **111** (1989), 381–385.
9. B. Parshall and Jian-pan Wang, *Quantum linear groups*, Memoirs of the A.M.S **439** (1991).
10. D. E. Radford, *Freeness (projectivity) criteria for Hopf algebras over Hopf subalgebras*, J. Pure Applied Algebra **11** (1977), 15–28.
11. M. E. Sweedler, *Hopf algebras*, Benjamin, New York, 1969.
12. M. Takeuchi, *A two parameter quantization of GL(n) (summary)*, Proc. Japan Acad. **66A** (1990), 112–113.

SCHOOL OF MATHEMATICAL SCIENCES, QUEEN MARY AND WESTFIELD COLLEGE, MILE END RD. LONDON E1 4NS, ENGLAND

ON COEFFICIENTS OF q IN KAZHDAN-LUSZTIG POLYNOMIALS

M. J. DYER

Dedicated to the memory of Roger Richardson

ABSTRACT. It is shown that the coefficients of q in Kazhdan-Lusztig polynomials $P_{x,y}$ or $Q_{x,y}$ for arbitrary Coxeter groups are non-negative. The coefficient is interpreted as the difference between the dimension of a vector space (associated to the interval $[x, y]$ and a reflection representation of the Coxeter group) and the cardinality of a set of vectors spanning this space.

INTRODUCTION

For crystallographic Coxeter groups, the Kazhdan-Lusztig polynomials $P_{x,y}$, $Q_{x,y}$ [8, 9] admit several important interpretations which make non-negativity of their coefficients obvious. This non-negativity of these coefficients for arbitrary Coxeter systems has been conjectured by Kazhdan and Lusztig. The coefficients in general have recently been given a conjectural representation-theoretic interpretation which would make their non-negativity clear if it was proven. In this note, ideas suggested by this conjectural interpretation are used to give a simple proof of non-negativity of the coefficient of q in $P_{x,y}$ and $Q_{x,y}$ for general Coxeter groups. The non-negativity of the coefficient of q in $P_{x,y}$ has been independently established by Tagawa [6, 7], using a more combinatorial approach.

STATEMENT OF RESULTS

Let (W, S) be a finitely generated Coxeter system in its standard reflection representation on a finite dimensional vector space U, and let $X = [u, v]$ be an interval in the Chevalley (Bruhat) order on W. Define $E = E_X = \{(x, y) \in X \times X \,|\, x \lessdot y\}$ where $x \lessdot y$ means x is an immediate predecessor of y.

For $x \lessdot y$ in X, let $\alpha = L_{x,y}$ be the unique positive root with $y = s_\alpha x_1$ where s_α is the reflection in α. For any $x < y \in X$ with $l(y) - l(x) = 2$, the closed interval $[x, y]$ has two coatoms, say u_i for $i = 1, 2$ and there are unique real numbers a_{ij} such that $L_{x,u_i} = \sum_{j=1}^{2} a_{ij} L_{u_j,y}$. Define $V_X = V$ as the quotient vector space $V := \bar{V}/\bar{U}$ where $\bar{V}$ is a $\mathbb{R}$-vector space with a basis consisting of symbols $\bar{\alpha}_{x,y}$ for $(x, y) \in E_X$ and $\bar{U}$ is the $\mathbb{R}$-subspace of $\bar{V}$ spanned by elements $\bar{\alpha}_{x,u_i} - \sum_{j=1}^{2} a_{ij}\bar{\alpha}_{u_j,y}$, for all elements $x < y$ of X with $l(y) - l(x) = 2$ and for u_i, a_{ij} as above. In this note, we prove the following.

Partially supported by N. S. F. grant DMS90-12836

Theorem.   (i)   *The coefficient of q in the Kazhdan-Lusztig polynomial $P_{x,y}$ (resp., $Q_{x,y}$) is equal to $c_{x,y} - d_{x,y}$ (resp., $a_{x,y} - d_{x,y}$) where $c_{x,y}$ (resp., $a_{x,y}$) is the number of coatoms (resp., atoms) of the interval $X = [x,y]$ and $d_{x,y} = \dim_E V_X$.*

(ii)   *The coefficient of q in $P_{x,y}$ (resp., $Q_{x,y}$) is non-negative.*

PROOF OF THE THEOREM

We introduce some additional notation concerning the reflection representation and Chevalley order of W (see [5] for more details). Let $(\cdot|\cdot)$ be the standard W-invariant, symmetric bilinear form on U. Denote the simple roots, roots, positive roots of W on U by Π, $\Phi = W\Pi$; Φ^+ respectively, and the reflection in a root α by $s_\alpha : v \mapsto v - 2(v|\alpha)\alpha$ (we assume $(\alpha|\alpha) = 1$ for $\alpha \in \Phi$).

Let $l = \pm l_0 : W \to \mathbb{Z}$ where $l_0 : W \to \mathbb{Z}$ is the usual length function on (W, S). For $v, w \in W$, write $v \lessdot w$ if $l(w) - l(v) = 1$ and $v = s_\alpha w$ for some $\alpha \in \Phi^+$; we then write $\alpha = L_{v,w}$. The reflexive transitive closure of the relation $\lessdot$ is a partial order on W to be denoted henceforward by $\leq$ (it is Chevalley order if $l = l_0$, reverse Chevalley order if $l = -l_0$). We will use the following "Z"-property of $\leq$; if $x, y \in W$ and $s \in S$ satisfy $sx > x$ and $sy < y$, then $x \leq y$ iff $sx \leq y$ iff $x \leq sy$.

For any $x < y$ in W with $l(y) - l(x) = 2$, it is well known that $[x, y]$ has two coatoms (see e.g. [1]), say u_i for $i = 1, 2$ and it's easily seen [4] that there are unique real numbers a_{ij} such that

$$L_{x,u_i} = \sum_{j=1}^{n} a_{ij} L_{u_j,y}. \tag{1}$$

Moreover, one has

$$a_{11}a_{22} - a_{12}a_{21} \neq 0, \qquad a_{12}a_{21} \neq 0 \tag{2}$$

For a closed interval $X = [x_0, y_0]$ in the order $\leq$, define $E = E_X = \{(v, w) \in X \times X \mid v < w\}$. We now define the $\mathbb{R}$-vector space $V = V_X$ as before i.e. V is the quotient ector space $V := \bar{V}/\bar{U}$ where $\bar{V}$ is a $\mathbb{R}$-vector space with a basis consisting of symbols $\bar{\alpha}_{x,y}$ for $(x,y) \in E_X$ and $\bar{U}$ is the $\mathbb{R}$-subspace of $\bar{V}$ spanned by elements

$$\bar{\alpha}_{x,u_i} - \sum_{j=1}^{n} a_{ij}\bar{\alpha}_{u_j,y} \tag{3}$$

for a_{ij} as in (1), for all elements $x < y$ of X with $l(y) - l(x) = 2$. We also define a function $\alpha = \alpha_X : E_X \to V_X$ by setting $\alpha_X(v, w) = \bar{\alpha}_{v,w} + \bar{U} \in V_X$ for any $(v, w) \in E$, and define $d_{x_0,y_0} = \dim_E V_{[x_0,y_0]} \in \mathbb{Z}$. The following lemma gives a recurrence formula which determines all d_{x_0,y_0}.

Lemma.   (i)   *For any $x_0 \in W, d_{x_0,x_0} = 0$*

(ii)   *If $s \in S$ satisfies $sx_0 > x_0$ and $sy_0 < y_0$, then*

$$d_{sx_0,y_0} = d_{x_0,sy_0} \quad and \quad d_{x_0,y_0} = \begin{cases} d_{x_0,sy_0} & \text{if } sx_0 \leq sy_0 \\ d_{x_0,sy_0} + 1 & \text{if } sx_0 \nleq sy_0 \end{cases}$$

Proof. First, observe that

$$\langle \alpha(z, y_0) \mid x_0 \le z \lessdot y_0 \rangle = V_X = \langle \alpha(x_0, z) \mid x_0 \lessdot z \le y_0 \rangle \tag{4}$$

where $\langle \cdot \rangle$ denotes $\mathbb{R}$-linear span. Indeed, the first equality is obvious from the definitions, and the second follows similarly using invertibility of the matrices $(a_{ij})_{i,j=1}^2$. For any maximal chain $x_0 \lessdot x_1 \lessdot \cdots \lessdot x_n = y_0$ in X, one has

$$V = \langle \alpha(x_{i-1}, x_i) \mid i = 1, \ldots, n \rangle. \tag{5}$$

To see this, note that (5) follows easily from the second part of (2) in case $l(y_0) - l(x_0) = 2$. In general, it follows if two maximal chains in X differ in a single place, the spans as in (5) for the two chains coincide. Since the order complex of the open interval $(x_0, y_0) = X \backslash \{x_0, y_0\}$ is a combinatorial sphere [1], any two maximal chains in X can be taken as the first and last of a series of maximal chains in X, of which any consecutive two differ in a single place. Hence the span on the right of (5) is independent of the maximal chain, and so must be all of V.

Suppose above that one has $x_0 \le z < sz \le y_0$, and $sx_0 > x_0$, for some $s = s_\beta$, $\beta \in \Pi$. By [1], one may choose the maximal chain so $x_j = z$ for some j, and $sx_i > x_i$ for $i = 0, \ldots, j$. For $i = 1, \ldots, j$ one has $L_{x_{i-1}, sx_{i-1}} = \beta = L_{x_i, sx_i}$ and so by (3) applied to the length two interval $[x_{i-1}, sx_i]$, one has $\alpha(x_{i-1}, sx_{i-1}) = \alpha(x_i, sx_i)$. In particular, one finds

$$\alpha(z, sz) = \alpha(x_0, sx_0). \tag{6}$$

Now (i) is trivial. Assume x_0, y_0, s are as in (ii). We prove the second equation of (ii). Write $s = s_\beta$ where $\beta \in \Pi$. Set $Y = [x_0, sy_0]$, $E' = E_Y$, $V' = V_Y$ and $\alpha' = \alpha_Y$. From the definitions, there is a unique $\mathbb{R}$-linear map $\theta : V' \to V$ with $\theta(\alpha'(x, y)) = \alpha(x, y)$ for $(x, y) \in E' \subseteq E$. Taking a maximal chain with $x_{n-1} = sy_0$ in (5), we get

$$V = \langle \mathrm{Im}\, \theta, \alpha(sy_0, y_0) \rangle \tag{7}$$

which implies that $d_{x_0, y_0} \le d_{x_0, sy_0} + 1$. Now since $sx_0 > x_0$ and $sx_{n-1} > x_{n-1}$, if $sx_0 \le sy_0$ then $\alpha(x_0, y_0) = \alpha(x_0, sx_0) \in \mathrm{Im}\, \theta$ by (6) and hence $d_{x_0, y_0} \le d_{x_0, sy_0}$ by (7) in that case.

To complete the proof of the second formula in (ii), it will clearly suffice to construct a vector space $\bar{V}$ and a map $\bar{\alpha} : E_X \to \bar{V}$ with the following properties (a)–(c);

(a) $\bar{V} = \langle \mathrm{Im}\, \bar{\alpha} \rangle$
(b) for $x \le y$ in X with $l(y) - l(x) = 2$, one has $\bar{\alpha}(x, u_i) = \sum_{j=1}^2 a_{ij} \bar{\alpha}(u_j, y)$ for $i = 1, 2$, where u_i and the a_{ij} are as in (1).
(c) $\bar{V}$ has dimension d_{x_0, sy_0} if $sx_0 \le sy_0$ and dimension $d_{x_0, sy_0} + 1$ if $sx_0 \not\le sy_0$.

To this end, if $sx_0 \le sy_0$ we set $\bar{V} = V'$ and $\bar{\beta} = \alpha'(x_0, sx_0) \in \bar{V}$. If $sx_0 \not\le sy_0$, we define $\bar{V} = V' \oplus \mathbb{R}\bar{\beta}$ where $\mathbb{R}\bar{\beta}$ is a one-dimensional space with basis element $\bar{\beta}$. Clearly $\bar{V}$ satisfies (c) in either case. Regard V' as a subspace of $\bar{V}$. Now by the definitions, there is a unique linear map $\Psi' : V' \to U$ such that $\Psi'(\alpha'(x, y)) = L_{x,y}$ for $(x, y) \in E'$. Note Ψ' extends to a $\mathbb{R}$-linear map $\Psi : \bar{V} \to U$ with $\Psi(\bar{\beta}) = \beta$, since in case $sx_0 \leq sy_0$, one has $\Psi'(\bar{\beta}) = \Psi'(\alpha'(x_0, sx_0) = L_{x_0,sx_0} = \beta$ (recall $s = s_\beta$). Define now an involutory linear map $\bar{s} : \bar{V} \to \bar{V}$ by $\bar{s}(v) = v - 2(\Psi(v) \mid \beta)\bar{\beta}$ for $v \in \bar{V}$, and observe that $\Psi \circ \bar{s} = \bar{s} \circ \Psi : \bar{V} \to V$.

We may now define the map $\bar{\alpha} : E_X \to \bar{V}$. Fix $(x,y) \in E_X$. We set $\bar{\alpha}(x,y) = \alpha'(x,y)$ if $(x,y) \in E'$ (i.e. if $y \leq sy_0$) and $\bar{\alpha}(x,y) = \bar{\beta}$ if $x = sy$ (note that if $(sx, x) \in E'$, then $\alpha'(sx, x) = \bar{\beta}$ by (6), so this is consistent). It remains to define $\bar{\alpha}(x,y)$ for $(x,y) \in E \backslash E'$ with $x \neq sy$. In that case, the Z-property implies that $(sx, sy) \in E'$, and we set $\bar{\alpha}(x,y) = \bar{s}(\alpha'(sx, sy))$. Clearly, (a) holds also.

The definitions immediately give that for $(x,y) \in E$, one has $\Psi(\bar{\alpha}(x,y)) = L_{x,y}$. Consider $(x,y) \in E$ with $sx < x$ and $sy < y$. We claim that $\bar{s}(\bar{\alpha}(x,y)) = \bar{\alpha}(sx, sy)$. By the definition, one need only check this in case $(x,y) \in E'$. The interval $[sx, y]$ has two atoms $u_1 = sy$, $u_2 = x$. Now $L_{sx,sy} = s(L_{x,y}) = L_{x,y} - 2(L_{x,y} \mid \beta)L_{sy,y}$ so by (3) for V', $\alpha'(sx, sy) = \alpha'(x,y) - 2(L_{x,y} \mid \beta)\alpha'(sy,y) = \bar{s}(\bar{\alpha}(x,y))$ as required.

We are now able to verify the claim (b). Consider $x < y$ in X with $l(y) - l(x) = 2$, and let u_1, u_2 be the two coatoms of $[x,y]$. If $y \leq sy_0$, then (b) holds as an equality of elements of V', by definition of V'. Otherwise, one must have $sy < y$ by the Z-property. If $sx > x$, one has say $u_1 = sx$, $u_2 = sy$. Then $L_{x,sx} = \beta = L_{sy,y}$ and $\bar{\alpha}(x, sx) = \bar{\beta} = \bar{\alpha}(sy, y)$, while $L_{x,y} = s(L_{sx,y}) = L_{sx,y} - 2(L_{sx,y} \mid \beta)L_{sy,y}$ and $\bar{\alpha}(x, sy) = \bar{s}(\bar{\alpha}(sx, y)) = \bar{\alpha}(sx, y) - 2(L_{sx,y} \mid \beta)\bar{\alpha}(sy, y)$ as required. Hence, we may assume $sx < s$ and $sy < y$. If $su_i < u_i$ for $i = 1, 2$, then multiplication by s gives a poset isomorphism $[x, y] \to [sx, sy]$ such that for $z \lessdot z'$ in $[x, y]$, one has $L_{z,z'} = s(L_{sz,sz'})$ and $\bar{\alpha}(z, z') = \bar{s}(\bar{\alpha}(sz, sz'))$. Hence in that case, (b) for $[x, y]$ follows by applying $\bar{s}$ to (b) for $[sx, sy]$, which is known. In the remaining case, we have $sx < x$ and say $su_2 > u_2$, hence $u_2 = sy$ by the Z-property. We must then have $su_1 < u_1$, and so $sx < su_1 < sy$ by the Z-property. Hence, the length two subinterval $[sx, sy]$ of X, for which (b) has been shown to hold, has coatoms $p_1 = x$ and $p_2 = su_1$. In particular, $\bar{\alpha}(sx, p_j) \in U' := \langle \bar{\alpha}(p_i, sy) \mid i = 1, 2 \rangle$ for $j = 1, 2$. Note $\bar{\alpha}(sx, x) = \bar{\beta} \in U'$ so U' is $\bar{s}$-invariant. One now sees that for $z \lessdot z'$ in $[x, y]$, $\bar{\alpha}(z, z') \in U'$, noting $\bar{\alpha}(u_1, y) = \bar{s}(\bar{\alpha}(p_2, sy), \bar{\alpha}(x, u_1) = \bar{s}(\bar{\alpha}(sx, p_2))$ and $\bar{\alpha}(u_2, y)\bar{\beta}$. On the other hand, $\Psi(\bar{\alpha}(z, z')) = L_{z,z'}$ and L_{x,u_1}, L_{x,u_2} are distinct postive roots, hence linearly independent. This shows that Ψ restricts to a linear isomorphism $\Psi'' : U' \to \langle L_{z,z'} \mid x \leq z \lessdot z' \leq y \rangle$. One has $L_{x,u_i} \sum_{j=1}^{2} a_{ij} L_{u_j, y}$; applying the inverse of Ψ'', recalling $\Psi''(\bar{\alpha}(z, z')) = L_{z,z'}$ gives (b) for $[x, y]$ as required.

We have now completed the proof of all claims (a)–(c), and hence the second equation in (ii). By symmetry (i.e. the same argument with l replaced by $-l$), one has also that d_{x_0,y_0} is equal to d_{sx_0,y_0} if $sx_0 \leq sy_0$ and to $d_{sx_0,y_0} + 1$ otherwise. Hence the first equation in (ii) follows.

Proof of the theorem. First, we recall the definition of (a variant of) the polynomials $R_{x,y}$ from [8]. Let $A = \mathbb{Z}[u, u^{-1}]$ be the ring of integral Laurent polynomials in an indeterminate u. It is known that there are unique elements $\tau_{x,y} \in A$ for $x \leq y$ in W such that $\tau_{x,x} = 1$ and for $s \in R$, $x, y \in W$ with $sx > x$ and $sy < y$ one has

$$\tau_{x,y} = (u - u^{-1})\tau_{x,sy} + \tau_{sx,sy} \quad \text{and} \quad \tau_{sx,y} = \tau_{x,sy}. \tag{8}$$

Here, we interpret $\tau_{v,w}$ as zero if $v \not\leq w$. (In case $l = l_0$, $q^{(l(y)-l(x))/2}\tau_{x,y}(q^{1/2}) = R_{x,y}$ as defined in [8]). Induction, using the Lemma, (8), and the Z-property, shows that

$$\tau_{x,y} = u^n - d_{x,y}u^{n-2} + \text{ lower powers of } u \qquad (9)$$

if $x \leq y$, where $n = l(y) - l(x)$. It is also known (see [8, 2]) that there are unique elements $p_{x,y} \in \mathbb{Z}[u]$ for $x \leq y$ in W satisfying $p_{x,x} = 1$ if $x \in W$, $p_{x,y} \in u\mathbb{Z}[u]$ if $x < y$, and

$$p_{x,y} = \sum_{w\in[x,y]} \tau_{x,w}p_{w,y}(u^{-1}) \qquad (10)$$

for $x \leq y$. For $x \leq y$, one sees from the definition and (10) that

$$p_{x,y} = u^n + (c_{x,y} - d_{x,y})u^{n-2} + \text{ lower powers of } u \qquad (11)$$

where $c_{x,y} = \#\{z \mid x \leq z \lessdot y\}$ is the number of coatoms of the interval $[x,y]$ and $n = l(y) - l(x)$ again. But (4) immediately implies that $d_{x,y} \leq c_{x,y}$ since $V_{[x,y]}$ is spanned by $\alpha_{[x,y]}(z,y)$ for $x \leq z < y$.

Finally, to complete the proof of the theorem one has only to note that $p_{x,y} = u^{l(y)-l(x)}P_{x,y}(u^{-2})$ if $l = l_0$ and $p_{x,y} = u^{l(y)-l(x)}Q_{y,x}(u^{-2})$ if $l = -l_0$ (see [8, 9] and [2, 1.2(iii)]).

Remark. In [3], certain generalizations of Chevalley order and its reverse were studied; they are defined as $\leq$ was in this paper, but using a more general length function l on W than just $\pm l_0$. Using results of [3], the preceding arguments apply mutatis mutandis to suitable intervals in these orders (precisely, closed intervals $[x,y]$ which are finite and for which every closed length two subinterval has cardinality four). Namely, for such an interval $[x,y]$, one defines the space $V_{[x,y]}$ and the elements $\tau_{x,y}$ and $p_{x,y}$ of A exactly as here (using of course the new order $\leq$), and the coefficient of $u^{l(y)-l(x)-2}$ in $p_{x,y}$ is $\#\{z \mid x \leq z \lessdot y\} - \dim V_{[x,y]} \geq 0$.

REFERENCES

1. A. Björner and M. Wachs, *Bruhat order of Coxeter groups and shellability*, Adv. in Math. **43** (1982), 87–100.
2. M. Dyer, *Hecke algebras and shellings of Bruhat intervals*, Comp. Math. **89** (1993), 91–115.
3. ———, *Hecke algebras and shellings of Bruhat intervals II; twisted Bruhat orders*, in Kazhdan-Lusztig theory and related topics (Providence, Rhode Island), Contemp. Math., vol. 139, Amer. Math. Soc., Providence, Rhode Island, 1992, pp. 141–165.
4. ———, *Algebras associated to Bruhat intervals and polyhedral cones*, in Finite dimensional algebras and related topics, Kluwer academic publishers, Dordrecht. Boston. London, 1994, pp. 95–122.
5. J. Humphreys, *Reflection groups and Coxeter groups*, Cambridge Studies in Advanced Math., 29, Cambridge University Press, 1990.
6. H. Tagawa, *On the first coefficients in q of the Kazhdan-Lusztig polynomials*, preprint.
7. ———, *On the non-negativity of the first coefficient of Kazhdan-Lusztig polynomials*, preprint.
8. D. Kazhdan and G. Lusztig, *Representations of Coxeter groups and Hecke algebras*, Invent. Math. **53** (1979), 165–184.

9. _____ , *Schubert varieties and Poincaré duality*, Proc. Symp. Pure Math. of Amer. Math. Soc. **36** (1980), 185–203.

DEPARTMENT OF MATHEMATICS, MAIL DISTRIBUTION CENTER, UNIVERSITY OF NOTRE DAME, INDIANA, 46556-5683

SPECTRAL ESTIMATES FOR POSITIVE ROCKLAND OPERATORS

A. F. M. TER ELST AND DEREK W. ROBINSON

Dedicated to the memory of Roger W. Richardson

ABSTRACT. Let $(\mathscr{H}, G, U)$ be an irreducible unitary representation of a homogeneous Lie group G and H a self-adjoint operator on $\mathscr{H}$ associated with a positive Rockland operator. We derive upper and lower bounds on the eigenvalue distribution of H in terms of volume estimates on the coadjoint orbit corresponding to the representation U. Hence we deduce bounds on the partition function $\beta \mapsto \mathrm{Tr}_{\mathscr{H}}(\exp(-\beta H))$. An application is given to the spectrum and eigenfunctions of the general anharmonic oscillator.

1. INTRODUCTION

Our purpose is to derive spectral estimates for Rockland operators H in each irreducible unitary representation of a homogeneous group G. These estimates are expressed in terms of the symbol of the differential operator H and are similar in spirit to the estimates of the spectra of quantum-mechanical Hamiltonians in terms of classical phase space integrals (see, for example, [26], [7]). The estimates extend recent results of Levy-Bruhl and Nourrigat [16] and Levy-Bruhl, Mohamed and Nourrigat [14] for sublaplacians on stratified groups and are related to Weyl's classical results on the asymptotic distribution of eigenvalues of the Laplacian on bounded regions. Our proofs partially rely upon the work of the above authors. Similar results for strongly elliptic operators have also been given by Manchon [18], [17] although his methods are quite different.

A differential operator H on a homogeneous group G is defined to be a Rockland operator if it is right-invariant, homogeneous and injective in each nontrivial irreducible unitary representation. The theory of Rockland operators began with Rockland's analysis of differential operators on the Heisenberg group [24]. Helffer and Nourrigat [10] proved that a Rockland operator on a graded group is hypoelliptic and in addition they derived several inequalities between the norm on the C^n-spaces and the operator norm. Then Miller [19] showed that one can replace a graded group by a homogeneous group in the Helffer–Nourrigat theorem. Subsequently, Folland and Stein [8] used the proof of an earlier theorem of Nelson and Stinespring [20] to deduce that a positive Rockland operator is essentially self-adjoint on the space $C_c^\infty(G)$. Moreover, they established that the closure generates a continuous semigroup with a kernel which is in the Schwartz space over the group. Hence it follows by a general structural theorem for nilpotent groups (see, for example, [2] Theorem 4.2.1) that the operators $\exp(-\beta H)$, $\beta > 0$, are trace class in each irreducible unitary representation. Our aim is to estimate the partition functions $\beta \mapsto Z(\beta) = \mathrm{Tr}(\exp(-\beta H))$ for each such representation. This problem is closely related to the estimation of the number $N(\lambda)$ of eigenvalues of H with values less than or equal to λ, because Z is the Abel transform of N.

Throughout the sequel we adopt the notation of [1], in which we used the general notation of [23], but to make this paper more self-contained we repeat the main definitions. Let G be a connected, simply connected, homogeneous group with Lie algebra $\mathfrak{g}$ and let $(\mathscr{X}, G, U)$ denote a strongly continuous, or weakly*, continuous representation of G on the Banach space $\mathscr{X}$ by bounded operators $g \mapsto U(g)$. If $a_i \in \mathfrak{g}$ then $A_i (= dU(a_i))$ will denote the generator of the one-parameter subgroup $t \mapsto U(\exp(-ta_i))$ of the representation. Let $(\gamma_t)_{t>0}$ be a family of **dilations** on $\mathfrak{g}$, i.e., a one-parameter group of automorphisms of the form

$$\gamma_t(a_i) = t^{w_i} a_i$$

for some basis $a_1, \ldots, a_d$ of $\mathfrak{g}$ and some positive numbers $w_1, \ldots, w_d$, which we call **weights**. We always assume that the smallest weight is at least one. Let $|||\cdot|||$ be a **homogeneous norm** on $\mathfrak{g}^*$, i.e., a norm such that $|||\gamma_t^*(l)||| = t\,|||l|||$ for all $l \in \mathfrak{g}^*$ and $t > 0$. A homogeneous norm can be constructed as follows. Let $\|\cdot\|$ be the dual norm on $\mathfrak{g}^*$ of a Euclidean norm $\|\cdot\|$ on $\mathfrak{g}$. Define $|||\cdot||| : \mathfrak{g}^* \to \mathbb{R}$ by

$$|||l||| = \inf\{\lambda > 0 : \|\gamma_{1/\lambda}^* l\| \leq 1\}.$$

One readily verifies that $|||\cdot|||$ is a homogeneous norm.

Next we introduce a multi-index notation. If $n \in \mathbb{N}_0$ let

$$J_n(d) = \bigoplus_{k=0}^{n} \{1, \ldots, d\}^k$$

and set

$$J(d) = \bigcup_{n=0}^{\infty} J_n(d).$$

Then if $\alpha = (i_1, \ldots, i_n) \in J(d)$ we denote the **Euclidean length** n of α by $|\alpha|$ and the **weighted length** by

$$\|\alpha\| = \sum_{k=1}^{n} w_{i_k}.$$

If $n \in \mathbb{N}$ we define $\mathscr{X}_n = \mathscr{X}_n(U) = \bigcap_{\alpha \in J_n(d)} D(A^\alpha)$ and

$$\|x\|_n = \max_{\substack{\alpha \in J(d) \\ |\alpha| \leq n}} \|A^\alpha x\|,$$

where $A^\alpha = A_{i_1} \ldots A_{i_n}$ if $\alpha = (i_1, \ldots, i_n)$. Similarly we define the weighted C^n-spaces

$$\mathscr{X}_n' = \mathscr{X}_n'(U) = \bigcap_{\substack{\alpha \in J(d) \\ \|\alpha\| \leq n}} D(A^\alpha)$$

for all $n \in \mathbb{R}$ with $n > 0$. Now, however, it can happen for a given n that there are no multi-indices α such that $\|\alpha\| = n$. Therefore the corresponding norms and seminorms are given by

$$\|x\|_n' = \|x\|_{U,n}' = \begin{cases} \max_{\substack{\alpha \in J(d) \\ \|\alpha\| \le n}} \|A^\alpha x\| & \text{if there exist } \alpha \in J(d) \text{ with } \|\alpha\| = n \\ 0 & \text{otherwise} \end{cases}$$

$$N_n'(x) = \begin{cases} \max_{\substack{\alpha \in J(d) \\ \|\alpha\| = n}} \|A^\alpha x\| & \text{if there exist } \alpha \in J(d) \text{ with } \|\alpha\| = n \\ 0 & \text{otherwise} \end{cases}$$

Note that if $b_1, \ldots, b_d$ is another basis for $\mathfrak{g}$ which satisfies $\gamma_i(b_i) = t^{v_i} b_i$ then the weighted C^n-space with respect to the basis $b_1, \ldots, b_d$ equals the space $\mathscr{X}_n'$, and, if there exists an $\alpha \in J(d)$ with $\|\alpha\| = n$ the norms are also equivalent. Moreover, let $\mathscr{X}_\infty = \mathscr{X}_\infty(U) = \bigcap_{n=1}^\infty \mathscr{X}_n$. It follows by a line by line extension of Lemma 2.4 of [3] that the Gårding space, and in particular the space $\mathscr{X}_\infty$, is dense in $\mathscr{X}_n'$ for all $n > 0$. The density is with respect to the weak, or weak*, topology corresponding to the continuity property of the representation. Further we let L denote the left regular representation on $L_2(G)$.

Let $m \in \langle 0, \infty \rangle$ and let $C: J(d) \to \mathbb{C}$ be such that $C(\alpha) = 0$ if $\|\alpha\| > m$ and there exists at least one $\alpha \in J(d)$ with $\|\alpha\| = m$ and $C(\alpha) \ne 0$. We call C a **form** of order m. We write $c_\alpha = C(\alpha)$. The **principal part** P of C is the form

$$P(\alpha) = \begin{cases} C(\alpha) & \text{if } \|\alpha\| = m, \\ 0 & \text{if } \|\alpha\| < m. \end{cases}$$

We say that C is **homogeneous** if $C = P$. The **formal adjoint** $C^\dagger$ of C is the function $C^\dagger: J(d) \to \mathbb{C}$ defined by

$$C^\dagger(\alpha) = (-1)^{|\alpha|} \overline{C(\alpha_*)},$$

where $\alpha_* = (i_n, \ldots, i_1)$ if $\alpha = (i_1, \ldots, i_n)$. We consider the operators

$$dU(C) = \sum_{\alpha \in J(d)} c_\alpha A^\alpha$$

with domain $D(dU(C)) = \mathscr{X}_m'$.

If P is the principal part of a form C we call P a **Rockland form** if the operator $dU(P)$ is injective on the space $\mathscr{X}_\infty(U)$ for every nontrivial irreducible unitary representation U of G. It follows then from the Helffer–Nourrigat theorem [10] that $dL(P)|_{C_c^\infty(G)}$ is a hypoelliptic operator. In fact the Helffer–Nourrigat theorem is formulated for graded groups. But it follows from Propositions 1.3 and 1.4 of [19] that the existence of a Rockland form ensures that the order m of P is an integer multiple of the smallest weight and all weights are rational multiples of this smallest weight. Therefore G is a graded group if one rescales the original weights by a large enough constant. (Actually there is a small gap in the proof of Proposition 1.3 in [19] where Miller applies his Lemma 1.2. For the operators that we consider we prove a stronger theorem in the spirit of Proposition 1.3 of [19]. This proof requires a lemma, Lemma 2.2, which also fills the gap in [19].)

A Rockland form P is called a **positive Rockland form** if $dL(P)$ is symmetric and $(\varphi, dL(P)\varphi) \geq 0$ for all φ in the Schwartz space on G (see [8], page 129). Throughout this paper we assume that C is a form of order m and that the principal part P of C is a positive Rockland form. We call $dL(P)$ a **positive Rockland operator**.

We study operators $dU(C)$ where U is a irreducible unitary representation. The irreducible unitary representations of a nilpotent Lie group are described by the Kirillov theory (see [13], [2], [22]). There is a one-to-one correspondence between the orbits in $\mathfrak{g}^*$ under the coadjoint action and the unitary dual of G. For an irreducible unitary representation U we denote by $\mathcal{O}_U$ the corresponding orbit in $\mathfrak{g}^*$ and let μ_U be the canonical invariant measure on $\mathcal{O}_U$ (see [2] Section 4.3).

At this point we can state a theorem which indicates the nature of our results.

Theorem 1.1. *Let $(\mathcal{H}, G, U)$ be an irreducible unitary representation of G and C a form of order m whose principal part P is a positive Rockland form. If there exists an $\omega > 0$ such that $dU(C) \geq \omega I$ then there is a $c > 0$ such that*

$$c^{-1} \int_{\mathcal{O}_U} d\mu_U(l)\, e^{-c\beta |||l|||} \leq \mathrm{Tr}_{\mathcal{H}}(e^{-\beta H}) \leq c \int_{\mathcal{O}_U} d\mu_U(l)\, e^{-c^{-1}\beta |||l|||}$$

for all $\beta > 0$. Moreover, these estimates are valid uniformly for all irreducible unitary representations whenever C is a positive Rockland form.

Note that the condition $dU(C) \geq \omega I$ automatically implies that $dU(C)$ is a positive operator, and hence a self-adjoint operator. The theorem automatically applies to positive Rockland forms because the estimate $dU(C) \geq \omega I$ is a direct consequence of the injectivity hypothesis. The ensuing uniformity over the irreducible representations will be a consequence of the proof. It relies upon a scaling argument. The estimates can be rephrased in terms of Euclidean integrals and symbols of differential operators by using more details of representation theory. These estimates, which will be derived in Section 4, are the direct analogue of the classical phase space estimates for quantum-mechanical partition functions.

The bounds for the partition function given by the theorem can be evaluated in greater detail in particular cases. As an illustration we consider spectral properties of the anharmonic oscillator in Section 5. We establish that the eigenvalues of the operator $P^{2j} + Q^{2k}$ satisfy the bounds $c_0^{-1} n^{2jk/(j+k)} \leq \lambda_n \leq c_0 n^{2jk/(j+k)}$ and the corresponding orthonormal eigenfunctions φ_n can be extended to entire functions on the complex plane satisfying growth bounds

$$|\varphi_n(x + iy)| \leq c^n e^{-a|x|^{(j+k)/j} + b|y|^{(j+k)/j}}$$

for some $a, b, c > 0$, independent of n.

2. POSITIVE ROCKLAND OPERATORS

In this section we prove some additional regularity theorems for operators $H = dU(C)$ associated with a (not necessarily unitary) representation $(\mathscr{X}, G, U)$ and a form C whose principal part is a positive Rockland form of order m. In particular we prove that H satisfies a Gårding inequality if U is a unitary representation. Next recall that Theorem 3.4 of [1] establishes that the closure $\overline{H}$ of H generates a continuous semigroup S which is holomorphic in the right half-plane and has a representation independent kernel K which depends only on the form C.

Proposition 2.1. *Let* $(\mathscr{X}, G, U)$ *be a (general) continuous representation of* G, C *a form of order* m *whose principal part is a positive Rockland form,* $H = dU(C)$ *and* S *the semigroup generated by* $\overline{H}$.

I.   *If* $n \in \mathbb{N}$ *and* $1 \leq k < mn$ *then* $D(\overline{H}^n) \subseteq \mathscr{X}_k'$ *and there exists* $c > 0$ *such that*

$$\|x\|_k' \leq \varepsilon^{mn-k}\|\overline{H}^n x\| + c\varepsilon^{-k}\|x\|$$

for all $x \in D(\overline{H}^n)$ *and* $\varepsilon \in \langle 0, 1]$. *In particular*

$$\mathscr{X}_\infty = \bigcap_{n=1}^{\infty} D(\overline{H}^n)$$

and S *maps into the smooth elements, i.e.,* $S_t \mathscr{X} \subseteq \mathscr{X}_\infty$ *for all* $t > 0$.

II.   *If* $k \in \mathbb{N}$ *then there exists* $c_k > 0$ *such that*

$$\|S_t x\|_k' \leq c_k t^{-k/m}\|x\|$$

for all $t \in \langle 0, 1]$ *and* $x \in \mathscr{X}$.

III. $\mathscr{X}_\infty$ *is a core for* $\overline{H}$.

Proof. Let $M, \rho \geq 0$ be such that $\|U(g)\| \leq Me^{\rho|g|'}$ for all $g \in G$, where $|\cdot|'$ is a homogeneous modulus on G. It follows as in Appendix A of [4] that the resolvent kernel $R_\lambda^{(n)}$ defined by

$$R_\lambda^{(n)}(g) = (n-1)!^{-1} \int_0^\infty dt\, e^{-\lambda t} t^{n-1} K_t(g)$$

belongs to $L_{1;nm-1}^{\rho'}(G)$ and $\|R_\lambda^{(n)}\|_{1;nm-1}^{\rho'} \leq c\lambda^{-(nm-k)/m}$ for all sufficiently large λ. Here K_t is the kernel of S_t (see [1]) and $L_{1;nm-1}^{\rho'}(G)$ is the space of weighted C^{nm-1}-vectors with respect to the left regular representation of G in $L_1(G; e^{\rho|g|'}dg)$ with norm $\|\cdot\|_{1;nm-1}^{\rho'}$. So $(\lambda I + \overline{H})^{-n} = U(R_\lambda^{(n)})$ maps $\mathscr{X}$ into $\mathscr{X}_k'$ and

$$\|(\lambda I + \overline{H})^{-n} x\|_k' \leq \|R_\lambda^{(n)}\|_{1;nm-1}^{\rho'}\|x\| \leq c\lambda^{-(nm-k)/m}\|x\|$$

for all $x \in \mathscr{X}$. The proposition now follows as at the end of the proof of Theorem 2.6 of [5]. Note that the constant c depends on the representation U only through the values of M and ρ.

The next lemma is slightly stronger than Lemma 1.2 of Miller [19] and should be used in Proposition 1.3 of [19].

Lemma 2.2. *Let $\mathfrak{g}$ be a homogeneous Lie algebra with dilations $(\gamma_t)_{t>0}$. Then there exist a basis $b_1, \ldots, b_d$ of $\mathfrak{g}$, $v_1, \ldots, v_d \geq 1$ and $d' \in \{1, \ldots, d\}$ such that $[\mathfrak{g}, \mathfrak{g}] \subset \mathrm{span}\{b_{d'+1}, \ldots, b_d\}$, and $\gamma_t(b_i) = t^{v_i} b_i$ for all $i \in \{1, \ldots, d\}$ and all $t > 0$. Moreover, $b_1, \ldots, b_{d'}$ is an algebraic basis for $\mathfrak{g}$.*

Proof. There exist $1 \leq u_1 < \ldots < u_n$ and non-trivial subspaces $\mathfrak{g}_{u_1}, \ldots, \mathfrak{g}_{u_n}$ of $\mathfrak{g}$ such that $\mathfrak{g} = \oplus_{i=1}^n \mathfrak{g}_{u_i}$ and $\gamma_t(a) = t^{u_i} a$ for all $i \in \{1, \ldots, n\}$, $t > 0$ and $a \in \mathfrak{g}_{u_i}$. Define $\mathfrak{g}_u = \{0\}$ if $u \notin \{u_1, \ldots, u_n\}$. Then $[\mathfrak{g}_u, \mathfrak{g}_v] \subseteq \mathfrak{g}_{u+v}$ for all $u, v \geq 1$. For all $i \in \{1, \ldots, n\}$ let $b_{i1}, \ldots, b_{id'_i}, b_{id'_i+1}, \ldots, b_{id_i}$ be a basis for $\mathfrak{g}_{u_i}$ such that $b_{id'_i+1}, \ldots, b_{id_i}$ is a basis for

$$\mathfrak{g}_{u_i} \cap \left(\mathrm{span} \bigcup_{u+v=u_i} [\mathfrak{g}_u, \mathfrak{g}_v] \right).$$

Then obviously $[\mathfrak{g}, \mathfrak{g}] \subseteq \mathrm{span}\{b_{ij} : i \in \{1, \ldots, n\}, \ j \in \{d'_i + 1, \ldots, d_i\}\}$ since $\mathfrak{g} = \oplus_{i=1}^n \mathfrak{g}_{u_i}$. Let $\mathfrak{h}$ be the Lie algebra generated by $\{b_{ij} : i \in \{1, \ldots, n\}, \ j \in \{1, \ldots, d'_i\}\}$. Then it follows by induction on N that $\oplus_{i=1}^N \mathfrak{g}_{u_i} \subseteq \mathfrak{h}$ and hence $\mathfrak{h} = \mathfrak{g}$ and $\{b_{ij} : i \in \{1, \ldots, n\}, \ j \in \{1, \ldots, d'_i\}\}$ is an algebraic basis for $\mathfrak{g}$. Now a basis with the required properties is given by the combination $b_{11}, \ldots, b_{1d'_1}, \ldots, b_{n1}, \ldots, b_{nd'_n}, b_{1d'_1+1}, \ldots, b_{1d_1}, \ldots, b_{nd'_n+1}, \ldots, b_{nd_n}$.

We will call a basis $b_1, \ldots, b_{d'}, \ldots, b_d$ an **adapted basis** for the homogeneous Lie algebra and $v_1, \ldots, v_d$ the corresponding weights if it satisfies the conclusion of Lemma 2.2. Note that the v_i are a permutation of the w_i.

Example 2.3. Let $\mathfrak{g}$ be the 4-dimensional homogeneous Lie algebra defined by

$$[a_1, a_2] = a_3 + a_4, \quad \gamma_t(a_1) = t a_1, \quad \gamma_t(a_2) = t a_2, \quad \gamma_t(a_3) = t^2 a_3, \quad \gamma_t(a_4) = t^2 a_4.$$

Then one can take $b_1 = a_1$, $b_2 = a_2$, $b_3 = a_3$ and $b_4 = a_3 + a_4$.

Lemma 2.4. *Let C be a form of order m whose principal part is a positive Rockland form and let $b_1, \ldots, b_{d'}, \ldots, b_d$ be an adapted basis with weights $v_1, \ldots, v_d$. Then m/v_i is even for all $i \in \{1, \ldots, d'\}$. In particular there exist $\alpha \in J(d)$ such that $\|\alpha\| = m/2$.*

Proof. We may assume that C is homogeneous of order m. Fix $j \in \{1, \ldots, d'\}$. Define $U \colon G \to \mathscr{L}(L_2(\mathbb{R}))$ by

$$\left(U(\exp(\sum_{n=1}^d \xi_n b_n)) f \right)(t) = f(t + \xi_j).$$

Then it follows from the inclusion $[\mathfrak{g}, \mathfrak{g}] \subseteq \mathrm{span}\{b_{d'+1}, \ldots, b_d\}$ and the Campbell–Baker–Hausdorff formula that U is a unitary representation. Then $B_j = D$, the differentiation operator, and $B_k = 0$ if $k \neq j$. Let $H = dU(C)$ and

$$c = \begin{cases} C(\alpha) & \text{if } m/v_j \in \mathbb{N} \text{ and } \alpha = (j, \ldots, j), \ \|\alpha\| = m, \\ 0 & \text{if } m/v_j \notin \mathbb{N}. \end{cases}$$

Then $H = c D^{m/v_j}$. Since $D(H) = L'_{2;m}(\mathbb{R})$ by Theorem 3.6 of [1] we have $c \neq 0$, so $m/v_j \in \mathbb{N}$. Moreover, H is self-adjoint since it is symmetric and the generator of a holomorphic semigroup by Theorem 3.4 of [1]. But as generator of a continuous semigroup it has to be lower semibounded. So cD^{m/v_j} must be lower semibounded and this is only possible if m/v_j is even.

It is now fairly standard to prove the Gårding inequality. An important tool is that an operator $dU(C)$ is self-adjoint if the representation U is unitary and $C = C^\dagger$.

Theorem 2.5. *Let C be a form of order m whose principal part is a positive Rockland form. Then there exist $p > 0$ and $q \in \mathbb{R}$ such that for each unitary representation U*

$$\mathrm{Re}(dU(C)x, x) \geq p(\|x\|'_{m/2})^2 - q\,\|x\|^2$$

for all $x \in D(dU(C))$. If the form C is homogeneous then there exists $p > 0$, where the value is independent of U, such that

$$\mathrm{Re}(dU(C)x, x) \geq pN'_{m/2}(x)^2$$

for all $x \in D(dU(C))$.

Proof. We may assume that the basis $a_1, \ldots, a_{d'}, \ldots, a_d$ is an adapted basis. Let C_1 be the form such that

$$dU(C_1) = \sum_{\substack{\alpha \in J(d) \\ \|\alpha\| \leq m/2}} (-1)^{|\alpha|} A^{(\alpha_*, \alpha)}$$

for any representation $(\mathscr{X}, G, U)$. Then the principal part P_1 of C_1 is a positive Rockland form. Indeed, if $(\mathscr{X}, G, U)$ is a nontrivial irreducible unitary representation, $x \in \mathscr{X}_\infty(U)$ and $dU(P_1)x = 0$ then $(dU(P_1)x, x) = 0$ and therefore $A^\alpha x = 0$ for all α with $\|\alpha\| = m/2$. Hence by Lemma 2.4 one deduces that $A_i x = 0$ for all $i \in \{1, \ldots, d'\}$. Since $a_1, \ldots, a_{d'}$ is an algebraic basis this implies that $x = 0$. So

$$\|dU(C_1)^{1/2}x\|^2 = (dU(C_1)x, x) = \sum_{\substack{\alpha \in J(d) \\ \|\alpha\| \leq m/2}} \|A^\alpha x\|^2$$

for all $x \in D(dU(C_1))$. Since $\mathscr{X}_\infty$ is a core for $\mathscr{X}'_{m/2}$ and $dU(C_1)$, by Proposition 2.1.III, it is also a core for $dU(C_1)^{1/2}$. It then follows that $D(dU(C_1)^{1/2}) = \mathscr{X}'_{m/2}$ with equivalent norms.

We may as well assume that the C is symmetric, i.e., $C = C^\dagger$. Then $dU(C)$ is self-adjoint. Since it is the generator of a semigroup it has to be lower semibounded. Let $\lambda > 0$ be such that $dU(C) + \lambda I \geq 0$. Then both $dU(C) + \lambda I$ and $dU(C_1)$ are positive self-adjoint operators with the same domain (see [1] Theorem 3.6), so by Kato [12] one deduces that $D((dU(C) + \lambda I)^{1/2}) = D(dU(C_1)^{1/2})$, with equivalent norms. Thus there exist $p, q > 0$ such that

$$\|x\|'_{m/2} \leq p\|dU(C)^{1/2}x\| + q\,\|x\|$$

for all $x \in \mathscr{X}'_{m/2}$. Then

$$2^{-1}(\|x\|'_{m/2})^2 - q^2\|x\|^2 \leq p^2\|dU(C)^{1/2}x\|^2 = p^2(dU(C)x, x)$$

for all $x \in D(dU(C))$.

The independence of p and q from the representation U follows because the kernels of the semigroups generated by $dU(C)$ and $dU(C_1)$ are independent of U and all constants involved can be expressed in terms of these kernels. Moreover, the Kato theorem involves only a global constant. If C is homogeneous then one can scale the lower order terms away (see [5] Corollary 3.4).

Corollary 2.6. *Suppose* $(\mathscr{X}, G, U)$ *is a unitary representation,* C *is a form whose principal part is a positive Rockland form of order* m. *Then*

I.   *For all* $n \in \mathbb{N}$ *and all large* $\lambda > 0$ *one has*

$$D((dU(C) + \lambda I)^{n/2}) = \mathscr{X}'_{nm/2}$$

with equivalent norms, with factors independent of the representation.

II.   *If* C *is homogeneous then* $dU(C)$ *is a positive self-adjoint operator. Moreover, for all* $n \in \mathbb{N}$ *the seminorms* $x \mapsto \|dU(C)^{n/2}x\|$ *and* $N'_{nm/2}$ *are equivalent, with factors independent of the representation.*

III.   *If* $n \in \mathbb{N}$ *and* $b_1, \ldots, b_{d'}, \ldots, b_d$ *is an adapted basis for* $\mathfrak{g}$ *with weights* $v_1, \ldots, v_d$ *then*

$$\mathscr{X}'_{nm/2} = \bigcap_{i=1}^{d'} D(B_i^{nm/(2v_i)}),$$

where $B_i = dU(b_i)$.

Proof. Statement I has been proved for $n = 1$ in the proof of Theorem 2.5. The general case can be dealt with similarly. The second statement follows again by scaling from the first. The proof of the third statement is analogous to the proof of Theorem 5.8.IV in [6].

3. Spectral Estimates

In this section we derive some preliminary estimates on the eigenvalue distributions of certain self-adjoint operators. Let H and H_0 be a self-adjoint operators satisfying $H \geq H_0$ in the sense of quadratic forms. If $\exp(-\beta H_0)$ is of trace-class for some $\beta > 0$ it follows that $\exp(-\beta H)$ is also of trace-class and

$$\mathrm{Tr}(e^{-\beta H}) \leq \mathrm{Tr}(e^{-\beta H_0}).$$

Moreover, if $N(\lambda)$ denotes the number of eigenvalues of H which are less than or equal to λ, counted according to multiplicities, and if $N_0(\lambda)$ is the corresponding measure for H_0 then

$$N(\lambda) \leq N_0(\lambda)$$

for all λ. Both these conclusions are direct consequences of the minimax theorem. Thus the eigenvalue density N of H and the trace of the semigroup generated by H can both be estimated from above by the introduction of a comparator H_0. Similarly they can be estimated from below with the aid of a comparator H_1 satisfying $H \leq H_1$. These various estimates are all closely related and we next give some general results of this nature which will be useful in the sequel.

First we describe two comparison results of Levy-Bruhl and Nourrigat [16] which allow the estimation of the eigenvalue density. These results are formulated in terms of a family of 'coherent states'. Thus we assume that $\mathscr{H}$ is a separable Hilbert space, that (X, ρ) is a σ-finite measure space and that there exists a measurable map $x \mapsto \psi_x$ from X into $\mathscr{H}$ satisfying the following properties:

(i) there is a $K > 0$ such that $\|\psi_x\| = K$ for all $x \in X$,

(ii) $\varphi = \int_X d\rho(x) (\varphi, \psi_x) \psi_x$, for all $\varphi \in \mathscr{H}$, in the weak sense.

The $\{\psi_x : x \in X\}$ are the coherent states.

Proposition 3.1. *Let H be a self-adjoint operator on $\mathscr{H}$ with compact resolvent and D a core of H. Further let h_0 be a positive measurable function over X with the property that $\rho(\{x \in X : h_0(x) \le \mu\}) < \infty$ for all $\mu > 0$. If*

$$\int_X d\rho(x) |(\varphi, \psi_x)|^2 h_0(x) \le (\varphi, H\varphi) + \lambda \|\varphi\|^2$$

for all $\varphi \in D$ and some $\lambda > 0$ then the eigenvalue density N of H satisfies

$$N(\lambda) \le 2K^2 \rho(\{x \in X : h_0(x) \le 4\lambda\}).$$

This statement is contained in Théorème 1.1 of [16] and we refer to this paper for the proof. It is based on the use of approximate spectral projections in the sense of Shubin [25]. There is a second complementary result which gives a lower bound on the eigenvalue density.

Proposition 3.2. *Let H be a self-adjoint operator on $\mathscr{H}$ with compact resolvent, D a core of H and h_1 a positive measurable function over X such that $\rho(\{x \in X : h_1(x) \le \mu\}) < \infty$ for all $\mu > 0$. Assume*

$$(\varphi, H\varphi) \le \int_X d\rho(x) |(\varphi, \psi_x)|^2 h_1(x)$$

for all $\varphi \in D$.

Then for all $r > 0$, $C > 0$ and $\alpha \ge 0$ there exists an $R > 0$ such that for all $\lambda > 0$ with

$$\int_X d\rho(x) |(\psi_x, \psi_y)| h(x, y)^{\alpha + 1/2} \le C(h_1(z)/\lambda)^\alpha,$$

where

$$h(x, y) = \max\left(h_1(x), h_1(y)\right) \Big/ \min\left(h_1(x), h_1(y)\right),$$

one has

$$N(R\lambda) \ge 2^{-1} K^2 \rho(\{x \in X : h_1(x) \le r\lambda\}).$$

Again this statement is contained in Théorème 1.1 of [16] and we refer to this paper for the proof.

Next we examine the relations between estimates on the eigenvalue density N of H and the trace of the semigroup generated by H. We do this in a general measure-theoretic setting. Specifically we compare the properties of two functions related to positive Borel measures μ, ν on $\mathbb{R}_+ = \langle 0, \infty \rangle$. These functions are defined by

$$Z_\mu(\beta) = \int_0^\infty d\mu(x) e^{-\beta x}$$

and
$$N_\mu(\lambda) = \int_{\langle 0,\lambda]} d\mu(x)$$

for all $\beta, \lambda > 0$, with similar definitions for Z_ν and N_ν. If the measures are purely atomic these functions are directly comparable with the traces and the eigenvalue densities discussed above. Note that

$$\beta \int_0^\infty d\lambda \, N_\mu(\lambda) \, e^{-\beta\lambda} = \beta \int_0^\infty d\mu(x) \int_x^\infty d\lambda \, e^{-\beta\lambda} = Z_\mu(\beta).$$

Thus Z_μ is the Abel transform of N_μ. This relationship allows one to relate ordering properties of the Z and N.

Lemma 3.3. *Let μ and ν be positive Borel measures on $\mathbb{R}_+$. If $N_\mu(\lambda) \leq N_\nu(\lambda)$ for all $\lambda > 0$ then $Z_\mu(\beta) \leq Z_\nu(\beta)$ for all $\beta > 0$.*

Proof. This follows because the Z are the Abel transforms of the N and the measures are positive.

This argument can also be adapted to give a version of the lemma which relates the small β behaviour of the Z to the large λ behaviour of the N.

Lemma 3.4. *Let μ and ν be non-zero positive Borel measures on $\mathbb{R}_+$. If $Z_\nu(\beta) < \infty$ for all $\beta > 0$, $c \in \mathbb{R}$ and*

$$\lim_{\lambda \to \infty} N_\mu(\lambda)/N_\nu(\lambda) = c$$

then

$$\lim_{\beta \to 0} Z_\mu(\beta)/Z_\nu(\beta) = c.$$

Proof. Let $\varepsilon > 0$ and choose $\delta > 0$ such that $(e^\delta - 1)(c + \varepsilon) \leq \varepsilon$. There exists $\beta_0 > 0$ such that $N_\mu(\delta/\beta) \leq (c + \varepsilon)N_\nu(\delta/\beta)$ for all $\beta \in \langle 0, \beta_0]$. Therefore

$$\begin{aligned}
\beta \int_0^{\delta\beta^{-1}} dx \, N_\mu(x) \, e^{-\beta x} &\leq N_\mu(\delta/\beta) \, \beta \int_0^{\delta\beta^{-1}} dx \, e^{-\beta x} \\
&\leq (c + \varepsilon)(1 - e^{-\delta}) \, N_\nu(\delta/\beta) \\
&\leq (c + \varepsilon)(1 - e^{-\delta}) \, N_\nu(\delta/\beta) \, e^\delta \beta \int_{\delta\beta^{-1}}^\infty dx \, e^{-\beta x} \\
&\leq (e^\delta - 1)(c + \varepsilon) \, \beta \int_{\delta\beta^{-1}}^\infty dx \, N_\nu(x) \, e^{-\beta x} \leq \varepsilon \, Z_\nu(\beta)
\end{aligned}$$

for all $\beta \in \langle 0, \beta_0]$. Alternatively,

$$\begin{aligned}
\beta \int_{\delta\beta^{-1}}^\infty dx \, N_\mu(x) \, e^{-\beta x} &= \int_\delta^\infty dx \, N_\mu(x/\beta) \, e^{-x} \\
&\leq (c + \varepsilon) \int_\delta^\infty dx \, N_\nu(x/\beta) \, e^{-x} \\
&= (c + \varepsilon)\beta \int_{\delta\beta^{-1}}^\infty dx \, N_\nu(x) \, e^{-\beta x} \leq (c + \varepsilon) \, Z_\nu(\beta).
\end{aligned}$$

Therefore

$$Z_\mu(\beta) \leq (c + 2\varepsilon) \, Z_\nu(\beta) \tag{1}$$

for all $\beta \in \langle 0, \beta_0]$. Since $Z_\mu(\beta), Z_\nu(\beta) > 0$ for all $\beta > 0$ it follows that $\lim_{\beta \to 0} Z_\mu(\beta)/Z_\nu(\beta) = 0$ if $c = 0$.

If $c \neq 0$ then it follows from (1) that $Z_\mu(\beta) < \infty$ for all $\beta > 0$ and one can interchange μ and ν and deduce that there exists $\beta_1 > 0$ such that

$$Z_\nu(\beta) \leq (c^{-1} + 2\varepsilon) Z_\mu(\beta)$$

for all $\beta \in \langle 0, \beta_1]$. So $\lim_{\beta \to 0} Z_\mu(\beta)/Z_\nu(\beta) = c$.

In some situations Lemma 3.4 has a converse. If $d\nu(x) = x^{\alpha-1}dx$ then one can establish that $\lim_{\beta \to 0} Z_\mu(\beta)/Z_\nu(\beta) = c$ implies $\lim_{\lambda \to \infty} N_\mu(\lambda)/N_\nu(\lambda) = c$ (see, for example, [26] pages 107–109). Another special case, with $d\nu(x) = \nu(x)\,dx$ and $\nu(x) \sim x^{-\alpha}(\log x)^{-1}$ as $x \to 0$, occurs in [27].

4. SPECTRA OF POSITIVE ROCKLAND OPERATORS

Let G be a connected simply connected homogeneous Lie group and C a positive Rockland form of order m. If U is a irreducible unitary representation of G then the operator $dU(C)$ has a discrete spectrum. For $\lambda > 0$ we denote by $N(\lambda, U, C)$ the number of eigenvalues (counted with multiplicity) of the operator $dU(C)$ which are less than or equal to λ.

Since any two homogeneous norms on $\mathfrak{g}^*$ are equivalent we may work with a specific one. For $\lambda > 0$ let

$$N_0(\lambda, U) = \mu_U(\{l \in \mathscr{O}_U : |||l||| \leq \lambda\}).$$

The next theorem extends results of [15], [16].

Theorem 4.1. *If C is a positive Rockland form of order m then there exists $c > 0$ such that*

$$c^{-1}N_0(\lambda, U) \leq N(\lambda^m, U, C) \leq c N_0(\lambda, U)$$

uniformly for all $\lambda > 0$ and all irreducible unitary representations U of G.

Proof. An irreducible unitary representation is represented in a one dimensional or an infinite dimensional Hilbert space. If the space is infinite dimensional then the representation is unitarily equivalent with a representation U in $L_2(\mathbb{R}^k)$ for some $k \in \mathbb{N}$ such that every infinitesimal generator $dU(a)$ has the following form: there exist polynomials $Y, X_1, \ldots, X_k \colon \mathbb{R}^k \times \mathfrak{g} \to \mathbb{R}$ such that

$$(dU(a)f)(x) = iY(x, a) + \sum_{j=1}^{k} X_j(x, a)\frac{\partial f}{\partial x_j}(x) \tag{2}$$

for all $a \in \mathfrak{g}$, $f \in \mathscr{S}(\mathbb{R}^k) = D^\infty(U)$ and $x \in \mathbb{R}^k$ (see [2] page 125 and Corollary 4.1.2). If the Hilbert space is one dimensional one has to make the obvious changes and $k = 0$. So it suffices to prove the theorem for representations U with infinitesimal generators of the form (2). Therefore, from now on we only consider this type of representation in the proof. Moreover $k \in \mathbb{N}$ will always be such that U is represented in $L_2(\mathbb{R}^k)$.

Next for every $x \in \mathbb{R}^k$ and $\xi \in \mathbb{R}^k$ define the symbol $l_{x,\xi}^U \colon \mathfrak{g} \to \mathbb{R}$ of the partial differential operator $dU(a)$ by

$$l_{x,\xi}^U(a) = Y(x,a) + \sum_{j=1}^{k} X_j(x,a)\,\xi_j.$$

Then

$$(x,\xi) \mapsto l_{x,\xi}^U$$

is a bijection from $\mathbb{R}^k \times \mathbb{R}^k$ onto $\mathscr{O}_U$. A proof can be given along the lines of proofs in [11], e.g., the proof of Proposition VIII.5.1, or [21] Theorem 2.13. Moreover,

$$\int_{\mathbb{R}^k \times \mathbb{R}^k} dx\, d\xi\, f(l_{x,\xi}^U) = (2^k k!)^{-1} \int_{\mathscr{O}_U} d\mu_U(l)\, f(l)$$

for every positive measurable function f on $\mathscr{O}_U$. Hence

$$N_0(\lambda, U) = \tau(\{(x,\xi) \in \mathbb{R}^{2k} : |||l_{x,\xi}^U||| \le \lambda\})$$

for all $\lambda > 0$, where τ denotes $(2^k k!)^{-1}$ times the Lebesgue measure on $\mathbb{R}^{2k}$. Let $\| \cdot \|$ be the dual norm on $\mathfrak{g}^*$ of a Euclidean norm $\| \cdot \|$ on $\mathfrak{g}$. For every representation U define $p(x,\xi,U) = \|l_{x,\xi}^U\|$ for all $x, \xi \in \mathbb{R}^k$. Next for all $\lambda > 0$ and all representations U (of the form (2)) define the representation U_λ by $U_\lambda(g) = U(\delta_\lambda(g))$ for all $g \in G$, where δ_λ is the dilation on G obtained via the exponential map. Then $p(x,\xi,U_\lambda) = \|\gamma_\lambda^* l_{x,\xi}^U\|$. Hence if $\mathscr{O}_U \ne \{0\}$ and $c, \lambda > 0$ then $|||l_{x,\xi}^U||| \le c\lambda$ if, and only if, $\|\gamma_{1/\lambda}^*(l_{x,\xi}^U)\| \le c$ or, if, and only if, $p(x,\xi,U_{1/\lambda}) \le c$. Here we have chosen for the homogeneous norm on $\mathfrak{g}^*$ the norm defined in the introduction. In particular:

$$N_0(c\lambda, U) = \tau(\{(x,\xi) \in \mathbb{R}^{2k} : p(x,\xi,U_{1/\lambda}) \le c\})$$

for all $c, \lambda > 0$.

Let w be the largest weight. Then

$$\mathscr{X}_n' \subseteq \mathscr{X}_n \subseteq \mathscr{X}_{nw}'$$

for all $n \in \mathbb{N}_0$. Let $n \in \mathbb{N}$ be such that $w \le nm$. By Corollary 2.6.I there exists $c_1 > 0$ such that

$$\|f\|_{U,nm}' \le c_1 \|dU(C)^n f\|$$

uniformly for all unitary representations U and $f \in \mathscr{X}_\infty(U)$. We next need the following proposition of Lévy-Bruhl and Nourrigat ([16] Proposition 4.1).

Proposition 4.2. *There exists $c > 0$ such that for every unitary representation U of G in $L_2(\mathbb{R}^k)$ there exists a continuous function $(x,\xi) \mapsto \psi_{x,\xi,U}$ from $\mathbb{R}^k \times \mathbb{R}^k$ into $\mathscr{S}(\mathbb{R}^k)$ such that $\|\psi_{x,\xi,U}\|_2 = (2\pi)^{-k/2}$ for all $x, \xi \in \mathbb{R}^k$,*

$$f = \int_{\mathbb{R}^{2k}} dx\, d\xi\, (\psi_{x,\xi,U}, f)\, \psi_{x,\xi,U}$$

for all $f \in L_2(\mathbb{R}^k)$, in the weak sense, and

$$\int_{\mathbb{R}^{2k}} dx\, d\xi\, p(x,\xi,U)^2 |(\psi_{x,\xi,U}, f)|^2 \le c\, \|f\|_{U,1}^2$$

uniformly for all $f \in \mathscr{X}_\infty(U)$.

Now with $c > 0$ the constant in Proposition 4.2 one has

$$\int_{\mathbb{R}^{2k}} dx\, d\xi\, p(x,\xi,U)^2 |(\psi_{x,\xi,U}, f)|^2 \le c\|f\|_{U,1}^2$$

$$\le c(\|f\|_{U,w}')^2 \le c(\|f\|_{U,nm}')^2 \le cc_1^2 \|dU(C)^n f\|^2$$

for all U and $f \in \mathscr{S}(\mathbb{R}^k)$. Next let S be the multiplication operator with the function $(x,\xi) \mapsto p(x,\xi,U)^2$ on $L_2(\mathbb{R}^{2k})$. Then the map $T\colon D(dU(C)^n) \to D(S)$ defined by $(Tf)(x,\xi) = (\psi_{x,\xi,U}, f)$ is continuous with norm bounded by cc_1^2. Hence by interpolation ([12]) it follows that T is bounded from $D(dU(C)^{1/2})$ into $D(S^{1/(2n)})$ and the norm is bounded by a constant which depends only on cc_1^2. Then with Theorem 2.5 one deduces that there exists a constant $c_2 > 0$ such that

$$\int_{\mathbb{R}^{2k}} dx\, d\xi\, p(x,\xi,U)^{1/n} |(\psi_{x,\xi,U}, f)|^2 \le c_2\Big((f, dU(C)f) + \|f\|^2\Big)$$

uniformly for all U and $f \in \mathscr{S}(\mathbb{R}^k)$. In particular, for all $\lambda > 0$ and U one obtains the inequalities

$$\int_{\mathbb{R}^{2k}} dx\, d\xi\, p(x,\xi,U_{1/\lambda})^{1/n} |(\psi_{x,\xi,U_{1/\lambda}}, f)|^2 \le c_2\Big((f, dU_{1/\lambda}(C)f) + \|f\|^2\Big)$$

$$= c_2\Big(\lambda^{-m}(f, dU(C)f) + \|f\|^2\Big)$$

and

$$\int_{\mathbb{R}^{2k}} dx\, d\xi\, c_2^{-1}\lambda^m p(x,\xi,U_{1/\lambda})^{1/n} |(\psi_{x,\xi,U_{1/\lambda}}, f)|^2 \le \Big((f, dU(C)f) + \lambda^m\|f\|^2\Big).$$

Then by Proposition 3.1 one obtains

$$N(\lambda^m, U, C) \le 2(2\pi)^{-k}\tau(\{(x,\xi) \in \mathbb{R}^{2k} : c_2^{-1}\lambda^m p(x,\xi,U_{1/\lambda})^{1/n} \le 4\lambda^m\})$$

$$= 2(2\pi)^{-k}\tau(\{(x,\xi) \in \mathbb{R}^{2k} : p(x,\xi,U_{1/\lambda}) \le (4c_2)^n\})$$

$$= 2(2\pi)^{-k} N_0((4c_2)^n \lambda, U)$$

for all $\lambda > 0$.

The next proposition gives an inequality in the opposite direction.

Proposition 4.3. *There exists $c > 0$ and, for all $y \in \mathbb{R}$, a $c_y > 0$ such that for every unitary representation U of G in $L_2(\mathbb{R}^k)$ there exists a continuous function $(x,\xi) \mapsto \psi_{x,\xi,U}$ from $\mathbb{R}^k \times \mathbb{R}^k$ into $\mathscr{S}(\mathbb{R}^k)$ with the following properties*

I. $\|\psi_{x,\xi,U}\|_2 = (2\pi)^{-k/2}$ *for all* $x,\xi \in \mathbb{R}^k$,

II. $f = \displaystyle\int_{\mathbb{R}^{2k}} dx\, d\xi\, (\psi_{x,\xi,U}, f)\, \psi_{x,\xi,U}$ *for all* $f \in L_2(\mathbb{R}^k)$, *in the weak sense,*

III. $|(f, dU(C)f)| \le c \displaystyle\int_{\mathbb{R}^{2k}} dx\, d\xi\, (1 + p(x,\xi,U))^{m+k} |(\psi_{x,\xi,U}, f)|^2$ *for all* $f \in \mathscr{S}(\mathbb{R}^k)$,

IV. $\displaystyle\int_{\mathbb{R}^{2k}} dy\, d\eta\, \Big(h_U(x,\xi,y,\eta)\Big)^y |(\psi_{x,\xi,U}, \psi_{y,\eta,U})| \le c_y\Big((1 + p(x,\xi,U))\Big)^k$ *uniformly for all* $x,\xi \in \mathbb{R}^k$, *where*

$$h_U(x,\xi,y,\eta) = \frac{1 + \max(p(x,\xi,U), p(y,\eta,U))}{1 + \min(p(x,\xi,U), p(y,\eta,U))}.$$

Proof. These properties follows from [16] Propositions 6.8 and 6.7.

In particular, if $\lambda > 0$, U is a representation in $\mathbb{R}^k$ and $f \in \mathscr{S}(\mathbb{R}^k)$ then

$$\lambda^{-m}|(f, dU(C)f)| = |(f, dU_{1/\lambda}(C)f)| \leq c\int_{\mathbb{R}^{2k}} dx\, d\xi (1 + p(x, \xi, U_{1/\lambda}))^{m+k} |(\psi_{x,\xi,U_{1/\lambda}}, f)|^2$$

and hence

$$|(f, dU(C)f)| \leq \int_{\mathbb{R}^{2k}} dx\, d\xi\, c\lambda^m (1 + p(x, \xi, U_{1/\lambda}))^{m+k} |(\psi_{x,\xi,U_{1/\lambda}}, f)|^2.$$

It then follows from Proposition 3.2, applied with $y = 2^{-1} + k(m+k)^{-1}$ that there exists $c_2 > 0$ such that

$$\int_{\mathbb{R}^{2k}} dy\, d\eta \left(\frac{1 + \max(p(x, \xi, U_{1/\lambda}), p(y, \eta, U_{1/\lambda}))}{1 + \min(p(x, \xi, U_{1/\lambda}), p(y, \eta, U_{1/\lambda}))} \right)^{\frac{k}{m+k} + \frac{1}{2}} |(\psi_{x,\xi,U_{1/\lambda}}, \psi_{y,\eta,U_{1/\lambda}})|$$

$$\leq c_2 \left(\frac{c\lambda^m (1 + p(x, \xi, U_{1/\lambda}))^{m+k}}{\lambda^m} \right)^{\frac{k}{m+k}}$$

uniformly for all U, all $x, \xi \in \mathbb{R}^k$ and $\lambda > 0$. Therefore, by Proposition 3.2 one deduces that there exists $c_3 > 0$ such that

$$N(c_3\lambda^m, U, C) \geq 2^{-1}(2\pi)^{-k}\tau(\{(x, \xi) \in \mathbb{R}^{2k} : c\lambda^m(1 + p(x, \xi, U_{1/\lambda}))^{m+k} \leq c2^{m+k}\lambda^m\})$$

$$= 2^{-1}(2\pi)^{-k}\tau(\{(x, \xi) \in \mathbb{R}^{2k} : p(x, \xi, U_{1/\lambda}) \leq 1\})$$

$$= 2^{-1}(2\pi)^{-k}N_0(\lambda, U)$$

for all $\lambda > 0$. This completes the proof of the theorem.

Corollary 4.4. *Let $(\mathscr{H}, G, U)$ be an irreducible unitary representation of G and C a form of order m whose principal part P is a positive Rockland form. If there exists an $\omega > 0$ such that $dU(C) \geq \omega I$ then there is a $c > 0$ such that*

$$c^{-1}N_0(c^{-1}\lambda, U) \leq N(\lambda^m, U, C) \leq cN_0(c\lambda, U)$$

for all $\lambda > 0$.

Proof. Since P is a Rockland form the operator $dU(P)$ is strictly positive and there exists an $\omega_0 > 0$ such that $\|x\|^2 \leq \omega_0(dU(P)x, x)$ for all $x \in \mathscr{X}_\infty$. Now $2P - C$ is a form whose principal part is a positive Rockland form. So by Theorem 2.5 there exists a $q > 0$ such that

$$(dU(2P - C)x, x) \geq -q\|x\|^2$$

for all $x \in \mathscr{X}_\infty$. Hence

$$(dU(C)x, x) \leq 2(dU(P)x, x) + q\|x\|^2 \leq (2 + q\omega_0)(dU(P)x, x)$$

for all $x \in \mathscr{X}_\infty$. By assumption one similarly has

$$(dU(P)x, x) \leq 2(dU(C)x, x) + q'\|x\|^2 \leq (2 + q'\omega^{-1})(dU(C)x, x)$$

for some $q' > 0$, uniformly for all $x \in \mathscr{X}_\infty$. Then by the minimax theorem one obtains

$$N((2 + q\omega_0)^{-1}\lambda, U, P) \leq N(\lambda, U, C) \leq N((2 + q'\omega^{-1})\lambda, U, P)$$

for all $\lambda > 0$. Now the corollary follows from Theorem 4.1.

These estimates for the eigenvalue density can be converted into estimates on the partition function of $H = dU(C)$ by the observation of Lemma 3.3. First for $\beta > 0$, $m > 1$ and U an irreducible unitary representation of G define

$$Z_0(\beta, m, U) = \int_{\mathscr{O}_U} d\mu_U(l)\, e^{-\beta|||l|||^m} = 2^k k! \int_{\mathbb{R}^k \times \mathbb{R}^k} dx\, d\xi\, e^{-\beta|||x,\xi|||^U}.$$

Corollary 4.5. *Let $(\mathscr{H}, G, U)$ be an irreducible unitary representation of G and C a form of order m whose principal part P is a positive Rockland form. If there exists an $\omega > 0$ such that $H = dU(C) \geq \omega I$ then there is a $c > 0$ such that*

$$c^{-1} Z_0(c\beta, m, U) \leq \mathrm{Tr}_{\mathscr{H}}(e^{-\beta H}) \leq c Z_0(c^{-1}\beta, m, U)$$

uniformly for all $\beta > 0$. Moreover, these estimates are valid, uniformly for all irreducible unitary representations whenever C is a positive Rockland form.

Proof. We shall prove that

$$Z_0(\beta, m, U) = \beta \int_0^\infty d\lambda\, N_0(\lambda^{1/m}, U)\, e^{-\beta\lambda}$$

for all $\beta > 0$ and irreducible unitary representations U. The equality follows from Fubini's theorem since

$$\beta \int_0^\infty d\lambda\, N_0(\lambda^{1/m}, U)\, e^{-\beta\lambda} = \beta \int_0^\infty d\lambda \int_{\mathscr{O}_U} d\mu_U(l)\, 1_{\{l \in \mathscr{O}_U : |||l||| \leq \lambda^{1/m}\}}(l)\, e^{-\beta\lambda}$$

$$= \beta \int_{\mathscr{O}_U} d\mu_U(l) \int_{|||l|||^m}^\infty d\lambda\, e^{-\beta\lambda}$$

$$= \int_{\mathscr{O}_U} d\mu_U(l)\, e^{-\beta|||l|||^m} = Z_0(\beta, m, U).$$

Now the corollary easily follows from Corollary 4.4 and the fact that $\beta \mapsto \mathrm{Tr}_{\mathscr{H}}(\exp(-\beta H))$ is the Abel transform of N. See Lemma 3.3.

The estimates of Corollary 4.5 are potentially useful for calculating the behaviour of the partition function for small β. In the next section this will be achieved for the simplest example, the anharmonic oscillator on the Heisenberg group. The general case appears more intractable and requires understanding the behaviour of $\lambda \mapsto N_0(\lambda, U)$ for large λ.

The large β behaviour of the partition function is much easier to establish. If λ_1 is the smallest eigenvalue of H then $\lim_{\beta \to \infty} e^{\beta\lambda_1} \mathrm{Tr}_{\mathscr{H}}(\exp(-\beta H)) = n_1$, where n_1 is the multiplicity of the eigenvalue λ_1. It is an interesting question whether $n_1 = 1$ for homogeneous C. For non-homogeneous C the spectrum of H need not be simple and in general $n_1 > 1$. An example is the operator

$$(P^2 + Q^2)^2 - 4(P^2 + Q^2) + 3I,$$

where P and Q are the usual self-adjoint operators in $L_2(\mathbb{R})$, see Section 5.

The simplicity of the lowest eigenvalue is often established by positivity arguments based on some variation of the Perron–Frobenius theorem. But this type of reasoning is in general not applicable in the present setting. Although the semigroup $t \mapsto \exp(-tH)$ has a kernel K it is usually not positive. It follows from [23], Chapter III, Section 5, that the semigroup kernel K^G on the group corresponding to the form C is positive if and only if C is second-order, in the unweighted sense, with real coefficients and with the principal coefficients satisfying an ellipticity condition. But positivity of K^G does not imply positivity of the kernel K on $\mathbb{R}^k \times \mathbb{R}^k$ corresponding to the semigroup generated by $H = dU(C)$ even for homogeneous, real, second-order C. An example is given by the operator $(P + Q)^2 + Q^2$, which is a sublaplacian for the algebraic basis $a_1 + a_2, a_2$ of the Heisenberg algebra, but as a second-order operator on $L_2(\mathbb{R})$ it is not even real.

5. THE ANHARMONIC OSCILLATOR

As an application we consider the general anharmonic oscillator.

Let G be the simply connected Heisenberg group, U the standard irreducible unitary representation of G in $L_2(\mathbb{R})$ and a_1, a_2, a_3 a basis in the Lie algebra $\mathfrak{g}$ of G such that $[a_1, a_2] = a_3$, $A_1 = -iP$, $A_2 = iQ$ and $A_3 = iI$, where P and Q are the self-adjoint operators in $L_2(\mathbb{R})$ such that $(Pf)(x) = if'(x)$ and $(Qf)(x) = xf(x)$ for all $f \in C_c^\infty(\mathbb{R})$ and $x \in \mathbb{R}$. Then

$$l_{x,\xi}^U(a_1) = \xi, \quad l_{x,\xi}^U(a_2) = x, \quad l_{x,\xi}^U(a_3) = 1.$$

Fix $j, k \in \mathbb{N}$. There are dilations $(\gamma_t)_{t>0}$ on $\mathfrak{g}$ such that $\gamma_t(a_1) = t^k a_1$, $\gamma_t(a_2) = t^j a_2$ and $\gamma_t(a_3) = t^{j+k} a_3$ for all $t > 0$. Let C be the positive Rockland form of order $m = 2jk$ such that

$$H = dU(C) = (-1)^j A_1^{2j} + (-1)^k A_2^{2k} = P^{2j} + Q^{2k}.$$

Define the modulus $||| \cdot |||_1 \colon \mathfrak{g}^* \to [0, \infty)$ by

$$|||l|||_1 = \left(l(a_1)^{2j} + l(a_2)^{2k} + l(a_3)^{2jk/(j+k)} \right)^{1/(2jk)}.$$

Then there exists a $c > 0$ such that

$$c^{-1} |||l|||_1 \le |||l||| \le c|||l|||_1$$

for all $l \in \mathfrak{g}^*$, so in Theorem 4.1 we may as well replace $||| \cdot |||$ by $||| \cdot |||_1$. Now $|||l_{x,\xi}^U|||_1^m = \xi^{2j} + x^{2k} + 1$. Then an elementary estimate shows that there exist $\Lambda > 0$ and $c > 0$ such that

$$c^{-1} \lambda^{1/\sigma} \le \tau(\{(x, \xi) \in \mathbb{R}^2 : |||l_{x,\xi}^U|||_1 \le \lambda\}) \le c\lambda^{1/\sigma}$$

for all $\lambda \ge \Lambda$, where τ now denotes the Lebesgue measure on $\mathbb{R}$ and $\sigma = 2jk/(j + k)$. So if $N(\lambda)$ denotes the number of eigenvalues of H which are less than or equal to λ one deduces that there exist $\Lambda > 0$ and $c > 0$ such that

$$c^{-1} \lambda^{1/\sigma} \le N(\lambda) \le c\lambda^{1/\sigma}$$

for all $\lambda \ge \Lambda$ and then, by increasing the value of c, for all $\lambda \ge \lambda_1$ where λ_1 is the strictly positive lowest eigenvalue of H. Next let $\lambda_1 \le \lambda_2 \le \dots$ be the eigenvalues of H, repeated according to multiplicity. Then for all small $\varepsilon > 0$ and $n \in \mathbb{N}$ with $\lambda_n > \Lambda$ one has

$$c^{-1}(\lambda_n - \varepsilon)^{1/\sigma} \le N(\lambda_n - \varepsilon) \le n \le N(\lambda_n) \le c\lambda_n^{1/\sigma},$$

so

$$(c^{-1}n)^\sigma \le \lambda_n \le (c\,n)^\sigma$$

for all $\lambda_n \ge \Lambda$. By increasing c one concludes that

$$c^{-1}n^\sigma \le \lambda_n \le cn^\sigma$$

for all $n \in \mathbb{N}$. This proves a conjecture in [6] page 40.

In [6] we proved that the eigenfunctions φ_n corresponding to H with eigenvalue λ_n belong to the Gel'fand–Shilov space $S_{j/(j+k)}^{k/(j+k)}$, which consists of all infinitely differentiable functions φ on $\mathbb{R}$ for which there exist $a, b, c > 0$ (depending on φ) such that

$$|x^r \varphi^{(s)}(x)| \le c\, a^r\, b^s\, r!^{j/(j+k)}\, s!^{k/(j+k)}$$

for all $r, s \in \mathbb{N}_0$ and $x \in \mathbb{R}$, or, equivalently, which consists of all functions φ on $\mathbb{R}$ which can be extended to entire functions (also denoted by φ) into the complex plane satisfying the growth bounds

$$|\varphi(x + iy)| \le ce^{-a|x|^{(j+k)/j} + b|y|^{(j+k)/j}}$$

for all $x, y \in \mathbb{R}$, for some constants $a, b, c > 0$, depending on φ, or, equivalently,

$$|\varphi(z)| \le ce^{b|z|^{(j+k)/j}}, \quad |\varphi(x)| \le ce^{-a|x|^{(j+k)/j}}$$

for all $z \in \mathbb{C}$ and $x \in \mathbb{R}$ (see [6] Section 7).

For the eigenfunctions $\varphi_1, \varphi_2, \ldots$ we next show that the constants a, b, c do not behave wildly if n varies. For the harmonic oscillator $P^2 + Q^2$ this was proved before by [29], equation (2).

Theorem 5.1. *Let $j, k \in \mathbb{N}$ and let $\lambda_1 \le \lambda_2 \le \ldots$ denote the eigenvalues of the operator $P^{2j} + Q^{2k}$, repeated according to multiplicity, with $\varphi_1, \varphi_2, \ldots$ a corresponding orthonormal basis of eigenfunctions. Then there exists $C > 0$ such that*

$$C^{-1}n^{2jk/(j+k)} \le \lambda_n \le Cn^{2jk/(j+k)}$$

for all $n \in \mathbb{N}$. Moreover, there exist $a, b, c > 0$ such that

$$|x^r \varphi_n^{(s)}(x)| \le c^n a^r b^s r!^{j/(j+k)} s!^{k/(j+k)}$$

uniformly for all $n \in \mathbb{N}$, $r, s \in \mathbb{N}_0$ and $x \in \mathbb{R}$.

Equivalently, each φ_n can be extended to an entire function and there exist $a, b, c > 0$ such that

$$|\varphi_n(x + iy)| \le c^n e^{-a|x|^{(j+k)/j} + b|y|^{(j+k)/j}}$$

uniformly for all $x, y \in \mathbb{R}$ and $n \in \mathbb{N}$, or, equivalently,

$$|\varphi_n(z)| \le c^n e^{b|z|^{(j+k)/j}}, \quad |\varphi_n(x)| \le c^n e^{-a|x|^{(j+k)/j}}$$

uniformly for all $z \in \mathbb{C}$, $x \in \mathbb{R}$ and $n \in \mathbb{N}$.

Proof. We have already proved the existence of the constant C for the eigenvalue estimates. Then for all $n \in \mathbb{N}$ one has

$$\|H^l \varphi_n\|_2 = \lambda_n^l \leq (Cn^{2jk/(j+k)})^l \leq (e^{2jk/(j+k)})^n C^l \, l!^{2jk/(j+k)}$$

for all $l \in \mathbb{N}_0$. If one now traces all the constants in [6], [28] §29.5 and [9] Section IV.3.3 one obtains the uniform estimates of the theorem.

Finally the partition function $Z(\beta) = \text{Tr}(\exp(-\beta H))$ can be bounded above and below by use of the eigenvalue estimates. One has

$$c^{-1}\beta \int_{\lambda_1}^{\infty} d\lambda \, \lambda^{\sigma} e^{-\beta\lambda} \leq \text{Tr}(e^{-\beta H}) \leq c\beta \int_{\lambda_1}^{\infty} d\lambda \, \lambda^{\sigma} e^{-\beta\lambda}$$

where again $\sigma = 2jk/(j+k)$. Alternatively one obtains bounds

$$c_\sigma^{-1}\beta^{-\sigma} e^{-\beta\lambda_1} \leq \text{Tr}(e^{-\beta H}) \leq c_\sigma \min(\beta^{-\sigma}, e^{-\beta\lambda_1})$$

by straightforward estimations of the integral.

REFERENCES

1. P. Auscher, A. F. M. ter Elst, and D. W. Robinson, *On positive Rockland operators*, Coll. Math. **67** (1994), 197–216.
2. L. Corwin and F. P. Greenleaf, *Representations of nilpotent Lie groups and their applications Part 1: Basic theory and examples*, Cambridge Studies in Advanced Mathematics, 18, Cambridge University Press, Cambridge, 1990.
3. A. F. M. ter Elst and D. W. Robinson, *Subcoercive and subelliptic operators on Lie groups: variable coefficients*, Publ. RIMS. Kyoto Univ. **29** (1993), 745–801.
4. _____, *Functional analysis of subelliptic operators on Lie groups*, J. Operator Theory **31** (1994), 277–301.
5. _____, *Subcoercivity and subelliptic operators on Lie groups II: The general case*, Potential Anal. **4** (1995), 205–243.
6. _____, *Weighted strongly elliptic operators on Lie groups*, J. Funct. Anal. **125** (1994), 548–603.
7. C. L. Fefferman, *The uncertainty principle*, Bull. Amer. Math. Soc. **9** (1983), 129–206.
8. G. B. Folland and E. M. Stein, *Hardy spaces on homogeneous groups*, Mathematical Notes, 28, Princeton University Press, Princeton, 1982.
9. I. M. Gel'fand and G. E. Shilov, *Generalized functions*, 2, Academic Press, New York, 1968.
10. B. Helffer and J. Nourrigat, *Caractérisation des opérateurs hypoelliptiques homogénes invariants á gauche sur un groupe de Lie nilpotent gradué*, Comm. Part. Diff. Eq. **4** (1979), 899–958.
11. _____, *Hypoellipticité maximale pour des opérateurs polynomes de champs de vecteurs*, Progress in Mathematics, 58, Birkhäuser, Boston, 1985.
12. T. Kato, *A generalization of the Heinz inequality*, Proc. Japan Acad. **37** (1961), 305–308.
13. A. A. Kirillov, *Unitary representations of nilpotent Lie groups*, Russian Math. Surveys **17** (1962), 53–104.
14. P. Lévy-Bruhl, A. Mohamed, and J. Nourrigat, *Spectral theory and representations of nilpotent groups*, Bull. Amer. Math. Soc **26** (1992), 299–303.
15. _____, *Etude spectrale d'opérateurs liés à des représentations de groupes nilpotents*, J. Funct. Anal. **113** (1993), 65–93.
16. P. Lévy-Bruhl and J. Nourrigat, *Etats cohérents, théorie spectrale et représentations de groupes nilpotents*, Ann. Sci. Ecole Norm. Sup. **27** (1994), 707–757.
17. D. Manchon, *Calcul symbolique sur les groupes de Lie nilpotents et applications*, J. Funct. Anal. **102** (1991), 206–251.

18. _____, *Formule de Weyl pour les groupes de Lie nilpotents*, J. reine angew. Math. **418** (1991), 77–129.
19. K. G. Miller, *Parametrices for hypoelliptic operators on step two nilpotent Lie groups*, Comm. Part. Diff. Eq. **5** (1980), 1153–1184.
20. E. Nelson and W. F. Stinespring, *Representation of elliptic operators in an enveloping algebra*, Amer. J. Math. **81** (1959), 547–560.
21. J. Nourrigat, L^2 *inequalities and representations of nilpotent groups*, World Scientific, 1991. Cours à l'école CIMPA-UNESCO d'Analyse Harmonique, Wuhan (China), to appear.
22. L. Pukanszky, *Leçons sur les représentations des groupes*, Dunod, Paris, 1967.
23. D. W. Robinson, *Elliptic operators and Lie groups*, Oxford Mathematical Monographs, Oxford University Press, Oxford, 1991.
24. C. Rockland, *Hypoellipticity for the Heisenberg group*, Trans. Amer. Math. Soc. **240** (1978), 1–52.
25. M. A. Shubin, *Pseudodifferential operators and spectral theory*, Springer Series in Sovjet Mathematics, Springer-Verlag, Berlin, 1987.
26. B. Simon, *Functional integration and quantum physics*, Pure and Applied Mathematics, 86, Academic Press, New York, 1979.
27. _____, *Nonclassical eigenvalue asymptotics*, J. Funct. Anal. **53** (1983), 84–98.
28. J. Wloka, *Grundräume und verallgemeinerte Funktionen*, Lecture Notes in Mathematics, vol. 82, Springer-Verlag, Berlin, 1969.
29. G.-Z. Zhang, *Theory of distributions of S type and pansions*, Chinese Math. **4** (1963), 211–221.

DEPARTMENT OF MATHEMATICS AND COMPUTING SCIENCE, EINDHOVEN UNIVERSITY OF TECHNOLOGY, P.O. BOX 513, 5600 MB EINDHOVEN, THE NETHERLANDS

CENTRE FOR MATHEMATICS AND ITS APPLICATIONS, SCHOOL OF MATHEMATICAL SCIENCES, AUSTRALIAN NATIONAL UNIVERSITY, CANBERRA, ACT 0200, AUSTRALIA

FACTORIZATION OF CERTAIN EXPONENTIALS IN LIE GROUPS

C. K. FAN AND G. LUSZTIG

Dedicated to the memory of Roger W. Richardson

ABSTRACT. The purpose of this paper is to verify a conjecture (proposition 1 below) of the second author [2, §11.4]. Our verification relies in part on the use of a computer and for this reason, is not entirely satisfactory.

Let G be a reductive algebraic group (over $\mathbb{C}$), and let $\mathfrak{g}$ be its Lie algebra. Assume $\mathfrak{g}$ is simple. Fix a Cartan subalgebra $\mathfrak{h}$ and a root decomposition of $\mathfrak{g}$. Let R (resp. $\check{R}$) be the set of roots (resp. coroots). Choose a set of simple roots $\Pi = \{\alpha_i\}_{i\in\mathbf{I}}$. Let $ht(\phi), \phi \in R$ denote the sum of the coefficients of ϕ when written as a linear combination of simple roots. Let $R^+ = \{\phi \in R \mid ht(\phi) > 0\}$. Let $A = (a_{ij})_{i,j\in\mathbf{I}}$ be the Cartan matrix defined by $a_{ij} = \alpha_j(\check{\alpha}_i)$, where $\check{\alpha}_i \in \check{R}$ and $\alpha_j \in R$. For each $i \in \mathbf{I}$, denote by $s_i : \mathfrak{h}^* \xrightarrow{\sim} \mathfrak{h}^*$ the simple reflection defined by $s_i(\alpha_j) = \alpha_j - a_{ij}\alpha_i$. The Weyl group W is the group generated by the s_i, $i \in \mathbf{I}$. Let $l(w)$, $w \in W$ be the smallest number n such that w is a product of n generators. Let $w_0 \in W$ be the unique element for which $l(w_0) = card(R^+)$. For each $i \in \mathbf{I}$, fix some $E_i \in \mathfrak{g}_{\alpha_i}$, $E_i \neq 0$. Let $\mathfrak{u}^+$ be the nilpotent subalgebra of $\mathfrak{g}$ generated by the E_i.

Proposition 1. *Choose* $(i_1,\ldots,i_N) \in \mathbf{I}^N$ *so that* $s_{i_1}s_{i_2}s_{i_3}\cdots s_{i_N}$ *is a reduced expression for* w_0. *Let* $h_k = ht(s_{i_1}s_{i_2}s_{i_3}\cdots s_{i_{k-1}}(\alpha_{i_k}))$. *For each* $i \in \mathbf{I}$, *let* $r_i = \sum_{i_k=i} h_k$. *Then*

$$\exp(h_1 E_{i_1})\exp(h_2 E_{i_2})\cdots\exp(h_N E_{i_N}) = \exp(\sum_{i\in\mathbf{I}} r_i E_i). \qquad (*)$$

We remark that in the simply laced case, $\sum_{i\in\mathbf{I}} r_i\alpha_i = \sum_{\phi\in R^+}\phi$.

In this paper, we assume that $\mathfrak{g}$ is of simply laced type. However, proposition 1 is true in the non-simply laced case without modification. The non-simply laced case can be reduced to the simply laced case by using standard techniques, see for example [2, §2.7].

We refer the reader to [2, §11.2] for a proof that the left hand side of $(*)$ is independent of the choice of reduced expression for w_0.

Case 1. Assume we are in type A_n. Let $\mathbf{I} = \{1,2,3,\ldots,n\}$, and label Π so that

---

G. L. supported in part by the National Science Foundation

215

$$a_{ij} = \begin{cases} 2 & \text{if } i = j, \\ -1 & \text{if } |i - j| = 1, \\ 0 & \text{otherwise.} \end{cases}$$

Now let $N = \frac{n(n+1)}{2}$ and $(i_1, \ldots, i_N) = (1, 2, 1, 3, 2, 1, \ldots, n, n - 1, \ldots, 3, 2, 1)$. Then $s_{i_1} s_{i_2} s_{i_3} \cdots s_{i_N}$ is a reduced expression for w_0. One can verify that $h_k = i_k$ and $r_i = i(n + 1 - i)$.

We proceed by induction on the rank. The formula is clearly true when $n = 1$. Assume the formula is true for all ranks less than n. To verify the formula for rank n, we must show that $\exp(\sum_{i \in I} r_i E_i) = \exp(\sum_{i \in I \setminus \{n\}} i(n - i) E_i) \, \Delta$, where $\Delta = \exp(n E_n) \cdots \exp(3 E_3) \exp(2 E_2) \exp(E_1)$.

We check this equation in the standard representation of $\mathfrak{g}$ on an $n + 1$ dimensional vector space V. Let $B = \{b_1, \ldots, b_{n+1}\}$ be a basis on which

$$E_i(b_j) = \begin{cases} b_{j-1} & \text{if } j = i + 1, \\ 0 & \text{otherwise.} \end{cases}$$

Direct computation yields the following formulae:

$$\exp(\sum_{i \in I} r_i E_i)(b_k) = \sum_{m=1}^{k} \frac{(k-1)!(n+1-m)!}{(m-1)!(n+1-k)!(k-m)!} b_m,$$

$$\exp(\sum_{i \in I \setminus \{n\}} i(n-i) E_i)(b_l) = \sum_{m=1}^{l} \frac{(l-1)!(n-m)!}{(m-1)!(n-l)!(l-m)!} b_m,$$

$$\Delta(b_k) = (k-1) b_{k-1} + b_k.$$

Now,

$$\exp(\sum_{i \in I \setminus \{n\}} i(n-i) E_i) \Delta(b_k) = (k-1) \sum_{m=1}^{k-1} \frac{(k-2)!(n-m)!}{(m-1)!(n-k+1)!(-m)!} b_m$$

$$= b_k + \sum_{m=1}^{k-1} \frac{(k-1)!(n-m)!((k-m) + (n-k+1))}{(m-1)!(n-k+1)!(k-m)!},$$

as desired.

Case 2. Assume we are in type D_n. Let $\mathbf{I} = \{1, 2, 3, \ldots, n-1, \overline{n-1}\}$. Label Π so that

$$a_{ij} = \begin{cases} 2 & \text{if } i = j, \\ -1 & \text{if } |i - j| = 1, \ \overline{n-1} \notin \{i, j\}, \\ -1 & \text{if } \{i, j\} = \{n-2, \overline{n-1}\}, \\ 0 & \text{otherwise.} \end{cases}$$

We proceed by induction on the rank. For $n = 2$ the proposition is clear. Assume $n \geq 3$ and the proposition is true for ranks smaller than n.

Let $N = n(n-1)$, and let $(i_1, \ldots, i_N)$ be the sequence $n-1, \overline{n-1}$ followed by $n-2, n-1, \overline{n-1}, n-2$, followed by $n-3, n-2, n-1, \overline{n-1}, n-2, n-3$, and so on, finishing with $1, 2, 3, \ldots, n-2, n-1, \overline{n-1}, n-2, \ldots, 3, 2, 1$. Then $s_{i_1} s_{i_2} s_{i_3} \cdots s_{i_N}$ is a reduced expression for the longest element in the Weyl group of type D_n. Furthermore, $s_{i_1} s_{i_2} s_{i_3} \cdots s_{i_{N-2n+2}}$ is a reduced expression for the longest element of the Weyl group of type D_{n-1} generated by $\{s_i\}_{i \in I \setminus \{1\}}$. Using this reduced expression, h_k is the sequence $1, 1$, followed by $3, 2, 2, 1$, followed by $5, 4, 3, 3, 2, 1$, and so on, finishing with $2n-3, 2n-4, \ldots, n-1, n-1, n-2, n-3, \ldots, 3, 2, 1$. We let $r_{i,n}$ denote the value of r_i as in the proposition, for the specific rank n. We have $r_{n-1,n} = r_{\overline{n-1},n} = \frac{n(n-1)}{2}$ and $r_{s,n} = s(2n - s - 1)$ for $s < n - 1$.

We must verify

$$\exp(\sum_{i \in I} r_{i,n} E_i) = \exp(\sum_{i \in I \setminus \{1\}} r_{i,n-1} E_i) \Delta$$

where Δ is given by:

$$\exp((2n-3)E_1) \cdots \exp((n-1)E_{n-1}) \exp((n-1)E_{\overline{n-1}}) \exp((n-2)E_{n-2}) \cdots \exp(E_1).$$

We use the representation of $\mathfrak{g}$ on a $2n$-dimensional vector space V with basis $B = \{b_0, \ldots, b_{n-1}, b'_0, \ldots b'_{n-1}\}$ on which the E_i act as follows:
For $i \neq \overline{n-1}$,

$$E_i(b_j) = \begin{cases} b_{j+1} & \text{if } i = j+1, \\ 0 & \text{otherwise.} \end{cases}$$

$$E_i(b'_j) = \begin{cases} b'_{j-1} & \text{if } i = j, \\ 0 & \text{otherwise.} \end{cases}$$

For $i = \overline{n-1}$, we have $E_{\overline{n-1}}(b_{n-2}) = b'_{n-1}$, $E_{\overline{n-1}}(b_{n-1}) = b'_{n-2}$, and $E_{\overline{n-1}}$ sends all other basis elements to zero.

Δ acts on the basis as follows:

$$b_k \mapsto b_k + \frac{(n-1)!}{k!}(b_{n-1} + b'_{n-1})$$
$$+ (n-1)(\sum_{l=k+1}^{n-2} \frac{2(l-1)!}{k!} b_l + \sum_{l=0}^{n-2} \frac{(2n-l-3)!}{k!} b'_l) \qquad (k < n-1)$$

$$b_{n-1} \mapsto b_{n-1} + \sum_{l=0}^{n-2} \frac{(2n-l-3)!}{(n-2)!} b'_l$$

$$b'_{n-1} \mapsto b'_{n-1} + \sum_{l=0}^{n-2} \frac{(2n-l-3)!}{(n-2)!} b'_l$$

$$b'_k \mapsto b'_k + 2(n-1) \sum_{l=0}^{n-2} \frac{(2n-l-3)!}{(2n-k-2)!} b'_l \qquad (k < n-1).$$

$\exp(\sum_{i \in I} r_{i,n} E_i)$ acts on the basis as follows:

$$b_k \mapsto \sum_{s=k}^{n-2} \frac{s!(2n-k-2)!}{k!(s-k)!(2n-s-2)!} b_s + \frac{(2n-k-2)!}{2(n-1-k)!k!}(b_{n-1} + b'_{n-1})$$

$$+ \sum_{s=0}^{n-2} \frac{(2n-s-2)!(2n-k-2)!}{2(2n-s-k-2)!k!s!} b'_s \qquad (k < n-1)$$

$$b_{n-1} \mapsto b_{n-1} + \frac{1}{2} \sum_{s=0}^{n-2} \frac{(2n-s-2)!}{(n-s-1)!s!} b'_s$$

$$b'_{n-1} \mapsto b'_{n-1} + \frac{1}{2} \sum_{s=0}^{n-2} \frac{(2n-s-2)!}{(n-s-1)!s!} b'_s$$

$$b'_k \mapsto \sum_{s=0}^{k} \frac{(2n-s-2)!k!}{(2n-k-2)!s!(k-s)!} b'_s \qquad (k < n-1).$$

The rest of the verification relies on the identity

$$\sum_{p=0}^{A-1} \frac{(p+q)!}{p!} = \frac{(q+A)!}{(A-1)!(q+1)}.$$

We omit further details.

Case 3. Now assume $\mathfrak{g}$ is of type E_6, E_7, or E_8. Here, we used the adjoint action of $\mathfrak{u}^+$ on the Borel subalgebra $\mathfrak{b} = \mathfrak{h} + \mathfrak{u}^+$. According to [3, §0.5], we may find a basis $B = \{X_\phi, t_i \mid \phi \in R^+, i \in I\}$ of $\mathfrak{b}$ such that the action of the E_i are as follows:

$$E_i(X_\phi) = \begin{cases} X_{\phi+\alpha_i} & \text{if } \phi, \phi + \alpha_i \in R^+, \\ 0 & \text{otherwise.} \end{cases}$$

$$E_i(t_j) = |a_{ij}| X_{\alpha_i}.$$

Note that this representation is faithful.

We utilized a computer to verify the formula in this case. One virtue of using the basis B is that all the structure constants are non-negative integers. We note that in E_8, the largest integer which appears in the final matrix is 41 digits long.

Note added February 13, 1996. Dale Peterson has informed us that he has a uniform proof of Proposition 1, which does not use computer calculations.

REFERENCES

1. N. Bourbaki, *Groupes et algèbres de Lie*, Chapitres 4, 5, et 6, Masson, Paris, 1981.
2. G. Lusztig, *Total positivity in reductive groups*, Lie Theory and Geometry: in honor of Bertram Kostant, Progress in Math., no. 123, Birkhauser, Boston, 1994, pp. 531–568.
3. ———, *Finite Dimensional Hopf Algebras Arising From Quantized Universal Enveloping Algebras*, J. Amer. Math. Soc. 3 (1990), 257–296.

DEPARTMENT OF MATHEMATICS, M.I.T., CAMBRIDGE, MASSACHUESETTS 02139

SYMMETRY BREAKING FOR EQUIVARIANT MAPS

MICHAEL FIELD

Dedicated to the memory of Roger W. Richardson

ABSTRACT. In this work we state and prove a number of foundational results in the local bifurcation theory of smooth equivariant maps. In particular, we show that stable one-parameter families of maps are generic and that stability is characterised by semi-algebraic conditions on the finite jet of the family at the bifurcation point. We also prove strong determinacy theorems that allow for high order forced symmetry breaking. We give a number of examples, related to earlier work of Field & Richardson, that show that even for finite groups we can expect branches of fixed or prime period two points with submaximal isotropy type. Finally, we provide a simplified proof of a result that justifies the use of normal forms in the analysis of the equivariant Hopf bifurcation.

CONTENTS

1. **Introduction**   **212**
   1.1. Acknowledgements

2. **Technical Preliminaries and Basic Notations**   **214**
   2.1. Generalities on groups
   2.2. Γ-sets and isotropy types
   2.3. Representations
   2.4. Isotropy types for representations
   2.5. Polynomial Invariants and Equivariants
   2.6. Smooth families of equivariant maps
   2.7. Normalized families

3. **Branching and invariant group orbits**   **217**
   3.1. Invariant group orbits and normal hyperbolicity
   3.2. Branches of invariant Γ-orbits
   3.3. The branching pattern
   3.4. Stabilities
   3.5. Branching conditions
   3.6. The signed indexed branching pattern
   3.7. Stable families
   3.8. Determinacy

1991 *Mathematics Subject Classification.* 58F14, 14E15, 14P05, 57S15, 58F36.
Research supported in part by NSF Grant DMS-9403624, Texas Advanced Research Program Award 003652026 and ONR Grant N00014-94-1-0317

4. Genericity theorems   **223**
4.1. Invariant and equivariant generators
4.2. The varieties Σ, Ξ
4.3. Geometric properties of Ξ
4.4. The sets C_τ
4.5. Stability theorems I: Weak regularity
4.6. Stability theorems II: Regular families
4.7. Determinacy
4.8. An invariant sphere theorem and Fiedler's theorem for maps

5. Examples for Γ finite   **236**
5.1. Preliminaries
5.2. Subgroups of the Weyl group of type B_n
5.3. Branches of fixed points
5.4. Branches of invariant orbits consisting of period two points

6. Strong determinacy   **238**
6.1. Strong determinacy theorem for maps
6.2. Proof of the Strong determinacy theorem for maps
6.3. Applications to normal forms

7. Equivariant Hopf bifurcation theorem for vector fields   **241**
7.1. Preliminaries
7.2. Equivariant Hopf bifurcation and normal forms

References   **244**

1. INTRODUCTION

Let Γ be a compact Lie group and V be an irreducible finite dimensional non-trivial representation space for Γ over $\mathbb{R}$ or $\mathbb{C}$. In Field & Richardson [22, 23, 24, 25], a theoretical framework was developed for the local analysis of symmetry breaking bifurcations[1] of one parameter families of smooth Γ-equivariant vector fields on V. This approach was developed further in [18, 12], where genericity and determinacy theorems were proved for bifurcation problems defined on general real or complex irreducible representations. Taken together, these results imply that there is a finite classification of branching patterns of stable families, that the stability of a family is determined by a finite jet at the bifurcation point (finite determinacy), and that branches persist generically under high order symmetry breaking perturbations (strong determinacy). We refer the reader to [11] for a discussion of some of these results and their proofs. Suffice it to say that techniques are typically geometric and depend on ideas from real algebraic geometry, equivariant transversality and resolution of singularities ("blowing-up").

Our aim in this work is to develop an analogous theory for smooth families of Γ-equivariant *maps*. Specifically, we study bifurcations of generic one-parameter

[1]We refer the reader to [28], [24, Introduction] for an overview and background on symmetry breaking and equivariant bifurcation theory.

families of Γ-equivariant maps defined on an irreducible representation (V, Γ). Rather than looking for branches of relative equilibria (or limit cycles), we search for branches of invariant group orbits. If Γ is *finite*, branches consist of fixed points or points of prime period two and some of our results extend work of Chossat & Golubitsky [8], Peckham & Kevrekidis [33], and Vanderbauwhede [38] to situations where the equivariant branching lemma does not apply. The reader should also note the important work by Ruelle [35] on bifurcations of equivariant maps (and vector fields). Using our results on families of maps we provide a simplified analysis of the effect of breaking normal form symmetries in the equivariant Hopf bifurcation. This approach avoids the use of the somewhat technical normal hyperbolicity results proved in [12, Appendix].

In more detail, we start in §2 with a review of basic notations and facts about group actions, representations and dynamics of equivariant maps. In §3, following [24], we cover the basic definitions of stable family, branching pattern and determinacy for equivariant maps. We conclude with the definition of strong determinacy which allows for forced symmetry breaking. In §4, we prove genericity and determinacy theorems for one-parameter families of equivariant maps (Theorems 4.5.3, 4.5.7, 4.6.5, 4.7.1). These results are proved using techniques based on equivariant transversality and stratified sets. Essentially, we show that our concept of genericity (or determinacy) can be formulated in terms of transversality conditions to stratified sets. Granted this, genericity and determinacy theorems follow easily using standard transversality theory. We conclude §4 with a version of the invariant sphere theorem [18] for maps (Theorem 4.8.1) and show how this can be used to prove a partial extension of Fiedler's Hopf bifurcation theorem [10] to maps. Using results from [25], we present in §5 a large class of examples based on the series of finite reflection groups $W(B_n)$, $n \geq 2$. In particular, we give many examples where there are stable submaximal branches of fixed points or points of prime period two. In §6, we prove a strong determinacy theorem for families of equivariant maps (Theorem 6.1.1). The methods used here are very similar to of [12, §§7–10] and depend on resolution of singularities arguments and, in the case of non-Abelian non-finite compact Lie groups, recent results of Schwarz [36] on the coherence of orbit strata. We conclude the section by showing how the strong determinacy theorem can be used to justify the use of normal forms in the analysis of period doubling bifurcations for equivariant maps (Theorems 6.3.3, 6.3.5). Finally, in §7, we show how our results on families of equivariant maps can be used to justify the use of normal forms in the equivariant Hopf bifurcation theorem for vector fields (Theorem 7.2.1). Using blowing-up arguments, we reduce the study of the $\Gamma \times S^1$-equivariant Hopf bifurcation to an analysis of a Γ-equivariant family of (blown-up) Poincaré maps.

1.1. Acknowledgements. We thank Marty Golubitsky for helpful conversations on period doubling and for telling us of the works by Vanderbauwhede and Peckam & Kevrekidis. Thanks also to Muriel Kœnig and Pascal Chossat for a number of stimulating conversations on the equivariant Hopf bifurcation which encouraged us to write down the simplified proof of the equivariant Hopf bifurcation theorem presented in §7.

2. TECHNICAL PRELIMINARIES AND BASIC NOTATIONS

As far as possible we shall follow the notational conventions of [12, 22].

2.1. Generalities on groups. Throughout, we shall be considering compact Lie groups Γ. If H is a (closed) subgroup of Γ, we let $N(H)$ denote the *normalizer* of H in Γ and $C(H)$ denote the *centralizer* of H in Γ. Obviously, $C(H) \subset N(H)$. We let H^0 denote the identity component of H.

2.2. Γ-sets and isotropy types. Let Γ be a group and X be a Γ-set. If $x \in X$, then $\Gamma \cdot x$ denotes the Γ-orbit of x and Γ_x denotes the isotropy subgroup of Γ at x. We refer to the conjugacy class (Γ_x) of Γ_x in Γ as the *isotropy type* or *orbit type* of x. We let $\mathscr{O}(X, \Gamma)$ denote the set of isotropy types for the Γ-set X. We abbreviate $\mathscr{O}(X, \Gamma)$ to $\mathscr{O}$ if X and Γ are implicit from the context. For $x \in \Gamma$, we let $\iota(x)$ denote the isotropy type of x. If $\tau \in \mathscr{O}(X, \Gamma)$, we let $X_\tau = \{x \in X \mid \iota(x) = \tau\}$ be the set of points of isotropy type τ. We let X^Γ denote the fixed point set for the action of Γ on X. We define the usual partial order on $\mathscr{O}(X, \Gamma)$ by "$\tau > \mu$, if there exists $H \in \tau$, $K \in \mu$ such that $H \supset K$, $H \neq K$".

2.3. Representations. Let V be a nontrivial (finite-dimensional) real representation space for Γ. We assume that V has a positive definite Γ-invariant inner product $(\,,\,)$ with associated norm $|\cdot|$ and regard Γ as acting on V by orthogonal transformations.

If V is a nontrivial complex representation space for Γ, we assume that V has a positive definite Γ-invariant hermitian inner product $<,>$ and regard (V, Γ) as a unitary representation. If we let $(\,,\,)$ denote the real part of $<,>$, then $(\,,\,)$ is a Γ-invariant inner product on V. We let J_V denote the complex structure on V defined by scalar multiplication by ι. If we regard $S^1 \subset \mathbb{C}$ as the group of complex numbers of unit modulus, we may extend the action of Γ on V to an action of $\Gamma \times S^1$ on V. The resulting representation $(V, \Gamma \times S^1)$ is complex and both $<,>$ and $(\,,\,)$ will be $\Gamma \times S^1$-invariant. We reserve the notation S^1 for the group of complex numbers of unit modulus and take the S^1-action on V defined by scalar multiplication.

Suppose that (V, Γ) is a real irreducible representation and let $L_\Gamma(V, V)$ denote the space of all Γ-equivariant $\mathbb{R}$-linear endomorphisms of V. We recall Frobenius' Theorem [30, 7.7] that $L_\Gamma(V, V)$ is isomorphic to either $\mathbb{R}$, $\mathbb{C}$ or $\mathbb{H}$ (the quaternions).

Definition 2.3.1. Let (V, Γ) be a nontrivial irreducible real representation.

(1) (V, Γ) is *absolutely irreducible* if $L_\Gamma(V, V) \cong \mathbb{R}$.

(2) (V, Γ) is *irreducible of complex type* if $L_\Gamma(V, V) \cong \mathbb{C}$.

(3) (V, Γ) is *irreducible of quaternionic type* if $L_\Gamma(V, V) \cong \mathbb{H}$.

Remark 2.3.2. If (V, Γ) is irreducible of complex type, we may give V the structure of a complex vector space so that (V, Γ) is irreducible as a complex representation. We take as complex structure on V any element of $L_\Gamma(V, V)$ whose square is $-I_V$. This choice is unique up to multiplication by $\pm I_V$. Elements of $L_\Gamma(V, V)$ will then be complex scalar multiples of the identity map of V. Similar remarks hold for the quaternionic case.

Suppose that (W, Γ) is absolutely irreducible. The action of Γ on W extends to a $\mathbb{C}$-linear action on the complexification $V = W \otimes_R \mathbb{C}$ of W. The representation (V, Γ) is then irreducible as a complex representation. More generally, we recall the following basic result [5] on complex representations.

Lemma 2.3.3. *Let (V, Γ) be an irreducible complex representation. Then one of the following three exclusive possibilities must occur.*

(R) (V, Γ) *is isomorphic to the complexification of an absolutely irreducible representation.*

(C) *If we regard (V, Γ) as a real representation, then (V, Γ) is of complex type.*

(Q) *If we regard (V, Γ) as a real representation, then (V, Γ) is of quaternionic type.*

Definition 2.3.4 ([12, §2]). Let (V, Γ) be a complex representation. We say that

(1) (V, Γ) is *complex irreducible* if (V, Γ) is nontrivial, irreducible and not of quaternionic type.

(2) (V, Γ) is *tangential* if (V, Γ) is complex irreducible and $\Gamma \supset S^1$.

Example 2.3.5. Let (V, Γ) be a complex representation. The natural action of S^1 on V commutes with Γ. Set $G = \Gamma \times S^1$. Then (V, G) is a complex representation of G. If (V, Γ) is complex irreducible, then (V, G) is tangential.

2.4. Isotropy types for representations. Let (V, Γ) be a finite dimensional real representation. It is well-known and elementary that the set $\mathcal{O}(V, \Gamma) = \mathcal{O}$ of isotropy types is finite. Obviously, we always have $(\Gamma) \in \mathcal{O}$. We define $\mathcal{O}^* = \mathcal{O} \setminus (\Gamma)$. If τ, $\mu \in \mathcal{O}$, it follows from linearity and slice theory that

$$\tau > \mu \text{ if and only if } V_\tau \subset \partial V_\mu$$

We say that an orbit type τ is *maximal* (respectively, *submaximal*) if (i) $\tau \neq (\Gamma)$ and (ii) $\mu > \tau$ implies $\mu = (\Gamma)$ (respectively, $\tau \neq (\Gamma)$ and τ is not maximal). Given $\tau \in \mathcal{O}$, choose $x \in V_\tau$ and let $N(\Gamma_x)$ denote the normalizer of Γ_x in Γ. Define

$$g_\tau = \dim(\Gamma \cdot x), \quad n_\tau = \dim(N(\Gamma_x)/\Gamma_x)$$

Of course, g_τ and n_τ depend only on τ and not on the choice of x in V_τ.

2.5. Polynomial Invariants and Equivariants. Let $P(V)$ denote the $\mathbb{R}$-algebra of $\mathbb{R}$-valued polynomial functions on V and $P(V, V)$ be the $P(V)$-module of all polynomial maps of V into V. For $k \in \mathbb{N}$, we let $P^k(V)$ (respectively, $P^{(k)}(V)$) denote the vector space of all homogeneous polynomials (respectively, polynomials) of degree k. We similarly define the spaces $P^k(V, V)$ and $P^{(k)}(V, V)$. If (V, Γ) is a finite dimensional real representation, we let $P(V)^\Gamma$ denote the $\mathbb{R}$-subalgebra of $P(V)$ consisting of invariant polynomials, and $P_\Gamma(V, V)$ denote the $P(V)^\Gamma$-module of Γ-equivariant polynomial endomorphisms of V. If (V, Γ) has the structure of a complex representation, then $P_\Gamma(V, V)$ may be given the structure of a complex vector space, with scalar multiplication defined by $P \mapsto \lambda P$, $\lambda \in \mathbb{C}$.

In the sequel, we assume that (V, Γ) is nontrivial and either absolutely irreducible or complex irreducible. We let $\mathscr{F} = \{F_1, \ldots, F_k\}$ be a minimal set of homogeneous generators for the $P(V)^\Gamma$-module $P_\Gamma(V,V)$ and let $\mathscr{F}_V$ denote the real vector subspace of $P_\Gamma(V,V)$ spanned by $\mathscr{F}$. Let $d_i = \text{degree}(F_i)$, $1 \le i \le k$. We order the F_i so that $d_1 \le d_2 \le \ldots \le d_k$. If (V,Γ) is absolutely irreducible, we may suppose that $F_1 = I_V$ and $d_i \ge 2$, $i > 1$. If (V,Γ) is complex irreducible, we may suppose that $F_1 = I_V$, $F_2 = J_V$ and $d_i \ge 2$, $i > 2$.

Lemma 2.5.1. *Suppose that (V,Γ) is a complex representation. We may choose $\mathscr{F}$ so that $\mathscr{F}_V$ is a complex vector subspace of $P_\Gamma(V,V)$. If we set $J = S^1 \cap \Gamma$, then $\mathscr{F}_V$ is invariant under the J-action defined by $\nu_g(F) = g \circ F = F \circ g$, $g \in J$. In particular, $\mathbb{R}^k$ inherits from $\mathscr{F}_V$ the natural structure of a complex J-representation.*

Proof. Let $\mathscr{H}$ be a minimal set of homogeneous generators for $P_\Gamma(V,V)$, regarded as a module over the complex valued invariants. Take $\mathscr{F} = \mathscr{H} \cup \imath\mathscr{H}$.

Remarks 2.5.2. (1) In the sequel we always assume that if (V,Γ) is complex irreducible then the set $\mathscr{F}$ of generators for $P_\Gamma(V,V)$ satisfies the hypotheses of Lemma 2.5.1. If $\exp(\imath\theta) \in S^1 \cap \Gamma$, we shall let $\nu_\theta : \mathbb{R}^k \to \mathbb{R}^k$ denote the linear isomorphism of $\mathbb{R}^k$ induced by multiplication by $\exp(\imath\theta)$. (2) Lemma 2.5.1 applies when $\Gamma \cap S^1 = \{e\}$. Suppose that $H \ne \Gamma$ is an isotropy group. Regard $N(H)/H$ as a subgroup of $O(V^H)$ and set $H^* = N(H)/H \cap S^1$. Since the S^1 action on V^H is a restriction of the S^1-action on V, and $F|V^H$ is $N(H)/H$-equivariant, $F \in \mathscr{F}_V$, it follows that $\mathscr{F}_V^H = \{F|V^H \mid F \in \mathscr{F}_V\}$ is H^*-invariant.

2.6. Smooth families of equivariant maps. Let $C_\Gamma^\infty(V \times \mathbb{R}, V)$ denote the space of smooth (that is, C^∞) Γ-equivariant maps from $V \times \mathbb{R}$ to V, where Γ acts on $V \times \mathbb{R}$ as $(v, \lambda) \mapsto (gv, \lambda)$. Let $C^\infty(V \times \mathbb{R})^\Gamma$ denote the space of smooth $\mathbb{R}$-valued invariant functions on $V \times \mathbb{R}$. We give $C_\Gamma^\infty(V \times \mathbb{R}, V)$ and $C^\infty(V \times \mathbb{R})^\Gamma$ the C^∞ topology. Let $f \in C_\Gamma^\infty(V \times \mathbb{R}, V)$. It follows either from the theory of closed ideals of differentiable functions (see [14]) or from Schwarz' theorem on smooth invariants (see [34]) that we may write

$$f_\lambda(x) = f(x, \lambda) = \sum_{i=1}^{k} f_i(x, \lambda) F_i(x),$$

where $f_i \in C^\infty(V \times \mathbb{R})^\Gamma$, $1 \le i \le k$.

2.7. Normalized families. Suppose $f \in C_\Gamma^\infty(V \times \mathbb{R}, V)$ and (V,Γ) is irreducible. Clearly, $x = 0$ is a fixed point of $f(x, \lambda) = x$. We refer to $x = 0$ as the *trivial branch* of fixed points for f.

Lemma 2.7.1. *Suppose that $f \in C_\Gamma^\infty(V \times \mathbb{R}, V)$, $\lambda_0 \in \mathbb{R}$ and $Df_{\lambda_0}(0)$ has no eigenvalues of unit modulus. We may choose a neighborhood U of $(0, \lambda_0)$ in $V \times \mathbb{R}$, such that if $(x, \lambda) \in U$ and $f_\lambda(x) \in \Gamma \cdot x$ then $x = 0$.*

Proof. Suppose that (V,Γ) is a complex representation (the proof when (V,Γ) is absolutely irreducible is similar). It follows from the irreducibility of (V,Γ) that $Df_\lambda(0) = \sigma_f(\lambda)I_V$, where $\sigma_f : \mathbb{R} \to \mathbb{C}$ is smooth. Without loss of generality, suppose $|\sigma_f(\lambda)| > 1$. It follows by continuity that we can choose a compact neighborhood J of λ_0 such that $|\sigma_f(\lambda)| > 1$ on J. Hence, for each $\lambda \in J$, $x = 0$

is a hyperbolic repelling point of f_λ. Let D_r denote the closed disc, center zero, radius r in V. Since J is compact, it follows from the mean value theorem that we can choose $r > 0$, $c > 1$, such that for all $x \in D_r$, $\lambda \in J$, we have $\|f(x)\| \geq c\|x\|$. Hence there are no fixed orbits for f in $U = D_r \times J$.

It follows from Lemma 2.7.1 that bifurcations of the trivial branch of fixed points only occur when $Df_\lambda(0)$ has an eigenvalue on the unit circle. In particular, note that our interest is in finding branches of invariant *group orbits* not just fixed points.

Just as in [22, 23, 24, 12], we restrict attention to families f that have a non-degenerate change of stability of the trivial branch of fixed points at $\lambda = 0$:

$$|\sigma_f(0)| = 1, \quad |\sigma_f(0)|' \neq 0 \tag{1}$$

Reparametrizing the bifurcation variable λ and noting that we shall only be interested in (generic) behavior of f near the origin of $V \times \mathbb{R}$, it is no loss of generality to restrict attention to the space

$$\mathscr{M}(V, \Gamma) = \{f \in C_\Gamma^\infty(V \times \mathbb{R}, V) \mid \sigma_f(\lambda) = \exp(\imath\omega(\lambda))(1 + \lambda)\},$$

where $\omega : \mathbb{R} \to \mathbb{R}$ is a smooth map. In case (V, Γ) is absolutely irreducible, we replace the term $\exp(\imath\omega(\lambda))(1 + \lambda)$ by $\pm 1 + \lambda$.

In the sequel, we refer to elements of the spaces $\mathscr{M}(V, \Gamma)$ as *normalized* families. If $\theta \in [0, 2\pi)$, we define $\mathscr{M}^\theta(V, \Gamma) = \{f \in \mathscr{M}(V, \Gamma) \mid \omega(0) = \theta\}$. If (V, Γ) is absolutely irreducible, we let $\mathscr{M}^+(V, \Gamma)$, $\mathscr{M}^-(V, \Gamma)$ denote the subspaces of $\mathscr{M}(V, \Gamma)$ corresponding to $\sigma_f(0) = 1$, $\sigma_f(0) = -1$ respectively.

3. Branching and invariant group orbits

After briefly reviewing some basic definitions about invariant group orbits for equivariant maps, we discuss branching and stability for 1-parameter families of equivariant maps. We assume the reader has some familiarity with the basic definitions and results about normal hyperbolicity described in Hirsch et al. [29].

3.1. Invariant group orbits and normal hyperbolicity. Let M be a smooth Γ-manifold. Suppose that $f : M \to M$ is a smooth and Γ-equivariant. We say that a Γ-orbit $\alpha \subset M$ is an f-*invariant orbit* if $f(\alpha) = \alpha$. It follows from equivariance that α is an f-invariant orbit if and only if there exists $x \in \alpha$ such that $f(x) \in \alpha$. If Γ (or α) is finite, each point of α is a periodic point of f and the stability condition we require is that α consists of hyperbolic periodic points. If Γ is infinite, the natural stability condition is to require that f is normally hyperbolic at α. We refer to [15] for basic properties of normally hyperbolic group orbits. As we shall be considering nondegenerate bifurcations off the trivial solution, we may and shall assume that maps are diffeomorphisms – at least on some neighborhood of invariant group orbits.

A spectral characterization of normal hyperbolicity for invariant group orbits is given in [15, 19]. We recall without proof the main definitions and results from [19] that we need.

Lemma 3.1.1 ([19, Proposition 5.2], [16, Lemma D]). *Let α be an f-invariant Γ-orbit and U be a Γ-invariant neighborhood of α. Suppose that $f(x) \in N(\Gamma_x)^0 \cdot x$, for some (hence any) $x \in \alpha$. Then there exists a smooth map $\chi : M \to \Gamma$ such that*

(1) $\chi(y) = e$, $y \in M \setminus U$.
(2) $\chi(y) \in C(\Gamma_y)^0$, *all* $y \in M$.
(3) *If we define* $\tilde{f} : M \to M$ *by* $\tilde{f}(y) = \chi(y)f(y)$, *then* $\tilde{f}$ *is equivariant and* $\tilde{f}|\alpha$ *is the identity map.*

Suppose that α is an f-invariant Γ-orbit of isotropy τ and $f(x) \in N(\Gamma_x)^0 \cdot x$, $x \in \alpha$. It follows from Lemma 3.1.1 that we may find a smooth Γ-equivariant map $\tilde{f}$ such that $\tilde{f}|\alpha$ is the identity and $\tilde{f}, f$ induce the same map on M/Γ. Since $\tilde{f}(x) = x$, $x \in \alpha$, $T_x\tilde{f} : T_xM \to T_xM$. Let spec$(\tilde{f}, x)$ denote the set of eigenvalues, with multiplicities, of $T_x\tilde{f}$. Then spec$(\tilde{f}, x)$ depends only on $\tilde{f}$ and is independent of $x \in \alpha$ [19]. We define spec$(\tilde{f}, \alpha)$ to be equal to spec$(\tilde{f}, x)$, any $x \in \alpha$.

Before giving the next definition, we recall that S^1 acts on $\mathbb{C}$ by scalar multiplication and that $\mathbb{C}/S^1 \cong \mathbb{R}^+$.

Definition 3.1.2. Let $f, \tilde{f}, \alpha$ be as above.

(1) The *(reduced) spectrum* SPEC(f, α) of f along α is the subset of $\mathbb{R}^+$ defined by
$$\text{SPEC}(f, \alpha) = \text{spec}(\tilde{f}, \alpha)/S^1$$
(2) The *index* of f along α, index(f, α), is defined to be the number of elements of SPEC(f, α) less than 1 (counting multiplicities).

Remarks 3.1.3. (1) It is shown in [15, 19] that the definition SPEC(f, α) depends only on f and α and not on the choice of $\tilde{f}$. Note also that the reduced spectrum may be defined even if $f(x) \in N(\Gamma_x) \cdot x$, as opposed to $N(\Gamma_x)^0 \cdot x$. We refer to [15] for details. (2) SPEC(f, α) contains at least g_τ elements equal to one.

Proposition 3.1.4 ([15]). *Let $f : M \to M$ be a smooth Γ-equivariant map and α be an f-invariant Γ-orbit of isotropy type τ. Then α is normally hyperbolic if and only if $1 \in$ SPEC(f, α) has multiplicity g_τ.*

We conclude this subsection by recalling the following lemma giving a decomposition of f into 'tangent and normal' components near an invariant Γ-orbit.

Lemma 3.1.5 ([19, Lemma 6.2]). *Let α be an f-invariant Γ-orbit and $\alpha(f) \geq 1$ be the smallest integer such that $f^{\alpha(f)}|\alpha$ is Γ-equivariantly isotopic to the identity. Suppose that U, U' are open Γ-invariant tubular neighborhoods of α such that $\overline{U}, \overline{f(U)} \subset U'$. Denote the corresponding families of slices determined by U, U' by $\mathscr{S} = \{\mathscr{S}_x | x \in \alpha\}$, $\mathscr{S}' = \{\mathscr{S}'_x | x \in \alpha\}$, respectively. There exist smooth maps $\rho : U \to \Gamma$, $h : U \to U'$ satisfying:*

(1) $f^{\alpha(f)}(y) = \rho(y)h(y)$, $y \in U$.
(2) $h : U \to U'$ *is an equivariant embedding.*
(3) $h : S_x \to S'_x$, *all* $x \in \alpha$.

(4) $\rho(y) \in C(\Gamma_y)$, all $y \in U$.

(5) α is normally hyperbolic if and only if each (any) $x \in \alpha$ is a hyperbolic fixed point of $h|S_x$.

3.2. Branches of invariant Γ-orbits. Next we discuss families of maps and branches of invariant group orbits. Most of what we say is a natural extension of the corresponding definitions for vector fields given in [24, 12]. We assume throughout that (V, Γ) is either absolutely irreducible or complex irreducible.

Definition 3.2.1. Given $f \in \mathcal{M}(V, \Gamma)$, let

$$\mathbf{I}(f) = \{(x, \lambda) \mid f_\lambda(x) \in \Gamma \cdot x\}$$
$$\mathbf{B}(f) = \{(x, \lambda) \in \mathbf{I}(f) \mid \Gamma \cdot x \text{ is not normally hyperbolic}\}$$

Both $\mathbf{I}(f)$ and $\mathbf{B}(f)$ are closed Γ-invariant subsets of $V \times \mathbb{R}$. We refer to $\mathbf{B}(f)$ as the *bifurcation* set of f and $\mathbf{D}(f) = (\mathbf{I}(f), \mathbf{B}(f))$ as the *bifurcation diagram* of f.

For each $\tau \in \mathcal{O}(V, \Gamma)$, choose $H \in \tau$ and set $\Delta_\tau = \Gamma/H$. Every Γ-orbit of isotropy type τ is smoothly Γ-equivariantly diffeomorphic to Δ_τ.

Definition 3.2.2. Let $f \in \mathcal{M}(V, \Gamma)$ and $\tau \in \mathcal{O}(V, \Gamma)$. A branch of invariant Γ-orbits (of isotropy type τ) for f at zero consists of a C^1 Γ-equivariant map

$$\phi = (\rho, \lambda) : [0, \delta] \times \Delta_\tau \to V \times \mathbb{R}$$

such that λ is independent of $u \in \Delta_\tau$ and

(1) $\phi(0, u) = (0, 0)$, all $u \in \Delta_\tau$.

(2) For all $t \in (0, \delta]$, $\alpha_t = \rho(t, \Delta_\tau)$ is an $f_{\lambda(t)}$-invariant Γ-orbit of isotropy τ.

(3) For every $u \in \Delta_\tau$, the map $\phi_u : [0, \delta] \to V \times \mathbb{R}$, $t \mapsto \phi(t, u)$ is a C^1-embedding.

If, in addition, we can choose $\delta > 0$ so that

(4) For all $t \in (0, \delta]$, $f_{\lambda(t)}$ is a normally hyperbolic at α_t,

we refer to ϕ as a branch of normally hyperbolic invariant Γ-orbits for f at zero.

Remark 3.2.3. Typically, parametrizations ϕ satisfying Definition 3.2.2 are smooth. In fact, if ϕ is smooth, satisfies (1,2) and ϕ_u has initial exponent $1 < p < \infty$, then we may define a new parametrization by $\tilde{\phi}(t, u) = \phi(t^{\frac{1}{p}}, u)$. Although $\tilde{\phi}$ is no longer smooth, it satisfies all the conditions of the definition. Subsequently, when we address the problem of constructing explicit parametrizations, we always construct smooth parametrizations with (minimal) finite initial exponent, possibly greater than one. Whether we work with smooth or C^1-parametrizations, the main point is that the direction of branching should be well-defined at $t = 0$.

Let $f \in \mathcal{M}(V, \Gamma)$. We regard two branches of invariant Γ-orbits for f as *equivalent* if they differ only by a (local) reparametrization. If ϕ is a branch, we let $[\phi]$ denote its equivalence class and we identify $[\phi]$ with the germ of the image of ϕ at the origin of $V \times \mathbb{R}$.

Example 3.2.4. Define $c_+, c_- : [0, \infty) \to V \times \mathbb{R}$ by $c_\pm(s) = (0, \pm s)$, $s \in [0, \infty)$. Then $c_\pm$ define two trivial branches of fixed points for any $f \in \mathcal{V}(V, \Gamma)$.

3.3. The branching pattern.

Definition 3.3.1 ([24, §1, §3]). Let $f \in \mathscr{M}(V, \Gamma)$. The *branching pattern* $\Xi(f)$ of f is the set of all equivalence classes of *non-trivial* branches of invariant orbits for f. Each point in $\Xi(f)$ is labelled with the isotropy type of the associated branch.

3.4. Stabilities. We refine our definition of the branching pattern to take account of stabilities. Suppose that $\phi = (\rho, \lambda)$ is a branch of normally hyperbolic invariant group orbits of isotropy type τ for $f \in \mathscr{M}(V, \Gamma)$. By continuity, $\text{index}(f_{\lambda(t)}, \rho(t, \Delta_\tau))$ is constant, $t > 0$, and we define the index of ϕ, $\text{index}(\phi)$, to be the common value of the indices of the non-trivial invariant orbits along the branch. We define $\text{index}([\phi]) = \text{index}(\phi)$ and note that $\text{index}([\phi])$ depends only on the equivalence class of ϕ. If all the branches of f are normally hyperbolic, then *index* is a well defined N-valued map on $\Xi(f)$.

Lemma 3.4.1. *Let $f \in \mathscr{M}(V, \Gamma)$ and suppose that ϕ is a branch of normally hyperbolic invariant Γ-orbits. Then ϕ is either a supercritical or subcritical branch.*

Proof. Similar to that of Lemma 3.4.2 [12]. $\quad\blacksquare$

If $[\phi] \in \Sigma(f)$ is a normally hyperbolic branch we define $\text{sign}([\phi])$ to be $+1$ if the branch is supercritical and -1 if it is subcritical. It follows from Lemma 3.4.1 that sign is well-defined on the set of normally hyperbolic branches.

3.5. Branching conditions. Following [24], we consider the following *branching conditions* on $f \in \mathscr{M}(V, \Gamma)$:

B1 There is a finite set $\phi_1, \ldots, \phi_{r+2}$ of branches of invariant Γ-orbits for f, with images $C_1, \ldots, C_{r+2}$, such that
  (1) $\Xi(f) = \{[\phi_1], \ldots, [\phi_r]\}$, $[\phi_{r+1}] = [c^+]$, $[\phi_{r+2}] = [c^-]$.
  (2) There is a neighborhood N of $(0,0)$ in $V \times \mathbb{R}$ such that if $(x, \lambda) \in N$ and $\Gamma \cdot x$ is f_λ-invariant then
$$\Gamma \cdot x \times \{\lambda\} \subset \cup_{j=1}^{r+2} C_j$$
  (3) If $i \neq j$, then $C_i \cap C_j = \{(0,0)\}$.
B2 Every $[\phi] \in \Xi(f)$ is a branch of normally hyperbolic f-invariant Γ-orbits.

Definition 3.5.1 ([24]). A family $f \in \mathscr{M}(V, \Gamma)$ is *weakly regular* if f satisfies the branching condition **B1**. If, in addition, f satisfies the branching condition **B2**, we say that f is *regular*.

Remark 3.5.2. If f is regular then $(0,0) \in V \times \mathbb{R}$ is an *isolated* point of $\mathbf{B}(f)$.

3.6. The signed indexed branching pattern.

Definition 3.6.1. Suppose $f \in \mathscr{M}(V, \Gamma)$ is regular. The *signed indexed branching pattern* $\Xi^*(f)$ of f consists of the set $\Xi(f)$, labelled by isotropy types, together with the maps index : $\Xi(f) \to \mathbb{N}$ and sign : $\Xi(f) \to \{\pm 1\}$.

Every regular family f has a well-defined signed indexed branching pattern $\Xi^*(f)$ which describes the stabilities of the f-invariant Γ-orbits on some neighborhood of zero.

Definition 3.6.2. Let $f, g \in \mathcal{M}(V, \Gamma)$ be weakly regular. We say that $\Xi(f)$ is isomorphic to $\Xi(g)$ if there is a bijection between $\Xi(f)$ and $\Xi(g)$ preserving isotropy type. If f, g are regular, we say that $\Xi^\star(f)$ is isomorphic to $\Xi^\star(g)$ if $\Xi(f)$ is isomorphic to $\Xi(g)$ by an isomorphism preserving the sign and index functions.

3.7. Stable families.

Definition 3.7.1 (cf. [24, §2]). A family $f \in \mathcal{M}(V, \Gamma)$ is *stable* if there is an open neighborhood U of f in $\mathcal{M}(V, \Gamma)$ consisting of regular families such that for every continuous path $\{f_t \mid t \in [0, 1]\}$ in U with $f_0 = f$ there exist a compact Γ-invariant neighborhood A of zero in $V \times \mathbb{R}$ and a continuous equivariant isotopy $K : A \times [0, 1] \to V \times \mathbb{R}$ of embeddings satisfying

(1) $K_0 = Id_A$.
(2) $K_t(\mathbf{I}(f) \cap A) = K_t(A) \cap \mathbf{I}(f_t)$, $t \in [0, 1]$.

Let $\mathcal{S}(V, \Gamma)$ denote the subset of $\mathcal{M}(V, \Gamma)$ consisting of stable families.

Proposition 3.7.2. (1) $\mathcal{S}(V, \Gamma)$ *is an open subset of* $\mathcal{M}(V, \Gamma)$.
(2) *If* f, f' *lie in the same connected component of* $\mathcal{S}(V, \Gamma)$ *then* f *and* f' *have isomorphic signed indexed branching patterns.*

Proof. We refer the reader to [24, §2].

3.8. Determinacy.
We conclude this section by extending the definitions of determinacy and strong determinacy given in [12, §3] to families of equivariant maps. As usual, if $f \in \mathcal{M}(V, \Gamma)$, $q \in \mathbb{N}$, we let $j^q f_0(0)$ denote the q-jet of f_0 at $(0, 0)$.

Definition 3.8.1. Γ-equivariant bifurcation problems on V are (generically) finitely determined if there exists $q \in \mathbb{N}$ and an open dense semi-algebraic subset $\mathcal{S}(q)$ of $P_\Gamma^{(q)}(V, V)$ such that if $f \in \mathcal{M}(V, \Gamma)$ and $j^q f_0(0) \in \mathcal{S}(q)$ then f is stable.

Remarks 3.8.2. (1) We say Γ-equivariant bifurcation problems on V are (generically) q-determined if q is the smallest positive integer for which we can find $\mathcal{S}(q)$ satisfying the conditions of Definition 3.8.1. For this value of q, we let $\mathcal{R}(q)$ denote the maximal semi-algebraic open subset of $P_\Gamma^{(q)}(V, V)$ satisfying the conditions of Definition 3.8.1. Granted this Definition of $\mathcal{R}(q)$, we shall say that f is q-*determined* if $j^q f_0(0) \in \mathcal{R}(q)$. (2) Let $f \in \mathcal{M}(V, \Gamma)$ be q-determined and set $Q = j^q f_0(0) - j^1 f_0(0)$. Define $J^Q \in \mathcal{M}(V, \Gamma)$ by $J^Q(x, \lambda) = Df_\lambda(0)(x) + Q(x)$. Then f and J^Q have isomorphic signed indexed branching patterns.

We may give refined definitions of stability and determinacy that allow for perturbations by maps which are only equivariant to some finite order.
Given $f \in C^\infty(V \times \mathbb{R}, V)$, $d \geq 1$, let $f^{[d]} = j^d f(0, 0) \in P^{(d)}(V \times \mathbb{R}, V)$.
Let H be a closed subgroup of Γ and let H act on $V \times \mathbb{R}$ and V by restriction of the action of Γ. For $d \geq 1$, define

$$\mathcal{M}_0^{[d]}[\Gamma : H] = \{f \in C_H^\infty(V \times \mathbb{R}, V) \mid f^{[d]} \in P_\Gamma^{(d)}(V \times \mathbb{R}, V)\}$$

If $H = \{e\}$, set $\mathcal{M}_0^{[d]}[\Gamma : H] = \mathcal{M}_0^{[d]}[\Gamma]$.

Definition 3.8.3. Let $H \subset \Gamma$ be a closed subgroup, N be a smooth compact H-manifold and $f \in \mathcal{M}_0^{[1]}[\Gamma : H]$. Let $1 \leq r \leq \infty$. A C^r-branch of normally hyperbolic invariant submanifolds of type N for f consists of a C^1 H-equivariant map $\phi = (\rho, \lambda) : [0, \delta] \times N \to V \times \mathbb{R}$ satisfying the following conditions:

(1) $\phi(0, x) = (0, 0)$, all $x \in N$.

(2) The map $\lambda : [0, \delta] \times N \to \mathbb{R}$ depends only on $t \in [0, \delta]$.

(3) For each $t \in (0, \delta]$, $\rho_t(N) = N_t$ is a normally hyperbolic submanifold of V for $f_{\lambda(t)}$.

(4) $\phi|(0, \delta] \times N$ is a C^r H-equivariant embedding and for all $(t, x) \in [0, \delta] \times N$, $\frac{\partial \phi}{\partial t}(t, x) \neq 0$.

Remarks 3.8.4. (1) We emphasize that we only require the manifolds $\rho_t(N)$ in Definition 3.8.3 to be C^r. Of course, in Definition 3.2.2, the invariant manifolds are Γ-orbits and therefore smoothly embedded. (2) In the usual way, we regard branches as equivalent if they differ only by a local C^r reparametrization. (3) Let $f \in \mathcal{M}_0^{[1]}[\Gamma : H]$ and $\phi : [0, \delta] \times N \to V \times \mathbb{R}$ be a C^r-branch of normally hyperbolic invariant submanifolds of type N for f. For $t > 0$, let $\{W^{ss}(N_t, x) \mid x \in N_t\}$ denote the strong stable foliation of $W^s(N_t)$. The the dimension of $W^{ss}(N_t, x)$ is independent of x and $t \in (0, \delta]$ and we define index$(\phi) = \dim(W^{ss}(N_t, x))$, $t \in (0, \delta]$. Finally, for possibly smaller $\delta > 0$, we may show that ϕ is either sub- or supercritical.

Suppose that $f \in \mathcal{M}(V, \Gamma)$ is regular. Choose a Γ-invariant neighborhood A of zero in $V \times \mathbb{R}$ such that

$$A \cap \mathbf{I}(f) = \bigcup_{i \in I} E_i$$

where each E_i is a (the image of) branch of normally hyperbolic Γ-orbits. We call $\mathscr{E} = \{E_i \mid i \in I\}$ a *local representation* of $\mathbf{I}(f)$ at zero. Given $E \in \mathscr{E}$, let $\tau(E)$ denote the isotropy type of the branch E.

Definition 3.8.5. Let $f \in \mathcal{M}(V, \Gamma)$ be stable and $\mathscr{E} = \{E_i \mid i \in I\}$ be a local representation of $\mathbf{I}(f)$ at zero. Let H be a closed subgroup of Γ and $d \in \mathbb{N}$. We say f is (d, H)-stable if there exists an open neighborhood U of f in $\mathcal{M}_0^{[d]}[G : H]$ such that for every continuous path $\{f_t \mid t \in [0, 1]\}$ in U with $f_0 = f$, there exists an H-invariant compact neighborhood A of zero in $V \times \mathbb{R}$ and a continuous H-equivariant isotopy $K : A \times [0, 1] \to V \times \mathbb{R}$ of embeddings such that

(1) $K_0 = Id_A$.

(2) For every $E \in \mathscr{E}$, $t \in [0, 1]$, $K_t(A \cap E)$ is a branch of normally hyperbolic submanifolds of type $\Delta_{\tau(E)}$ for f_t.

Remarks 3.8.6. (1) If $H = \{e\}$ in Definition 3.8.5, we say f is *strongly d-stable*. (2) In (2) of Definition 3.8.5, we implicitly assume that the branch is C^r for some $r \geq 1$. The differentiability class does not play a major role in our results and the strong determinacy theorem that we prove holds for all r, $1 \leq r < \infty$.

Definition 3.8.7. We say Γ-equivariant bifurcation problems on V are (generically) *strongly determined* if there exist $d \in \mathbb{N}$ and an open dense semi-analytic subset $\mathscr{S}(d) \subset P_\Gamma^{(d)}(V,V)$ such that if $f \in \mathscr{M}(V,\Gamma)$ and $j^d f_0(0) \in \mathscr{S}(d)$ then f is strongly d-stable.

Remarks 3.8.8. (1) We shall say that Γ-equivariant bifurcation problems on V are (generically) strongly d-determined if d is the smallest positive integer for which we can find $\mathscr{S}(d)$ satisfying the conditions of Definition 3.8.7. For this value of d, we let $\mathscr{N}(d)$ denote the maximal semi-analytic open subset of $P_\Gamma^{(d)}(V,V)$ satisfying the conditions of Definition 3.8.7. We say that f is *strongly d-determined* if $j^d f_0(0) \in \mathscr{N}(d)$. (2) Let H a closed subgroup of Γ. We say that Γ-equivariant bifurcation problems on V are (generically) strongly H-determined if there exist $d \in \mathbb{N}$ and an open and dense semi-analytic subset $\mathscr{S}(d) \subset P_\Gamma^{(d)}(V,V)_0$ such that if $f \in \mathscr{M}(V,\Gamma)$ and $j^d f_0(0) \in \mathscr{S}(d)$ then f is (d,H)-stable. Modulo statements about H-equivariance of isotopies (see Definition 3.8.5), it is clear that strong-determinacy implies strong H-determinacy for all closed subgroups H of Γ.

4. GENERICITY THEOREMS

In this section, we prove our basic genericity and determinacy theorems for Γ-equivariant bifurcation problems on V. The approach we use is broadly similar to that followed in [18, Appendix], [22, §5] and, in particular, [12, §4]. Our results hold for general absolutely or complex irreducible representations of compact Lie groups. We assume known basic facts about semi-algebraic sets and Whitney stratified sets. We refer the reader to [22], [12, §4] for a brief review of these topics and to Costi [9], Mather [32] and Gibson et al. [26] for more detailed presentations.

4.1. Invariant and equivariant generators. Let $\mathscr{F} = \{F_1, \ldots, F_k\}$ be a minimal set of homogeneous generators for the $P(V)^\Gamma$-module $P_\Gamma(V,V)$ and $\mathscr{P} = \{p_1, \ldots, p_\ell\}$ be a minimal set of homogeneous generators for the $\mathbb{R}$-algebra $P(V)^\Gamma$. Set $P = (p_1, \ldots, p_\ell) : V \to \mathbb{R}^\ell$ and note that $P : V \to \mathbb{R}^\ell$ may be regarded as the *orbit map* of V onto $V/\Gamma \subset \mathbb{R}^\ell$.

4.2. The varieties Σ, Ξ. Define $F : V \times \mathbb{R}^k \to V$ by $F(x,t) = \sum_{i=1}^k t_i F_i(x)$ and let

$$\Sigma = \{(x,t) \in V \times \mathbb{R}^k \mid F(x,t) \in T_x N(\Gamma_x) \cdot x\}$$

$$\Xi = \{(x,t) \in V \times \mathbb{R}^k \mid P(x) = P(F(x,t))\}$$

Obviously Ξ is a Γ-invariant real algebraic subset of $V \times \mathbb{R}^k$ and $F(x,t) \in \Gamma \cdot x$ if and only if $(x,t) \in \Xi$. Moreover, if $(x,t) \in \Xi$, then $F(x,t) \in N(\Gamma_x) \cdot x$.

Remarks 4.2.1. (1) The variety Σ was defined in [22, 18, 12] and plays a basic role in the codimension 1 bifurcation theory of smooth Γ-equivariant vector fields. We refer to these works, especially [12, §4], for properties of Σ. Much of what we do in this section will be a relatively straightforward extension of this theory to Ξ. (2) If (V,Γ) is complex irreducible, then we assume (see Lemma 2.5.1, Remarks 2.5.2) that $\mathscr{F}$ is chosen so that $\mathscr{F}_V$ has the structure of a complex vector space. In particular, k will be even and we may identify $\mathbb{R}^k$ with $\mathbb{C}^m$, where $k = 2m$.

Example 4.2.2. Suppose that Γ is finite. Define $\Xi^+ = \{(x,t) \in V \times \mathbb{R}^k \mid F(x,t) = x\}$. Obviously, $\Xi^+ \subset \Xi$. However, it is not generally true that $\Xi^+ = \Xi$. For example, if $\Gamma = \mathbb{Z}_2$ acts non-trivially on $\mathbb{R}$, then $P(x) = x^2$ and $F(x,t) = tx$. Hence Ξ is the zero variety of $x^2(t^2 - 1)$ while Ξ^+ is the zero variety of $x(t - 1)$.

For each $\tau \in \mathcal{O}(V,\Gamma)$, we let Ξ_τ denote the subset of Ξ consisting of points of isotropy type τ. Clearly, Ξ is the disjoint union over $\mathcal{O}(V,\Gamma)$ of the sets Ξ_τ. Since (V,Γ) is a non-trivial irreducible representation, $\Xi_{(\Gamma)} = \{0\} \times \mathbb{R}^k$.

4.3. Geometric properties of Ξ. Suppose that $H \in \tau \in \mathcal{O}^*$. Then $N(H)/H$ acts on the fixed point space V^H.

Remark 4.3.1. For future reference, note that if $-I \in \Gamma$ and $H \in \tau \in \mathcal{O}^*$, then $-I \in N(H)/H$, where we regard $N(H)/H$ as a subgroup of the orthogonal group of V^H. Similarly, if (V,Γ) is complex irreducible and $\Gamma \supset S^1$, then $N(H)/H \supset S^1$, where the S^1 action on V^H is the restriction of the action of S^1 on V. In general, $N(H)/H$ may contain $-I$ or S^1 without the same being true for Γ.

Lemma 4.3.2. Suppose that $(x_0, t_0) \in \Xi_\tau^H$. Then $D_2 F_{(x_0,t_0)} : \mathbb{R}^k \to V^H$ is surjective.

Proof. Since $x_0 \in V_\tau^H$, $\{F_1(x_0), \ldots, F_k(x_0)\}$ spans V^H. Since the matrix of $D_2 F_{(x_0,t_0)}$ equals $[F_1(x_0), \ldots, F_k(x_0)]$, $D_2 F_{(x_0,t_0)}$ is onto.

Lemma 4.3.3. For each $\tau \in \mathcal{O}(V,\Gamma)$, Ξ_τ is a Γ-invariant smooth semi-algebraic submanifold of $V \times \mathbb{R}^k$ of dimension $k + g_\tau$.

Proof. Since $\Xi_\tau = \Xi \cap (V \times \mathbb{R}^k)_\tau$, it is obvious that Ξ_τ is a Γ-invariant semi-algebraic subset of $V \times \mathbb{R}^k$. In order to show that Ξ_τ is smooth, it suffices to show that the map $G : V^H \times \mathbb{R}^k \to V^H/N(H)$ defined by $G(x,t) = P(F(x,t)) - P(x)$ is a submersion at (x_0, t_0). This follows from Lemma 4.3.2, since the orbit map restricts to a submersion $P : V_\tau^H \to V_\tau^H/N(H)$ [2].

Lemma 4.3.4. Let $\gamma, \tau \in \mathcal{O}(V,\Gamma)$. Then

(1) $\Xi_\tau \cap \overline{\Xi}_\gamma = \emptyset$ if $\gamma > \tau$.
(2) $dim(\Xi_\tau \cap \overline{\Xi}_\gamma) < g_\tau + k$, if $\gamma < \tau$.

Proof. Similar to that of Lemma 4.3.4 [12].

In future we regard $\mathbb{R}^k$ as embedded in $V \times \mathbb{R}^k$ as the subspace $\{0\} \times \mathbb{R}^k$. We let $\mathbb{R}^{k-1}$, $\mathbb{R}^{k-2}$ denote the subspaces of $\mathbb{R}^k$ defined by $t_1 = 0$ and $t_1, t_2 = 0$ respectively.

If (V,Γ) is absolutely irreducible, we let $\mathbf{C}_+$, $\mathbf{C}_- \subset \mathbb{R}^k$ respectively denote the hyperplanes defined by $t_1 = +1$, $t_1 = -1$. We set $\mathbf{C} = \mathbf{C}_+ \cup \mathbf{C}_-$. If (V,Γ) is complex irreducible, we let $\mathbf{C} \subset \mathbb{R}^k$ denote the cylinder $t_1^2 + t_2^2 = 1$. If (V,Γ) is complex, we let $\mathbf{C}_\theta$ denote the codimension 2 subspace $\{\theta\} \times \mathbb{R}^{k-2}$.

Lemma 4.3.5. Suppose that $\tau \in \mathcal{O}^*$. Then $\partial \Xi_\tau \cap \mathbb{R}^k \subset \mathbf{C}$.

Proof: We prove in case (V, Γ) is complex irreducible. Set $(h_1, \ldots, h_\ell)(x, t) = P(F(x, t)) - P(x)$. Since the invariant p_1 is the square of the Euclidean norm on V and $F(x, t) = (t_1 + \imath t_2)x + O(\|x\|^2)$, it follows that

$$h_1(x, t) = (t_1^2 + t_2^2 - 1)\|x\|^2 + O(\|x\|^3) \tag{2}$$

Suppose that (x^n, t^n) is a sequence of points of Ξ_τ converging to the point $(0, t) \in V \times \mathbb{R}^k$. Substituting in (2), dividing by $\|x^n\|^2$, and letting $n \to \infty$, we deduce that $t_1^2 + t_2^2 = 1$.

Example 4.3.6. Take the standard action of SO(2) on $\mathbb{C}$. A basis for the SO(2)-equivariant polynomial maps of $\mathbb{C}$ is given by $F_1(z) = z$, $F_2(z) = \imath z$. Let (e) denote the conjugacy class of the identity element. The variety Ξ is the zero set of $((t_1^2 - 1) + t_2^2)|z|^2$. Consequently, $\Xi_{(\mathrm{SO}(2))} = \mathbb{R}^2$ and $\Xi_{(e)} = \{(z, t) \mid z \neq 0, t_1^2 + t_2^2 = 1\}$. Obviously, $\partial\Xi_{(e)}$ meets $\mathbb{R}^2$ along the circle $t_1^2 + t_2^2 = 1$.

Lemma 4.3.7. *Let* $H \in \tau \in \mathcal{O}^*$. *Suppose* (x^n, t^n) *is a sequence of points in* Ξ_τ^H *converging to* $(0, t) \in \{0\} \times \mathbb{R}^k$, (γ^n) *is a sequence of points of* $N(H)/H \subset O(V^H)$ *converging to* $\gamma \in N(H)/H$ *and that for* $n \geq 1$ *we have*

$$\sum_{j=1}^{k} t_j^n F_j(x^n) = \gamma^n x^n$$

Then

(a) *If* (V, Γ) *is absolutely irreducible, then* $t_1 = \pm 1$ *and* $\gamma^2 = I$. *If* $t_1 = +1$, *then* $\gamma = I \in N(H)/H$. *If* $t_1 = -1$ *and* $-I \in N(H)/H$, *then* $\gamma = -I$.

(b) *If* (V, Γ) *is complex irreducible, then* $t_1 + \imath t_2 = \exp(\imath\theta)$, *for some* $\theta \in [0, 2\pi)$. *If* $\theta = 2\pi/p$, *for some* $p \in \mathbb{N}$, *then* $\gamma^p = I$. *If* $\exp(\imath\theta) \in S^1 \cap N(H)/H$, *then* $\gamma = \exp(\imath\theta)$.

Proof. Given $t \in \mathbb{R}^k$, define $f_t : V \to V$ by $f_t(x) = F(x, t)$. Provided $t_1 \neq 0$, f_t will be a Γ-equivariant diffeomorphism on some Γ-invariant neighborhood D of $0 \in V$. Restrict f_t to the $N(H)/H$-space V^H. Following [12, §9], [20], we (polar) blow-up V^H along the non-principal $N(H)/H$-orbit strata. If $\Pi : \tilde{V}^H \to V^H$ denotes the blowing-down map, we set $E = \Pi^{-1}(0)$. Let $\tilde{f}_t$ denote the lift of $f_t|D$ to $\tilde{V}^H \cap \Pi^{-1}(D)$. The restriction of $\tilde{f}_t$ to E equals the lift of $Df_t(0)$ to E. Suppose now that (V, Γ) is absolutely irreducible and the hypotheses of the lemma are satisfied with $t_1 = 1$. It follows that $\tilde{f}_{t_n}|E$ converges to the identity map. Since $N(H)/H$ acts freely on $E \subset \tilde{V}^H$, we see that $\gamma = I$. If $t_1 = -1$, then the same argument proves that $\tilde{f}_{t_n}^2|E$ must converge to the identity and so $\gamma^2 = I$. If $-I \in N(H)/H$, then $f_{-t_n}(x^n) = (-\gamma^n)(x^n)$, where $-\gamma^n \in N(H)/H$. Hence $-\gamma = I$, since $-t_1^n \to 1$. The other cases are similarly proved. $\blacksquare$

Remark 4.3.8. Let (V, Γ) be an absolutely irreducible representation. With the notation of Lemma 4.3.7, suppose that $x^n/|x^n| \to u$, where u lies in the unit sphere of V^H. If $t_1 = -1$ then u must be fixed by $-\gamma$ and so $-\gamma$ must restrict to the identity map on the line through u. In particular $\gamma(\mathbb{R}u) = \mathbb{R}u$. Consequently, if $N(H)/H$ acts freely on the projective space of V^H, then we can never have $t_1 = -1$. Similar remarks hold for the case of complex irreducible representations.

Example 4.3.9. Let $\Gamma = \mathbf{D}_3$ denote the dihedral group of order 6 acting in the standard way on $\mathbb{C} = \mathbb{R}^2$. As bases for $P(\mathbb{C})^\Gamma$, $P_\Gamma(\mathbb{C}, \mathbb{C})$, we take $\{|z|^2, \text{Re}(z^3)\}$ and $\{z, \bar{z}^2\}$ respectively. The action of Γ on $\mathbb{C}$ has three isotropy types: $\tau_0 = (\Gamma)$, $\tau_1 = (\mathbb{Z}_2)$ and $\tau_2 = (e)$. It is easy to verify directly that $\overline{\Xi}_{\tau_1}$ meets $\mathbb{R}^2$ along the line $t_1 = 1$. On the other hand $\overline{\Xi}_{\tau_2} \cap \mathbb{R}^2$ consists of the line $t_1 = -1$ together with the isolated point $(1, 0)$. Indeed, set $z(\rho) = i\rho \exp(i\rho)u(\rho)$. Using the implicit function theorem together with the defining equations for Ξ_{τ_2}, it is not hard to show that for all $t_2 \in \mathbb{R}$, we can find smooth maps $u, t_1 : [0, \delta] \to \mathbb{R}(> 0)$, such that $t_1(0) = -1$ and $F(t_1(\rho), t_2, z(\rho)) = z(\rho)$, $\rho \in [0, \delta]$. Note that this curve of points of period two is tangent to the line in $\mathbb{C}$ on which $z \mapsto \bar{z}$ acts as minus the identity. Later, we prove a general version of this result (see Lemma 4.4.2).

We conclude this subsection with an elementary lemma that will be useful later.

Lemma 4.3.10. *Let* $\mathfrak{m} = \{q \in P(V)^\Gamma \mid q(0) = 0\}$ *and suppose*

$$G(x) = \sum_{i=1}^k q_i(x) F_i(x),$$

where $q_i \in \mathfrak{m}$, $1 \le i \le k$. *If we define* $\tilde{F} : V \times \mathbb{R}^{k+1} \to V$ *by* $\tilde{F}(x, t) = F(x, t) + t_{k+1} G(x)$, *and let* $\tilde{\Xi} = \{(x, t) \mid P(\tilde{F}(x, t)) = P(x)\}$, *then for all* $\tau \in \mathcal{O}$ *we have*

$$\partial \tilde{\Xi}_\tau \cap \mathbb{R}^{k+1} = \partial \Xi_\tau \cap \mathbb{R}^k \times (\{0\} \times \mathbb{R})$$

Proof. The result follows easily by observing that $(x, (t_1, \ldots, t_{k+1})) \in \tilde{\Xi}_\tau$ if and only if $(x, (t_1 + t_{k+1} q_1(x), \ldots, t_k + t_{k+1} q_k(x))) \in \Xi_\tau$.

Let $\mathcal{S}$ denote the canonical (minimal) semi-algebraic stratification of Ξ (see [32] or [12, §4]).

Theorem 4.3.11 ([22, Theorem 5.10]). *The stratification* $\mathcal{S}$ *induces a semi-al-gebraic Whitney stratification of each* Ξ_τ. *In particular, each* Ξ_τ *is a union of* $\mathcal{S}$*-strata.*

Proof. The proof is similar to that of [22, Theorem 5.10] or [12, Theorem 4.3.7] and we shall not repeat the details.

4.4. The sets C_τ. Given $\tau \in \mathcal{O}$, define $C_\tau = \mathbb{R}^k \cap \overline{\Xi}_\tau$. Clearly $C_{(\Gamma)} = \mathbb{R}^k$. If $\tau \ne (\Gamma)$, then it follows from Lemma 4.3.5 that $C_\tau \subset \mathbf{C}$. If (V, Γ) is absolutely irreducible, we define $C_\tau^+ = C_\tau \cap \mathbf{C}_+$ and $C_\tau^- = C_\tau \cap \mathbf{C}_-$. (Note that C_τ^- may be empty – Example 4.3.9 – but C_τ^+ always contains $(1, 0, \ldots, 0)$). If (V, Γ) is complex irreducible, then for each point θ lying on the circle $t_1^2 + t_2^2 = 1$, we define $C_\tau^\theta = \mathbf{C}_\theta \cap C_\tau$.

Remark 4.4.1. We recall from [22, 18, 12] that for $\tau \in \mathcal{O}^*$, we define $A_\tau = \mathbb{R}^k \cap \overline{\Sigma}_\tau$. The sets A_τ are closed semi-algebraic conical subsets of the hyperplane $t_1 = 0$. If (V, Γ) is complex irreducible and tangential then A_τ is invariant under translations by vectors $(0, s, 0, \ldots, 0) \in \mathbb{R}^k$, $s \in \mathbb{R}$. The same is true if $H \in \tau$ and $S^1 \subset N(H)/H$.

Lemma 4.4.2. *Let (V, Γ) be absolutely irreducible. Let $H \in \tau \in \mathcal{O}$.*

(1) *If $\dim(V^H) = 1$, then C_τ^+ is the hyperplane $t_1 = 1$.*

(2) *If there exists $\gamma \in \Gamma \cap N(H)$, $\mathbf{u} \in V_\tau^H$ such that the fixed point space of the map $-\gamma : V^H \to V^H$ is the line $\mathbb{R}\mathbf{u}$, then C_τ^- is the hyperplane $t_1 = -1$.*

Proof. Statement (1) is well-known and is essentially just a reformulation of the equivariant branching lemma (see [11, Example 6.8]). It remains to prove (2). It suffices to show that there is an open and dense subset $\mathcal{R}$ of $\mathbb{R}^{k-1}$ such that if $(t_2, \ldots, t_k) \in \mathcal{R}$, then there is a smooth solution $(x(\rho), (t_1(\rho), t_2, \ldots, t_k))$, $\rho \in [0, \rho]$, to $F(x, t) = \gamma x$ with $x(0) = 0$, $t_1(0) = -1$. It follows from Lemma 4.3.10 that if we add the cubic $|x|^2 x$ to our generating set $\mathscr{F}$ and define

$$\tilde{F}(x, \tilde{t}) = \tilde{F}(x, (t, t_{k+1})) = F(x, t) + t_{k+1}|x|^2 x,$$

then it suffices to show that there is an open and dense subset $\tilde{\mathcal{R}}$ of $\mathbb{R}^k$ such that if $(t_2, \ldots, t_{k+1}) \in \tilde{\mathcal{R}}$, we can find a smooth solution $(x(\rho), (t_1(\rho), t_2, \ldots, t_k, t_{k+1}))$ of $\tilde{F}(x, t) = \gamma x$ satisfying the conditions listed previously.

Write $\sum_{j=2}^{k+1} t_j F_j(x) = Q(x) + C(x) + H(x)$, where $Q(x)$ is the sum of the quadratic terms, $C(x)$ is the sum of the cubic terms and $H(x)$ is the sum of the remaining higher order terms. Let W be the orthogonal complement of $\mathbb{R}\mathbf{u}$ in V^H. Denote the orthogonal projections of V^H on $\mathbb{R}\mathbf{u}$ and W by π_u and π_W respectively. Let $j \in \mathbb{N}$ be the smallest integer such that $\pi_W F_j(\mathbf{u}) \neq 0$ and set $b = d_j$. Note that $j > 1$, since $F_1(\mathbf{u}) = \mathbf{u}$, and that $j \in \{2, \ldots, k\}$, since $\{F_1(\mathbf{u}), \ldots, F_k(\mathbf{u})\}$ span V^H. Let $t_1 = -1 + c\rho^2$, where $c = \pm 1$, and $(t_2, \ldots, t_{k+1}) \in \mathbb{R}^k$, and look for a solution to $\tilde{F}(x, \tilde{t}) = \gamma x$ of the form $x(\rho) = (\rho q(\rho)\mathbf{u}, \rho^b \hat{w}(\rho))$, where $q : [0, \delta] \to \mathbb{R}$, $\hat{w} : [0, \delta] \to W$ and $q(0) > 0$, $\hat{w}(0) \neq 0$. Since $b > 1$, $x(\rho)$ will be tangent to $\mathbb{R}\mathbf{u}$ at $\rho = 0$. Substituting in the equation $\tilde{F}(x, \tilde{t}) = \gamma x$, we find the following conditions on $x(\rho)$ for it to be a solution.

$$-\rho q(\rho) = (-1 + c\rho^2)\rho q(\rho) + \rho^2 q \pi_u[q Q(\mathbf{u}) + \rho^{b-1} A(\mathbf{u}, \hat{w}) + q^2 \rho C(\mathbf{u})] + O(|\rho|^4)$$

$$\rho^b \gamma \hat{w}(\rho) = (-1 + c\rho^2)\rho^b \hat{w}(\rho) + q^b \rho^b \pi_W F_b(\mathbf{u}) + O(|\rho|^{1+b}),$$

where $A(\mathbf{u}, \hat{w}) = 2DQ(\mathbf{u})(\hat{w})$. Since Q is even and $\gamma \mathbf{u} = -\mathbf{u}$, we have $\pi_u(Q(\mathbf{u})) = 0$. It follows from Lemma 4.3.7 that $\gamma^2 = I$ and so $\gamma|W = I_W$. Dividing the second equation by ρ^b and setting $\rho = 0$, we see that

$$\hat{w}(0) = q(0)^b \pi_W F_b(\mathbf{u})/2.$$

Now suppose $b = 2$ (the proof if $b > 2$ is similar but easier). Dividing the first equation by ρ^3 and substituting for $\hat{w}(0)$ we find that

$$cq(0) + q(0)^3 \Lambda(\mathbf{u}) = 0, \tag{3}$$

where $\Lambda(u) = \pi_u(A(\mathbf{u}, \pi_W F_b(\mathbf{u})/2)) + C(\mathbf{u})$ is independent of $\hat{w}(0)$. Since $\tilde{F}$ contains the term $t_{k+1}|x|^2 x$, there is an open and dense subset $\tilde{\mathcal{R}}$ of $\mathbb{R}^k$ such that, if $(t_2, \ldots, t_{k+1}) \in \tilde{\mathcal{R}}$, then $\Lambda(\mathbf{u}) \neq 0$. From (3), $q(0)^2 = -c/\Lambda(\mathbf{u})$ and so we can choose c so that there is a unique positive value of $q(0)$ satisfying the equation.

The construction of a smooth solution $x(\rho) = (\rho q(\rho)\mathbf{u}, \rho^b \hat{w}(\rho))$ with these initial values of $q, \hat{w}$ is now a routine application of the implicit function theorem.

Remark 4.4.3. An alternative proof of Lemma 4.4.2(2) can be based on Liapunov-Schmidt reduction and the equivariant branching lemma. See Vanderbauwhede [38], Peckham & Kevrekidis [33] and also [27, Lecture 2].

As a corollary of Theorem 4.3.11 and the definition of the sets C_τ, we have

Proposition 4.4.4. *Each semi-algebraic set C_τ inherits a Whitney regular stratification $\mathscr{C}_\tau$ from $\mathscr{S}$.*

In the sequel we shall denote the stratification $\mathscr{C}_{(\Gamma)}$ of $\Xi_{(\Gamma)} = \mathbb{R}^k$ by $\mathscr{B}$. As usual, we denote the union of the i-dimensional strata of $\mathscr{B}$ by $\mathscr{B}_i$, $i \geq 0$. It follows from our constructions that

$$\mathscr{B}_k = \mathbb{R}^k \setminus \bigcup_{\tau \neq (\Gamma)} C_\tau \tag{4}$$

$$\mathscr{B}_i \subset \mathbf{C}, \, i < k \tag{5}$$

In the next few paragraphs, we undertake a more careful analysis of the sets C_τ. The way we do this is to start by showing that if $\exp(\imath\theta) \in N(H)/H$ we can reduce the study of C_τ^θ to that of C_τ^+. We then show how the structure of C_τ^+ can sometimes be given in terms of the corresponding sets A_τ (see Remark 4.4.1).

Our first result shows how the set C_τ transform under change of generating set for $P_\Gamma(V, V)$. Very similar results can be found in [3, §5], [15, 13].

Lemma 4.4.5. *Let $\mathscr{F} = \{F_1, \ldots, F_k\}$, $\mathscr{G} = \{G_1, \ldots, G_k\}$ be two minimal sets of homogeneous generators for $P_\Gamma(V, V)$. Let $\mathscr{B}^{\mathscr{F}}$, $\mathscr{B}^{\mathscr{G}}$ denote the corresponding stratifications of $\mathbb{R}^k$. There is a linear strata preserving isomorphism $L : \mathbb{R}^k \to \mathbb{R}^k$ mapping $\mathscr{B}^{\mathscr{F}}$ onto $\mathscr{B}^{\mathscr{G}}$.*

Proof. Since $\mathscr{F}$ is a minimal set of homogeneous generators for $P_\Gamma(V, V)$, there exist $\alpha_{ij} \in P(V)^\Gamma$ such that

$$G_i(x) = \sum_{j=1}^k \alpha_{ij}(x) F_j(x), \quad 1 \leq i \leq k.$$

It follows from [14, Lemma 3.4] that the map $L(t) = (\sum_{i=1}^k t_i \alpha_{ij}(0))$ is a linear isomorphism of $\mathbb{R}^k$. Define $H : V \times \mathbb{R}^k \to V \times \mathbb{R}^k$ by $H(x, t) = (x, h_1(x, t), \ldots, h_k(x, t))$, where $h_j(x, t) = \sum_{i=1}^k t_i \alpha_{ij}(x)$, $1 \leq j \leq k$. Then there is a Γ-invariant open neighborhood U of $\mathbb{R}^k$ such that $H|U$ is a Γ-equivariant analytic embedding onto an open neighborhood of $\mathbb{R}^k$. Let $\bar{F} : V \times \mathbb{R}^k \to \mathbb{R}^\ell$ denote the map defined by $\bar{F}(x, t) = P(\sum_{j=1}^k t_j F_j(x)) - P(x)$ and similarly define $\bar{G}$. It follows from our constructions that $\bar{G} = \bar{F} \circ H$. Hence, $H(\bar{G}^{-1}(0) \cap U) = \bar{F}^{-1}(0) \cap H(U)$. Since $H|U$ is an analytic isomorphism and the stratifications $\mathscr{S}^{\mathscr{F}}$, $\mathscr{S}^{\mathscr{G}}$ are minimal, it follows that H is strata preserving. In particular, $H|\mathbb{R}^k = L$ will map $\mathscr{B}^{\mathscr{F}}$ onto $\mathscr{B}^{\mathscr{G}}$.

Lemma 4.4.6. *Let $H \in \tau \in \mathcal{O}^\star$.*

(a) *If (V, Γ) is absolutely irreducible and $-I \in N(H)/H$, then $C_\tau^- = -C_\tau^+$.*

(b) *Let (V, Γ) be complex irreducible. Suppose that $\mathscr{F}$ is chosen so that $\mathscr{F}_V^H$ is $S^1 \cap N(H)/H$-invariant (see Remarks 2.5.2). If $\exp(\imath\theta) \in N(H)/H$, then $\exp(\imath\theta)(C_\tau^0) = C_\tau^\theta$. In particular, if $S^1 \subset N(H)/H$, then S^1 acts freely on C_τ and $\exp(\imath\theta)(C_\tau^\phi) = C_\tau^{\theta+\phi}$, all $\theta, \phi \in [0, 2\pi)$.*

Proof. The result follows easily from Lemma 4.3.7.

The next lemma is a straightforward application of standard existence and regularity theory for ordinary differential equations.

Lemma 4.4.7. *We may construct an open Γ-invariant neighborhood U of $\mathbf{C} \subset V \times \mathbb{R}^k$ and a smooth Γ-equivariant map $\Phi : U \times (-2, 2) \to V$ such that for each $((x, t), s) \in U \times (-2, 2)$, $\Phi((x, t), s)$ is the solution of the ordinary differential equation $\frac{dx}{ds} = F(x, t)$ with initial condition (x, t).*

Following the notation of Lemma 4.4.7, we define $\tilde{F} : U \to V$ by $\tilde{F}(x, t) = \Phi((x, t), 1)$. That is, $\tilde{F}$ is the time-one map defined by the solutions of $\frac{dx}{ds} = F(x, t)$. Note that $\tilde{F}$ is a smooth equivariant map.

Lemma 4.4.8. *If (V, Γ) is absolutely irreducible, there exist smooth invariant functions $f_j : U \to \mathbb{R}$, $2 \le j \le k$ such that $f_j(0, t) = O(|t_1|)$, $j \ge 2$, and*

$$\tilde{F}(x, t) = \exp(t_1)[x + \sum_{i=2}^{k}(t_j + f_j(x, t))F_j(x)].$$

Lemma 4.4.9. *Suppose that (V, Γ) is tangential. There exist smooth invariant functions $f_j : U \to \mathbb{R}$, $3 \le j \le k$, satisfying*

(a) *$f_j(x, t)$ is independent of t_2, $j \ge 3$.*
(b) *$f_j(0, t) = O(|t_1|)$, $j \ge 3$.*
(c) *$\tilde{F}(x, t) = \exp(t_1 + \imath t_2)[x + \sum_{i=3}^{k}(t_j + f_j(x, t))F_j(x)]$.*

Proofs of Lemmas 4.4.8, 4.4.9. Lemma 4.4.8 is an elementary exercise in ordinary differential equations, the first step of which is to make the transformation $x(t) = \exp(t_1 t)u(t)$. The proof of Lemma 4.4.9 is similar. In this case, we make the transformation $x(t) = \exp((t_1 + \imath t_2)t)u(t)$ and use the S^1-equivariance to obtain an equation $u' = H(u, t)$, where H is independent of t_2. We omit details.

Lemma 4.4.10. *Suppose that (V, Γ) is either absolutely irreducible or tangential. Let $\tilde{\Xi} = \{(x, t) \in U \mid P(\tilde{F}(x, t)) = P(x)\}$. If $\tau \in \mathcal{O}^*$, then $\tilde{\Xi}_\tau$ is smooth near $\mathbf{C}$ and $C_\tau = \partial\tilde{\Xi}_\tau \cap \mathbf{C}$.*

Proof. Suppose first that (V, Γ) is absolutely irreducible. Noting that $\tilde{F}$ agrees with F at terms of lowest order, the result follows easily by a standard 'invariance' argument similar to that used in the proof of Lemma 4.4.5 (see also [3, §5], [13, Proposition]). If (V, Γ) is tangential, we start by observing that the S^1-equivariance implies that $\tilde{\Xi} = \{(x, t) \mid P(\tilde{F}(x, (t_1, 0, t_3, \dots, t_k))) = P(x)\}$. The result then follows, just as before, using the expression for $\tilde{F}$ given by Lemma 4.4.9.

For $x \in V$, let $d(x) = \dim(V^{\Gamma x})$. Given $\tau \in \mathcal{O}$, $x \mapsto d(x)$ is constant on V_τ and we set $d_\tau = d(x)$, $x \in V_\tau$. If $t \in \mathbb{R}^k$, we let $t - 1$ be the point with coordinates $(t_1 - 1, t_2, \ldots, t_k)$. We similarly define $t + 1$ and, more generally, $C \pm 1$ for any subset C of $\mathbb{R}^k$.

Proposition 4.4.11. *Let $H \in \tau \in \mathcal{O}^*$.*

(a) *If (V, Γ) is absolutely irreducible, then $C_\tau^+ = \in A_\tau + 1$. In particular, $C_\tau^+ - 1$ is a closed semi-algebraic conical subset of $\mathbb{R}^k$ and $k - d_\tau + n_\tau \leq \dim(C_\tau^+) \leq k - 1$. If $-I \in N(H)/H$, then $C_\tau^- = A_\tau - 1$.*

(b) *Suppose that (V, Γ) is complex irreducible and set $A_\tau^0 = \{t \in A_\tau \mid t_2 = 0\}$. If $\exp(\imath\theta) \in S^1 \cap N(H)/H$, then $C_\tau^\theta = \exp(\imath\theta)(A_\tau^0 + 1)$.*

(c) *If (V, Γ) is tangential or $S^1 \subset N(H)/H$, then*

$$C_\tau = \{\exp(\imath s)(t + 1) \mid t \in A_\tau^0, s \in \mathbb{R}\}$$

and $k - d_\tau + n_\tau \leq \dim(C_\tau) \leq k - 1$.

Proof. If Γ is finite, (a,b,c) follow easily from Lemma 4.3.7 and [12, Lemma 4.3.9]. Suppose Γ is not finite and (V, Γ) is absolutely irreducible. It follows from Lemmas 4.4.5, 4.4.6, and [12, Lemma 4.3.9] that it suffices to show that $C_\tau^+ - 1 = A_\tau$. Now $C_\tau^+ = \partial \tilde{\Xi}_\tau \cap \mathbb{R}^k \cap \mathbf{C}_+$ (Lemma 4.4.10) and so we may work with the variety $\tilde{\Xi}$. Suppose $t \in C_\tau^+$, $H \in \tau$ and (x^n, t^n), (γ^n) are sequences in $\tilde{\Xi}_\tau^H$, $N(H)/H$ such that

$$\tilde{F}(x^n, t^n) = \gamma^n x^n \quad \text{and}$$

$$((x^n, t^n), \gamma^n) \to ((0, t), I) \in (\{0\} \times \mathbb{R}^k) \times N(H)/H$$

Denote the Lie algebra of $N(H)/H$ by $\mathfrak{k}$. We may choose a sequence $(k^n) \subset \mathfrak{k}$ so that $k^n \to 0$ and $\gamma^n = \exp(k^n)$ for sufficiently large n (that is, if $\gamma^n \in (N(H)/H)^0$). It follows from the definition of $\tilde{F}$ and equivariance, that

$$F(x^n, t^n - 1) = \frac{d}{ds} \exp(sk^n)x^n|_{s=0}.$$

Hence $(x^n, t^n) \in \tilde{\Xi}$ if and only if $(x^n, t^n - 1) \in \Sigma$. Letting $n \to \infty$, it follows that $C_\tau^+ - 1 = A_\tau$. In particular, $C_\tau^- + 1$ is conical. Similar arguments apply when (V, Γ) is complex or tangential.

Remarks 4.4.12. (1) For the last part of (a), it suffices to assume that $F_1, \ldots, F_k$ restrict to odd (possibly zero) maps on V^H. It follows that the restriction of any homogeneous invariant to V^H is either zero or of even degree. Since invariants separate Γ-orbits, it follows that $-I \in N(H)/H$. (2) In general, C_τ^- may be empty. If $C_\tau^- \neq \emptyset$ and the conditions of (a) do not hold, we may ask whether $C_\tau^- + 1$ is a conical subset of $\mathbb{R}^k$. This seems possible, at least if there are no quadratic equivariants.

4.5. Stability theorems I: Weak regularity. Suppose $f \in \mathcal{M}(V,\Gamma)$ and write

$$f(x,\lambda) = \sum_{j=1}^{k} f_j(x,\lambda) F_j(x)$$

where $f_j \in C^\infty(V \times \mathbb{R})^\Gamma$, $1 \leq j \leq k$. Define smooth maps $\mathrm{graph}_f : V \times \mathbb{R} \to V \times \mathbb{R}^k$ and $\gamma_f : \mathbb{R} \to \mathbb{R}^k$ by

$$\mathrm{graph}_f(x,\lambda) = (x, (f_1(x,\lambda), \ldots, f_k(x,\lambda)))$$
$$\gamma_f(\lambda) = (f_1(0,\lambda), \ldots, f_k(0,\lambda))$$

Since $f = F \circ \mathrm{graph}_f$, it follows that $\mathbf{I}(f) = \mathrm{graph}_f^{-1}(\Xi)$.

In the sequel, we frequently talk about transversality of maps to semi-algebraic sets. If no stratification is specified, it will always be understood that transversality is meant with respect to the canonical semi-algebraic stratification of the set. More generally, if a semi-algebraic set C comes with a Whitney stratification $\mathcal{S}$ (not necessarily canonical), we say that a map is transverse to $\mathcal{S}$ if it is transverse to C, given the stratification $\mathcal{S}$. We use the notation $f \pitchfork C$ (or $f \pitchfork \mathcal{S}$) to indicate that f is transverse to C (or $\mathcal{S}$).

Define

$$\mathscr{L}_\Gamma(V) = \{f \in \mathcal{M}(V,\Gamma) \mid \mathrm{graph}_f \pitchfork \Xi \text{ at } \lambda = 0\}$$
$$\mathscr{L}_\Gamma^\pm(V) = \{f \in \mathscr{L}_\Gamma(V) \mid f_1(0,0) = \pm 1\}$$
$$\mathscr{L}_\Gamma^\theta(V) = \{f \in \mathscr{L}_\Gamma(V) \mid f_1(0,0) + \imath f_2(0,0) = \exp(\imath\theta)\}$$

The methods of Bierstone [3] and Field [14] may be used to show that $\mathscr{L}_\Gamma(V)$, together with the subsets defined above, are independent of the choices of generators for $P(V)^\Gamma$, $P_\Gamma(V,V)$ and coefficient functions $f_1, \ldots, f_k$ implicit in the definition of graph_f. Just as in [12, §4], it may be shown that the map γ_f is uniquely defined once $\mathcal{S}$ has been chosen.

Proposition 4.5.1. *Let $f \in \mathcal{M}(V,\Gamma)$. The following conditions on f are equivalent.*

(1) $f \in \mathscr{L}_\Gamma(V)$.
(2) $\gamma_f \pitchfork \mathscr{B}$ at $\lambda = 0$.
(3) $\gamma_f(0) \in \mathscr{B}_i$, where $i \geq k - 1$.

Proof. Since $f \in \mathcal{M}(V,\Gamma)$, $\gamma_f(0) \in \mathbf{C}$ and $\gamma_f \pitchfork \mathbf{C}$ at $\lambda = 0$. We prove the equivalence of (1) and (2) (see [12, Proposition 4.4.3] for details on the analogous results for vector fields). Since $\mathscr{B} \subset \mathscr{S}$, and all strata of $\mathscr{S}$ which meet $\mathbb{R}^k$ lie in $\mathscr{B}$, it follows by the Whitney regularity of the stratification $\mathscr{S}$ that $\mathrm{graph}_f \pitchfork \mathscr{S}$ at $\lambda = 0$ if and only if $\mathrm{graph}_f \pitchfork \mathscr{B}$ at $\lambda = 0$. Noting the definition of graph_f, we see that $\mathrm{graph}_f \pitchfork \mathscr{B}$ at $\lambda = 0$ if and only if $\gamma_f \pitchfork \mathscr{B}$ at $\lambda = 0$. Hence (1) and (2) are equivalent.

As an immediate consequence of Proposition 4.5.1 and standard properties of maps transversal to Whitney stratified sets we have

Proposition 4.5.2.   (1) $\mathscr{L}_\Gamma(V)$ *is an open and dense subset of $\mathcal{M}(V,\Gamma)$.*

(2) *Every $f \in \mathscr{L}_\Gamma(V)$ is weakly regular.*

Theorem 4.5.3. *Let $f \in \mathscr{L}_\Gamma(V)$. Then*

(1) *If $codim(C_\tau) \geq 2$, the germ of $\mathbf{I}(f)$ at zero contains no points of isotropy type τ.*

(2) *If $codim(C_\tau) = 1$ and $\gamma_f(0) \in C_\tau$, there is a branch of invariant group orbits of isotropy type τ for f at zero.*

(3) *The map $\gamma_f : \mathbb{R} \to \mathbb{R}^k$ is transverse to the canonical stratification of C_τ for all $\tau \in \mathcal{O}$.*

Similar results hold if we replace C_τ by $C_\tau^\pm$. Finally, if (V, Γ) is complex irreducible and $C_\tau^\theta \neq \emptyset$ for only finitely many values of θ, then $codim(C_\tau) \geq 2$ and so the germ of $\mathbf{I}(f)$ at zero contains no points of isotropy type τ.

Proof. The only result that is not immediate by transversality is (2). That is, we have to prove the branch is C^1. If $H \in \tau$ and $N(H)/H$ is finite, we follow the method used in the proofs of [22, Proposition 5.16] or [12, Theorem 4.4.5]. Otherwise, we use methods based on blowing-up [12, §9].

Remark 4.5.4. Just as for vector fields, it follows from our results that if $f \in \mathscr{L}_\Gamma(V)$ and $f(x, \lambda) = \sum_{i=1}^k f_i(x, \lambda) F_i(x)$, then the branching pattern $\Xi(f)$ is determined completely by $(f_1(0,0), f_2(0,0), \ldots, f_k(0,0))$. In particular, the branching pattern will be determined by the d_k-jet of f_0 at the origin.

Definition 4.5.5 (cf. [22, 24, 12]). Let $\tau \in \mathcal{O}^*$. If $\sigma \in \{+, -\}$ and (V, Γ) is absolutely irreducible, we say that τ is *σ-symmetry breaking* (respectively, *generically σ-symmetry breaking*) if there exists a non-empty open (respectively, open and dense) subset U of $\mathscr{L}_\Gamma(V)$ such that for every $f \in U$, the germ of $\mathbf{I}(f)$ at zero contains points of isotropy type τ. If (V, Γ) is complex irreducible, we say that τ is *symmetry breaking* (respectively, *generically symmetry breaking*) if there exists a non-empty open (respectively, open and dense) subset U of $\mathscr{L}_\Gamma(V)$ such that for every $f \in U$, the germ of $\mathbf{I}(f)$ at zero contains points of isotropy type τ.

As an immediate consequence of our results so far, we have

Proposition 4.5.6. *Let $\tau \in \mathcal{O}^*$ and suppose that (V, Γ) is absolutely irreducible and $\sigma \in \{+, -\}$. Then τ is a σ-symmetry breaking isotropy type if and only if $codim(C_\tau^\sigma) = 1$. Moreover, τ is generically σ-symmetry breaking if and only if $C_\tau^\sigma = \mathbf{C}_\sigma$. A similar result holds if (V, Γ) is complex irreducible.*

It follows from Proposition 4.5.2 that $\mathscr{L}_\Gamma(V)$ is an open and dense subset of $\mathscr{M}(V, \Gamma)$ consisting of weakly regular families. The next theorem gives a weak version of stability for families in $\mathscr{L}_\Gamma(V)$.

Theorem 4.5.7. *Let $f \in \mathscr{L}_\Gamma(V)$ and $\{f_t \mid t \in [0,1]\}$ be a continuous path in $\mathscr{L}_\Gamma(V)$ such that $f_0 = f$. Then there exists a compact Γ-invariant neighborhood A of $(0,0) \in V$ and a continuous Γ-equivariant isotopy $K : A \times [0,1] \to V \times \mathbb{R}$ of embeddings such that*

(1) $K_0 = I_V$.

(2) $K_t(A \cap \mathbf{I}(f)) = \mathbf{I}(f_t) \cap K_t(A)$, $t \in [0,1]$.

Proof. The result follows from Thom's first isotopy lemma (see [12, Theorem 4.4.11] for further details).

As an immediate corollary of Theorem 4.5.7, we have

Proposition 4.5.8. *Let* $\tau \in \mathcal{O}^*$ *and suppose that* (V, Γ) *is absolutely irreducible and* $\sigma \in \{+, -\}$. *Then* τ *is* σ-*symmetry breaking if and only if there exists* $f \in \mathscr{L}_\Gamma^\sigma(V)$ *such that* $\Xi(f)$ *contains a branch of isotropy type* τ. *Further* τ *is generically* σ-*symmetry breaking if and only if for every* $f \in \mathscr{L}_\Gamma^\sigma(V)$, $\Xi(f)$ *contains a branch of isotropy type* τ. *Similar results hold for the case of complex irreducible representations.*

4.6. Stability theorems II: Regular families. In this section we extend results of Field [18, 12] and show that $\mathscr{S}(V, \Gamma)$ is an open and dense subset of $\mathscr{M}(V, \Gamma)$ for all absolutely irreducible and complex irreducible representations (V, Γ). Most of what we say is a straightforward extension of the theory and methods in [18, 12] and so we shall only give brief details of proofs.

Let $\tau \in \mathcal{O}(V, \Gamma)$. Define

$$Z_0(\tau) = \{(x, y) \in V \times V \mid x \in V_\tau, \, y \in N(\Gamma_x) \cdot x\}$$

and set

$$Z_0 = \text{closure}(\bigcup_{\tau \in \mathcal{O}} Z_0(\tau))$$

Lemma 4.6.1 ([12, Lemma 4.5.2]). *Each set* $Z_0(\tau)$ *is a smooth* Γ-*invariant semi-algebraic subset of* $V \times V$, $\tau \in \mathcal{O}$. *In particular,* Z_0 *is semi-algebraic.*

Proof. We prove that $Z_0(\tau)$ is a Γ-invariant semi-algebraic set. Let $H \in \tau$ and set $Z_0(H) = Z_0(\tau) \cap (V^H \times V^H)$. Since $Z_0(\tau) = \Gamma \cdot Z_0(H)$, it suffices to prove that $Z_0(H)$ is a smooth semi-algebraic subset of $V^H \times V^H$. But $Z_0(H) = \{(x, y) \mid Q(x) = Q(y)\}$, where $Q : V^H \to V^H/N(H)$ denotes the orbit map, and so $Z_0(H)$ is semi-algebraic.

For each $\tau \in \mathcal{O}$, let $L_\tau(V, V)$ denote the semi-algebraic subset of $L(V, V)$ consisting of maps that have at least $g_\tau + 1$ eigenvalues of unit modulus (counting multiplicities).

We define sets $Z_1(\tau) \subset V_\tau \times V_\tau \times L(V, V)$ and $Z_1 \subset V \times V \times L(V, V)$ by

$$Z_1(\tau) = \{((x, y), A) \mid \exists \gamma \in N(\Gamma_x) \text{ such that } y = \gamma x \text{ and } \gamma^{-1} A \in L_\tau(V, V)\}$$

$$Z_1 = \overline{(\bigcup_{\tau \in \mathcal{O}} Z_1(\tau))}$$

Lemma 4.6.2. (1) *For every* $\tau \in \mathcal{O}$, $Z_1(\tau)$ *is a* Γ-*invariant semi-algebraic subset of* $V \times V \times L(V, V)$.
  (2) Z_1 *is a closed* Γ-*invariant semi-algebraic subset of* $V \times V \times L(V, V)$.

Proof. Let $H \in \tau$ and define $\tilde{Z}_1(H)$ to be the semi-algebraic subset of $V_\tau^H \times N(H) \times L(V, V)$ consisting of points (x, γ, A) such that $\gamma^{-1} A \in L_\tau(V, V)$. It follows that $\tilde{Z}_1(\tau) = \Gamma \cdot \tilde{Z}_1(H)$ is a Γ-invariant semi-algebraic set. But now there is a natural semi-algebraic map of $\tilde{Z}_1(\tau)$ onto $Z_1(\tau)$ defined by mapping (x, γ, A) to $((x, \gamma x), A)$. Hence $Z_1(\tau)$ is semi-algebraic, proving (1). Statement (2) follows from (1) and the fact that closures and finite unions of semi-algebraic sets are semi-algebraic.

Let $J^1(V, V)$ denote the space of 1-jets of maps from V to V. Note that $J^1(V, V)$ inherits the structure of a Γ-space from V. Recall that $J^1(V, V) \cong V \times V \times L(V, V)$ and that if $f \in C^\infty_\Gamma(V, V)$ then $j^1 f(x) = (x, f(x), Df(x))$ under this isomorphism.

From now on, we regard Z_1 and $Z_1(\tau)$ as defining semi-algebraic subsets of $J^1(V, V)$.

We turn next to families of maps. The jet space $J^1(V \times \mathbb{R}, V)$ is isomorphic (as a Γ-representation) to $J^1(V, V) \oplus (\mathbb{R} \times V)$. Thus, if $f \in C^\infty_\Gamma(V \times \mathbb{R}, V)$, then

$$j^1 f(x, \lambda) = ((x, \lambda), f(x, \lambda), Df(x, \lambda)) \in J^1(V \times \mathbb{R}, V)$$
$$= ([x, f(x, \lambda), D_1 f(x, \lambda)], [\lambda, D_2 f(x, \lambda)]) \in J^1(V, V) \times (\mathbb{R} \times V)$$

Let $\Pi : J^1(V \times \mathbb{R}, V) \to J^1(V, V)$ denote the associated projection map. Set $Z_1^1 = \Pi^{-1}(Z_1)$. Since Z_1^1 may be identified with $Z_1 \times (\mathbb{R} \times V)$, Z_1^1 is a closed semi-algebraic subset of $J^1(V \times \mathbb{R}, V)$.

Lemma 4.6.3. *Let $f \in C^\infty_\Gamma(V \times \mathbb{R}, V)$. Then*

(1) $\mathbf{I}(f) = \{(x, \lambda) \mid f_\lambda(x) \in Z_0\} = \{(x, \lambda) \mid graph_f(x, \lambda) \in \Xi\}$.
(2) $\mathbf{B}(f) = (j^1 f)^{-1} Z_1^1$.

Proof. Similar to that of [12, Lemma 4.6.9]. ∎

Just as in [12, §4], [18, Appendix], we may express genericity conditions on maps or families in terms of equivariant jet transversality conditions. We briefly summarize the main results that we shall need on equivariant jet transversality. (We refer the reader to [4], [12, §4] for more details.) Suppose that (V_i, Γ), $i = 1, 2$, are Γ-representations and that Q is a Γ-invariant closed semi-algebraic subset of $J^1(V_1, V_2)$. If $f \in C^\infty_\Gamma(V_1, V_2)$, and $A \subset V_1$ is compact and Γ-invariant, we write "$j^1 f \pitchfork_\Gamma Q$ on A" to signify that $j^1 f : V_1 \to J^1(V_1, V_2)$ is in equivariant general position to Q on A. We recall from Bierstone [4] that $\{f \mid j^1 f \pitchfork_\Gamma Q \text{ on } A\}$ is an open and dense subset of $C^\infty_\Gamma(V_1, V_2)$. Moreover, the usual isotopy and stability theorems hold (for precise statements, see [4, Theorems 7.6, 7.7, 7.8]).

As an immediate consequence of our definitions, we have

Lemma 4.6.4 (cf. [12, Lemma 4.6.10]). *Let $f \in C^\infty_\Gamma(V, V)$. The following conditions on f are equivalent:*

(1) *All f-invariant Γ-orbits are normally hyperbolic.*
(2) $j^1 f(V) \cap Z_1 = \emptyset$.
(3) $j^1 f \pitchfork_\Gamma Z_1$ *on V.*

Theorem 4.6.5. *Let*

$$\mathscr{S}_0(V, \Gamma) = \{f \in \mathscr{L}_\Gamma(V) \mid^1 f \pitchfork_\Gamma Z_1^1 \text{ at } (x, \lambda) = (0, 0)\}$$

Then

(a) $\mathscr{S}_0(V, \Gamma)$ *is an open and dense subset of $\mathscr{M}(V, \Gamma)$.*
(b) $\mathscr{S}_0(V, \Gamma) \subset \mathscr{S}(V, \Gamma)$.

Proof. Statement (a) follows from Bierstone's jet transversality theorem. The proof of (b) is very similar to that of the corresponding Theorem 4.6.11 in [12]. We start by observing that if $f \in \mathscr{S}_0(V, \Gamma)$, then we can find a compact Γ-invariant neighborhood A of $(0, 0) \in V \times \mathbb{R}$ such that $j^1 f \pitchfork_\Gamma Z_1^1$ on A. Using the definitions of equivariant general position and Z_1^1, it may be shown that $(j^1 f | A)^{-1}(Z_1^1)$ has the structure of a Whitney regular stratified set. It suffices to prove that the origin is an *isolated point* in $(j^1 f | A)^{-1}(Z_1^1)$. If not, there is a continuous non-constant arc in $(j^1 f | A)^{-1}(Z_1^1)$, with initial point at the origin. We derive a contradiction using the openness of equivariant transversality together with Lemma 4.6.4.

Remark 4.6.6. Using methods similar to those in [4, 14], one can show that $\mathscr{S}_0(V, \Gamma)$ is defined independently of choices of generating sets.

4.7. Determinacy. In this section, we indicate how the determinacy theorems of [12, §4], [18, Appendix] can be extended from vector fields to smooth maps. As the methods are completely analogous to those used in [12, 18], we only give very brief details.

We regard Z_0 as a subset of $J^1(V, V)$ and define $Z_0^1 = \Pi^{-1}(Z_0) \subset J^1(V \times \mathbb{R}, V)$. Clearly $Z_1^1 \subset Z_0^1$. It follows that $f \in \mathscr{S}_0(V, \Gamma)$ if and only if

$$j^1 f \pitchfork_\Gamma Z_0^1 \text{ and } j^1 f \pitchfork_\Gamma Z_1^1 \text{ at } (0, 0)$$

Just as in [12, 18], we construct a Whitney regular stratification $\mathscr{T}$ of Z_0^1 that induces a Whitney regular stratification of Z_1^1 (we do not know whether this stratification always coincides with the canonical stratification of Z_1^1). We define

$$\mathscr{S}_1(V, \Gamma) = \{f \mid f \pitchfork_\Gamma \mathscr{T} \text{ at } (0, 0)\},$$

and note that $\mathscr{S}_1(V, \Gamma) \subset \mathscr{S}(V, \Gamma)$. membership in $\mathscr{S}_1(V, \Gamma)$. Using this, together with the standard technique for proving density (see the proof of [12, Theorem 4.7.10]) we may prove

Theorem 4.7.1. *Let (V, Γ) be an absolutely irreducible or complex irreducible representation. There exists $q > 0$ such that Γ-equivariant bifurcation problems on V are q-determined.*

4.8. An invariant sphere theorem and Fiedler's theorem for maps. Given $f : V \times \mathbb{R} \to V$ and $a \in \mathbb{R}$, we define $f^a(x, \lambda) = f(x, \lambda) + a|x|^2 x$. Using techniques based on the persistence of normally hyperbolic sets, similar to those used in the proof of [18, Theorem 5.2], we may prove the following *invariant sphere theorem* for maps (see also [17, §4]).

Theorem 4.8.1. *Let (V, Γ) be irreducible and suppose $\dim_\mathbb{R}(V) = m$. Let $r > 0$. Assume that $P_\Gamma^2(V, V) = \{0\}$ and let $f \in \mathscr{M}(V, \Gamma)$. Then we may find $a_0 \in \mathbb{R}$ such that if $a \leq a_0$, then there exist $\varepsilon > 0$, a neighborhood U of $0 \in V$, and a continuous family $S_\lambda : S^{m-1} \to U$, $\lambda \in [0, \varepsilon]$, satisfying the following properties.*

(1) *$S_\lambda : S^{m-1} \to U$ is a Γ-equivariant C^r-embedding, $\lambda > 0$, and S_0 is the zero map.*

(2) *If we set $S(\lambda) = S_\lambda(S^{m-1})$, then $S(\lambda)$ is f_λ^a-invariant.*

(3) *If $x \in U \setminus \{0\}$, then the f_λ^a-orbit of x is forward asymptotic to $S(\lambda)$.*

Example 4.8.2 (cf. [10], [12, §11]). Suppose (V, Γ) is complex irreducible and let $\dim_R(V) = 2n$. Set $G = \Gamma \times S^1$ and consider the tangential representation (V, G). If $f \in \mathscr{L}_G(V)$, it follows from Theorem 4.5.3 that $f^a \in \mathscr{L}_G(V)$ and $\Xi(f) = \Xi(f^a)$ for all $a \in \mathbb{R}$. Since $P_G^2(V, V) = \{0\}$, we may choose $a_0 \in \mathbb{R}$ so that if $a \leq a_0$ then f^a spawns a branch $\{S(\lambda) \mid \lambda \in [0, \varepsilon]\}$ of G- and f^a-invariant C^1-embedded spheres. Set $f^a = g$. Since S^1 acts freely on each $S(\lambda)$, g_λ-induces a C^1 G-equivariant map on the $(2n - 2)$-dimensional complex projective space $S(\lambda)/S^1$, $\lambda \in (0, \varepsilon]$. Let $\tau \in \mathscr{O}(V, G)$ be a maximal isotropy type and let $H \in \tau$. For each $\lambda \in (0, \varepsilon]$, g_λ restricts to an $N(H) \times S^1$-equivariant C^1-mapping of $S(\lambda) \cap V^H$ and so induces a C^1-mapping g_λ^H on the associated projective space. Each g_λ^H is clearly $N(H) \times S^1$-equivariantly homotopic to the identity map. Since the Euler characteristic of every complex projective space is nonzero, it follows that the Lefschetz number $L(g_\lambda^H) \neq 0$. Hence, g_λ^H has at least one fixed point for each $\lambda \in (0, \varepsilon]$. Consequently, g_λ has an invariant G-orbit of isotropy type τ for each $\lambda \in (0, \varepsilon]$. It follows that $C_\tau = \mathbb{C}$. That is, every G-maximal isotropy type is generically symmetry breaking. Using the strong determinacy theorem (see §6), we may ask what happens if we break symmetry from G to Γ. Unlike the case of vector fields [12, §11], complicated dynamics may appear [6].

5. EXAMPLES FOR Γ FINITE

In this section, Γ is always assumed finite. We discuss some examples of symmetry breaking closely related to those studied in Field and Richardson [22, 25].

5.1. Preliminaries. Suppose (V, Γ) is complex irreducible. It follows from Theorem 4.5.3 that $\mathrm{codim}(C_\tau) \geq 2$ for all $\tau \in \mathscr{O}$, $\tau \neq (\Gamma)$. Consequently, no branches of invariant orbits bifurcate off the trivial branch for generic maps in $\mathscr{M}(V, \Gamma)$. Hence we may assume (V, Γ) is absolutely irreducible. Since Γ is finite it follows that if $f \in \mathscr{L}^+(V, \Gamma)$ then $\Xi(f)$ consists of branches of *fixed points*. If $f \in \mathscr{L}^-(V, \Gamma)$, then $\Xi(f)$ consists of branches of Γ-orbits consisting of points of prime period 2 for f.

5.2. Subgroups of the Weyl group of type B_n. Assume $n \geq 3$. Let S_n denote the group of $n \times n$ permutation matrices and Δ_n denote the group of diagonal matrices with entries ± 1. The Weyl group $W(B_n)$ is the semi-direct product $\Delta_n \rtimes S_n$. Set $W_n = W(B_n)$ and recall that $(\mathbb{R}^n, W_n)$ is absolutely irreducible.

Definition 5.2.1 (cf. [25, Conditions 13.5.1-2]). Suppose $n \geq 3$. Let $\mathscr{W}_n$ denote the set of representations $(\mathbb{R}^n, \Gamma)$ satisfying

(1) Γ is a subgroup of W_n.
(2) $(\mathbb{R}^n, \Gamma)$ is absolutely irreducible.
(3) $P_\Gamma^2(\mathbb{R}^n, \mathbb{R}^n) = \{0\}$.

Example 5.2.2. If H is a subgroup of $O(n)$, we let H' denote the determinant one subgroup of H. We recall from [25, §14], that if T is a transitive subgroup of S_n then, with the single exception of the tetrahedral group $\mathbb{T} = \Delta_3' \rtimes \mathbb{Z}_3$, we have

$$(\mathbb{R}^n, \Delta_n \rtimes T), \ (\mathbb{R}^n, \Delta_n' \rtimes T), \ (\mathbb{R}^n, (\Delta_n \rtimes T)') \in \mathscr{W}_n.$$

5.3. Branches of fixed points. Let $\mathscr{E}$ denote the set of non-zero vectors $\varepsilon \in \mathbb{R}^n$ such that $\varepsilon_i \in \{0, +1, -1\}$, $1 \le i \le n$. Given $(\mathbb{R}^n, \Gamma) \in \mathscr{W}_n$, let $\mathscr{O}_S = \{\iota(\varepsilon) \mid \varepsilon \in \mathscr{E}\}$. It was shown in [25] that an isotropy type τ was symmetry breaking for 1-parameter families of vector fields if and only if $\tau \in \mathscr{O}_S$. In view of this result and Lemma 4.4.11(1), we have the following simple characterization of $+$-symmetry breaking isotropy types for representations in the class $\mathscr{W}_n$.

Proposition 5.3.1. *Let $(\mathbb{R}^n, \Gamma) \in \mathscr{W}_n$. Then $\tau \in \mathscr{O}$ is $+$-symmetry breaking if and only if $\tau \in \mathscr{O}_S$.*

As an easy consequence of results in [24, 25], together with Proposition 4.4.11, we have the following determinacy result.

Lemma 5.3.2. *There is a maximal open and dense semi-algebraic subset $\mathscr{R}$ of $P_\Gamma^3(\mathbb{R}^n, \mathbb{R}^n)$ such that*

(1) *If $f \in \mathscr{M}^+(\mathbb{R}^n, \Gamma)$ and $D^3 f_0(0) \in \mathscr{R}$, then f is stable.*
(2) *If $\mu \in \mathbb{R} \setminus \{0\}$, then $\mu\mathscr{R} = \mathscr{R}$.*

5.4. Branches of invariant orbits consisting of period two points.

Lemma 5.4.1. *Let $(\mathbb{R}^n, \Gamma) \in \mathscr{W}_n$.*

(1) *Let $H \in \tau \in \mathscr{O}_S$. Then τ is $-$-symmetry breaking if $-I \in N(H)/H$.*
(2) *If $-I \in \Gamma$, then $\tau \in \mathscr{O}$ is $-$-symmetry breaking if and only if $\tau \in \mathscr{O}_S$.*

Proof. The result follows immediately from Lemma 4.4.6(a).

Lemma 5.4.2. *Let $(\mathbb{R}^n, \Gamma) \in \mathscr{W}_n$ and suppose $-I \in \Gamma$. Let $\mathscr{R} \subset P_\Gamma^3(\mathbb{R}^n, \mathbb{R}^n)$ be the subset given by Lemma 5.3.2. If $f \in \mathscr{M}(\Gamma, \mathbb{R}^n)$ and $D^3 f_0(0) \in \mathscr{R}$, then f is stable. In particular, Γ-equivariant bifurcation problems on $\mathbb{R}^n$ are 3-determined.*

Proof. It suffices to show that if $f \in \mathscr{M}^-(\mathbb{R}^n, \Gamma)$ and $D^3 f_0(0) \in \mathscr{R}$, then f is stable. Suppose that $F_2, \ldots, F_r$ are the cubic equivariants in $\mathscr{F}$ and set $F_{r+1}(x) = |x|^2 x$. Then $F_2, \ldots, F_{r+1}$ define a basis for $P_\Gamma^3(\mathbb{R}^n, \mathbb{R}^n)$. Suppose $f \in \mathscr{M}^-(\mathbb{R}^n, \Gamma)$. Then

$$D^3 f_0(0)(x) = \sum_{i=2}^{r+1} a_i F_i(x),$$

where $a_2, \ldots, a_{r+1} \in \mathbb{R}$. A simple computation verifies that

$$D^3 f_0^2(0)(x) = \sum_{i=2}^{r+1} -2a_i F_i(x).$$

Hence, by Lemma 5.3.2(2), $D^3 f_0^2(0) \in \mathscr{R}$ if and only if $D^3 f_0(0) \in \mathscr{R}$.

We extend our results to cover the case where $-I \notin \Gamma$. Let $\mathbb{Z}_2 \subset O(n)$ be the subgroup generated by $-I$ and set $\Gamma^2 = \Gamma \times \mathbb{Z}_2$. Note that if $-I \in \Gamma$, then $(-I, -I) \in \Gamma^2$ fixes every point of $\mathbb{R}^n$. If we identify Γ^2 with its image in $O(n)$, then $(\mathbb{R}^n, \Gamma^2) \in \mathscr{W}_n$. Let $\mathscr{O}_S^2$ be the set of +-symmetry breaking isotropy types for Γ^2. If $\tilde{H} \in \tau \in \mathscr{O}_S^2$, then $(\tilde{H} \cap \Gamma) \in \mathscr{O}_S$. This construction defines a natural map $\Pi : \mathscr{O}_S^2 \to \mathscr{O}_S$. Note that if x has Γ^2-isotropy τ, then $\Pi(\tau)$ is the Γ-isotropy of x.

Proposition 5.4.3. *Suppose* $(\mathbb{R}^n, \Gamma) \in \mathscr{W}_n$.

(1) *If τ is --symmetry breaking then $\tau \in \mathscr{O}_S$.*
(2) *If $D^3 f_0(0) \in \mathscr{R}$, then f is stable.*
(3) *If $\tau \in \mathscr{O}_S$, τ will be --symmetry breaking if there exists $\eta \in \Pi^{-1}(\tau)$ such that $\eta \neq \tau$. (That is, if we choose $x \in V_\eta$, then $[\Gamma_x^2 : \Gamma_x] = 2$.)*

Proof. If $-I \in \Gamma$, the result follows from Lemma 5.4.2 so we may suppose $-I \notin \Gamma$. Since $\Gamma^2 \in \mathscr{W}_n$ and $-I \in \Gamma^2$, Lemma 5.4.2 applies to $(\mathbb{R}^n, \Gamma^2)$. Since $(\mathbb{R}^n, \Gamma^2)$ and $(\mathbb{R}^n, \Gamma)$ have the *same* cubic equivariants and quadratic equivariants are trivial, (1,2) of the Proposition follow easily. It remains to prove (3). Let τ, η satisfy the conditions of (3). Suppose that $f \in \mathscr{M}^-(\mathbb{R}^n, \Gamma^2)$, $D^3 f_0(0) \in \mathscr{R}$ and f has a curve $\phi = (\rho, \lambda)$ of $-I$-invariant points of prime period two and isotropy type η. Denote the initial direction $\rho'(0)$ of the curve by $\pm u \in S^{n-1}$ (we define the initial direction only up to ± 1 to allow for the reverse parametrization of the branch). Since $D^3 f_0(0) \in \mathscr{R}$, it is easy to verify that the initial direction u depends only on $D^3 f_0(0)$. Consequently, if we take any Γ-equivariant perturbation of f by terms of order at least four, the resulting perturbed curve $\tilde{\phi}$ of period two points will have the *same* initial direction $\pm u$. It follows from our hypotheses on τ, η, that there exists $\gamma \notin \Gamma_u$ such that $-\gamma \in \Gamma_u^2$. By Γ-equivariance, $\gamma \tilde{\phi}$ must also be a branch of points of period two. Since $\gamma \tilde{\phi}$ has the same initial direction $\pm u$ as $\tilde{\phi}$, it follows that $\gamma \tilde{\phi} = \tilde{\phi}$ (with reverse parametrization) and so τ is --symmetry breaking. Conversely, every --symmetry breaking isotropy type arises in this way. $\blacksquare$

Using our results in combination with the techniques of [24], we may easily verify strong 3-determinacy for representations in $\mathscr{W}_n$. In summary, we have proved

Theorem 5.4.4. *Let $(\mathbb{R}^n, \Gamma) \in \mathscr{W}_n$. Then Γ-equivariant bifurcation problems on $\mathbb{R}^n$ are strongly 3-determined.*

Example 5.4.5. Let $(\mathbb{R}^5, \Gamma) \in \mathscr{W}_5$, where $\Gamma = \Delta_5' \rtimes \mathbb{Z}_5$. It follows from [25, §14] and Proposition 5.3.1 that all isotropy types in $\mathscr{O}^*(\mathbb{R}^5, \Gamma)$ are +-symmetry breaking. It follows either directly or using Proposition 5.4.3 that the maximal isotropy types $\iota(1, \ldots, 1, \pm 1)$ are *not* --symmetry breaking. On the other hand, the trivial isotropy type $\iota(1, 1, 1, 1, 0)$ is --symmetry breaking. In this case, branches of invariant orbits will be tangent to the Γ-orbit of the plane $x_5 = 0$. All of the remaining isotropy types satisfy the conditions of Lemma 5.4.1 and so are also --symmetry breaking.

Remark 5.4.6. Additional examples of symmetry breaking for maps for the standard representation of S_{n+1} on $\mathbb{R}^n$ are implicit in [25, §17] and explicit in [1].

6. Strong determinacy

6.1. Strong determinacy theorem for maps. Just as for vector fields, we may prove a strong determinacy theorem for one parameter families of equivariant maps.

Theorem 6.1.1. *Let (V,Γ) be either absolutely or complex irreducible. Then Γ-equivariant bifurcation problems on V are strongly determined. In particular, there exists $d \in \mathbb{N}$ and an open and dense semi-analytic subset $\mathcal{N}(d)$ of $P_{\Gamma}^{(d)}(V,V)$ such that if $f \in \mathcal{M}(V,\Gamma)$ and $j^d f_0(0) \in \mathcal{N}(d)$ then*

(1) *f is strongly determined.*

(2) *If H is a closed subgroup of Γ then f is (d, H)-stable.*

6.2. Proof of the Strong determinacy theorem for maps. As the proof of Theorem 6.1.1 is similar to that of the corresponding result for vector fields, we shall only sketch the main techniques.

We start by restricting to the set $\mathcal{M}_\omega(V,\Gamma) \subset \mathcal{M}(V,\Gamma)$ of real-analytic families and assume that (V,Γ) is absolutely irreducible or tangential. Using methods based on resolution of singularities, it can be shown that we can find $d, N \in \mathbb{N}$, and an open and dense semi-algebraic subset $\mathcal{R}^1$ of $P_{\Gamma}^{(d)}(V,V)$ such that if we define

$$\mathcal{M}_1(V,\Gamma) = \{ f \in \mathcal{M}_\omega(V,\Gamma) \mid j^d f_0(0) \in \mathcal{R}^1 \}$$

then, for all $p \in \mathbb{N}$, the p-jet at zero of solution branches of $f \in \mathcal{M}_1(V,\Gamma)$ depends analytically on $j^{p+N} f(0,0)$. (Full details of this construction are given in [12, §10].) If Γ is finite, we may use this parametrization theorem, in combination with methods based on Newton-Puiseaux series, to obtain estimates on eigenvalues of the linearization along branches of invariant group orbits. A routine application of Tougeron's implicit function theorem [37] then yields strong determinacy for smooth maps. If Γ is not finite, we have to work a little harder. First of all we blow-up along orbit strata using recent results of Schwarz on the coherence of the orbit stratification (see [36] and [12, §9]). In this way, we desingularize the branch. Next we use the tangential and normal form for the family given by Lemma 3.1.5 and apply the same arguments used for the Γ-finite case to the normal component to obtain eigenvalue estimates along the branch. We obtain strong determinacy using persistence results on families of normally hyperbolic manifolds. (Proofs of these results are in [12, Appendix].) Finally, we extend our strong determinacy result from tangential to complex irreducible representations using equivariant normal forms (see [12, §9.18]).

6.3. Applications to normal forms. We conclude this section by briefly describing how we can use Theorem 6.1.1 to justify normal form computations.

First of all, suppose that (V,Γ) is absolutely irreducible and that $-I \notin \Gamma$. Set $\Gamma^2 = \Gamma \times \mathbb{Z}_2$. It follows from Theorem 6.1.1 that we can find $d \in \mathbb{N}$ such that Γ^2-equivariant bifurcation problems on V are strongly (d, Γ)-determined. Suppose that $f \in \mathcal{M}^-(V, \Gamma^2)$ and f is strongly (d, Γ^2)-determined. Let $f' \in \mathcal{M}^-(V, \Gamma)$ satisfy $j^d f_0'(0) = j^d f_0(0)$. We regard f' as a perturbation of f breaking symmetry from Γ^2 to Γ. It follows from the strong determinacy theorem that each branch of normally hyperbolic invariant Γ^2-orbits in $\Xi^*(f)$ will persist as a branch of Γ-invariant normally hyperbolic submanifolds for f'. Typically, some of these branches will be branches of Γ-orbits (and so will appear in $\Xi^*(f')$), others will not be Γ-orbits. If Γ is finite, each branch for f' will consist of hyperbolic points of prime period two.

Example 6.3.1. Let $\Gamma = \Delta_5' \rtimes \mathbb{Z}_5$ (Example 5.4.5). Then Γ^2-equivariant bifurcation problems are strongly $(3, \Gamma)$-determined (Theorem 5.4.4). It is easy to verify directly that if $f \in \mathcal{L}_\Gamma^-(\mathbb{R}^5)$, then f has branches of points of prime period two tangent to the axes $\mathbb{R}(1, 1, 1, 1, \pm1)$. However, the period two points are not related by Γ-symmetries. In this example, $(\mathbb{R}^5, \Gamma)$, $(\mathbb{R}^5, \Gamma^2)$ have the same cubic equivariants and fourth order terms are required to break symmetry from Γ^2 to Γ.

Let $P_\Gamma^{(d)}(V \times \mathbb{R}, V)_0$ denote the subset of $P_\Gamma^{(d)}(V \times \mathbb{R}, V)$ consisting of polynomial maps with linear term $(\lambda - 1)I_V$. We similarly define $P_{\Gamma^2}^{(d)}(V \times \mathbb{R}, V)_0$. The next result follows from the theory of equivariant normal forms [28, Chapter XVI, §5] (see also the proof of [12, Lemma 9.18.3]).

Lemma 6.3.2. *Let $d \in \mathbb{N}$. There is a polynomial submersion*

$$N_d : P_\Gamma^{(d)}(V \times \mathbb{R}, V)_0 \to P_{\Gamma^2}^{(d)}(V \times \mathbb{R}, V)_0,$$

such that if $f \in \mathcal{M}^-(V, \Gamma)$ then $N_d(j^d f(0, 0))$ is the Γ^2-equivariant normal form of f to order d. Moreover, if $p > d$, $N_d(N_p(j^p f(0, 0))) = N_d(j^d f(0, 0))$. In particular, N_d restricts to the identity map on $P_{\Gamma^2}^{(d)}(V \times \mathbb{R}, V)_0$.

Suppose that Γ-equivariant bifurcation problems on V are p-determined, Γ^2-equivariant bifurcation problems on V are q-determined and Γ^2-equivariant bifurcation problems on V are strongly (d, Γ)-determined. It is easy to see that $d \geq p, q$.

Theorem 6.3.3. *We may construct an open and dense semi-analytic subset $\mathcal{N}$ of $P_\Gamma^{(d)}(V \times \mathbb{R}, V)$ such that if $f \in \mathcal{M}^-(V, \Gamma)$ satisfies $j^d f(0, 0) \in \mathcal{N}$ then*
  (1) *$f \in \mathcal{S}(V, \Gamma)$.*
  (2) *$\tilde{f} = N_d(j^d f(0, 0)) \in \mathcal{S}(V, \Gamma^2)$.*
  (3) *Every branch of invariant normally hyperbolic Γ^2-orbits of $\tilde{f}$ persists as a branch of normally hyperbolic Γ-invariant submanifolds manifolds for f and every branch of invariant Γ-orbits of f arises via such a perturbation.*

Proof. Let $\mathcal{R}, \mathcal{R}^2$ be the open and dense semi-algebraic subsets of $P_\Gamma^{(d)}(V \times \mathbb{R}, V)$, $P_{\Gamma^2}^{(d)}(V \times \mathbb{R}, V)$ that respectively determine stable maps for Γ- and Γ^2-equivariant bifurcation problems on V. let $\mathcal{D}$ be the semi-analytic subset of $P_{\Gamma^2}^{(d)}(V \times \mathbb{R}, V)$ that determines the strongly (d, Γ)-stable mappings in $\mathcal{M}^-(V, \Gamma^2)$. We define

$$\mathcal{N} = \mathcal{R} \cap N_d^{-1}(\mathcal{R}^2 \cap \mathcal{D})$$

Since N_d is a polynomial submersion, $\mathcal{N}$ is an open and dense semi-analytic subset of $P_\Gamma^{(d)}(V \times \mathbb{R}, V)$. The theorem follows.

We have somewhat similar results if (V, Γ) is a complex representation such that $S^1 \not\subset \Gamma$. In this case, we set $G = \Gamma \times S^1$. Let $d \in \mathbb{N}$. Let $P_\Gamma^{(d)}(V \times \mathbb{R}, V)_0$ be the set of polynomial maps which have linear term $(1 + \lambda) \exp(i\omega) I_V$, where $\omega \in [0, 2\pi)$. Let $P_\Gamma^{(d)}(V \times \mathbb{R}, V)_*$ denote the open and dense semi-algebraic subset of $P_\Gamma^{(d)}(V \times \mathbb{R}, V)_0$ defined by requiring that $\exp(i\omega)$ is not a qth root of unity, $1 \leq q \leq d$. We similarly define $P_G^{(d)}(V \times \mathbb{R}, V)_*$.

Lemma 6.3.4. *Let $d \in \mathbb{N}$. There is a polynomial submersion*

$$N_d : P_\Gamma^{(d)}(V \times \mathbb{R}, V)_* \to P_G^{(d)}(V \times \mathbb{R}, V)_*,$$

such that if $f \in \mathcal{M}(V, \Gamma)$ then $N_d(j^d f(0, 0))$ is the G-equivariant normal form of f to order d. Moreover, if $p > d$, $N_d(N_p(j^p f(0, 0))) = N_d(j^d f(0, 0))$. In particular, N_d restricts to the identity map on $P_G^{(d)}(V \times \mathbb{R}, V)_*$.

Suppose G-equivariant bifurcation problems on V are strongly (d, Γ)-determined. We have

Theorem 6.3.5. *We may construct an open and dense semi-analytic subset $\mathcal{N}$ of $P_\Gamma^{(d)}(V \times \mathbb{R}, V)$ such that if $f \in \mathcal{M}^-(V, \Gamma)$ satisfies $j^d f(0, 0) \in \mathcal{N}$ then*

(1) $f \in \mathscr{S}(V, \Gamma)$.

(2) $\tilde{f} = N_d(j^d f(0, 0)) \in \mathscr{S}(V, G)$.

(3) *Every branch of normally hyperbolic G-orbits of $\tilde{f}$ persists as a branch of normally hyperbolic Γ-invariant submanifolds manifolds for f. Moreover, every branch of invariant Γ-orbits of f arises as such a perturbation.*

Remark 6.3.6. The residual dynamics on branches when we break normal form symmetry may, of course, be complicated (see Broer et al. [6]).

7. EQUIVARIANT HOPF BIFURCATION THEOREM FOR VECTOR FIELDS

One of the applications of the theory developed in [12] was a proof of a variant of Fiedler's equivariant Hopf bifurcation theorem based on strong determinacy and equivariant normal forms [12, Theorem 11.2.1]. The proof of the general strong determinacy theorem given in [12] depends on rather technical and delicate results on persistence of branches of normally hyperbolic group orbits under symmetry breaking perturbations. It is clear, however, that simpler proofs should be available if we restrict attention to symmetry breaking perturbations which only break normal form symmetry. In this section, we outline a relatively simple proof of a version of the strong determinacy theorem that applies to the normal form analysis of the equivariant Hopf bifurcation. Our proof avoids most normal hyperbolicity issues.

7.1. Preliminaries. Suppose that (V, Γ) is a complex irreducible representation. Let X be a smooth Γ-equivariant vector field on V and α be a relative equilibrium of X. We recall [12, §3], [15, 19] that we may define the *reduced Hessian* $\mathrm{HESS}(X, \alpha)$ of X along α and that $\mathrm{HESS}(X, \alpha)$ is a subset of $\mathbb{C}/\imath\mathbb{R}$. The orbit α is normally hyperbolic for X if and only if the multiplicity of 0 in $\mathrm{HESS}(X, \alpha)$ is equal to the dimension of α. If α is an equilibrium orbit, $\mathrm{HESS}(X, \alpha)$ is, up to translations by pure imaginary numbers, the set of eigenvalues (counting multiplicities) of the Hessian of X along α. If $X = X^N + X^T$ is a tangent and normal decomposition of X on a neighborhood of α (see [31, 19]), then $\mathrm{HESS}(X^N, \alpha) = \mathrm{HESS}(X, \alpha)$. In the sequel, we typically work with the tangent and normal decomposition.

Let $\mathscr{V}(V, \Gamma)$ denote the space of normalized Γ-equivariant vector fields on V. We recall that if $f \in \mathscr{V}(V, \Gamma)$ then $Df_\lambda(0) = (\lambda + \imath)I_V$, $\lambda \in \mathbb{R}$.

The next result follows from [12, Theorem 9.18.1] (cf. [28, Chapter XVI, §11]).

Proposition 7.1.1 (Estimates on eigenvalues). *There exist* $d \in \mathbb{N}$, $\nu > 0$ *and an open and dense semi-analytic subset* $\mathscr{R}$ *of* $P_\Gamma^{(d)}(V, V)$ *such that if* $f \in \mathscr{V}(V, G)$ *and* $j^d f_0(0) \in \mathscr{R}$ *then*

(1) $f \in \mathscr{S}(V, \Gamma)$.

(2) *If* $\mathfrak{b} \in \Sigma(f)$ *is a branch of relative equilibria of isotropy type* τ, *there exists a parametrization* $\Psi = (\phi, \lambda) : [0, \delta] \times \Delta_\tau \to V \times \mathbb{R}$ *of* $\mathfrak{b}$ *and* $C = C(f) > 0$ *such that if* $t \in (0, \delta]$ *and* $\mu(t) \in \mathrm{HESS}(f_{\lambda(t)}, \phi(t, \Delta_\tau))$ *is nonzero, then*

$$|Re(\mu(t))| \geq Ct^\nu$$

Remark 7.1.2. The first step of the proof of Proposition 7.1.1 depends on choosing $\mathscr{R}$ and $d \in \mathbb{N}$ so that if f is analytic and $j^d f_0(0) \in \mathscr{R}$, then we can choose parametrizations of branches so that initial exponents along the branch are locally constant (as functions of f). This is done in [12, §7] and we may suppose $\mathscr{R}$ is semi-algebraic. In order to obtain estimates on eigenvalues along the branch, we use blowing-up techniques and results based on Newton-Puiseaux series. Typically, we now have to allow $\mathscr{R}$ to be semi-analytic rather than semi-algebraic. Finally, using the tangent and normal form, we extend estimates to smooth families (see [12, §8]).

7.2. Equivariant Hopf bifurcation and normal forms. Continuing with our assumptions on (V, Γ), set $G = \Gamma \times S^1$ and consider the representation (V, G). Let $d \in \mathbb{N}$, $\mathscr{R} \subset P_G^{(d)}(V, V)$ and ν satisfy the conditions of Proposition 7.1.1 for (V, G).

Theorem 7.2.1. *There exists* $\tilde{d} \geq d$ *such that* (V, G) *is strongly* $(\tilde{d}, \Gamma)$*-determined.*

Proof. Let $\mu \in \mathscr{O}(V, G)$ be a symmetry breaking isotropy type. The conjugacy class of Γ_x is constant on V_μ and so μ determines a unique isotropy type $\tau \in \mathscr{O}(V, \Gamma)$. In particular, $V_\mu \subset V_\tau$ (see also [21, §3] and note that in general V_μ may not be an open subset of V_τ). If G-orbits of isotropy type μ are Γ-orbits, it is easy to see that normally hyperbolic branches of relative equilibria of isotropy type μ persist when we break symmetry from G to Γ. Therefore, we assume that G-orbits of isotropy type μ are not Γ-orbits. It follows that if α is a G-orbit of isotropy type μ, then G/Γ is diffeomorphic to S^1 and $G_x/\Gamma_x \cong \mathbb{Z}_p$, for some $p \geq 1$.

Following [12, Lemma 9.3.5], we successively polar blow-up $V \times \mathbb{R}$ along (the strict transforms) of the G-orbit strata $V_\rho \times \mathbb{R}$, $\rho > \mu$. In this way, we obtain an analytic G-equivariant map $\Pi : W \to V \times \mathbb{R}$ such that $W_\mu = \Pi^{-1}(\overline{V}_\mu \times \mathbb{R})$ is a closed submanifold of W. In addition, W_μ will be a submanifold of W_τ – the set of points of Γ-isotropy type τ. Indeed, if we order $\mathscr{O}(V, G)$ so that $\rho > \mu$ if $W_\rho \subset \partial W_\tau$, we may and shall assume that W_τ is a closed submanifold of W. The blowing-down map Π restricts to a local finite-to-one analytic diffeomorphism on the complement of $\Pi^{-1}(\cup_{\rho > \tau}(V \times \mathbb{R})_\rho)$. Every $f \in \mathscr{V}(V, G)$ lifts uniquely to a smooth G-equivariant vector field on $\tilde{f}$ on W. If $j^d f_0(0) \in \mathscr{R}$, then every branch $\mathfrak{b} \in \Sigma(f)$ of isotropy type μ lifts to a branch $\tilde{\mathfrak{b}} \subset W_\mu$ of normally hyperbolic relative equilibria of $\tilde{f}$. If we let $\tilde{\Psi} : [0, \delta] \times \Delta_\mu \to W_\mu \subset W$ denote the lift of the parametrization of $\mathfrak{b}$ given by Proposition 7.1.1, then the estimates of Proposition 7.1.1 hold for $\tilde{\Psi}$ (since Π is a local analytic diffeomorphism off $\Pi^{-1}(\cup_{\rho > \tau}(V \times \mathbb{R})_\rho)$). The map $\tilde{\Psi}$ is a G-equivariant embedding. We set $Z_t = \tilde{\Psi}(t, \Delta_\mu)$, $t \in [0, \delta]$, and note that Z_t will be a smooth family of G-orbits of G-isotropy type μ and Γ-isotropy type τ. We regard the dynamics on Z as Γ-equivariant and define a local Poincaré section $D \subset W$ for Z_0. We recall that D will be a smoothly Γ-equivariantly embedded submanifold of codimension 1 which intersects Z_0 transversally along a Γ-orbit (we refer to [15, 19] for details). It follows by transversality that D will be a Poincaré section for Z_t, $t \in [0, \delta']$, where $0 < \delta' \leq \delta$. Choosing a sufficiently small $D' \subset D$, we may define an associated family of Poincaré maps $P_{\lambda(t)} : D' \to D$, $t \in [0, \delta']$, such that $P_{\lambda(t)}$ has an invariant Γ-orbit $Z_t \cap D$ for each $t \in [0, \delta']$. If $z \in D'$, $P_{\lambda(t)}(z)$ will be defined as the first point of intersection of the forward $\tilde{f}_{\lambda(t)}$-trajectory through z with D. If $z \in Z_t \cap D$, then $P_{\lambda(t)}(z) = \exp(2\pi i/p)z$. Set $I_t = Z_t \cap D$. It follows from our hyperbolicity conditions on $\tilde{\mathfrak{b}}$ that I_t is a branch of normally hyperbolic invariant Γ-orbits for the family of Γ-equivariant diffeomorphisms $P_{\lambda(t)}$, $t \in [0, \delta']$. Moreover, our estimates on elements of $\text{HESS}(\tilde{f}_{\lambda(t)}, Z_t)$ exponentiate to estimates on elements of $\text{SPEC}(P_t, I_t)$. Specifically, if $\mu(t) \in \text{SPEC}(P_{\lambda(t)}, I_t)$ is not equal to one, then there exists $C > 0$ such that

$$|1 - |\mu(t)|| \geq Ct^\nu, \quad t \in [0, \delta']$$

It is now a straightforward application of the techniques used to study families of Γ-equivariant maps to show that the branch I_t of invariant Γ-orbits will persist, as a family of normally hyperbolic invariant Γ-orbits, under sufficiently small high order Γ-equivariant perturbations of the family P. Moreover, we can choose an S^1-invariant horn neighborhood H of the original branch I_t such that $H \cap \Pi^{-1}(\cup_{\rho > \tau}(V \times \mathbb{R})_\rho) = I_0$ and require that perturbed families lie within H. Finally, we may choose $\tilde{d} \geq d$ (independent of f) such that if $f \in \mathscr{M}(V, \Gamma)$ and $j^{\tilde{d}} f'(0,0) = j^{\tilde{d}} f(0,0)$, then $\tilde{f}'$ defines a family of Poincaré maps $P'_{\lambda(t)}$ with a corresponding branch of invariant normally hyperbolic Γ-orbits contained in H. This branch determines in the usual way a family of Γ- and $\tilde{f}'$-invariant normally hyperbolic submanifolds of W_τ and again we may require that the family is contained in H. Blowing-down by Π we obtain the required family of Γ- and f'-invariant normally hyperbolic submanifolds of f'.

Remarks 7.2.2. (1) Note that it is somewhat easier to prove *persistence* of the branch in W_τ. Once we have defined the Poincaré maps, restrict to the free Γ-manifold W_τ and reduce to the orbit manifold W_τ/Γ. The branch of invariant Γ-orbits drops down to a branch of fixed points. Somewhat similar techniques were used in [7]. (2) Note that the advantage of working with the Poincaré maps is that they are Γ-invariant, even when we break normal form symmetries, and so we can characterize normal hyperbolicity in terms of spectral conditions on eigenvalues of linearizations. This approach is not open to us if we do not work with flow or map invariant Γ-orbits as the behavior *tangent* to the manifold then becomes critical.

REFERENCES

1. D. G. Aronson, M. Golubitsky, and M. Krupa, *Coupled arrays of Josephson Junctions and bifurcation of maps with S_N symmetry*, Nonlinearity **4** (1991), 861–902.
2. E. Bierstone, *Lifting isotopies from orbit spaces*, Topology **14** (1975), 245–252.
3. _____, *General position of equivariant maps*, Trans. Amer. Math. Soc. **234** (1977), 447–466.
4. _____, *Generic equivariant maps*, Real and Complex Singularities, Oslo 1976, Sijthoff and Noordhoff International Publ., Leyden, 1977, Proc. Nordic Summer School/NAVF Sympos. Math., pp. 127–161.
5. T. Bröcker and T. tom Dieck, *Representations of Compact Lie groups*, Graduate Texts in Mathematics, Springer, New York, 1985.
6. H. W. Broer, G. B. Huitema, F. Takens, and B. L. T. Braaksma, *Unfoldings and bifurcation of quasi-periodic tori*, Mem. Amer. Math. Soc. **83** (1990), no. 421.
7. P. Chossat and M. J. Field, *Geometric analysis of the effect of symmetry breaking perturbations on an $O(2)$ invariant homoclinic cycle*, Fields Institute Communications, **4** (1995), 21–42.
8. P. Chossat and M. Golubitsky, *Iterates of maps with symmetry*, SIAM J. Math. Anal. **19** (1988), no. 6, 1259–1270.
9. M. Coste, *Ensembles semi-algébriques*, Géométrie Algebébrique Réelle et Formes Quadratiques, Springer Lecture Notes in Math., no. 959, 1982, pp. 109–138.
10. B. Fiedler, *Global Bifurcation of Periodic Solutions with Symmetry*, Springer Lecture Notes in Math., no. 1309, Springer-Verlag, New York-London, 1988.
11. M. J. Field, *Geometric methods in bifurcation theory*, Pattern formation and symmetry breaking in PDEs, Fields Institute Communications, **5** (1995).
12. _____, *Symmetry breaking for compact Lie groups*, Mem. Amer. Math Soc., to appear, 1996.
13. _____, *Stratifications of equivariant varieties*, Bull. Austral. Math. Soc. **16** (1977), no. 2, 279–296.
14. _____, *Transversality in G-manifolds*, Trans. Amer. Math. Soc. **231** (1977), 429–450.
15. _____, *Equivariant Dynamical Systems*, Trans. Amer. Math. Soc. **259** (1980), 185–205, **26** (1982), 161–180.
16. _____, *Isotopy and stability of equivariant diffeomorphisms*, Proc. London Math. Soc. **46** (1983), no. 3, 487–516.
17. _____, *Equivariant Dynamics*, Contemp. Math **56** (1986), 69–95.

18. _____, *Equivariant Bifurcation Theory and Symmetry Breaking*, J. Dynamics and Diff. Eqns. **1** (1989), no. 4, 369–421.
19. _____, *Local structure of equivariant dynamics*, Singularity Theory and its Applications, II, (M. Roberts and I. Stewart, eds.), Springer Lecture Notes in Math., no. 1463, 1991, pp. 168–195.
20. _____, *Blowing-up in equivariant bifurcation theory*, Dynamics, Bifurcation and Symmetries: New Trends and New Tools (Amsterdam) (P. Chossat and J.-M. Gambaudo, eds.), NATO ARW Series, Kluwer, Amsterdam, 1994, pp. 111–122.
21. _____, *Determinacy and branching patterns for the equivariant Hopf bifurcation*, Nonlinearity **7** (1994), 403–415.
22. M. J. Field and R. W. Richardson, *Symmetry Breaking and the Maximal Isotropy Subgroup Conjecture for Reflection Groups*, Arch. for Rational Mech. and Anal. **105** (1989), no. 1, 61–94.
23. _____, *Symmetry breaking in equivariant bifurcation problems*, Bull. Amer. Math. Soc. **22** (1990), no. 1, 79–84.
24. _____, *Symmetry breaking and branching patterns in equivariant bifurcation theory I*, Arch. Rational Mech. Anal. **118** (1992), 297–348.
25. _____, *Symmetry breaking and branching patterns in equivariant bifurcation theory II*, Arch. Rational Mech. Anal. **120** (1992), 147–190.
26. C. Gibson, K. Wirthmüller, A. A. du Plessis, and E. Looijenga, *Topological stability of smooth mappings*, Springer Lecture Notes in Math., no. 553, 1976.
27. M. Golubitsky, *Genericity, bifurcation and Symmetry*, Patterns and Dynamics in Reactive Media (New York) (R. Aris, D. G. Aronson, and H. L. Swinney, eds.), The IMA Volumes in Math. and its Applic., no. 37, Springer-Verlag, New York, 1991, pp. 71–87.
28. M. Golubitsky, D. G. Schaeffer, and I. N. Stewart, *Singularities and groups in bifurcation theory*, vol. II, Appl. Math. Sci., no. 69, Springer-Verlag, New York, 1988.
29. M. W. Hirsch, C. C. Pugh, and M. Shub, *Invariant Manifolds*, Springer Lect. Notes Math., no. 583, 1977.
30. N. Jacobson, *Basic algebra I*, W. H. Freeman, San Francisco, 1974.
31. M. Krupa, *Bifurcations of relative equilibria*, SIAM J. Math. Anal. **21** (1990), no. 6, 1453–1486.
32. J. Mather, *Stratifications and mappings*, Proceedings of the Dynamical Systems Conference, Salvador, Brazil (M. Peixoto, ed.), Academic Press, New York, San Francisco, London, 1973.
33. B. B. Peckam and I. G. Kevrekidis, *Period doubling with higher-order degeneracies*, SIAM J. Math. Anal **22** (1991), no. 6, 1552–1574.
34. V. Poenaru, *Singularités C∞ en Présence de Symétrie*, Springer Lect. Notes in Math., no. 510, Springer-Verlag, New York, 1976.
35. D. Ruelle, *Bifurcation in the presence of a symmetry group*, Arch. Rational Mech. Anal. **51** (1973), no. 2, 136–152.
36. G. W. Schwarz, *Algebraic quotients of compact group actions*, preprint, 1994.
37. J. C. Tougeron, *Ideaux de fonctions differentiable*, Erge. der Math. und ihrer Gren., no. 71, Springer-Verlag, Berlin, Heidelberg, New-York, 1972.
38. A. Vanderbauwhede, *Equivariant period doubling*, Advanced Topics in the Theory of Dynamical Systems (G. Fusco, M. Ianelli, and L. Salvadori, eds.), Notes Rep. Math. Sci. Engrg., no. 6, 1989, pp. 235–246.

DEPARTMENT OF MATHEMATICS, UNIVERSITY OF HOUSTON, HOUSTON, TX 77204-3476, USA
E-mail address: mf@uh.edu

LOW DIMENSIONAL REPRESENTATIONS OF REDUCTIVE GROUPS ARE SEMISIMPLE

JENS CARSTEN JANTZEN
Dedicated to the memory of Roger Richardson

Let G be a connected reductive algebraic group over an algebraically closed field k of characteristic $p \neq 0$. The purpose of this note is to prove the following two results:

(A) *Any G-module of dimension $\leq p$ is semisimple.*

and

(B) *If G is defined over a finite field $\mathbf{F}_q \subset k$ and does not have a component of type A_1, then any $kG(\mathbf{F}_q)$-module of dimension $\leq p$ is semisimple.*

Early in 1986 Serre asked me whether (B) was true in the case of a prime field. In an answer I gave a proof where I forgot to check one condition. It contained as auxiliary results what is contained below in 1.2–1.6. (However, I have replaced my proof of Lemma 1.4 by a reference to a much nicer argument by Serre.) In 1992 I heard that Larsen was interested in something like (A) and later on I got his preprint [11] that mentions (A) as a conjecture and proves a similar semisimplicity result involving a worse bound. Unfortunately, at that point I had forgotten about my 1986 results. It was only when Serre reminded me in 1993 of our correspondence in connection with his work [13], that I realized that I had all the necessary ingredients to prove (A) and (B).

We prove our results first in the case where G is almost simple and simply connected. In the last section we extend them to all general reductive groups. The assumptions and notations from 1.1 will be in force throughout Sections 1 and 2.

1. THE ALGEBRAIC GROUPS

1.1.   Let G be a connected semisimple group with maximal torus T over an algebraically closed field k of characteristic $p \neq 0$. Denote by $X = X(T)$ the group of characters of T and by R the root system of G with respect to T. Assume that R is indecomposable and that G is simply connected. Let W be the Weyl group of R and let W_p be the affine Weyl group generated by W and the translations by all $p\alpha$ with $\alpha \in R$.

We choose a positive system R^+ in R, denote the simple roots (for this choice) by $\alpha_1, \alpha_2, \ldots, \alpha_n$ using the same numbering as in the tables in [5]. Denote the largest short root by α_0, and the half sum of the positive roots by ρ. For each root

Research in part supported by an NSF-grant

α let $\alpha^\vee$ be the corresponding coroot. We denote the set of dominant weights by X^+ and set for all integers $r > 0$

$$X_r = \{\lambda \in X \mid 0 \leq \langle \lambda, \alpha_i^\vee \rangle < p^r \quad \text{for all } i\}.$$

For all $\lambda \in X^+$ let $L(\lambda)$ be the simple module with highest weight λ and $H^0(\lambda)$ the G–module induced from the one dimensional representation given by λ of a suitable Borel subgroup of G as in [8], II.2.1.

An arbitrary dominant weight $\lambda \in X^+$ has a p–adic expansion $\lambda = \sum_{i=1}^r p^{i-1}\lambda_i$ with all $\lambda_i \in X_1$ (for a suitable integer $r > 0$). Then Steinberg's tensor product theorem says that

$$L(\lambda) \simeq L(\lambda_1) \otimes L(\lambda_2)^{[1]} \otimes \cdots L(\lambda_r)^{[r-1]}, \tag{1}$$

cf. [8], II.3.17. We have in particular

$$\dim L(\lambda_i) \leq \dim L(\lambda) \qquad \text{for all } i. \tag{2}$$

This shows: If $\dim L(\lambda) \leq p$, then $\dim L(\lambda_i) \leq p$ for all p.

Lemma 1.2. *If $\lambda \in X_1$ with $\dim L(\lambda) \leq p$, then*

$$\langle \lambda, \alpha^\vee \rangle < p \qquad \text{for all } \alpha \in R^+ \tag{1}$$

and

$$\sum_{\alpha > 0} \langle \lambda, \alpha^\vee \rangle \leq \dim L(\lambda) - 1. \tag{2}$$

Proof. We shall call (for any $\alpha > 0$) the direct sum of all weight spaces of $L(\lambda)$ corresponding to weights of the form $\lambda - i\alpha$ with $i \in \mathbb{Z}$ the α–string through λ. This sum is a submodule for the subgroup G_α of G generated by the root subgroups corresponding to α and $-\alpha$. This subgroup is isomorphic to SL_2, cf. [8], II.1.3. The α–string through λ contains (as a composition factor) the simple G_α–module with highest weight λ restricted to $T \cap G_\alpha$. If $\langle \lambda, \alpha^\vee \rangle < p$, then this simple module has dimension equal to $1 + \langle \lambda, \alpha^\vee \rangle \leq p$. So this is then a lower bound for the dimension of the α–string.

In order to prove (1) we use induction over the height of $\alpha^\vee$. If the height is 1, then α is a simple root α_i, and (1) holds by the assumption that $\lambda \in X_1$. Otherwise there are two positive roots β_1, β_2 with $\alpha^\vee = \beta_1^\vee + \beta_2^\vee$. (Apply [5], Ch. VI, §1, Cor. 1 de la Prop. 19 to $R^\vee$.) We have by induction $\langle \lambda, \beta_i^\vee \rangle < p$ for both i. The discussion above shows that the β_i–string through λ has dimension at least $1 + \langle \lambda, \beta_i^\vee \rangle$. Distinct α–strings have only the λ weight space in common, so we get

$$\langle \lambda, \alpha^\vee \rangle = \langle \lambda, \beta_1^\vee \rangle + \langle \lambda, \beta_2^\vee \rangle \leq \dim L(\lambda) - 1 < p$$

as desired.

In order to deduce (2) from (1) we use the same argument as in the last step. We get now from (1) that the α–string through λ has dimension at least $1 + \langle \lambda, \alpha^\vee \rangle$ for all $\alpha \in R^+$. Since distinct α–strings have only the λ weight space in common, we get $1 + \sum_{\alpha > 0} \langle \lambda, \alpha^\vee \rangle \leq \dim L(\lambda)$, i.e., (2).

1.3. Let $\lambda \in X_1$ with $\dim L(\lambda) \le p$. If $\lambda \ne 0$, then there is a simple root α_i with $\langle \lambda, \alpha_i^\vee \rangle > 0$. We get then for any root $\alpha > 0$ with α_i in the support of α that $\langle \lambda, \alpha^\vee \rangle > 0$. If R is not of type A_1 or A_2, then there is at least one root $\alpha > 0$ with α_i in the support of α and $\alpha \ne \alpha_0$. (Take, e.g., $\alpha = \alpha_i + \alpha_j$ where j is linked to i in the Dynkin diagram of R.) In 1.2(2) we get from this α and from α_i contributions ≥ 2 on the left hand side, hence

$$\langle \lambda, \alpha_0^\vee \rangle \le \dim L(\lambda) - 3 \qquad \text{for } \lambda \ne 0, \text{ type } \ne A_1, A_2. \tag{1}$$

We can improve the bound for R of type G_2 where we have three α as above and get

$$\langle \lambda, \alpha_0^\vee \rangle \le \dim L(\lambda) - 5 \qquad \text{for } \lambda \ne 0, \text{ type } = G_2, \tag{2}$$

whereas for type A_2 we get only

$$\langle \lambda, \alpha_0^\vee \rangle \le \dim L(\lambda) - 2 \qquad \text{for } \lambda \ne 0, \text{ type } = A_2. \tag{3}$$

We have $\langle \lambda, \alpha_0^\vee \rangle > 0$ in all cases since $\lambda \ne 0$. So $\dim L(\lambda) \le p$ implies in the case of (3) that $p - 3 > 0$, hence $p \ge 5$. We argue similarly in the other cases and get: *If G has a nontrivial irreducible representation of dimension $\le p$, then*

$$\begin{aligned} p &\ge 5 \qquad \text{for } R \text{ of type } \ne A_1, A_2, \\ p &\ge 7 \qquad \text{for } R \text{ of type } G_2, \\ p &\ge 3 \qquad \text{for } R \text{ of type } A_2. \end{aligned} \tag{4}$$

Note that one gets rid of the restriction to $\lambda \in X_1$ using Steinberg's tensor product theorem, cf.1.1(2). One can give a different proof of (4) (and get stronger bounds), if one observes that $p \ge |W\lambda|$ and uses the tables in [5] in order to estimate $|W\lambda|$. Finally, one can also use tables for the smallest dimension of a non-trivial representation of G, e.g. in [10], 5.4.13.

Lemma 1.4. *If $\lambda \in X_1$ with $\dim L(\lambda) \le p$ and $\lambda \ne 0$, then $\langle \lambda + \rho, \alpha_0^\vee \rangle \le p$.*

Proof. This follows immediately from Lemma 1.2 and the following inequality, discovered by Serre, cf. [13], 2.4, Prop. 5: One has for all dominant weights $\lambda \in X^+$ with $\lambda \ne 0$

$$\langle \lambda + \rho, \alpha_0^\vee \rangle \le 1 + \sum_{\alpha > 0} \langle \lambda, \alpha^\vee \rangle. \tag{1}$$

1.5. Set

$$\overline{C} = \{\nu \in X \mid 0 \le \langle \nu + \rho, \alpha^\vee \rangle \le p \quad \text{for all } \alpha \in R^+\}. \tag{1}$$

This is the closure of the "first dominant alcove" for W_p (denoted by $\overline{C}_{\mathbb{Z}}$ in [8], II.5.5). So Lemma 1.4 says: If $\lambda \in X_1$ with $\dim L(\lambda) \le p$ and $\lambda \ne 0$, then $\lambda \in \overline{C}$. This implies, cf. [8], II.5.6, that the simple module $L(\lambda)$ is equal to the module $H^0(\lambda)$, hence that $\dim L(\lambda)$ is given by Weyl's dimension formula

$$\dim H^0(\lambda) = \prod_{\alpha > 0} \frac{\langle \lambda + \rho, \alpha^\vee \rangle}{\langle \rho, \alpha^\vee \rangle}. \tag{2}$$

This allows us (in principle) to determine all λ with this property. Using 1.1(1), (2) we see that we can find (in principle) all irreducible G–modules of dimension less or equal to p.

1.6. Any $\lambda \in X_1$ with $\dim L(\lambda) = p$ satisfies $\langle \lambda + \rho, \alpha_0^\vee \rangle = p$, since we have for all factors in the numerator in 1.5(2)

$$\langle \lambda + \rho, \alpha^\vee \rangle \leq \langle \lambda + \rho, \alpha_0^\vee \rangle \leq p.$$

On the other hand, suppose that $\lambda \in X_1$ satisfies the assumption of Lemma 1.4 (i.e., $\lambda \neq 0$ and $\dim L(\lambda) \leq p$). If now $\langle \lambda + \rho, \alpha_0^\vee \rangle = p$, then $\dim L(\lambda) = p$, because the numerator in 1.5(2) is divisible by p and the denominator isn't since

$$\langle \rho, \alpha^\vee \rangle \leq \langle \rho, \alpha_0^\vee \rangle < \langle \lambda + \rho, \alpha_0^\vee \rangle \leq p$$

for all $\alpha > 0$.

It is now easy to determine all $\lambda \in X_1$ with $\dim L(\lambda) = p$. We have just seen that necessarily $\langle \lambda + \rho, \alpha_0^\vee \rangle = p$. Lemma 1.2 and 1.4(1) imply that

$$p = \langle \lambda + \rho, \alpha_0^\vee \rangle = 1 + \sum_{\alpha > 0} \langle \lambda, \alpha^\vee \rangle = \dim H^0(\lambda).$$

It is easy to inspect the tables in [5] to check where one gets equality. One can also observe that the proof of Lemma 1.2 shows that all weight spaces in $L(\lambda) = H^0(\lambda)$ have dimension equal to 1 and then use the classification of such modules in [12], 6.1. Both methods show that the complete list of all $\lambda \in X_1$ with $\dim L(\lambda) = p$ is

$$
\begin{array}{lll}
\lambda = (p-1)\omega_1 & \text{for type } A_1, & p \text{ arbitrary;} \\
\lambda = \omega_1 & \text{for type } A_n, n \geq 2, & p = n+1; \\
\lambda = \omega_n & \text{for type } A_n, n \geq 2, & p = n+1; \\
\lambda = \omega_1 & \text{for type } B_n, n \geq 2, & p = 2n+1; \\
\lambda = \omega_1 & \text{for type } G_2, & p = 7.
\end{array}
$$

If we allow arbitrary $\lambda \in X^+$, we get by Steinberg's tensor product theorem in addition the Frobenius twists of the representations above.

Lemma 1.7. *Suppose that R is not of type A_1. Let $\lambda, \lambda' \in X^+$ with $\langle \lambda + \rho, \alpha_0^\vee \rangle < p$ and $\langle \lambda' + \rho, \alpha_0^\vee \rangle < p$. If $\lambda \neq \lambda'$, then $\mathrm{Ext}_G^1(L(\lambda + p\mu), L(\lambda' + p\mu')) = 0$ for all $\mu, \mu' \in X^+$.*

Proof. Let h be the Coxeter number of R. If $p < h$ then there is not a weight $\nu \in X^+$ with $\langle \nu + \rho, \alpha_0^\vee \rangle < p$, if $p = h$ then there is only one such weight (namely $\nu = 0$), cf. [8], II.6.2(10). So our assumption implies $p > h$.

Let $\mu, \mu' \in X^+$. Consider first the case $\lambda = 0$. We have by [8], II.10.17(3) an isomorphism

$$\mathrm{Ext}_G^1(L(p\mu), L(\lambda' + p\mu')) \simeq \mathrm{Hom}_G(L(\mu), H^1(G_1, L(\lambda'))^{[-1]} \otimes L(\mu')).$$

Note that the $\oplus$ in [8] is a misprint and should be replaced by $\otimes$. Here G_1 is the first Frobenius kernel of G (equivalent to the restricted Lie algebra of G) and we have rewritten $\mathrm{Ext}_{G_1}(L(0), L(\lambda'))$ as $H^1(G_1, L(\lambda'))$. The exponent $[-1]$ denotes a Frobenius "untwist". This formula shows that is enough to show that

$$H^1(G_1, L(\nu)) = 0$$

for all $\nu \in X^+$ with $\langle \nu + \rho, \alpha_0^\vee \rangle < p$. Well, we have $L(\nu) = H^0(\nu)$ for these ν. There is in [9] a list of all $\nu \in X_1$ with $H^1(G_1, H^0(\nu)) \neq 0$. Since $p > h$ we can exclude the special p from [9], 2.2, and get from [9], 4.1, that $\nu = p\omega_i - \alpha_i$ for some simple root α_i. We get then

$$\langle \nu + \rho, \alpha_0^\vee \rangle = p \langle \omega_i, \alpha_0^\vee \rangle + h - 1 - \langle \alpha_i, \alpha_0^\vee \rangle$$

and this term is greater or equal to p since $\langle \alpha_i, \alpha_0^\vee \rangle \leq h - 1$ for all possible root systems except for the excluded type A_1. So the claim for the ν with $\langle \nu + \rho, \alpha_0^\vee \rangle < p$ follows.

Let us now turn to the general case. We can assume that $\lambda + p\mu$ and $\lambda' + p\mu'$ are conjugate under the dot action of the affine Weyl group W_p, since otherwise the two modules belong to different blocks for G and do not extend, cf. [8], II.7.1/2. The weights $\lambda + p\mu$ and $0 + p\mu$ are in the interior of the same alcove for W_p. So there is a translation functor T with $TL(\lambda + p\mu) \simeq L(p\mu)$, cf. [8], II.7.15. It is an equivalence of categories of the corresponding blocks, cf. [8], II.7.9, and induces isomorphisms of Ext groups. There is a weight $\lambda'' \in X^+$ with $\langle \lambda'' + \rho, \alpha_0^\vee \rangle < p$ such that $TL(\lambda' + p\mu') \simeq L(\lambda'' + p\mu')$. We have

$$\mathrm{Ext}_G^1(L(\lambda + p\mu), L(\lambda' + p\mu')) \simeq \mathrm{Ext}_G^1(L(p\mu), L(\lambda'' + p\mu')).$$

So we get the desired vanishing result from the case $\lambda = 0$ as soon as we know that $\lambda'' \neq 0$. Well, there is an element w in the Weyl group W and a weight $\nu \in \mathbb{Z}R$ with $\lambda' + p\mu' = w(\lambda + \rho) - \rho + p(\mu + \nu)$. We have $w \neq 1$ since $\lambda' \neq \lambda$. We have now $\lambda'' + p\mu' = w(\rho) - \rho + p(\mu + \nu)$, hence $\lambda'' = w(\rho) - \rho + p(\mu + \nu - \mu')$. Now $p > h$ and $w \neq 1$ imply $w(\rho) - \rho \notin pX$, and thus $\lambda'' \neq 0$ as desired.

Lemma 1.8. *Suppose that R is not of type A_1. Let L and L' be simple G-modules with $\dim L \leq p$ and $\dim L' \leq p$. Then $\mathrm{Ext}_G^1(L, L') = 0$.*

Proof. We can assume that L and L' are not isomorphic, cf. [8], II.2.12(1). So at least one of these modules has dimension > 1. Now 1.3(4) implies in particular that $p > 2$, since we exclude the type A_1.

There are $\lambda, \lambda' \in X_1$ and $\mu, \mu' \in X^+$ with $L \simeq L(\lambda + p\mu)$ and $L' \simeq L(\lambda' + p\mu')$. Steinberg's tensor product theorem implies that $L \simeq L(\lambda) \otimes L(\mu)^{[1]}$ and $L' \simeq L(\lambda') \otimes L(\mu')^{[1]}$. So the dimensions of $L(\lambda)$, $L(\lambda')$, $L(\mu)$, and $L(\mu')$ are less than or equal to p. If $\lambda = \lambda'$, then [8], II.10.17(2) and II.12.9, show (since $p > 2$) that

$$\mathrm{Ext}_G^1(L, L') \simeq \mathrm{Ext}_G^1(L(\mu), L(\mu')).$$

We can then use induction on the weights to get the desired vanishing.

Suppose now that $\lambda \neq \lambda'$. If $\langle \lambda + \rho, \alpha_0^\vee \rangle < p$ and $\langle \lambda' + \rho, \alpha_0^\vee \rangle < p$, then we can apply Lemma 1.7. If $\langle \lambda + \rho, \alpha_0^\vee \rangle < p = \langle \lambda' + \rho, \alpha_0^\vee \rangle$ (or vice versa), then $\lambda + p\mu$ and $\lambda' + p\mu'$ are not conjugate under the dot action of W_p and the Ext group vanishes by the linkage principle. If $\langle \lambda + \rho, \alpha_0^\vee \rangle = p = \langle \lambda' + \rho, \alpha_0^\vee \rangle$, then the discussion at the beginning of 1.6 implies $\dim L(\lambda) = p = \dim L(\lambda')$. (Note: since $\lambda \neq \lambda'$, one of these, say λ, is not equal to 0. Then $\langle 0 + \rho, \alpha_0^\vee \rangle < \langle \lambda + \rho, \alpha_0^\vee \rangle = p$ implies that also $\lambda' \neq 0$.) We get therefore $L \simeq L(\lambda)$ and $L' \simeq L(\lambda')$. Now λ and λ' are two distinct weights in the closure of one alcove for W_p. So they are not conjugate under the dot action of W_p and the Ext group vanishes by the linkage principle.

Proposition 1.9. *Any G–module of dimension $\leq p$ is semisimple.*

Proof. If R is not of type A_1, then we can apply Lemma 1.8. So let us assume that R is of type A_1. Since G is not solvable, any module of dimension ≤ 2 is semisimple. So we can assume that $p > 2$. We have to show: If L and L' are simple G-modules with $\dim L + \dim L' \leq p$, then $\mathrm{Ext}_G^1(L, L') = 0$. Proceed as in the proof of Lemma 1.8 and use the notations from that proof. If $\lambda = \lambda'$, we can again use induction since also $\dim L(\mu) + \dim L(\mu') \leq p$. Suppose now $\lambda \neq \lambda'$. Since we can assume that $\lambda + p\mu$ and $\lambda' + p\mu'$ are conjugate under the dot action of the affine Weyl group, there has to be an integer i $(0 \leq i < p-1)$ with $\lambda = i\omega_1$ and $\lambda' = (p-i-2)\omega_1$. We have then $\dim L(\lambda) + \dim L(\lambda') = (i+1) + (p-i-1) = p$, hence $L = L(\lambda)$ and $L' = L(\lambda')$. Again the linkage principle implies the vanishing of the Ext group.

Remarks. (1) This proves the conjecture made in [11], 3.7.

(2) For G of type A_1 the module $H^0(p\omega_1)$ has dimension $p + 1$. Its socle is $L(p\omega_1) \simeq L(\omega_1)^{[1]}$ and has dimension 2. (The factor module is isomorphic to $L((p - 2)\omega_1)$ of dimension $p - 1$.) So $H^0(p\omega_1)$ is not semisimple.

2. THE FINITE GROUPS

Lemma 2.1. *Suppose that R is not of type A_1. Let L and L' be simple G–modules with highest weights in X_r with $\dim L + \dim L' \leq p$. If L and L' are not both trivial, then each weight λ of $L \otimes L'$ satisfies*

$$\langle \lambda, \alpha_0^\vee \rangle \leq \begin{cases} p^r - 3p^{r-1} - 3, & \text{if } R \text{ is of type } G_2; \\ p^r - 2p^{r-1} - 2, & \text{otherwise.} \end{cases} \tag{1}$$

Proof. The existence of L and L' implies that $p \geq 5$ (and $p \geq 7$ for the type G_2). This follows from 1.3(4) unless R is of type A_2. In that case one of the two modules has dimension at least 3, so $\dim L + \dim L' \leq p$ yields $p \geq 4$, hence $p \geq 5$.

Denote the highest weights of L and L' by μ and μ'. We have then $\lambda \leq \mu + \mu'$ for each weight λ of $L \otimes L'$, hence

$$\langle \lambda, \alpha_0^\vee \rangle \leq \langle \mu + \mu', \alpha_0^\vee \rangle$$

So we may assume that $\lambda = \mu + \mu'$. Consider the p–adic expansions

$$\mu = \sum_{i=0}^{r-1} p^i \mu_i \quad \text{and} \quad \mu' = \sum_{i=0}^{r-1} p^i \mu_i'$$

with all μ_i and μ_i' in X_1. Note that Steinberg's tensor product theorem implies for all i, cf. 1.1(2)

$$\dim L(\mu_i) \leq \dim L(\mu) \leq p - 1; \tag{2}$$

similarly for the μ_i'.

We want to show first for all i that

$$\langle \mu_i + \mu_i', \alpha_0^\vee \rangle \leq \begin{cases} p - 3, & \text{for type } A_2; \\ p - 6, & \text{for type } G_2; \\ p - 4, & \text{otherwise.} \end{cases} \tag{3}$$

In case $\mu_i = \mu_i' = 0$ this follows from our bound on p. In case $\mu_i \neq 0 = \mu_i'$ we apply 1.3(1)–(3) and use (2). If $\mu_i \neq 0 \neq \mu_i'$ we use the same formulas; for example in the case A_2 we get

$$\langle \mu_i + \mu_i', \alpha_0^\vee \rangle \leq \dim L(\mu_i) + \dim L(\mu_i') - 4.$$

We observe for each type that

$$\dim L(\mu_i) + \dim L(\mu_i') \leq \dim L(\mu) + \dim L(\mu') \leq p,$$

cf. (2). We get thus (3), even with a strict inequality (for $\mu_i \neq 0 \neq \mu_i'$).

If $r = 1$, then (1) follows directly from (3) except for R of type A_2. In that case we have a bound equal to $p - 4$ in (1) whereas (3) yields only $p - 3$. If $\mu \neq 0 \neq \mu'$ we can apply the last remark in the preceding paragraph and get a strict inequality in (3), hence the desired inequality in (1). Suppose now that $\mu \neq 0 = \mu'$ and that $\langle \mu, \alpha_0^\vee \rangle = p - 3$. Write μ as a linear combination of the fundamental weights: $\mu = (a-1)\omega_1 + (b-1)\omega_2$ with $a, b > 0$. Now $\langle \mu, \alpha_0^\vee \rangle = p - 3$ implies $a + b = p - 1$ and $\dim L \leq p - 1$ implies by Weyl's dimension formula (recall 1.4/5) $ab(a + b) \leq 2(p - 1)$, hence $ab \leq 2$ and $p - 1 = a + b \leq 3$, a contradiction.

Suppose now that $r > 1$. We have

$$\langle \mu + \mu', \alpha_0^\vee \rangle = \sum_{i=0}^{r-1} p^i \langle \mu_i + \mu_i', \alpha_0^\vee \rangle \leq \sum_{i=0}^{r-1} p^i (p - m)$$

where m is 3, 6, or 4 (as in (3)). One checks easily for all $n > 0$ that

$$(p - m)\frac{p^n - 1}{p - 1} \leq p^n - m.$$

This yields

$$\langle \mu + \mu', \alpha_0^\vee \rangle \leq \sum_{i=0}^{r-2} p^i (p - m) + p^{r-1}(p - m) = (p - m)\frac{p^{r-1} - 1}{p - 1} + p^r - mp^{r-1}$$

$$\leq p^{r-1} - m + p^r - mp^{r-1} = p^r - (m - 1)p^{r-1} - m.$$

The last term is in each case smaller than the right hand side in (1).

2.2. Suppose now that G and T are defined and split over the prime field $\mathbf{F}_p$. Let σ be a (possibly trivial) automorphism of the Dynkin diagram of R. We extend σ to R and X. For each p-power $q = p^r$ let $G(q)$ be the group of points over $\mathbf{F}_q$ of G twisted by σ.

Proposition. *Let $r \geq 1$ be an integer and set $q = p^r$. Let V be a finite dimensional G-module such that each weight λ of V satisfies*

$$\langle \lambda, \alpha_0^\vee \rangle \leq \begin{cases} p^r - 3p^{r-1} - 3, & \text{if } R \text{ is of type } G_2; \\ p^r - 2p^{r-1} - 2, & \text{otherwise.} \end{cases} \tag{1}$$

Then the natural restriction map $H^1(G, V) \to H^1(G(q), V)$ is an isomorphism.

Proof. By [6], Lemma 5.6 (resp. by the remark in the middle of p. 250 in [3]) it is enough to show (for a suitable Borel subgroup B of G) that the natural restriction map from $H^1(B, V) = 0$ to $H^1(B(q), V)$ is an isomorphism. So it is enough (in case $\sigma = 1$) to check for each weight λ of V that λ satisfies the isomorphism condition in [6], 5.5 for $n = 0, 1$ and the injectivity condition in [6], 5.4 for $n = 2$. For $\sigma \neq 1$ we have to take the analogous conditions in [3], p. 251.

Note that (1) implies that each weight λ of V satisfies for all roots α

$$|\langle \lambda, \alpha^\vee \rangle| \leq \begin{cases} p^r - 3p^{r-1} - 3, & \text{if } R \text{ is of type } G_2; \\ p^r - 2p^{r-1} - 2, & \text{otherwise.} \end{cases} \tag{2}$$

Well, we can replace λ in (2) by any conjugate of λ under the Weyl group W. So we can assume that λ is dominant. We have then

$$0 \leq \langle \lambda, \alpha^\vee \rangle \leq \langle \lambda, \alpha_0^\vee \rangle$$

for all $\alpha \in R^+$ and can apply (1).

We want to show first: If λ is a weight of V and if α is a root with $\lambda - p^i\alpha \in (q - \sigma)X$ for some integer i with $0 \leq i < r$, then $\lambda - p^i\alpha = 0$. This will imply the isomorphism condition in [6], 5.5 or in [3] for $n = 1$ (where only positive roots are considered). Well, suppose that $\lambda - p^i\alpha = (q - \sigma)\nu$ with $\nu \neq 0$. Then there is a root β with $\langle \nu, \beta^\vee \rangle \neq 0$. Choose β such that $|\langle \nu, \beta^\vee \rangle|$ is maximal for $\beta \in R$. We can replace β by $-\beta$ and assume that $\langle \nu, \beta^\vee \rangle > 0$. We have then

$$\langle (p^r - \sigma)\nu, \beta^\vee \rangle = p^r \langle \nu, \beta^\vee \rangle - \langle \nu, \sigma^{-1}\beta^\vee \rangle \geq (p^r - 1)\langle \nu, \beta^\vee \rangle$$

by the choice of β. Now $\langle \nu, \beta^\vee \rangle > 0$ implies

$$p^r - 1 \leq \langle \lambda - p^i\alpha, \beta^\vee \rangle.$$

We have $|\langle \alpha, \beta^\vee \rangle| \leq 2$ (resp. ≤ 3 for type G_2) and $p^i \leq p^{r-1}$, hence

$$\langle \lambda, \beta^\vee \rangle \geq \begin{cases} p^r - 3p^{r-1} - 1, & \text{for } G_2; \\ p^r - 2p^{r-1} - 1, & \text{otherwise.} \end{cases}$$

This contradicts (2). We get therefore $\nu = 0$ and $\lambda = p^i\alpha$.

We show similarly: If λ is a weight of V with $\lambda \in (q - \sigma)X$, then $\lambda = 0$. This yields the isomorphism condition in [6], 5.5 (resp. [3]) for $n = 0$.

Let us now turn to the injectivity condition for $n = 2$ in [6], 5.4 (or [3]). There are two cases to be considered. Suppose first that λ is a weight of V and that α is a root with $\lambda = p^j \alpha$ for some $j > 0$. Then $\lambda \neq 0$ since $\alpha \neq 0$. We can suppose that λ is dominant (using conjugation under the Weyl group) and get then

$$\langle \lambda, \alpha_0^\vee \rangle = p^j \langle \alpha, \alpha_0^\vee \rangle \geq p^j$$

hence $p^j < q$ by (1). This is even stronger than the condition in [6], 5.4 where only $p^j \leq q$ is required.

Suppose now that λ is a weight of V and that α, β are roots with $\alpha \neq -\beta$ and $\lambda = p^i \alpha + p^j \beta$ for some $i, j \geq 0$. We have to show that $i, j < r$. Suppose first that $i = j$. Then $\lambda = p^i(\alpha + \beta) \neq 0$; as above we can suppose that λ is dominant and get $\langle \lambda, \alpha_0^\vee \rangle \geq p^i$, hence $i < r$ by (1), as desired. Suppose now that say $i > j$ (hence $i \geq 1$). We can suppose (using conjugation under the Weyl group) that α is dominant, hence $\langle \alpha, \alpha_0^\vee \rangle \geq 2$. We get

$$2p^i \leq \langle p^i \alpha, \alpha_0^\vee \rangle = \langle \lambda, \alpha_0^\vee \rangle - p^j \langle \beta, \alpha_0^\vee \rangle \leq \langle \lambda, \alpha_0^\vee \rangle + \begin{cases} 3p^{i-1}, & \text{for } G_2; \\ 2p^{i-1}, & \text{otherwise.} \end{cases}$$

This yields

$$\langle \lambda, \alpha_0^\vee \rangle \geq \begin{cases} p^i + (p-3)p^{i-1}, & \text{for } G_2; \\ p^i + (p-2)p^{i-1}, & \text{otherwise.} \end{cases}$$

Now (1) yields $i < r$.

Corollary 2.3. *Let λ, μ be dominant weights with*

$$\langle \lambda + \mu, \alpha_0^\vee \rangle \leq \begin{cases} p^r - 3p^{r-1} - 3, & \text{if } R \text{ is of type } G_2; \\ p^r - 2p^{r-1} - 2, & \text{otherwise.} \end{cases} \tag{1}$$

Then the natural restriction map

$$\text{Ext}_G^1(L(\mu), L(\lambda)) \to \text{Ext}_{G(q)}^1(L(\mu), L(\lambda))$$

is an isomorphism.

Proof. This is an immediate consequence of the proposition and the isomorphism

$$\text{Ext}_G^1(L(\mu), L(\lambda)) \simeq H^1(G, L(\mu)^* \otimes L(\lambda))$$

and its analogue for $G(q)$. We use also: If $L(\mu)^* \simeq L(\mu^*)$, then

$$\langle \mu^*, \alpha_0^\vee \rangle = \langle \mu, \alpha_0^\vee \rangle.$$

Remark. For $p \geq 3(h-1)$ (where h is the Coxeter number) the corollary follows also from Theorem 2.8 in [1] in the split case and from Theorem 3.4(i) in [1] in the twisted case. In fact, one can then replace the right hand side in (1) by the (better) bound $p^r - p^{r-1} - 2$.

Proposition 2.4. *Suppose that R is not of type A_1. Each $kG(p^r)$-module of dimension $\leq p$ is semisimple.*

Proof. Set $q = p^r$. We have to show: Whenever L and L' are simple $kG(q)$–modules with $\dim L + \dim L' \leq p$, then $\text{Ext}^1_{G(q)}(L, L') = 0$. We can assume that L and L' are simple G–modules with a highest weight in X_r. We can also assume that not both L and L' are trivial. (We have excluded all cases where $G(q)$ is not equal to its derived group.) We know by Lemma 1.8 that $H^1(G, L \otimes L') = 0$. So we want to show that the natural restriction map from $H^1(G, L \otimes L') = 0$ to $H^1(G(q), L \otimes L')$ is an isomorphism. This follows from Lemma 2.1 and Proposition 2.2.

Remark. The corresponding statement does not hold for type A_1 since there

$$\text{Ext}^1_{G(p)}(L(r\omega_1), L((p - r - 3)\omega_1)) \neq 0 \qquad \text{for all } r \text{ with } 0 \leq r \leq p - 3,$$

cf. [2], p. 49. (The dimensions add up to $p - 1$.) On the other hand, for $r > 1$ Corollary 4.5 in [2] implies that a nonsplit extension of two simple $kG(p^r)$–modules has dimension at least $p + 1$. The restriction of the module $H^0(p\omega_1)$ (mentioned in 1.9, Remark 2) to $G(p^r)$ yields an example of such an extension.

3. The General Case

3.1. We want to extend our results to arbitrary reductive groups. Let k be again an algebraically closed field of characteristic $p \neq 0$. Consider for any algebraic group G over k and any integer n the property:
(Sn) Each rational G–module of dimension $\leq n$ is semisimple.

Lemma. (a) *If G satisfies* (Sn) *and if H is a normal (closed) subgroup of G, then G/H satisfies* (Sn).
(b) *If G_1 and G_2 are algebraic groups over k satisfying* (Sn), *then their direct product satisfies* (Sn).
(c) *Let G be an algebraic group over k and H a closed subgroup of G that has finite index in G. If H satisfies* (Sn) *and if the index $(G : H)$ is prime to p, then G satisfies* (Sn).

Proof. (a) This is obvious.
(b) This follows from the following (well known) result: If a finite dimensional $(G_1 \times G_2)$–module M is semisimple as a G_1–module and as a G_2–module, then M is semisimple as a $(G_1 \times G_2)$–module. Since it seems to be difficult to find a reference, let me indicate a proof:
    The image of the group algebra kG_i in $\text{End}_k(M)$ is a semisimple k–algebra; denote it by A_i. Then the action of the group algebra $k(G_1 \times G_2)$ on M factors through $A_1 \otimes_k A_2$. This is again a semisimple algebra, so M is a semisimple as a $(G_1 \times G_2)$–module.
(c) This follows from the more general result: Any G–module that is semisimple over H is semisimple over G. Indeed, we can construct a G–linear projection on a submodule from an H–linear one: One averages using a system of representatives for G/H. (Again this is well known, but it is difficult to find a convenient reference. For G finite one can quote [7], I.19.5(ix). It says that each G–module is (G, H)–projective which easily implies the result.)

Proposition 3.2. *If G is a connected reductive algebraic group over k, then any rational G-module of dimension $\leq p$ is semisimple.*

Proof. The claim holds by Proposition 1.9 for G almost simple and simply connected. It then follows by Lemma 3.1.b for direct products of such groups, i.e., for G semisimple and simply connected. By Lemma 3.1.a it holds also for homomorphic images of these groups, hence for all G semisimple.

Any reductive G is a homomorphic image of the direct product of its derived group (which is semisimple and connected) with its connected center (which is a torus). Each rational module for a torus is semisimple. So we get the claim for arbitrary G using 3.1.b/a again.

Remark. Using Lemma 3.1.c one can extend the result to non-connected G if $(G : G^0)$ is prime to p.

Proposition 3.3. *Let G be a connected reductive algebraic group over k defined over a finite field $\mathbf{F}_q \subset k$. If G has no component of type A_1, then any $kG(\mathbf{F}_q)$-module of dimension $\leq p$ is semisimple.*

Proof. Suppose for the sake of simplicity that k is the algebraic closure of $\mathbf{F}_q$. Let F be the Frobenius endomorphism of G with respect to $\mathbf{F}_q$ so that $G(\mathbf{F}_q)$ is the group G^F of fixed points of F in G.

Consider first the case where G is semisimple and simply connected. Then G is the direct product of almost simple groups that are again simply connected. The Frobenius endomorphism permutes these factors. The product over all factors in an orbit under F is defined over $\mathbf{F}_q$. By Lemma 3.1.b (now applied to finite groups regarded as algebraic groups) it is enough to prove the result for each of these products. So we may assume that F permutes the almost simple factors transitively. In that case suppose that there are r factors and that G_1 is one on them. The other factors are then the $F^i G_1$ with $0 < i < r$, we have $F^r G_1 = G_1$ and G^F is isomorphic to $G_1^{F^r}$. We can regard G_1 as defined over $\mathbf{F}_{q^r}$ with Frobenius endomorphism F^r. Now $G_1(\mathbf{F}_{q^r})$ satisfies the claim by Proposition 2.4, hence so does the isomorphic group $G(\mathbf{F}_q)$.

Assume now that G is semisimple, but not necessarily simply connected. Then there is a simply connected semisimple group G_0 defined over $\mathbf{F}_q$ with an isogeny (defined over $\mathbf{F}_q$) onto the derived group of G, cf. [14], 2.6.1. Now $G_0(\mathbf{F}_q)$ satisfies the claim, hence its image (say H) in $G(\mathbf{F}_q)$. Lemma 3.1.c implies then that it also holds for $G(\mathbf{F}_q)$, since the index $(G(\mathbf{F}_q) : H)$ is prime to p. [Since $G_0(\mathbf{F}_q)$ and $G(\mathbf{F}_q)$ have the same order, cf. [4], 16.8, this index is equal to the order of the kernel of the map from $G_0(\mathbf{F}_q)$ to $G(\mathbf{F}_q)$ which is contained in a maximal torus of G_0, hence has order prime to p.]

Consider now the general case. The derived group (say G_1) of G and the connected center (say G_2) of G are both defined over $\mathbf{F}_q$. There is an obvious central isogeny (defined over $\mathbf{F}_q$) from $G_1 \times G_2$ onto G. Arguing as in the preceding paragraph we see that it is enough to prove the claim for $G_1 \times G_2$. By Lemma 3.1.b it suffices to look at the two factors. We have proved the claim for $G_1(\mathbf{F}_q)$ above. On the other hand, G_2 is a torus, so $G_2(\mathbf{F}_q)$ has order prime to p and all $kG_2(\mathbf{F}_q)$-modules are semisimple.

Remark. If G has a component of type A_1, then one can prove a similar result with p replaced by $p - 2$.

REFERENCES

1. H. H. Andersen, *Extensions of simple modules for finite Chevalley groups*, J. Algebra **111** (1987), 388–403.
2. H. H. Andersen, J. Jørgensen, and P. Landrock, *The projective indecomposable modules of* $SL(2, p^n)$, Proc. London Math. Soc. (3) **46** (1983), 38–52.
3. G. Avrunin, *Generic cohomology for twisted groups*, Trans. Amer. Math. Soc. **268** (1981), 247–253.
4. A. Borel, *Linear Algebraic Groups*, Benjamin, 1969, [2nd ed.: Graduate Texts in Mathematics, vol. 126, New York, 1991 (Springer)].
5. N. Bourbaki, *Groupes et algèbres de Lie*, Hermann, Paris, 1968, Chap. 4, 5 et 6.
6. E. Cline, B. Parshall, L. Scott, and W. van der Kallen, *Rational and generic cohomology*, Invent. math. **39** (1977), 143–163.
7. C. W. Curtis and I. Reiner, *Methods of Representation Theory with Applications to Finite Groups and Orders*, vol. I, Wiley, New York, 1981.
8. J. C. Jantzen, *Representations of Algebraic Groups*, Pure and Applied Mathematics, vol. 131, Academic Press, Orlando, Fla., 1987.
9. _____, *First cohomology groups for classical Lie algebras*, Representation Theory of Finite Groups and Finite-Dimensional Algebras (Basel etc.) (G. O. Michler and C. M. Ringel, eds.), Proc. Bielefeld, vol. 95, Birkhäuser, Basel etc., 1991, pp. 289–315.
10. P. Kleidman and M. Liebeck, *The Subgroup Structure of the Finite Classical Groups*, London Mathematical Society Lecture Notes Series, vol. 129, Cambridge Univ. Press, Cambridge etc., 1990.
11. M. Larsen, *On the semisimplicity of low-dimensional representations of semisimple groups in characteristic p*, J. Algebra **173** (1995), 219–236.
12. G. M. Seitz, *The maximal subgroups of classical algebraic groups*, Memoirs Amer. Math. Soc. **67** (1987), no. 365.
13. J. P. Serre, *Sur la semi-simplicité des produits tensoriels de représentations de groupes*, Invent. math. **116** (1994), 513–530.
14. J. Tits, *Classification of algebraic semisimple groups*, Algebraic Groups and Discontinuous Subgroups (A. Borel and G. D. Mostow, eds.), Proc. Symp. Pure Math., vol. 9, Amer. Math. Soc., Providence, 1966, pp. 33–62.

DEPARTMENT OF MATHEMATICS, UNIVERSITY OF OREGON, EUGENE, OR 97403, U. S. A.
CURRENT ADDRESS:
MATEMATISK INSTITUT, AARHIS UNIVERSITET, 8000 AARHUS C, DENMARK

GROSSES CELLULES POUR LES VARIÉTÉS SPHÉRIQUES

D. LUNA

À la mémoire de Roger W. Richardson

Soit G un groupe réductif connexe (le corps de base k étant algébriquement clos et de caractéristique nulle), et soit B un sous- groupe de Borel de G. Une G-variété algébrique X est appelée "sphérique", si X est normale et irréductible, et si B a une orbite ouverte (et donc dense) dans X. Un sous-groupe algébrique H de G est appelé "sphérique", si l'espace homogène G/H l'est. Par exemple, les sous-groupes paraboliques et les sous-groupes symétriques (ceux qui sont composés des points fixes d'un automorphisme involutif) sont sphériques.

Pour chaque groupe réductif G se pose le problème de la classification (à conjugaison près) de ses sous-groupes sphériques, en termes combinatoires (c'est-à-dire essentiellement en termes du système de racines de G).

Soit X une G-variété algébrique. On dit que X est "sobre" (mot employé ici comme synonyme de simple), si X est une variété complète et si G n'a qu'une seule orbite fermée dans X. On dit que X est "magnifique", si X est sphérique sobre et si tout diviseur irréductible de X qui est stable par B et qui contient une orbite de G, est stable par G. Un sous-groupe algébrique H de G est appelé "sobre", si le groupe $N_G(H)/H$ est fini.

Tous les sous-groupes d'isotropie d'une variété sphérique sobre sont sphériques sobres ; inversement, quel que soit le sous-groupe sphérique sobre H de G, l'espace homogène G/H peut être plongé comme orbite ouverte dans une G-variété magnifique X, unique à isomorphisme près (ces résultats sont en grande partie dus à F. Pauer, voir [10] et [14]) ; on appelle ce X le "plongement magnifique" de G/H. Il s'ensuit que classifier les sous-groupes sphériques sobres de G (à conjugaison près), revient à classifier les G-variétés magnifiques (à isomorphisme près).

Cela étant rappelé, voici un résumé du contenu de cet article. Au §1, pour toute G-variété sphérique sobre X et pour tout sous- groupe parabolique Q de G, on introduit une "grosse cellule" X_Q, sous-ensemble ouvert dans X et stable par Q (généralisant la grosse cellule de la "décomposition de Bruhat" de G en doubles classes sous Q), puis on étudie la structure de ces grosses cellules. Au §2, on propose une manière pour attaquer le problème de la classification des sous-groupes sphériques. On explique d'abord pourquoi, lorsqu'on veut classifier tous les sous-groupes sphériques, il faut commencer par ceux qui sont sobres. On introduit ensuite pour tout sous-groupe sphérique sobre H des données combinatoires (le "triplet" de H). On montre quelques propriétés de ces triplets (s'appuyant pour cela sur les résultats du §1), puis on explique le rôle qu'ils semblent pouvoir jouer dans la classification des sous-groupes sphériques sobres (cette partie n'est qu'esquissée). Enfin au §3, on établit des liens entre le formalisme des triplets, et les résultats des articles [5] et [8] concernant les diviseurs et la dualité entre

267

courbes et diviseurs dans les variétés sphériques (voir aussi [3]). La plupart des
énoncés du §3 m'ont été fournis par M. Brion.
Des conversations avec Th. Vust, F. Pauer, M. Brion et F. Knop (et la lecture
de leurs articles) m'ont fait beaucoup avancer dans la compréhension des variétés
sphériques ; je voudrais ici les en remercier.

1. GROSSES CELLULES.

Soient G un groupe réductif connexe, T un tore maximal de G, et B un sous-
groupe de Borel de G contenant T. On note $\Xi(T)$ le groupe des caractères de T.
Soit $R \subset \Xi(T)$ le système des racines de G, et soit S la base de R associée à B.
Soit $S' \subset S$. On désigne par $G_{S'}$ le sous-groupe parabolique de G contenant B
et associé à S', et par $G_{-S'}$ le sous-groupe parabolique opposé à $G_{S'}$ contenant T.
On pose $G^{S'} = G_{S'} \cap G_{-S'}$; c'est un sous-groupe de Levi de $G_{S'}$ et de $G_{-S'}$. On
désigne par $C^{S'}$ le centre connexe de $G^{S'}$. En particulier, on a $B = G_\emptyset$, $B_- = G_{-\emptyset}$
est le sous-groupe de Borel de G opposé à B contenant T, et $C^\emptyset = T$.
Toute G-variété algébrique sobre normale est projective (c'est une conséquence
d'un résultat bien connu de Sumihiro, voir [15]). Rappelons aussi que B n'a qu'un
nombre fini d'orbites dans toute G-variété sphérique (voir [4], [19], [18] ou [12]).

1.1. Soit X une G-variété sphérique sobre. Désignons par z l'unique point fixe
de B_- dans X, et par $Z = G \cdot z$ l'unique orbite fermée de G dans X.
Pour tout $S' \subset S$, on pose $X_{S'} = \{x \in X, \overline{G_{S'} \cdot x} \supset Z\}$. On appellera ce
sous-ensemble de X la "grosse cellule" de X associée à S'. L'ensemble $X_{S'}$ est
ouvert dans X, est stable par $G_{S'}$, et $G_{S'} \cdot z$ est l'unique orbite fermée de $G_{S'}$ dans
$X_{S'}$; inversement, ces propriétés caractérisent $X_{S'}$. On notera $X^{S'}$ l'ensemble
des points fixes de $C^{S'}$ dans $X_{S'}$; c'est une sous-$G^{S'}$-variété de X.
L'ensemble $X_\emptyset = \{x \in X, \overline{B \cdot x} \supset Z\}$ (qu'on note aussi X_B) est déjà bien connu
dans la théorie des variétés sphériques : il s'agit de la "carte affine canonique"
de X en Z (voir par exemple [5], p. 398) ; dans ce cas $X^\emptyset = \{z\}$. Pour S'
quelconque, on vérifie sans peine que $X_{S'} = G_{S'} \cdot X_B$.

1.2. Soit X une G-variété algébrique complète et irréductible. On note $\Xi(T)$
l'ensemble des morphismes de groupes algébriques $\lambda : k^* \to T$. À tout $\lambda \in$
$\Xi_*(T)$, on peut associer (voir [2]) une "décomposition cellulaire" de X : c'est la
décomposition de X en réunion disjointe des X_Y, où Y parcourt l'ensemble des
composantes connexes de $X^{\lambda(k^*)}$ et où

$$X_Y = \{x \in X, \lim_{t \to 0} \lambda(t)x \in Y\}.$$

Désignons par X^λ la "cuvette" (en anglais : sink) de cette décomposition (c'est-
à-dire la composante connexe Y de $X^{\lambda(k^*)}$ telle que X_Y est ouvert dans X), et
posons $X_\lambda = X_{X^\lambda}$ (la "cellule ouverte" de la décomposition). Notons G^λ le
commutant de $\lambda(k^*)$ dans G, et posons

$$G_\lambda = \{g \in G, \lim_{t \to 0} \lambda(t)g\lambda(t)^{-1} \in G^\lambda\}.$$

Le groupe G_λ laisse stable X_λ et le groupe G^λ laisse stable X^λ. Notons π_λ
à la fois le morphisme de variétés algébriques $X_\lambda \to X^\lambda$ défini par $\pi_\lambda(x) =$

$\lim_{t\to 0} \lambda(t)x$, $(x \in X_\lambda)$, et aussi le morphisme de groupes algébriques $G_\lambda \to G^\lambda$ défini par $\pi_\lambda(g) = \lim_{t\to 0} \lambda(t)g\lambda(t)^{-1}$, $(g \in G_\lambda)$; on a $\pi_\lambda(g \cdot x) = \pi_\lambda(g) \cdot \pi_\lambda(x)$ quels que soient $g \in G_\lambda$ et $x \in X_\lambda$.

On dira qu'un $\lambda \in \Xi_*(T)$ est adapté à $S' \subset S$, s'il vérifie $\langle \lambda, \alpha \rangle > 0$ quel que soit $\alpha \in S \setminus S'$ et $\langle \lambda, \alpha \rangle = 0$ quel que soit $\alpha \in S'$. Pour tout λ adapté à S', on a $G_\lambda = G_{S'}$ et $G^\lambda = G^{S'}$.

1.3. Soit X une G-variété sphérique sobre, et soit $S' \subset S$.

Proposition. (1) *Pour tout* $\lambda \in \Xi_*(T)$ *adapté à* S', *on a* $X_\lambda = X_{S'}$ *et* $X^\lambda = X^{S'}$.

(2) *Tout élément de* $G^u_{-S'}$ *(le radical unipotent de* $G_{-S'}$*) laisse fixe tout élément de* $X^{S'}$.

(3) *La* $G^{S'}$-*variété* $X^{S'}$ *est sphérique et sobre.*

(4) *Si* X *est une* G-*variété magnifique, alors* $X^{S'}$ *est une* $G^{S'}$-*variété magnifique.*

Preuve. Soit $\lambda \in \Xi_*(T)$ adapté à S'. Commençons par montrer que X^λ est une variété normale. Pour tout $x \in X^\lambda$, on sait qu'il existe des voisinages ouverts affines V dans X_λ, stables par k^* (où k^* opère dans X_λ à travers λ). Il suffit de montrer que V^λ est normale. L'opération de k^* dans V se reflète en une graduation $k[V] = \bigoplus_{n \in \mathbb{Z}} k[V]_n$. De $V \subset X_\lambda$, on déduit que $k[V]_n = 0$ pour $n < 0$, et que $k[V^\lambda]$ s'identifie à $k[V]_0$. Puisque X est normale, V l'est aussi, donc $k[V]$ est intégralement clos, d'où il suit que $k[V]_0 = k[V^\lambda]$ est intégralement clos, ce qui signifie que V^λ est normale.

Posons $B^\lambda = B \cap G^\lambda$; c'est un sous-groupe de Borel de G^λ et on a $\pi_\lambda(B) = B^\lambda$. Si $X_{(B)}$ désigne l'orbite ouverte de B dans X, alors $X_{(B)}$ est contenue dans X_λ, et $\pi_\lambda(X_{(B)})$ est une orbite ouverte de B^λ dans X^λ, ce qui montre que X^λ est une G^λ-variété sphérique.

On pose

$$G_{\lambda-1} = \left\{ g \in G, \lim_{t\to 0} \lambda(t)^{-1}g\lambda(t) \in G^\lambda \right\} ;$$

c'est un sous-groupe parabolique de G dont le radical unipotent est $G^u_{\lambda-1} = \left\{ g \in G, \lim_{t\to 0} \lambda(t)^{-1}g\lambda(t) = 1 \right\}$. Soit $g \in G^u_{-S'} = G^u_{\lambda-1}$ et soit $x \in X^\lambda$. On a manifestement $\lim_{t\to 0} \lambda^{-1}(t) \cdot (g \cdot x) = x$. Puisque X^λ est la "cuvette" de la décomposition de Bialynicki-Birula associée à λ, X^λ est aussi la "source" de la décomposition de Bialynicki-Birula associée à λ^{-1}. Il s'ensuit que $g.x = x$.

Montrons maintenant que X^λ est une G^λ-variété sobre. Il est clair que X^λ est une variété complète. Posons $B^\lambda_- = B_- \cap G^\lambda$; c'est un sous-groupe de Borel de G^λ et on a $B_- = G^u_{\lambda-1} \cdot B^\lambda_-$. De là et de ce qui précède, on déduit que tout point de X^λ fixé par B^λ_- est aussi fixé par B_-. Comme B_- n'a qu'un seul point fixe dans X, B^λ_- n'a qu'un seul point fixe dans X^λ. Il s'ensuit bien que G^λ n'a qu'une seule orbite fermée dans X^λ.

L'ensemble X_λ est ouvert dans X et est stable par $G_\lambda = G_{S'}$. Puisque X^λ est une G^λ-variété sobre, on voit que $G_\lambda \cdot z$ est la seule orbite fermée de G_λ dans X_λ.

Il s'ensuit que $X_\lambda = X_{S'}$ et que $X^\lambda \supset X^{S'}$. Puisque X^λ est une G^λ-variété sobre, $C^{S'}$ (qui est le centre connexe de G^λ) ne peut opérer que trivialement dans X^λ (sinon, puisque X^λ est une variété projective, l'ensemble des points de X^λ fixés par $C^{S'}$ posséderait au moins deux composantes connexes contenant des orbites fermées de G^λ distinctes, ce qui n'est pas possible puisque X^λ est une G^λ-variété sobre). Il s'ensuit que $X^\lambda = X^{S'}$.

Enfin, supposons que X est une G-variété magnifique. Soit D' un diviseur de X^λ stable par B^λ. Notons D l'adhérence dans X de $\pi_\lambda^{-1}(D')$; c'est un diviseur de X stable par B tel que $D \cap X^\lambda = D'$. Posons $Z^\lambda = Z \cap X^\lambda$ (où Z désigne l'unique orbite fermée de G dans X) ; c'est l'unique orbite fermée de G^λ dans X^λ. Si D' contient une orbite de G^λ, alors D' contient aussi Z^λ. Par suite, D contient $B \cdot Z^\lambda$ qui est dense dans Z, donc D contient Z, d'où il suit que D est stable par G. Il s'ensuit que D' est stable par G^λ, ce qui montre bien que X^λ est une G^λ-variété magnifique.

La proposition précédente montre en particulier que les cellules ouvertes X_λ des décompositions cellulaires associées aux $\lambda \in \Xi_*(T)$ adaptés à S' ne dépendent pas de λ (car elles coïncident avec $X_{S'}$). De même, les morphismes $\pi_\lambda : X_\lambda = X_{S'} \to X^\lambda = X^{S'}$ sont indépendants du choix de λ (et on peut par conséquent les noter $\pi_{S'} : X_{S'} \to X^{S'}$) : en effet, pour tout $x \in X_{S'}, \pi_{S'}(x)$ est l'unique point de $X^{S'}$ adhérent à $C^{S'} \cdot x$.

1.4. Terminons ce § par une variante des résultats de 1.3, valable pour toute variété sphérique (non nécessairement magnifique). Rappelons qu'une G-variété sphérique X est dite "toroïdale", si tout diviseur irréductible de X qui est stable par B et qui contient une orbite de G, est stable par G.

Pour tout groupe algébrique H, notons H^r le radical de H.

Proposition. *Soit X une G-variété sphérique (pouvant être singulière et incomplète), soit Q un sous-groupe parabolique de G, et soit Y une composante irréductible de X^{Q^r}. Alors Y est une Q/Q^r-variété sphérique (en particulier, Y est une variété normale) ; de plus, si X est toroïdale, Y l'est aussi.*

Preuve. L'énoncé étant "local", on peut se ramener au cas où X est projective. De plus, toute sous-G-variété irréductible de X étant sphérique, on peut supposer que $G \cdot Y$ est dense dans X. Enfin, quitte à remplacer Q par l'un de ses conjugués dans G, on peut supposer que $Q = G_{S'}$, avec $S' \subset S$.

Soit $\lambda \in \Xi_*(T)$ adapté à S'. Puisque pour tout $g \in G_{S'}, \lim_{t \to 0} \lambda(t) g \lambda(t)^{-1} \in G^{S'} \subset G_{-S'}$, on a $\lim_{t \to 0} \lambda(t) \cdot x \in G^{S'} \cdot Y \subset Y$, quel que soit $x \in G_{S'} \cdot Y$. Comme $G_{S'} \cdot Y = G_{S'} G_{-S'} \cdot Y$ est dense dans X, on en déduit que $Y = X^\lambda$, la "cuvette" de la décomposition de Bialynicki-Birula de X associée à λ. On raisonne alors comme dans la première partie de la preuve de 1.3 pour montrer que Y est normale et sphérique, et comme dans la dernière partie de la preuve de 1.3 pour montrer que X toroïdale implique Y toroïdale.

2. Triplet associé à un sous-groupe sphérique.

Conservons les notations du §1. Rappelons que G est un groupe réductif connexe, que B est un sous-groupe de Borel de G, et que T est un tore maximal de B. Un sous-groupe de G qui contient le centre de G, sera appelé un sous-groupe "adjoint" de G.

2.1. Soit H un sous-groupe sphérique de G. Quitte à remplacer H par l'un de ses conjugués dans G, on peut supposer BH ouvert dans G, ce que nous ferons désormais. On note $\Delta_{G/H}$ l'ensemble des orbites de codimension 1 de B dans G/H ; on appelle $\Delta_{G/H}$ l'ensemble des couleurs de H (et de G/H).

Le groupe $N_G(H)/H = \mathrm{Aut}_G(G/H)$ opère dans $\Delta_{G/H}$. Si cette opération est fidèle, on dit que H est un sous-groupe (sphérique) "très sobre" de G. Comme $\Delta_{G/H}$ est un ensemble fini, la composante connexe neutre $(N_G(H)/H)^0$ opère trivialement dans $\Delta_{G/H}$. Il s'ensuit que tout sous-groupe sphérique très sobre est sobre ; par ailleurs tout tel sous-groupe est visiblement aussi adjoint (mais un sous-groupe sphérique sobre adjoint n'est pas toujours très sobre, comme le montre l'exemple $G = PGL_n$ et $H = PSO_n$, n pair et ≥ 4).

Si H est un sous-groupe sphérique de G, désignons par $\overline{H}$ le sous-groupe des éléments de $N_G(H)$ qui opèrent trivialement dans $\Delta_{G/H}$. Il est clair que le morphisme naturel $G/H \to G/\overline{H}$ induit une bijection $\Delta_{G/H} \leftrightarrow \Delta_{G/\overline{H}}$, et que $\overline{H}$ est un sous-groupe sphérique très sobre de G (en fait c'est le plus petit sous-groupe sphérique très sobre de G contenant H). On appelle $\overline{H}$ la clôture très sobre de H.

On voit que le problème de la classification des sous-groupes sphériques se scinde en deux : il faut classifier les sous-groupes sphériques très sobres ; puis, pour chaque sous-groupe sphérique très sobre H, il faut classifier les sous-groupes sphériques ayant H comme clôture très sobre.

Puisque pour tout sous-groupe sphérique H de G, $\overline{H}/H$ est diagonalisable (voir [10]), le deuxième problème est plus facile que le premier. Par exemple, les sous-groupes sphériques de G dont la clôture très sobre est un sous-groupe parabolique de G, sont ceux qui contiennent un sous-groupe unipotent maximal de G (pour une description de ces sous-groupes "horosphériques", voir [13]).

Dans la suite de cet article, nous n'allons considérer que le premier problème.

2.2. Désignons par U le radical unipotent de B. Si M est un T-module, notons $\Xi(T, M)$ l'ensemble des $\gamma \in \Xi(T)$ qui sont caractères d'un vecteur propre de T dans M.

Soit X une variété sphérique de G. Désignons par $k(X)$ le corps des fonctions rationnelles sur X, et par $k(X)^U$ le sous-corps des invariants de U dans $k(X)$. Posons $\Xi_X = \Xi(T, k(X)^U)$; c'est un sous-réseau de $\Xi(T)$. Le rang de Ξ_X s'appelle le rang de (la variété sphérique) X. Ce qui précède s'applique en particulier à G/H, où H est un sous-groupe sphérique de G. Le rang de $\Xi_{G/H}$ s'appelle alors aussi le corang de H dans G.

Supposons X sphérique et sobre, et désignons par Z l'unique orbite fermée de G dans X. Rappelons que $X_B = \{x \in X, \overline{B \cdot x} \supset Z\}$ est la "carte canonique" de X ; c'est un ouvert affine, dont on connaît bien la structure (voir [9]). Posons

$\Xi_X^- = \Xi(T, k[X_B^U])$ (où $k[X_B]$ désigne l'algèbre des fonctions régulières sur X_B).
Si H est un sous-groupe sphérique sobre de G, posons $\Xi_{G/H}^- = \Xi_X^-$, où X est
le plongement magnifique de G/H. On sait que $\Xi_{G/H}^-$ engendre dans l'espace
vectoriel $\Xi_{G/H} \otimes_{\mathbb{Z}} \mathbb{Q}$ un cône convexe simplicial d'intérieur non vide $C_{G/H}$, et que
$\Xi_{G/H}^- = C_{G/H} \cap \Xi_{G/H}$ (voir [7] et [14]).

2.3. Si X est une G-variété magnifique de rang 1, alors Ξ_X est (comme groupe)
isomorphe à $\mathbb{Z}$. Désignons par γ_X le générateur de Ξ_X tel que $-\gamma_X \in \Xi_X^-$.

On dit qu'une G-variété X est adjointe, si le centre de G opère trivialement
dans X. Désignons par Σ_G l'ensemble des γ_X, où X parcourt les différentes G-
variétés magnifiques adjointes de rang 1. On appelle Σ_G l'ensemble des racines
sphériques de G.

On connaît tous les sous-groupes sphériques sobres de corang 1 des groupes
réductifs (voir [1] et [6]), donc on connaît aussi toutes les G-variétés magnifiques
de rang 1. En particulier, on connaît Σ_G pour tout groupe réductif G ; c'est un
ensemble fini (contenu, par définition, dans Ξ_R, le réseau radiciel de G).

À titre d'exemple, explicitons le cas du groupe SL_{n+1}. On utilisera la
numérotation usuelle $\alpha_1, \alpha_2, \ldots, \alpha_n$ des racines simples de SL_{n+1}. L'ensemble
$\Sigma_{SL_{n+1}}$ comprend alors les éléments suivants :

(1) les $\alpha_i + \alpha_{i+1} + \cdots + \alpha_{i+j}$ (pour $1 \le i$ et $0 \le j \le n-i$) ;
(2) les $2\alpha_i$ (pour $1 \le i \le n$) ;
(3) les $\alpha_i + \alpha_{i+j}$ (pour $1 \le i$ et $2 \le j \le n-i$) ;
(4) les $\alpha_i + 2\alpha_{i+1} + \alpha_{i+2}$ (pour $1 \le i \le n-2$).

Attention, Σ_G n'est pas contenu dans R. Mais on sait que toute racine sphérique
γ peut s'écrire comme somme d'éléments de S (voir [7]). On notera S_γ l'ensemble
des racines de S nécessaires pour exprimer γ.

2.4. Soit H un sous-groupe sphérique sobre adjoint de G. On va associer à H un
sous-ensemble $\Sigma_{G/H}$ de Σ_G, qu'on appellera l'ensemble des "racines sphériques"
de H.

Soit X le plongement magnifique de G/H. Notons r le rang de X (qui est
aussi le corang de H dans G). On sait que X contient exactement r sous-variétés
magnifiques $X_1, \ldots, X_r$ de rang 1 (voir par exemple [14]). On pose $\Sigma_{G/H} =$
$\{\gamma_1, \ldots, \gamma_r\}$, où $\gamma_1, \ldots, \gamma_r \in \Sigma_G$ sont les racines sphériques associées à $X_1, \ldots, X_r$.

On peut caractériser $\Sigma_{G/H}$ aussi comme l'ensemble des éléments primitifs $\gamma \in$
$\Xi_{G/H}$ tels que $-\gamma$ est situé sur une arête de $C_{G/H}$ (où $C_{G/H}$ est le cône introduit au
n° 2.2). De la structure bien connue de X_B résulte que X est lisse si et seulement
si le réseau engendré par $\Sigma_{G/H}$ coïncide avec $\Xi_{G/H}$.

Mentionnons ici l'article [11], où est introduit un système de racines (noté Δ_X
dans [11]), par une approche assez différente de celle qu'on suit ici, possédant
les propriétés suivantes : l'ensemble $\Sigma_{G/H}$ défini ci-dessus est une base de ce
système de racines, et si $N_G(H)/H$ est trivial, alors $\Xi_{G/H}$ est le réseau radiciel
de ce système de racines (d'où en particulier une preuve de la lissité de X, si
$N_G(H)/H$ est trivial).

2.5. Soit H un sous-groupe sphérique sobre adjoint de G. On vient d'associer à H son ensemble des couleurs $\Delta_{G/H}$ et son ensemble des racines sphériques $\Sigma_{G/H}$ (qui est un sous-ensemble de Σ_G). Définissons maintenant un accouplement naturel $\rho_{G/H} : \Delta_{G/H} \times \Sigma_{G/H} \to \mathbb{Z}$. Soit $D \in \Delta_{G/H}$ et soit $\gamma \in \Sigma_{G/H}$. On pose $\rho_{G/H}(D,\gamma) = v_D(f_\gamma)$, où v_D est la valuation naturelle du corps $k(G/H)$ qu'on peut associer à D, et où $f_\gamma \in k(G/H)^U$ est une fonction rationnelle (unique à constante multiplicative près), qui est vecteur propre de T de caractère γ.

On appellera $\Delta_{G/H}, \Sigma_{G/H}, \rho_{G/H}$ le "triplet" associé à H. Lorsqu'on veut interpréter ce triplet comme invariant **combinatoire** attaché à H, on considère $\Delta_{G/H}$ comme ensemble "abstrait" (oubliant qu'il est l'ensemble des orbites de codimension 1 de B dans G/H).

À titre d'exemple, considérons le cas particulier où $G = \mathbf{G} \times \mathbf{G}$, $\mathbf{G}$ étant un groupe semi-simple adjoint, et où H est la diagonale de $\mathbf{G} \times \mathbf{G}$. Dans ce cas, l'ensemble S (la base du système de racines de $\mathbf{G}$) est en bijection naturelle avec $\Delta_{G/H}$ et $\Sigma_{G/H}$, et modulo cette bijection, $\rho_{G/H}$ est donné par la matrice de Cartan de $\mathbf{G}$ (pour cela, et pour des résultats analogues concernant les espaces homogènes symétriques, voir [20]).

2.6. Soit H un sous-groupe sphérique sobre et adjoint de G. Désignons par X le plongement magnifique de G/H, et notons x_H le point $H/H \in G/H \subset X$.

Soit $S' \subset S$. Revenons aux notations $G^{S'}, G_{S'}, X^{S'}, X_{S'}$, et $\pi_{S'}$ du §1. Désignons par $H^{S'}$ le groupe d'isotropie de $G^{S'}$ en $\pi_{S'}(x_H)$ (comme on suppose BH ouvert dans G, on a $x_H \in X_{S'}$). Pour abréger, on utilisera aussi les notations G', H', X' pour $G^{S'}, H^{S'}, X^{S'}$ et on pose $x'_H = \pi_{S'}(x_H)$. D'après 1.3, H' est un sous-groupe sphérique sobre et adjoint de G', et X' est le plongement magnifique de G'/H'.

Le tore maximal T de G est également un tore maximal de G'. Notons R' le système des racines de G'. L'ensemble S' peut s'interpréter aussi comme la base de R' correspondante au sous-groupe de Borel $B' = B \cap G'$ de G'. Notons $\Delta_{G/H}(S')$ l'ensemble des éléments de $\Delta_{G/H}$ qui, considérés comme diviseurs de X, ne sont pas stables par $G_{S'}$.

Proposition. (1) On a $\Sigma_{G'/H'} = \Sigma_{G/H} \cap \Xi_{R'}$ (où $\Xi_{R'}$ désigne le réseau radiciel de R') ;

    (2) $\Delta_{G'/H'}$ est en bijection naturelle avec $\Delta_{G/H}(S')$;

    (3) modulo cette bijection, l'accouplement $\Sigma_{G'/H'} \times \Delta_{G'/H'} \to \mathbb{Z}$ est la restriction de l'accouplement $\Sigma_{G/H} \times \Delta_{G/H} \to \mathbb{Z}$.

Preuve. Soit X_B la carte canonique de X. Soit $\lambda \in \Xi_*(T)$ adapté à S'. Puisque $X_B \subset X_\lambda$, on a $\langle \lambda, \gamma \rangle \geq 0$ quel que soit $\gamma \in \Xi(T, k[X_B]^U)$. On sait que $X_B \cong P^u \times W$, où W est une T-variété, et où P est le sous-groupe parabolique de G donné par $P = \{g \in G, gBH = BH\}$ (voir [9]). Soit $X'_{B'}$ la carte canonique de X'. Posons $U' = U \cap G'$; c'est le radical unipotent de B'. On a $X'_{B'} = X' \cap X_B$, d'où il suit que $X'_{B'} \cong (P^u \cap G') \times W^{\lambda(k^*)}$. On en déduit que

$$\Xi(T, k[X'_{B'}]^{U'}) = \Xi(T, k[W^{\lambda(k^*)}]) =$$
$$\{\gamma \in \Xi(T, k[W]), \langle \lambda, \gamma \rangle = 0\} = \{\gamma \in \Xi(T, k[X_B]^U), \langle \lambda, \gamma \rangle = 0\}.$$

De cela et de la caractérisation des racines sphériques donnée au n° 2.4, il résulte que $\Sigma_{G'/H'}$ s'identifie à l'ensemble des $\gamma \in \Sigma_{G/H}$ tels que $\langle \lambda, \gamma \rangle = 0$ (condition qui signifie : $\gamma \in \Xi_{R'}$).

Pour tout $D' \in \Delta_{G'/H'}$ (considéré comme diviseur B'-stable et non G'-stable de X'), l'adhérence dans X de $\pi_{S'}^{-1}(D')$ est un diviseur B-stable et non $G_{S'}$-stable de X, d'où un $D \in \Delta_{G/H}(S')$. L'application inverse est celle qui envoie $D \in \Delta_{G/H}(S')$ sur $D' = D \cap X' \in \Delta_{G'/H'}$ (voir 1.3).

Soit $D' \in \Delta_{G'/H'}$ et notons $D \in \Delta_{G/H}(S')$ l'élément correspondant. Soit $\gamma \in \Sigma_{G'/H'}$ et soit $f_{-\gamma} \in k[X'_{B'}] \subset k(X') = k(G'/H')$ un vecteur propre de B' de valeur propre $-\gamma$. Alors $f_{-\gamma} \circ \pi_{S'}$ est une fonction rationnelle sur G/H, vecteur propre de B de valeur propre $-\gamma$ (modulo l'identification $\Xi(B') = \Xi(T) = \Xi(B)$). Puisque $\pi_{S'} : G_{S'} \cdot x_H \to G' \cdot x_{H'}$ est lisse, on a $\rho(D', \gamma) = -v_{D'}(f_{-\gamma}) = -v_D(f_{-\gamma} \circ \pi_{S'}) = \rho(D, \gamma)$, ce qui prouve la dernière assertion de la proposition.

2.7. La proposition précédente, jointe à un examen cas par cas des variétés magnifiques adjointes de rang ≤ 2, conduit à de nombreuses conditions restrictives et à une compréhension plus approfondie des triplets. Nous ne ferons pas cette étude ici de façon systématique. Nous allons seulement éclairer brièvement les liens qui existent entre S et $\Delta_{G/H}$.

Les variétés sphériques magnifiques sous SL_2 (et plus généralement, sous un groupe réductif connexe de rang semi-simple1) sont de quatre types (c'est-à-dire classes d'isomorphisme) :

(1) type a : $\mathbf{P}_1 \times \mathbf{P}_1$;
(2) type a' : $\mathbf{P}_2$;
(3) type b : $\mathbf{P}_1$;
(4) type p : un point.

Soit H un sous-groupe sphérique sobre, soit X le plongement magnifique de G/H et soit $\alpha \in S$. D'après 1.3 appliqué à $S' = \{\alpha\}, G^{\{\alpha\}}$ est un groupe réductif de rang semi- simple 1 , et $X^{\{\alpha\}}$ est une $G^{\{\alpha\}}$-variété sphérique sobre. Par conséquent, pour tout $\alpha \in S, X^{\{\alpha\}}$ est isomorphe à l'une des quatre variétés ci-dessus. Désignons par $S^a_{G/H}$ (resp. $S^{a'}_{G/H}, S^b_{G/H}, S^p_{G/H}$) l'ensemble des $\alpha \in S$ tels que $X^{\{\alpha\}}$ est de type a (resp. a', b, p).

Le sous-groupe parabolique de G contenant B et associé à $S^p_{G/H}$ n'est rien d'autre que $P = \{g \in G, gBH = BH\}$.

De 2.6, et d'une inspection des quatre cas ci-dessus, on obtient que $\Delta_{G/H}(\{\alpha\})$ contient deux éléments si $\alpha \in S^a_{G/H}$, un élément si $\alpha \in S^{a'}_{G/H} \cup S^b_{G/H}$, et aucun élément si $\alpha \in S^p_{G/H}$. Si $\alpha \in S^{a'}_{G/H} \cup S^b_{G/H}$, notons D_α l'unique élément de $\Delta_{G/H}(\{\alpha\})$. Du fait que les $G_{\{\alpha\}}$, $\alpha \in S$ engendrent le groupe G, on déduit que $\Delta_{G/H} = \bigcup_{\alpha \in S} \Delta_{G/H}(\{\alpha\})$.

Bien entendu, cette réunion n'est pas disjointe. On peut montrer que les trois ensembles

$$\bigcup_{\alpha \in S^a_{G/H}} \Delta_{G/H}(\{\alpha\}), \quad \bigcup_{\alpha \in S^{a'}_{G/H}} \Delta_{G/H}(\{\alpha\}), \quad \bigcup_{\alpha \in S^b_{G/H}} \Delta_{G/H}(\{\alpha\})$$

sont disjoints (autrement-dit, à tout $D \in \Delta_{G/H}$ on peut associer un type a, a' ou b). Mais si α et β sont deux éléments distincts de $S^a_{G/H}$, il peut arriver que $\Delta_{G/H}(\{\alpha\}) \cap \Delta_{G/H}(\{\beta\})$ contienne un élément. De même, si α et β sont deux éléments distincts de $S^b_{G/H}$, il peut arriver que $D_\alpha = D_\beta$ (on peut montrer que cela n'arrive que si α et β sont orthogonales, et que $\alpha + \beta \in \Sigma_{G/H}$).

2.8. Soit Q un sous-groupe parabolique de G. On dit qu'un sous- groupe H de G peut s'obtenir par induction à partir de Q, si quitte à remplacer H par l'un de ses conjugués dans G, on a $Q^r \subset H \subset Q$ (où Q^r est le radical de Q). On peut écrire des sorites (assez longs) sur l'induction des sous-groupes sphériques très sobres (par exemple, on peut caractériser en termes du triplet de H les $S' \subset S$ tels que H peut s'obtenir par induction à partir de $G_{-S'}\cdots$). On dira qu'un sous-groupe H de G est réduit, s'il n'est induit par aucun sous-groupe parabolique propre de G. On peut montrer que, pour qu'un sous-groupe sphérique très sobre H soit réduit et ne contienne aucun sous-groupe distingué non trivial de G, il faut et il suffit que son triplet vérifie $\bigcup_{\gamma \in \Sigma_{G/H}} S_\gamma = S$.

Voici quelques questions pour lesquelles je n'ai pour le moment pas encore de réponse complète (autrement formulé, voici l'annonce d'un programme de recherches qui n'est pas encore terminé) :

Pour tout sous-groupe sphérique très sobre réduit H de G, le triplet $\Sigma_{G/H}$, $\Delta_{G/H}$, $\rho_{G/H}$ détermine-t-il H à conjugaison près ? Quelles sont les conditions qu'il faut demander à un triplet "abstrait" composé d'un ensemble "abstrait" Δ, d'un sous-ensemble Σ de Σ_G, et d'un accouplement $\rho : \Delta \times \Sigma \to \mathbb{Z}$, pour qu'il provienne d'un H comme ci-dessus (c'est-à-dire, pour qu'il existe un H comme ci-dessus, tel que $\Sigma_{G/H} = \Sigma$, et tel que l'accouplement $\rho_{G/H} : \Delta_{G/H} \times \Sigma_{G/H} \to \mathbb{Z}$ s'identifie à l'accouplement $\rho : \Delta \times \Sigma \to \mathbb{Z}$, modulo une bijection $\Delta_{G/H} \leftrightarrow \Delta$) ?

Mentionnons ici l'article [16], dans lequel est donnée la classification des sous-groupes sphériques sobres résolubles (et de leurs triplets).

2.9. Indiquons enfin brièvement le lien (immédiat) qui existe entre les triplets introduits au n° 2.5, et le formalisme (voir [14]) de la théorie des plongements des espaces homogènes sphériques.

Soit H un sous-groupe sphérique sobre adjoint de G. L'espace vectoriel sur $\mathbb{Q}$ dans lequel vivent les "éventails colorés" est $\mathbb{Q}_{G/H} = \mathrm{Hom}_{\mathbb{Z}}(\Xi_\Sigma, \mathbb{Q})$, où Ξ_Σ désigne le sous-réseau du réseau radiciel de R engendré par $\Sigma_{G/H}$. Désignons par $\gamma^*(\gamma \in \Sigma_{G/H})$ la base de $\mathbb{Q}_{G/H}$ duale de la base γ ($\gamma \in \Sigma_{G/H}$) de Ξ_Σ. Alors le "cône des valuations G-invariantes" est le cône convexe engendré par les $-\gamma^*(\gamma \in \Sigma_{G/H})$, et l'application $\rho_{G/H} : \Delta_{G/H} \to \mathbb{Q}_{G/H}$ est donnée par $\rho_{G/H}(D) = \sum_{\gamma \in \Sigma_{G/H}} \rho_{G/H}(D, \gamma)\gamma^*$.

3. Compléments.

On conserve les notations des § précédents. Soit H un sous-groupe sphérique très sobre de G, et soit X le plongement magnifique de G/H.

On supposera dans la suite que X est lisse. Cette hypothèse est probablement toujours remplie (ce que rend plausible l'article [11], où est prouvé que X est lisse lorsque $N_G(H)/H$ est trivial).

3.1. Si D est un diviseur de X, on notera $[D]$ son image dans $\text{Pic}(X)$, le groupe de Picard de X. Dans [5] est montré que $\text{Pic}(X)$ est un $\mathbb{Z}$-module libre ayant comme base les $[D]$, $D \in \Delta_{G/H}$.

Rappelons que X_B désigne la carte canonique de X. Pour tout $\gamma \in \Sigma_{G/H}$, on choisit $f_{-\gamma} \in k[X_B]$ un vecteur propre de B de caractère $-\gamma$ (unique à constante non nulle près), et on note D^γ le diviseur irréductible de X tel que $D^\gamma \cap X_B = (f_{-\gamma})^{-1}(0)$; c'est un diviseur de X stable par G. On obtient ainsi une bijection de $\Sigma_{G/H}$ sur l'ensemble des diviseurs irréductibles G-stables de X.

Proposition. *Pour tout* $\gamma \in \Sigma_{G/H}$*, on a* $[D^\gamma] = \sum\limits_{D \in \Delta_{G/H}} \rho(D, \gamma)[D]$.

Preuve. Par définition de v_D (la valuation qui est associée à $D \in \Delta_{G/H}$), on a $\text{div}_X(f_{-\gamma}) = D^\gamma + \sum\limits_{D \in \Delta_{G/H}} v_D(f_{-\gamma})D$. Par définition de ρ on a $v_D(f_{-\gamma}) = -\rho(D, \gamma)$.

3.2. Revenons aux notations de 2.7, qu'on va compléter. Pour tout $\alpha \in S$, posons $B_-^{\{\alpha\}} = G^{\{\alpha\}} \cap B_-$; c'est un sous-groupe de Borel de $G^{\{\alpha\}}$.

(1) Soit $\alpha \in S_{G/H}^a$. Alors $X^{\{\alpha\}}$ est isomorphe à $\mathbf{P}_1 \times \mathbf{P}_1$. Désignons par z_α^+ et z_α^- les deux points fixes de T dans $X^{\{\alpha\}} \setminus Z$ (rappelons que Z est l'unique orbite fermée de G dans X). Notons D_α^+ et D_α^- les deux éléments de $\Delta_{G/H}(\{\alpha\})$ (pour les distinguer, on suppose que D_α^+ contient z_α^+ et D_α^- contient z_α^-). Notons c_α, c_α^+, c_α^- les trois courbes de $X^{\{\alpha\}}$ qui sont stables par $B_-^{\{\alpha\}}$ (pour les distinguer, on suppose que c_α est contenue dans Z, que c_α^+ contient z_α^+, et que c_α^- contient z_α^-).

(2) Soit $\alpha \in S_{G/H}^{a'}$. Alors $X^{\{\alpha\}}$ est isomorphe à $\mathbf{P}_2$. Désignons par z_α' le point fixe de T dans $X^{\{\alpha\}} \setminus Z$. Notons D_α l'unique élément de $\Delta_{G/H}(\{a\})$, et notons c_α et c_α' les deux courbes de $X^{\{\alpha\}}$ qui sont stables par $B_-^{\{\alpha\}}$ (pour les distinguer, on suppose que c_α est contenue dans Z, et que c_α' contient z_α').

(3) Soit $\alpha \in S_{G/H}^b$. Alors $X^{\{\alpha\}}$ est isomorphe à $\mathbf{P}_1$. Notons D_α l'unique élément de $\Delta_{G/H}(\{a\})$, et posons $c_\alpha = X^{\{\alpha\}}$.

Puisque le radical unipotent de $G_{-\{\alpha\}}$ opère trivialement dans $X^{\{\alpha\}}$, les courbes $c_\alpha, c_\alpha^+, c_\alpha^-, c_\alpha'$ ci-dessus sont aussi stables par B_-. Dans [8] est montré qu'inversement toute courbe stable par B_- dans X est égale à l'une de ces courbes.

Si c est une courbe (complète) de X, désignons par $[c]$ son image dans le groupe des 1-cycles de X modulo équivalence rationnelle. Si $c \in Z_1(X)$ est un 1-cycle et si $D \in \text{div}(X)$ est un diviseur de X, on note $(c.D)$ l'entier obtenu par l'accouplement naturel $Z_1(X) \times \text{div}(X) \to \mathbb{Z}$.

Lemme 1. (1) *Si* $\alpha \in S_{G/H}^a$*, alors* $[c_\alpha] = [c_\alpha^+] + [c_\alpha^-]$.

(2) *Si* $\alpha \in S_{G/H}^{a'}$*, alors* $[c_\alpha] = 2[c_\alpha']$.

Preuve. Ces relations sont déjà (trivialement) vraies dans $X^{\{\alpha\}}$ (il suffit de les constater dans $\mathbf{P}_1 \times \mathbf{P}_1$ et $\mathbf{P}_2$).

Lemme 2. (1) *Si* $\alpha \in S^a_{G/H}$, *alors*

$$(c^+_\alpha \cdot D^+_\alpha) = 1 \text{ et } (c^+_\alpha \cdot D) = 0, \text{ pour tout } D \in \Delta_{G/H} \setminus \{D^+_\alpha\} \ ;$$
$$(c^-_\alpha \cdot D^-_\alpha) = 1 \text{ et } (c^+_\alpha \cdot D) = 0, \text{ pour tout } D \in \Delta_{G/H} \setminus \{D^-_\alpha\} \ .$$

(2) *Si* $a \in S^{a'}_{G/H}$, *alors*

$$(c'_\alpha \cdot D_\alpha) = 1 \text{ et } (c'_\alpha \cdot D) = 0, \ \text{ pour tout } D \in \Delta_{G/H} \setminus \{D_\alpha\}.$$

(3) *Si* $a \in S^b_{G/H}$, *alors*

$$(c_\alpha \cdot D_\alpha) = 1 \text{ et } (c'_\alpha \cdot D) = 0, \ \text{ pour tout } D \in \Delta_{G/H} \setminus \{D_\alpha\}.$$

Preuve. On a $c^+_\alpha \cap D^+_\alpha = \{z^+_\alpha\}$ (intersection transverse) et $c^+_\alpha \cap D = \emptyset$ quel que soit $D \in \Delta_{G/H} \setminus \{D^+_\alpha\}$, d'où la première assertion du lemme 2. Les autres assertions se vérifient de façon analogue.

Lemme 3. *Soit* $\gamma \in \Sigma_{G/H}$.

(1) *Si* $\alpha \in S^a_{G/H}$, *alors* $(c^+_\alpha \cdot D^\gamma) = \rho(D^+_\alpha, \gamma)$ *et* $(c^-_\alpha \cdot D^\gamma) = \rho(D^-_\alpha, \gamma)$.
(2) *Si* $\alpha \in S^{a'}_{G/H}$, *alors* $(c'_\alpha \cdot D^\gamma) = \rho(D_\alpha, \gamma)$.
(3) *Si* $a \in S^b_{G/H}$, *alors* $(c_\alpha \cdot D^\gamma) = \rho(D_\alpha, \gamma)$.

Preuve. Le lemme 3 découle aussitôt de 3.1 et du lemme 2 ci-dessus.

3.3. Soit Y une variété projective lisse dans laquelle opère un tore algébrique T.

Soit c une courbe dans Y, isomorphe à $\mathbf{P}_1$ et stable par T. On suppose que T opère non trivialement dans c, et on note y_0 et y_∞ les deux points fixes de T dans c. On note α le caractère (unique) de T possédant la propriété suivante : il existe une opération fidèle de k^* dans c, telle que $\lim\limits_{t \to 0} t \cdot y = y_0$ quel que soit $y \in c \setminus \{y_\infty\}$, et telle que l'opération de T dans c se factorise par α.

Soit L un fibré en droites T-linéarisé sur Y. Soit β_0 (resp. β_∞) le caractère de l'opération de T dans la fibre de L au-dessus de y_0 (resp. y_∞). Notons $(c \cdot L)$ l'entier obtenu par l'accouplement naturel $Z_1(Y) \times \text{Pic}(Y) \to \mathbb{Z}$. Le résultat suivant est démontré dans [8] (voir p 377).

Lemme. *Le caractère* $\beta_\infty - \beta_0$ *est un multiple de* α, *et on a* $(c \cdot L) = (\beta_\infty - \beta_0)/\alpha$.

Voici une légère variante de cet énoncé. Soit D un diviseur de Y, stable par T et contenant y_0 et y_∞. Supposons D défini par une équation locale en y_0 (resp. en y_∞), stable par T de caractère γ_0 (resp. γ_∞). Alors on a $(c \cdot D) = (\gamma_0 - \gamma_\infty)/\alpha$.

En effet, le fibré en droites $L(D)$ associé à D est alors naturellement T-linéarisé, et le caractère de T dans la fibre de $L(D)$ au-dessus de y_0 (resp. y_∞) est $-\gamma_0$ (resp. $-\gamma_\infty$). Par conséquent, $(c \cdot D) = (c \cdot L(D)) = (\gamma_0 - \gamma_\infty)/\alpha$.

3.4. Si $\alpha \in S$, on note $\alpha^\vee$ la racine duale de α.

Proposition. *Soit* $\gamma \in \Sigma_{G/H}$.

(1) *Si* $\alpha \in S^a_{G/H}$, *alors* $\rho(D^+_\alpha, \gamma) + \rho(D^-_\alpha, \gamma) = \alpha^\vee(\gamma)$.

(2) *Si* $\alpha \in S^{a'}_{G/H}$, *alors* $\rho(D_\alpha, \gamma) = \frac{1}{2}\alpha^\vee(\gamma)$.

(3) *Si* $\alpha \in S^b_{G/H}$, *alors* $\rho(D_\alpha, \gamma) = \alpha^\vee(\gamma)$.

Preuve. Pour tout $\alpha \in S^a_{G/H} \cup S^{a'}_{G/H} \cup S^b_{G/H}$, nous avons introduit dans 3.2 une courbe c_α, stable par T et isomorphe à $\mathbf{P}_1$. Soit z l'unique point fixe de B^- dans X. Notons s_α l'élément d'ordre 2 du groupe de Weyl $W = N_G(T)/T$ associé à α, et aussi un élément de $N_G(T)$ "au-dessus" de s_α. Si $y_0 = s_\alpha(z)$ et $y_\infty = z$, alors y_0 et y_∞ sont les points fixes de T dans c_α, et α est aussi le caractère de T introduit en 3.3.

Considérons le diviseur D^γ de X qui est stable par G. On a $z \in D^\gamma \cap X_B$ et $D^\gamma \cap X_B = (f_{-\gamma})^{-1}(0)$, où $f_{-\gamma} \in k[X_B]$ est un vecteur propre de B de caractère $-\gamma$. Il s'ensuit que $s_\alpha(-\gamma) = -\gamma + \alpha^\vee(\gamma)\alpha$ est le caractère d'une équation locale de D^γ en $s_\alpha(z)$. De 3.3 résulte alors que $(c_\alpha \cdot D^\gamma) = \alpha^\vee(\gamma)$.

Les assertions de la proposition découlent aussitôt de cette dernière formule et de 3.2 (lemmes 1 et 3).

3.5. Rappelons que P désigne le sous-groupe parabolique de G contenant B et associé à $S^p_{G/H}$ (on peut aussi le définir par $P = \{g \in G, gBH = BH\}$). On note $\mathrm{Lie}(P^u)$ l'algèbre de Lie du radical unipotent de P.

Voici un procédé pour retrouver certaines données du triplet de H, en termes de l'opération de T dans le fibré tangent de X.

Proposition.   (1) *Les caractères de* T *dans (l'espace tangent)* $T_z X$ *sont ceux de* $\mathrm{Lie}(P^u)$ *et les* $\gamma, \gamma \in \Sigma_{G/H}$.

(2) *Pour tout* $\alpha \in S^a_{G/H}$, *les caractères de* T *dans* $T_{z^+_\alpha} X$ *(resp. dans* $T_{z^-_\alpha} X$*) sont ceux de* $\mathrm{Lie}(P^u)$, $-\alpha$, *et les* $\gamma - \rho(D^+_\alpha, \gamma)\alpha, \gamma \in \Sigma_{G/H} \setminus \{\alpha\}$ *(resp. les* $\gamma - \rho(D^-_\alpha, \gamma)\alpha, \gamma \in \Sigma_{G/H} \setminus \{\alpha\}$*).*

Preuve. On sait que $\bigcap_{\gamma \in \Sigma_{G/H}} D^\gamma = G \cdot z$ (intersection qui est transverse), et que $D^\gamma \cap X_B = (f_{-\gamma})^{-1}(0)$, où $f_{-\gamma} \in k[X_B]$ est un vecteur propre de B de caractère $-\gamma$. Il s'ensuit que les caractères dans $T_z X$ sont ceux de $T_z G \cdot z \cong \mathrm{Lie}(P^u)$ et les $\gamma, \gamma \in \Sigma_{G/H}$.

Pour tout $\gamma \in \Sigma_{G/H} \setminus \{\alpha\}$, désignons par $h_{-\gamma^*}$ une équation locale de D^γ au voisinage de z^+_α, stable par T de caractère $-\gamma^*$. D'après 3.2 et 3.3, on a $\rho(D^+_\alpha, \gamma) = (c^+_\alpha \cdot D^\gamma) = (\gamma - \gamma^*)/\alpha$, d'où il suit que $\gamma^* = \gamma - \rho(D^+_\alpha, \gamma)\alpha$. Puisque $\bigcap_{\gamma \in \Sigma_{G/H} \setminus \{\alpha\}} D^\gamma = \overline{G \cdot z^+_\alpha}$ (intersection transverse), les caractères de T dans $T_{z^+_\alpha} X$ sont ceux de $T_{z^+_\alpha} G \cdot z^+_\alpha$ (c'est-à-dire ceux de $\mathrm{Lie}(P^u)$ et $-\alpha$), et les $\gamma^* = \gamma - \rho(D^+_\alpha, \gamma)\alpha, \gamma \in \Sigma_{G/H} \setminus \{\alpha\}$. L'assertion pour z^-_α se démontre de façon analogue.

3.6. Ce qui précède permet d'expliciter le diviseur canonique de (toute G-variété magnifique lisse) X. Pour un résultat plus général, valable pour toute G-variété sphérique (non nécessairement lisse ni complète), voir [3].

Notons σ_p la somme des poids de T dans $\text{Lie}(P^u)$. Définissons une application $r : \Delta_{G/H} \to N$ par :

$r(D) = 1$ si D est de type a ou a' ; et

$r(D) = \alpha^\vee(\sigma_p)$ si D est de type b et si $D = D_\alpha$ avec $\alpha \in S^b_{G/H}$ (dans ce cas on a $r(D) \geq 2$).

Désignons par ω_X le fibré en droites canonique de X. Puisque les $[D]$, $D \in \Delta_{G/H}$ forment une base de $\text{Pic}(X)$ (voir [5]), il existe des entiers $m^D_{\omega_X}$ uniques tels que $[\omega_X] = \sum\limits_{D \in \Delta_{G/H}} m^D_{\omega_X}[D]$.

Proposition. (1) *Pour tout* $D \in \Delta_{G/H}$, *on a* $m^D_{\omega_X} = -r(D) - \sum\limits_{\gamma \in \Sigma_{G/H}} \rho(D, \gamma)$.

(2) *Le diviseur* $-\sum\limits_{\gamma \in \Sigma_{G/H}} D^\gamma - \sum\limits_{D \in \Delta_{G/H}} r(D) D$ *est un diviseur canonique de* X.

Preuve. Pour abréger, posons $\Sigma = \Sigma_{G/H}$, $\Delta = \Delta_{G/H}$, et $\sigma_\Sigma = \sum\limits_{\gamma \in \Sigma} \gamma$.

D'après 3.5, le caractère de l'opération de T dans la fibre de ω_X au-dessus de z est $-\sigma_p - \sigma_\Sigma$, et celui au-dessus de z^+_α (resp. z^-_α) est $-\sigma_p - \sigma_\Sigma + (1 + \sum\limits_{\gamma \in \Sigma} \rho(D^+_\alpha, \gamma))\alpha$ (resp. $-\sigma_p - \sigma_\Sigma + (1 + \sum\limits_{\gamma \in \Sigma} \rho(D^-_\alpha, \gamma))\alpha$).

Supposons D de type b, et soit $\alpha \in S^b_{G/H}$ tel que $D = D_\alpha$. De 3.2 (lemme 2) et de 3.3 on obtient :

$$m^D_{\omega_X} = (c_\alpha \cdot \omega_X) = \{-\sigma_p - \sigma_\Sigma - s_\alpha(-\sigma_p - \sigma_\Sigma)\}/\alpha$$
$$= -\alpha^\vee(\sigma_p) - \alpha^\vee(\sigma_\Sigma) = -r(D) - \sum\limits_{\gamma \in \Sigma} \rho(D, \gamma).$$

Supposons D de type a', et soit $\alpha \in S^{a'}_{G/H}$ tel que $D = D_\alpha$. De 3.2 (lemmes 1 et 2) et de 3.3 on obtient :

$$m^D_{\omega_X} = (c'_\alpha \cdot \omega_X) = \frac{1}{2}(c_\alpha \cdot \omega_X) = -\frac{1}{2}\alpha^\vee(\sigma_p) - \frac{1}{2}\alpha^\vee(\sigma_\Sigma)$$
$$= -1 - \sum\limits_{\gamma \in \Sigma} \rho(D'_\alpha, \gamma) = -\rho(D) - \sum\limits_{\gamma \in \Sigma} \rho(D, \gamma).$$

Enfin, supposons D de type a. Soit $\alpha \in S^a_{G/H}$ tel que $D = D^+_\alpha$. De 3.2 (lemme 2) et de 3.3 on obtient :

$$m^D_{\omega_X} = (c^+_\alpha \cdot \omega_X) = \left\{-\sigma_p - \sigma_\Sigma - \left(-\sigma_p - \sigma_\Sigma + \left(1 + \sum\limits_{\gamma \in \Sigma} \rho(D^+_\alpha, \gamma)\right)\alpha\right)\right\}/\alpha$$
$$= -1 - \sum\limits_{\gamma \in \Sigma} \rho(D, \gamma) = -r(D) - \sum\limits_{\gamma \in \Sigma} \rho(D, \gamma).$$

Si $D = D^-_\alpha$, on raisonne de la même façon.

La deuxième assertion de la proposition résulte de la première et de 3.1.

REFERENCES

1. D. N. Akhiezer, *Equivariant completion of homogeneous algebraic varieties by homogeneous divisors*, Ann. Global Anal. geom. **1** (1983), 49–78.

2. A. Bialynicki-Birula, *Some theorems on actions of algebraic groups*, Ann. of Math. **98** (1980), 480–497.

3. M. Brion, *Curves and divisors in spherical varieties*, publié dans ce volume.

4. _____, *Quelques propriétés des espaces homogènes sphériques*, Manuscr. Math. **55** (1986), 191–198.

5. _____, *Groupe de Picard et nombres caractéristiques des variétés sphériques*, Duke Math. J. **58** (1989), 397–424.

6. _____, *On spherical varieties of rank one*, CMS Conf. Proc. **10** (1989), 31–41.

7. _____, *Vers une généralisation des espaces symétriques*, J. of Algebra **134** (1990), 115–143.

8. _____, *Variétés sphériques et théorie de Mori*, Duke Math. J. **72** (1993), 369–404.

9. M. Brion, D. Luna, and Th. Vust, *Espaces homogènes sphériques*, Invent. Math. **84** (1986), 617–632.

10. M. Brion and F. Pauer, *Valuations des espaces homogènes sphériques*, Comment. Math. Helv. **62** (1987), 265–285.

11. F. Knop, *Automorphisms, root systems and compactifications of homogeneous varieties*, à paraître.

12. _____, *On the set of orbits for a Borel subgroup*, à paraître.

13. _____, *Weylgruppe und momentabbildung*, Invent. Math. **99** (1990), 40–54.

14. _____, *The Luna-Vust theorey of spherical embeddings*, Proc. of the Hyderabad conference on algebraic groups (1989) (Madras), Manoj Prakashan, 1991, pp. 225–249.

15. F. Knop, H. Kraft, D. Luna, and Th. Vust, *Local properties of algebraic group actions, Algebraische Transformationsgruppen und Invariantentheorie*, DMV Seminar (H. Kraft, P. Slodowy, and T. Springer, eds.), vol. 13, Birkhäuser, 1989, pp. 63–76.

16. D. Luna, *Sous-groupes sphériques résolubles*, à paraître.

17. _____, *Adhérence d'orbite et invariants*, Invent. Math. **29** (1975), 231–238.

18. T. Matsuki, *Orbits on flag manifolds*, Proc. of the international congress of mathematicians, Kyoto 1990, II, Springer-Verlag, 1991, pp. 807–813.

19. E. B. Vinberg, *Complexity of action of reductive groups*, Funct. Anal. Appl. **20** (1986), 1–11.

20. Th. Vust, *Plongements d'espaces symétriques algébriques : une classification*, Ann. Sc. Norm. Super. Pisa, Serie IV **XVII** Fasc. 2 (1990), 165–194.

UNIVERSITÉ DE GRENOBLE I, INSTITUT FOURIER, LABORATOIRE DE MATHÉMATIQUES, ASSOCIÉ AU CNRS (URA 188), B.P. 74, 38402 ST MARTIN D'HÈRES CEDEX (FRANCE)

TOTAL POSITIVITY AND CANONICAL BASES

GEORGE LUSZTIG
To the memory of Roger Richardson

Let G be a connected split reductive algebraic group over $\mathbf{R}$; we identify G with the group of its $\mathbf{R}$-points. In [4] we have defined (in terms of an "épinglage" of G) the subset $G_{\geq 0}$ (resp. $G_{>0}$) of totally positive (resp. totally strictly positive) elements of G. Both these subsets are closed under multiplication, $G_{\geq 0}$ is the closure of $G_{>0}$ and $G_{>0}$ is the interior of $G_{\geq 0}$.

Although the definition of these subsets was elementary, many of their properties were proved in a non-elementary way, using canonical bases and their positivity properties [2],[3]. Another relationship between canonical bases and total positivity was found in [3, 42.2] (see also [4]); namely we have shown that one can naturally index canonical bases in the corresponding enveloping algebras in terms of the geometry of $G_{\geq 0}$ (where $\mathbf{R}$ is replaced by $\mathbf{R}(\epsilon)$).

This paper contains two results which strengthen the relationship between canonical bases and the semigroups $G_{\geq 0}$. (For simplicity we assume that G is simply laced, but the results can be extended to the general case by standard methods.)

To explain our first result, we recall that the canonical basis of a simple (finite dimensional) G-module can be parametrized in a combinatorial way in two different ways, by regarding it either as a highest weight module or as a lowest weight module. The problem arises to describe combinatorially the (rather complicated) relation between the indices in the two parametrizations. Surprisingly, it turns out that the necessary combinatorics also appears in an entirely different setting, namely in the geometry of $G_{\geq 0}$ over $\mathbf{R}(\epsilon)$.

Our second result is that $G_{\geq 0}$ and $G_{>0}$ admit a global definition, namely they can be defined by explicit inequalities, provided by certain canonical basis elements.

It is a pleasure to thank the Institut des Hautes Etudes Scientifiques for its hospitality during my visit (January 1995), when most of the work on this paper was done.

1. Recollections on total positivity

1.1. Let K be one of the fields $\mathbf{R}$ or $\mathbf{R}(\epsilon)$, where ϵ is an indeterminate. The subset $K_{>0}$ of K is defined as follows. If $K = \mathbf{R}$ the definition is the standard one. If $K = \mathbf{R}(\epsilon)$, $K_{>0}$ consists of the $f \in K - \{0\}$ with a power series expansion $f = a_s\epsilon^s + a_{s+1}\epsilon^{s+1} + \dots$ such that $a_s \in \mathbf{R}_{>0}, a_{s+1} \in \mathbf{R}, a_{s+2} \in \mathbf{R}, \dots$ (we then

Supported in part by the National Science Foundation

set $|f| = s$). (An equivalent condition for f is that for any sufficiently small $\epsilon_0 \in \mathbf{R}_{>0}$, the specialization $f(\epsilon_0)$ is a well defined number in $\mathbf{R}_{>0}$.) We set $K_{\geq 0} = K_{>0} \cup \{0\}$.

1.2. We will often identify an algebraic variety over K with the set of its K-rational points. Let G be a connected, split, simply laced, reductive algebraic group over K with a fixed épinglage (as in [4, 1.1]). Thus, we have a fixed K-split maximal torus T of G, a pair B^+, B^- of opposed Borel subgroups containing T, with unipotent radicals U^+, U^- and fixed imbeddings $x_i : K \rightarrow U^+, y_i : K \rightarrow U^-$ (indexed by $i \in I$) subject to a certain requirement (*loc. cit.*). In particular, there are unique isomorphisms $\Psi : G \xrightarrow{\sim} G^{opp}$, $\Pi : G \xrightarrow{\sim} G^{opp}$ (the opposite group structure) such that

$$\Psi(x_i(a)) = y_i(a), \Psi(y_i(a)) = x_i(a) \quad \text{for all } i \in I, a \in K \text{ and}$$
$$\Psi(t) = t \quad \text{for all } t \in T,$$
$$\Pi(x_i(a)) = x_i(a), \Pi(y_i(a)) = y_i(a) \quad \text{for all } i \in I, a \in K \text{ and}$$
$$\Pi(t) = t^{-1} \quad \text{for all } t \in T.$$

1.3. Let Y (resp. X) be the free abelian group of all homomorphisms of algebraic groups $K^* \rightarrow T$ (resp. $T \rightarrow K^*$). We write the operation in these groups as addition. Let $\langle , \rangle : Y \times X \rightarrow \mathbf{Z}$ be the standard pairing. For $i \in I$, let $\alpha_i \in X$ be the simple root such that $tx_i(a)t^{-1} = x_i(\alpha_i(t)a)$ for all $a \in K, t \in T$; let $\check{\alpha}_i \in Y$ be the corresponding simple coroot.

Let X^+ be the set of all $\lambda \in X$ such that $\langle \check{\alpha}_i, \lambda \rangle \in \mathbf{N}$ for all $i \in I$.

1.4. Let NT be the normalizer of T in G and let $W = NT/T$ be the Weyl group of G. For $i \in I$, let $s_i \in W$ be the simple reflection corresponding to x_i, y_i. Let w_0 be the longest element of W with respect to the set of generators $\{s_i | i \in I\}$; let n be the length of w_0.

Let $i \mapsto i^*$ be the involution of I defined by $w_0 s_i w_0^{-1} = s_{i^*}$ for all $i \in I$.

For any $w \in W$, we denote by $\dot{w}$ a representative of w in NT.

Let $\mathbf{H}$ be the set of all sequences $\mathbf{i} = (i_1, i_2, \ldots, i_n)$ in I such that $s_{i_1} s_{i_2} \ldots s_{i_n} = w_0$.

1.5. Recall the following definitions from [4].

$U^+_{\geq 0}$ is the submonoid with 1 of U^+ generated by the elements $x_i(a)$ for various $i \in I$ and $a \in K_{\geq 0}$.

$U^-_{\geq 0}$ is the submonoid with 1 of U^- generated by the elements $y_i(a)$ for various $i \in I$ and $a \in K_{\geq 0}$.

$T_{>0}$ is the subgroup of T generated by the elements $\chi(a)$ for various $\chi \in Y, a \in K_{>0}$.

$G_{\geq 0}$ is the submonoid $U^+_{\geq 0} T_{>0} U^-_{\geq 0} = U^-_{\geq 0} T_{>0} U^+_{\geq 0}$ of G.

$U^+_{>0}$ is the subset of U^+ consisting of all products $x_{i_1}(a_1) x_{i_2}(a_2) \ldots x_{i_n}(a_n)$ where $\mathbf{i} = (i_1, i_2, \ldots, i_n)$ is a fixed element of $\mathbf{H}$ and $(a_1, a_2, \ldots, a_n) \in K^n_{>0}$. (This is a submonoid of $U^+_{\geq 0}$ independent of the choice of $\mathbf{i}$.)

$U_{>0}^-$ is the subset of U^- consisting of all products $y_{i_1}(a_1)y_{i_2}(a_2)\ldots y_{i_n}(a_n)$ where $\mathbf{i} = (i_1, i_2, \ldots, i_n)$ is a fixed element of $\mathbf{H}$ and $(a_1, a_2, \ldots, a_n) \in K_{>0}^n$. (This is a submonoid of $U_{>0}^-$ independent of the choice of $\mathbf{i}$.)

$G_{>0}$ is the submonoid $U_{>0}^+ T_{>0} U_{>0}^- = U_{>0}^- T_{>0} U_{>0}^+$ of $G_{\geq 0}$.

1.6. Let $\mathscr{B}$ be the set of all Borel subgroups of G defined over K. For $B \in \mathscr{B}$ and $g \in G$ we shall sometime write ^gB instead of gBg^{-1}.

2. COMBINATORICS

2.1. We begin by the definition of a graph introduced in [2] in connection with the parametrization of canonical bases. The present definition differs slightly from that in [2]: whereas in [2] we only used positive integers, we now allow arbitrary integers.

Let $\hat{\mathbf{H}}$ be the set of all pairs $(\mathbf{i}, \mathbf{c})$ where $\mathbf{i} \in \mathbf{H}$ and $\mathbf{c} \in \mathbf{Z}^n$. We regard $\hat{\mathbf{H}}$ as the set of vertices of a graph in which $(\mathbf{i}, \mathbf{c}), (\mathbf{i}', \mathbf{c}')$ form an edge if one of the conditions (a),(b) below is satisfied.

(a) $\mathbf{i}'$ is obtained from $\mathbf{i}$ by replacing two consecutive indices i, j (with $s_i s_j = s_j s_i$) by j, i and $\mathbf{c}'$ is obtained from $\mathbf{c}$ by replacing the two consecutive coordinates (a, b) of $\mathbf{c}$ (corresponding to the positions in which i, j appear in $\mathbf{i}$) by (b, a).

(b) $\mathbf{i}'$ is obtained from $\mathbf{i}$ by replacing three consecutive indices i, j, i (with $s_i s_j s_i = s_j s_i s_j$) by j, i, j and $\mathbf{c}'$ is obtained from $\mathbf{c}$ by replacing the three consecutive coordinates (a, b, c) of $\mathbf{c}$ (corresponding to the positions in which i, j, i appear in $\mathbf{i}$) by (a', b', c'), where $a' = b + c - \min(a, c), b' = \min(a, c), c' = a + b - \min(a, c)$.

Note that conditions (a),(b) above are symmetric, hence the edges are well defined. Let $\hat{\mathbf{H}}^\dagger$ be the set of connected components of the graph $\hat{\mathbf{H}}$.

The connected component of $\hat{\mathbf{H}}$ containing $(\mathbf{i}, \mathbf{h})$ is denoted by $[\mathbf{i}, \mathbf{c}]$.

From [4, 9.5(b)], it follows that

(c) for any $\mathbf{i}$, the map $\mathbf{c} \mapsto [\mathbf{i}, \mathbf{c}]$ is a bijection $\mathbf{Z}^n \xrightarrow{\sim} \hat{\mathbf{H}}^\dagger$.

2.2. For $i \in I$, we define a function $g_i : \hat{\mathbf{H}}^\dagger \to \mathbf{Z}$ by $g_i[\mathbf{i}, \mathbf{c}] = c_1$ where $\mathbf{i}$ begins with $i_1 = i$ and $\mathbf{c}$ begins with c_1. From [3, 42.1.14], we see that g_i is well defined.

For $i \in I$, we define a function $g_i' : \hat{\mathbf{H}}^\dagger \to \mathbf{Z}$ by $g_i'[\mathbf{i}, \mathbf{c}] = c_n$ where $\mathbf{i}$ begins with $i_1 = i$ and $\mathbf{c}$ ends with c_n. Again, g_i' is well defined.

2.3. For any $(i, p) \in I \times \mathbf{Z}$ we define a bijection $T_{i,p} : \hat{\mathbf{H}}^\dagger \xrightarrow{\sim} \hat{\mathbf{H}}^\dagger$ by

$$T_{i,p}[\mathbf{i}, \mathbf{c}] = [\mathbf{i}, \mathbf{c}']$$

where $\mathbf{i} = (i_1, i_2, \ldots, i_n), i_1 = i, \mathbf{c}' = \mathbf{c} + (p, 0, 0, \ldots, 0)$. (This is map is well defined, by an argument similar to the one in [3, 42.1.14].)

We have $T_{i,p}T_{i,p'} = T_{i,p+p'}$ for any $i \in I$ and $p, p' \in \mathbf{Z}$. Moreover, $T_{i,0} = 1$.

2.4. For any function $f : I \to \mathbf{Z}$, we define a bijection $S_f : \hat{\mathbf{H}} \xrightarrow{\sim} \hat{\mathbf{H}}$ by $S_f(\mathbf{i}, \mathbf{c}) = (\mathbf{i}, \mathbf{c} + (f(i_1), f(i_2), \ldots, f(i_n)))$. It is easily verified that this map takes any edge of the graph $\hat{\mathbf{H}}$ to an edge of the graph $\hat{\mathbf{H}}$ hence it defines a bijection $S_f : \hat{\mathbf{H}}^\dagger \xrightarrow{\sim} \hat{\mathbf{H}}^\dagger$. Clearly,

(a) $S_f S_{f'} = S_{f+f'}$ for any $f, f' : I \to \mathbf{Z}$ and $S_0 = 1$.

2.5. The elements $(\mathbf{i}, (0, 0, \ldots, 0))$ of $\hat{\mathbf{H}}$ (for various $\mathbf{i} \in \mathbf{H}$) form a single connected component of the graph $\hat{\mathbf{H}}$. This component may be regarded as an element $h_0 \in \hat{\mathbf{H}}^\dagger$. For any $f : I \to \mathbf{Z}$, we set $h_f = S_f(h_0) \in \hat{\mathbf{H}}^\dagger$. Note that h_f is the connected component of $(\mathbf{i}, (f(i_1), f(i_2), \ldots, f(i_n))) \in \hat{\mathbf{H}}$, for any $\mathbf{i} \in \mathbf{H}$.

2.6. It is clear that $T_{i,p} S_f = S_f T_{i,p}$ for any $i \in I, p \in \mathbf{Z}, f : I \to \mathbf{Z}$. (Equality of maps $\hat{\mathbf{H}}^\dagger \to \hat{\mathbf{H}}^\dagger$.)

2.7. We now define a new graph, whose set of vertices is $\hat{\mathbf{H}}^\dagger$ and in which two vertices are joined when one is obtained from the other by applying $T_{i,p}$ for some $i \in I$ and some $p \in \mathbf{Z} - \{0\}$.

Proposition 2.8. *This graph is connected.*

The subset

$$\hat{\mathbf{H}}_+ = \{(\mathbf{i}, \mathbf{c}) \in \hat{\mathbf{H}} | \mathbf{c} \in \mathbf{N}^n\}$$

of $\hat{\mathbf{H}}$ is clearly a union of connected components of $\hat{\mathbf{H}}$. We denote by $\hat{\mathbf{H}}_+^\dagger$ the set of connected components of $\hat{\mathbf{H}}$ that are contained in $\hat{\mathbf{H}}_+$.

We regard $\hat{\mathbf{H}}_+^\dagger$ as a full subgraph of $\hat{\mathbf{H}}^\dagger$. From [3, 42.1.15] it follows that this subgraph is connected. Now let ζ, ζ' be two elements of $\hat{\mathbf{H}}^\dagger$. If $f : I \to \mathbf{Z}$ is a constant function whose value is a sufficiently large integer > 0, then $S_f(\zeta), S_f(\zeta')$ belong to $\hat{\mathbf{H}}_+^\dagger$ hence are in the same connected component of $\hat{\mathbf{H}}^\dagger$. But S_f is an automorphism of the graph $\hat{\mathbf{H}}^\dagger$ (see 2.6). Hence ζ, ζ' are in the same connected component of $\hat{\mathbf{H}}^\dagger$.

Theorem 2.9. *There exists a unique bijection* $\Phi : \hat{\mathbf{H}}^\dagger \xrightarrow{\sim} \hat{\mathbf{H}}^\dagger$ *such that*

(a) $\Phi(h_0) = h_0$, *and*
(b) $\Phi \circ T_{i,p} = T_{i^\bullet, -p} \circ \Phi$

for any $i \in I, p \in \mathbf{Z}$.
We have $\Phi \circ \Phi = 1$.

The existence of Φ will be proved in §3. The uniqueness of Φ follows immediately from 2.8. Clearly, Φ^{-1} satisfies conditions like (a),(b); by uniqueness, we have $\Phi^{-1} = \Phi$.

Theorem 2.10. *For any* $f : I \to \mathbf{Z}$, *we define* $f^* : I \to \mathbf{Z}$ *by* $f^*(i) = f(i^*)$. *We have*

(a) $\Phi(h_{f^\bullet}) = h_{-f}$;
(b) $\Phi \circ S_{f^\bullet} = S_{-f} \circ \Phi$.

The proof will be given in 4.11.

3. THE MAP ϕ

3.1. Given $\mathbf{i} \in \mathbf{H}, k \in [0, n]$ let $f_{\mathbf{i},k} : K_{>0}^n \to \mathscr{B}$ be the map defined by

$$f_{\mathbf{i},k}(a_1, a_2, \ldots, a_n) = {}^{x_{i_1}(a_1)x_{i_2}(a_2)\cdots x_{i_k}(a_k)y_{i_n}(a_n)y_{i_{n-1}}(a_{n-1})\cdots y_{i_{k+1}}(a_{k+1})\dot{s}_{i_{k+1}}\dot{s}_{i_{k+2}}\cdots\dot{s}_{i_n}} B^-.$$

Proposition 3.2. (a) *The map $f_{\mathbf{i},k}$ is injective.*
(b) *The image of $f_{\mathbf{i},k}$ is independent of $\mathbf{i}, k$.*
(c) *For any $\mathbf{i} \in \mathbf{H}$ and any $k \in [0, n-1]$, there exists a bijection $g_{\mathbf{i},k} : K_{>0}^n \xrightarrow{\sim} K_{>0}^n$ such that $f_{\mathbf{i},k} = f_{\mathbf{i},k+1} \circ g_{\mathbf{i},k}$.*

Assume first that (c) is known. Then (a),(b) are proved by descending induction on k; the beginning of the induction ($k = n$) is provided by [4, 2.7], while the induction step is provided by (c).

It remains to verify (c). For $s \in [k + 1, n]$, we define a map $f'_s : K_{>0}^n \to \mathscr{B}$ by

$$f'_s(a_1, a_2, \ldots, a_n) =$$

$${}^{x_{i_1}(a_1)x_{i_2}(a_2)\cdots x_{i_k}(a_k)\cdot y_{i_n}(a_n)y_{i_{n-1}}(a_{n-1})\cdots y_{i_{s+1}}(a_{s+1})\cdot x_{i_{k+1}}(a_{k+1})\cdot y_{i_s}(a_s)\cdots y_{i_{k+2}}(a_{k+2})\dot{s}_{i_{k+2}}\cdots\dot{s}_{i_n}} B^-.$$

If $s > k + 1$, in the expression for f'_s, we can try to interchange $x_{i_{k+1}}(a_{k+1})$ and $y_{i_s}(a_s)$. This can be done without trouble, if $i_{k+1} \neq i_s$, while if $i_{k+1} = i_s$, this can be done at the expense of modifying a_{k+1}, a_s and introducing a factor in T (which then can be moved to the right until it is absorbed by B^-). (We use the standard commutation formulas [4, 1.3].) We see that

$$f'_s = f'_{s-1} \quad \text{if } i_{k+1} \neq i_s;$$

while, if $i_{k+1} = i_s$,

$$f'_s(a_1, a_2, \ldots, a_{k+1}, a_{k+2}, \ldots, a_s, \ldots, a_n)$$

$$= f_{s-1}(a'_1, a'_2, \ldots, a'_{k+1}, a'_{k+2}, \ldots, a'_s, \ldots, a'_n)$$

where $a'_p = a_p$ for $p = 1, \ldots, k$ and for $p = s + 1, \ldots, n$;

$$a'_s = a_s/(1 + a_{k+1}a_s), \quad a'_{k+1} = a_{k+1}(1 + a_{k+1}a_s);$$

$$a'_p = a_p(1 + a_{k+1}a_s)^{-\langle \check{\alpha}_{i_s}, \alpha_{i_p} \rangle} \quad \text{for } p = k + 2, \ldots, s - 1.$$

(These formulas may be inverted:

$$a_p = a'_p \quad \text{for } p = 1, \ldots, k \text{ and for } p = s + 1, \ldots, n;$$

$$a_s = a'_s(1 + a'_{k+1}a'_s), \quad a_{k+1} = a'_{k+1}/(1 + a'_{k+1}a'_s);$$

$$a_p = a'_p(1 + a'_{k+1}a'_s)^{\langle \check{\alpha}_{i_s}, \alpha_{i_p} \rangle} \quad \text{for } p = k + 2, \ldots, s - 1.)$$

Thus, for $s \in [k + 2, n]$, there exists a bijection $g'_s : K_{>0}^n \xrightarrow{\sim} K_{>0}^n$ such that $f'_{s-1} = f'_s g'_s$. Note that $f'_n = f_{\mathbf{i},k+1}$. Hence $f'_{k+1} = f_{\mathbf{i},k+1}g'_n g'_{n-1} \cdots g'_{k+2}$. We have

$$f'_{k+1}(a_1, a_2, \ldots, a_n)$$

$$= {}^{x_{i_1}(a_1)x_{i_2}(a_2)\cdots x_{i_k}(a_k)\cdot y_{i_n}(a_n)y_{i_{n-1}}(a_{n-1})\cdots y_{i_{k+2}}(a_{k+2})x_{i_{k+1}}(a_{k+1})\dot{s}_{i_{k+2}}\cdots\dot{s}_{i_n}} B^-.$$

By a standard identity in SL_2, we have

$$x_{i_{k+1}}(a_{k+1}) = y_{i_{k+1}}(1/a_{k+1})\dot{s}_{i_{k+1}}t y_{i_{k+1}}(1/a_{k+1})$$

for some $t \in T$. Note also that

$$\dot{s}_{i_n}^{-1} \ldots \dot{s}_{i_{k+2}}^{-1} t y_{i_{k+1}}(1/a_{k+1}) \dot{s}_{i_{k+2}} \ldots \dot{s}_{i_n} \in B^-$$

(since $s_{i_{k+1}} s_{i_{k+2}} \ldots s_{i_n}$ is a reduced expression in W). Hence

$$x_{i_{k+1}}(a_{k+1}) \dot{s}_{i_{k+2}} \cdots \dot{s}_{i_n} B^- = y_{i_{k+1}}(1/a_{k+1}) \dot{s}_{i_{k+1}} t y_{i_{k+1}}(1/a_{k+1}) \dot{s}_{i_{k+2}} \cdots \dot{s}_{i_n} B^-$$

$$= y_{i_{k+1}}(1/a_{k+1}) \dot{s}_{i_{k+1}} \dot{s}_{i_{k+2}} \cdots \dot{s}_{i_n} B^-$$

so that

$$f'_{k+1}(a_1, a_2, \ldots, a_n) = f_{i,k}(a'_1, a'_2, \ldots a'_n)$$

where $a'_{k+1} = 1/a_{k+1}$ and $a'_p = a_p$ for $p \neq k+1$. Hence there exists a bijection $g'_{k+1} : K^n_{>0} \xrightarrow{\sim} K^n_{>0}$ such that $f_{i,k} = f'_{k+1} g'_{k+1}$. It follows that

$$f_{i,k} = f_{i,k+1} g'_n g'_{n-1} \cdots g'_{k+2} g'_{k+1}.$$

Then $g_{i,k} = g'_n g'_{n-1} \cdots g'_{k+2} g'_{k+1}$ is as required. The proposition is proved.

Proposition 3.3. (a) *There exists a unique bijection $\phi : U^+_{>0} \xrightarrow{\sim} U^-_{>0}$ such that $\phi(u)^{-1} B^+ \phi(u) = u^{-1} B^- u$ for all $u \in U^+_{>0}$.*
(b) *There exists a unique bijection $\phi' : U^+_{>0} \xrightarrow{\sim} U^-_{>0}$ such that $\phi'(u) B^+ \phi'(u)^{-1} = u B^- u^{-1}$ for all $u \in U^+_{>0}$.*
(c) *For any $u \in U^+_{>0}$, we have $\phi(u) = \Psi(\phi'^{-1}(\Psi(u)))$, where Ψ is as in 1.2.*

We prove (b). Choose some $i \in H$. Applying 3.2 with $k = 0$ and with $k = n$ we see that the (injective) map $U^-_{>0} \to \mathscr{B}$ given by $u \mapsto u B^+ u^{-1}$ has the same image as the (injective) map $U^+_{>0} \to \mathscr{B}$ given by $u \mapsto u B^- u^{-1}$. This clearly implies (b).

We prove (a). We define $\phi : U^+_{>0} \xrightarrow{\sim} U^-_{>0}$ in terms of ϕ' in (a), by the formula in (c). This proves the existence part of (a); the uniqueness is immediate. Now both (a), (c) follow.

3.4. When $K = \mathbf{R}$, the previous proposition could be also deduced directly from [4, 8.7]; then the case $K = \mathbf{R}(\epsilon)$ could be deduced from the case $K = \mathbf{R}$ by specialization. There is, however, an advantage in the proof based on 3.2. Namely, that proof shows that, for any $i, i' \in H$, the bijection $K^n_{>0} \to K^n_{>0}$ given by $(a_1, a_2, \ldots, a_n) \mapsto (a'_1, a'_2, \ldots a'_n)$ where

$$\phi'(x_{i_1}(a_1) x_{i_2}(a_2) \ldots x_{i_n}(a_n)) = y_{i'_1}(a'_1) y_{i'_2}(a'_2) \ldots y_{i'_n}(a'_n))$$

is such that the a'_k are rational functions in $a_1, a_2, \ldots, a_n$ of a very special kind, that is, they are obtained from $a_1, a_2, \ldots, a_n$ by performing a finite sequence of multiplications, divisions and additions (in some order), but no differences. The same holds for the inverse of ϕ' and for ϕ and its inverse. (This holds, since our bijections are compositions of simpler bijections which have the same kind of property.) Note that the property above is independent of the choice of i, i'; this follows from the transformation rules [4, 2.5].

For example, if $G = SL_2$ and $I = \{1\}$, then $\phi(x_1(a)) = y_1(1/a)$. If $G = SL_4$ and $I = \{1, 2, 3\}$ with the standard notation, we have

$$\phi(x_2(p) x_1(q) x_3(r) x_2(v) x_1(u) x_3(w)) = y_2(p') y_1(q') y_3(r') y_2(v') y_1(u') y_3(w')$$

where

$$p' = \frac{uw}{rpq}, \qquad q' = \frac{q}{u(u+q)}, \qquad r' = \frac{r}{w(r+w)},$$

$$v' = \frac{(r+w)(u+q)}{qvr}, \qquad u' = \frac{1}{u+q}, \qquad w' = \frac{1}{r+w}.$$

3.5. Let $u \in U_{>0}^+$. By [4, 2.7(d)], we can write uniquely $u = vmv'$ where $v \in U^-, v' \in U^-, m \in T\dot{w}_0$. Using the definitions, we see that

(a) $v' = \phi(u), v = \phi'(u)$.

Lemma 3.6. *Assume that* $(i_1, i_2, \ldots, i_n) \in \mathbf{H}$ *is given. There exist functions* $p_k : K_{>0}^{n-1} \to K_{>0}, k = 1, 2, \ldots, n$, *such that*

$$\phi(x_{i_1}(a_1)x_{i_2}(a_2)\ldots x_{i_n}(a_n))$$
$$= y_{i_1^*}(p_1(a_2, a_3, \ldots, a_n)a_1^{-1})y_{i_2^*}(p_2(a_2, a_3, \ldots, a_n))\ldots y_{i_n^*}(p_n(a_2, a_3, \ldots, a_n))$$

for all $(a_1, a_2, \ldots, a_n) \in K_{>0}^n$.

Let $a_1, a_2, \ldots, a_n$ be elements in $K_{>0}$. Throughout this proof, $a_2, a_3, \ldots, a_n$ will be fixed, but a_1 will be allowed to vary.

Let $x = x_{i_1}(a_1)x_{i_2}(a_2)\ldots x_{i_n}(a_n), v = x_{i_2}(a_2)\ldots x_{i_n}(a_n)$. We have $v \in B^- \dot{s}_{i_1}^{-1}\dot{w}_0 B^-$ (see [4, 2.7(d)]) hence we can write uniquely $v = v_1 m v_2$ where

$$v_1 \in U^- \cap \dot{s}_{i_1}^{-1} U^- \dot{s}_{i_1}, v_2 \in U^-, m \in T\dot{s}_{i_1}^{-1}\dot{w}_0.$$

A standard identity in SL_2 shows that $x_{i_1}(a_1) = y_{i_1}(a_1^{-1})m_1 y_{i_1}(a_1^{-1})$ where $m_1 \in T\dot{s}_{i_1}$. Thus,

$$x = x_{i_1}(a_1)v = y_{i_1}(a_1^{-1})m_1 y_{i_1}(a_1^{-1})v_1 m v_2.$$

Now the subgroup $U^- \cap \dot{s}_{i_1}^{-1}U^-\dot{s}_{i_1}$ is normalized by $\{y_{i_1}(a)|a \in K\}$, hence $y_{i_1}(a_1^{-1})v_1 = v_1' y_{i_1}(a_1^{-1})$ for some $v_1' \in U^- \cap \dot{s}_{i_1}^{-1}U^-\dot{s}_{i_1}$. From the definitions we see that there is a unique element $c \in K - \{0\}$ (depending only on m, hence constant as a function of a_1) such that

$$m^{-1}y_{i_1}(a)m = y_{i_1^*}(ca)$$

for all $a \in K$. In particular, this holds for $a = a_1$. Thus, we have

$$x = y_{i_1}(a_1^{-1})m_1 v_1' y_{i_1}(a_1^{-1})mv_2 = y_{i_1}(a_1^{-1})m_1 v_1' m y_{i_1^*}(ca_1^{-1})v_2 = u'm_1 mu''$$

where

$$u' = y_{i_1}(a_1^{-1})m_1 v_1' m_1^{-1} \in U^-, u'' = y_{i_1^*}(ca_1^{-1})v_2 \in U^-, m_1 m \in T\dot{w}_0.$$

By 3.5(a), we have $u'' = \phi(x)$. Since $\phi(x) \in U_{>0}^-$, we have $u'' = y_{i_1^*}(b_1)y_{i_2^*}(b_2)\ldots$ $y_{i_n^*}(b_n)$ for well defined elements $b_1, b_2, \ldots, b_n$ in $K_{>0}$. We deduce that

(a) $v_2 = y_{i_1^*}(ca_1^{-1})^{-1}y_{i_1^*}(b_1)y_{i_2^*}(b_2)\ldots y_{i_n^*}(b_n) = y_{i_1^*}(b_1')y_{i_2^*}(b_2)\ldots y_{i_n^*}(b_n),$

where $b_1' = -ca_1^{-1} + b_1$. From (a) we see that $b_1', b_2, b_3, \ldots, b_n$ are uniquely determined by v_2 (by the same argument as in the proof of [4, 2.7(a)]), hence they are constant as functions of a_1. We have

$$u'' = y_{i_1^*}(b_1' + ca_1^{-1})y_{i_2^*}(b_2)\ldots y_{i_n^*}(b_n).$$

Since c is also constant as a function of a_1, we see that, to conclude the proof, it is enough to show that $b_1' = 0$.

Since $b_1 \in K_{>0}$, we have $b_1' + ca_1^{-1} \in K_{>0}$. Here b_1', c are constant and a_1 can take any value in $K_{>0}$. This clearly implies that $b_1' \in K_{>0} \cup \{0\}$ and $c \in K_{>0}$.

Assume that $b_1' \in K_{>0}$. We will show that this leads to a contradiction. By specializing ϵ to a small $\epsilon_0 \in \mathbf{R}_{>0}$ (if $K = \mathbf{R}(\epsilon)$) we see that it is enough to consider the case where $K = \mathbf{R}$. The equality $x^{-1}B^-x = u''^{-1}B^+u''$ can be written as

$$x_{i_n}(a_n)^{-1}\ldots x_{i_1}(a_1)^{-1}B^- = {}^{y_{i_n^*}(b_n)^{-1}\ldots y_{i_2^*}(b_2)^{-1}y_{i_1^*}(b_1'+ca_1^{-1})^{-1}}B^+.$$

We consider the limit of both sides (in the real flag manifold $\mathscr{B}$) as $a_1 \to \infty$ (through values in $\mathbf{R}_{>0}$). Note that $x_{i_1}(a_1)^{-1}B^-x_{i_1}(a_1)$ tends to $\dot{s}_iB^-\dot{s}_i^{-1}$ and $y_{i_1^*}(b_1' + ca_1^{-1})$ tends to $y_{i_1^*}(b_1')$, hence in the limit we obtain the equality

(b) $x_{i_n}(a_n)^{-1}\ldots x_{i_2}(a_2)^{-1}\dot{s}_i B^- = {}^{y_{i_n^*}(b_n)^{-1}\ldots y_{i_2^*}(b_2)^{-1}y_{i_1^*}(b_1')^{-1}}B^+.$

The relative position of B^+ with the left hand side of (b) is s_iw_0; since $b_1', b_2, \ldots, b_n$ are in $\mathbf{R}_{>0}$, the right side is in $\mathscr{B}_{>0}$ (see [4, 8.8]) hence the relative position of B^+ with the right hand side of (b) is w_0 (see [4, 8.14]). This is a contradiction. The lemma is proved.

3.7.   *In the remainder of this section we assume that $K = \mathbf{R}(\epsilon)$ (unless otherwise specified).*

For any $(\mathbf{i}, \mathbf{c}) \in \mathbf{H}$, let $U_{\mathbf{i},\mathbf{c}}^+$ be the set of all elements of U^+ of the form $x_{i_1}(a_1)x_{i_2}(a_2)\ldots x_{i_n}(a_n)$ where $a_1, a_2, \ldots, a_n$ are in $K_{>0}$ and $|a_1| = c_1, |a_2| = c_2, \ldots, |a_n| = c_n$.

Similarly, let $U_{\mathbf{i},\mathbf{c}}^-$ be the set of all elements of U^- of the form $y_{i_1}(a_1)y_{i_2}(a_2)\ldots y_{i_n}(a_n)$ where $a_1, a_2, \ldots, a_n$ are in $K_{>0}$ and $|a_1| = c_1, |a_2| = c_2, \ldots, |a_n| = c_n$.

For fixed $\mathbf{i} \in \mathbf{H}$, we have partitions

$$U_{>0}^+ = \sqcup_{\mathbf{c}\in\mathbf{Z}^n}U_{\mathbf{i},\mathbf{c}}^+, \quad U_{>0}^- = \sqcup_{\mathbf{c}\in\mathbf{Z}^n}U_{\mathbf{i},\mathbf{c}}^-$$

(see [4, 2.7]). These partitions are independent of the choice of $\mathbf{i}$ (see [4, 9.5]). The subsets of $U_{>0}^+$ (resp. of $U_{>0}^-$) which make up this partition are called the w_0-zones of $U_{>0}^+$ (resp. of $U_{>0}^-$).

The map $(\mathbf{i}, \mathbf{c}) \mapsto U_{\mathbf{i},\mathbf{c}}^+$ is constant on each connected component of $\hat{\mathbf{H}}$ (see [4, 9.4]) and it induces a bijection between $\hat{\mathbf{H}}^\dagger$ and the set of w_0-zones of $U_{>0}^+$. Similarly, the map $(\mathbf{i}, \mathbf{c}) \mapsto U_{\mathbf{i},\mathbf{c}}^-$ is constant on each connected component of $\hat{\mathbf{H}}$ and it induces a bijection between $\hat{\mathbf{H}}^\dagger$ and the set of w_0-zones of $U_{>0}^-$.

Lemma 3.8. *Let* $\mathbf{i} = (i_1, i_2, \ldots, i_n) \in \mathbf{H}$ *and let* $\mathbf{i}^* = (i_1^*, i_2^*, \ldots, i_n^*) \in \mathbf{H}$.

(a) Let $\mathbf{c} = (c_1, c_2, \ldots, c_n) \in \mathbf{Z}^n, \mathbf{c}' = (c_1', c_2', \ldots, c_n') \in \mathbf{Z}^n$ and let $p \in \mathbf{Z}$. Let $_p\mathbf{c} = \mathbf{c} + (p, 0, 0, \ldots, 0), _{-p}\mathbf{c}' = \mathbf{c}' - (p, 0, 0, \ldots, 0)$. If $\phi(U_{\mathbf{i},\mathbf{c}}^+) \subset U_{\mathbf{i}^*,\mathbf{c}'}^-$, then $\phi(U_{\mathbf{i},_p\mathbf{c}}^+) \subset U_{\mathbf{i}^*,_{-p}\mathbf{c}'}^-$.

(b) If $\mathbf{0} = (0, 0, \ldots, 0) \in \mathbf{Z}^n$, then $\phi(U_{\mathbf{i},\mathbf{0}}^+) \subset U_{\mathbf{i}^*,\mathbf{0}}^-$.

We prove (a). Let $p_k : K_{>0}^{n-1} \to K_{>0}, k = 1, 2, \ldots, n$ be as in 3.6. By assumption, we have

(c) $(a_1, a_2, \ldots, a_n) \in K_{>0}^n, |a_1| = c_1, |a_2| = c_2, \ldots |a_n| = c_n \implies |p_1(a_2, \ldots a_n) a_1^{-1}| = c_1'$ and $|p_k(a_2, \ldots a_n)| = c_k'$ for $k = 2, \ldots, n$.

Now let $(a_1, a_2, \ldots, a_n) \in K_{>0}$ be such that $|a_1| = c_1 + p, |a_2| = c_2, \ldots |a_n| = c_n$. Then $a_1\epsilon^{-p} \in K_{>0}$ satisfies $|a_1\epsilon^{-p}| = c_1$. Applying (c) to $(a_1\epsilon^{-p}, a_2, \ldots, a_n) \in K_{>0}^n$, we obtain

$$|p_1(a_2, \ldots a_n)a_1^{-1}\epsilon^p| = c_1'$$

(that is, $|p_1(a_2, \ldots a_n)a_1^{-1}| = c_1' - p$) and $|p_k(a_2, \ldots a_n)| = c_k'$ for $k = 2, \ldots, n$. This shows that $\phi(U_{\mathbf{i},_p\mathbf{c}}^+) \subset U_{\mathbf{i}^*,_{-p}\mathbf{c}'}^-$ and (a) is proved.

We prove (b). We fix $(a_1, a_2, \ldots, a_n) \in K_{>0}^n$ such that $|a_1| = |a_2| = \cdots = |a_n| = 0$. Let $(b_1, b_2, \ldots, b_n) \in K_{>0}^n$ be defined by

$$\phi(x_{i_1}(a_1)x_{i_2}(a_2)\ldots x_{i_n}(a_n)) = y_{i_1^*}(b_1)y_{i_2^*}(b_2)\ldots y_{i_n^*}(b_n).$$

Then

$$x_{i_n}(a_n)^{-1}\ldots x_{i_1}(a_1)^{-1}B^- = y_{i_n^*}(b_n)^{-1}\ldots y_{i_1^*}(b_1)^{-1}B^+.$$

We may specialize ϵ to an arbitrary $\epsilon_0 \in \mathbf{R}_{>0}$ bounded above by a fixed small number in $\mathbf{R}_{>0}$. We then get an equality of Borel subgroups (over $\mathbf{R}$)

(d) $x_{i_n}(a_{n,\epsilon_0})^{-1}\ldots x_{i_1}(a_{1,\epsilon_0})^{-1}B^- = y_{i_n^*}(b_{n,\epsilon_0})^{-1}\ldots y_{i_1^*}(b_{1,\epsilon_0})^{-1}B^+$

depending on the parameter ϵ_0. Since $|a_k| = 0$, we have $\lim_{\epsilon_0 \to 0} a_{k,\epsilon_0} = d_k$ where $d_k \in \mathbf{R}_{>0}$. (Here $k = 1, \ldots, n$.). Hence the left hand side of (a) has a limit (in $\mathscr{B}$) as ϵ_0 tends to 0, namely $x_{i_n}(d_n)^{-1}\ldots x_{i_1}(d_1)^{-1}B^-$; by 3.3(a), this is of the form

(e) $y_{i_n^*}(f_n)^{-1}\ldots y_{i_1^*}(f_1)^{-1}B^+$

for some well defined $(f_1, \ldots, f_n) \in \mathbf{R}_{>0}^n$. Thus, the right hand side of (d) converges to (e) as ϵ_0 tends to 0. Since the map $U^- \to \mathscr{B}$ given by $u \mapsto u^{-1}B^+u$ is homeomorphism of U^- onto its image and the map $(g_1, \ldots, g_n) \to y_{i_1^*}(g_1)\ldots y_{i_n^*}(g_n)$ is a homeomorphism of $\mathbf{R}_{>0}^n$ onto $U_{>0}^-$, it follows that $\lim_{\epsilon_0 \to 0} b_{k,\epsilon_0} = f_k \in \mathbf{R}_{>0}$. (Here $k = 1, \ldots, n$.) This clearly forces $|b_k| = 0$ for all k. It follows that $y_{i_1^*}(b_1)y_{i_2^*}(b_2)\ldots y_{i_n^*}(b_n) \in U_{\mathbf{i}^*,\mathbf{0}}^-$. This proves (b).

Theorem 3.9. (a) The bijection $\phi : U_{>0}^+ \xrightarrow{\sim} U_{>0}^-$ maps any w_0-zone of $U_{>0}^+$ onto a w_0-zone of $U_{>0}^-$.

(b) We define a bijection $\Phi : \hat{\mathbf{H}}^\dagger \xrightarrow{\sim} \hat{\mathbf{H}}^\dagger$ by the commutative diagram (of bijections)

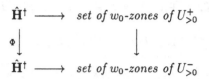

where the horizontal maps are as in 3.7 and the right vertical map is induced
by ϕ (see (a)).
Then Φ satisfies the conditions 2.9(a),(b).

We prove (a). We transport the graph structure on $\hat{\mathbf{H}}^\dagger$ (see 2.7) to the set of
w_0-zones of $U^+_{>0}$ by means of the bijection in 3.7. We obtain a graph whose set
of vertices is the set of w_0-zones of $U^+_{>0}$. Let $\mathscr{J}$ be the set consisting of those
w_0-zones Z of $U^+_{>0}$ such that $\phi(Z)$ is contained in a w_0-zone of $U^-_{>0}$. By 3.8(b),
$\mathscr{J}$ is non-empty and by 3.8(a), $\mathscr{J}$ is a union of connected components of our
graph. Since our graph is connected (see 2.8), we see that $\mathscr{J}$ is the full set of
w_0-zones of $U^+_{>0}$. Since ϕ is a bijection, (a) follows.
Now (b) follows immediately from 3.8 and the definitions. The theorem is
proved.

3.10. (a) Part (a) of the theorem follows also directly from the remarks in 3.4
(without using 3.6).
(b) Theorem 2.9 clearly follows from Theorem 3.9. Using the remarks in 3.4,
we deduce that the map Φ in 2.9 is of a very special kind. More precisely, let
us fix $\mathbf{i}, \mathbf{j}$ in $\mathbf{H}$. Then $\Phi[\mathbf{i}, \mathbf{c}] = [\mathbf{j}, \mathbf{c}']$ where the coordinates of $\mathbf{c}'$ can be obtained
from the coodinates of $\mathbf{c}$ by performing a finite succession of operations of the
form $x + y, x - y, \min(x, y)$. Hence Φ is a piecewise linear function.

4. CANONICAL BASES

In this section, we assume that $K = \mathbf{R}$.

4.1. Let $\mathbf{U}$ be the enveloping algebra of the Lie algebra of G over K. For
$i \in I$, let $e_i, f_i \in \mathbf{U}$ be the differentials of x_i, y_i (in the Lie algebra of G) regarded
as elements of $\mathbf{U}$. Let $\mathbf{U}^+$ (resp. $\mathbf{U}^-$) be the subalgebra of $\mathbf{U}$ generated by
the e_i (resp. the f_i). Let $\mathfrak{f}$ be the associative K-algebra with 1 defined by the
generators $\theta_i (i \in I)$ and the Serre relations so that we have algebra isomorphisms
$\mathfrak{f} \xrightarrow{\sim} \mathbf{U}^+, x \mapsto x^+$ and $\mathfrak{f} \xrightarrow{\sim} \mathbf{U}^-, x \mapsto x^-$, defined by $\theta_i^+ = e_i, \theta_i^- = f_i$.
Let $\mathbf{B}$ be the basis of the K-vector space $\mathfrak{f}$ obtained by specializing at $v = 1$ the
canonical basis [2] of the corresponding quantized enveloping algebra and then
extending scalars from $\mathbf{Q}$ to K.

4.2. For $i \in I$, let $\tilde{\phi}_i : \mathbf{B} \to \mathbf{B}$ and $\tilde{\epsilon}_i : \mathbf{B} \to \mathbf{B} \cup \{0\}$ be the operators defined
by Kashiwara [1] (here we follow the notation of [3, 17.3]). (Strictly speaking,
these operators are defined on the limit as $q = \infty$ of the q-analogue of $\mathbf{B}$; but
this limit is in natural bijection with $\mathbf{B}$.)

4.3. In [2] (see also [3, §42]) we have defined an explicit bijection

(a) $u : \hat{\mathbf{H}}^\dagger_+ \xrightarrow{\sim} \mathbf{B}$.

(For any fixed $\mathbf{i} \in \mathbf{H}$, the map $\mathbf{N}^n \mapsto \mathbf{B}$ given by $\mathbf{c} \mapsto u[\mathbf{i}, \mathbf{c}]$ is a bijection.) We have

(b) $u(h_0) = 1$.

Using [3, 42.1.15, 17.3.7], we see that, for any $\zeta \in \hat{\mathbf{H}}^\dagger_+$, we have

(c) $\tilde{\phi}_i u(\zeta) = u(T_{i,1}\zeta)$;
(d) $\tilde{\epsilon}_i u(\zeta) = u(T_{i,-1}\zeta)$ if $g_i(\zeta) \geq 1$;
(e) $\tilde{\epsilon}_i u(\zeta) = 0$ if $g_i(\zeta) = 0$.

Note that properties (b),(c) characterize u.

4.4. Let $\lambda \in X^+$. There exists a simple algebraic G-module Λ_λ of finite dimension over K with a non-zero vector η such that $x_i(a)\eta = \eta$ for all $i \in I, a \in K$, $t\eta = \lambda(t)\eta$ for all $t \in T$. Then Λ_λ is naturally a (simple) $\mathbf{U}$-module. Let

$$_\lambda\mathbf{B} = \{b \in \mathbf{B} | b^-\eta \neq 0\}.$$

By [2, §8], $\{b^-\eta | b \in {}_\lambda\mathbf{B}\}$ is a (canonical) basis of the K-vector space Λ_λ. We denote it by $\underline{\mathbf{B}}_\lambda$. Let

$$\hat{\mathbf{H}}_\lambda = \{\zeta \in \hat{\mathbf{H}}^\dagger_+ | g'_{i\bullet}(\zeta) \leq f(i) \quad \forall i\}$$

where $f : I \to \mathbf{N}$ is defined by

$$f(i) = \langle \check{\alpha}_i, \lambda \rangle.$$

From [2, §8] one can deduce that:

(a)   *the bijection $u : \hat{\mathbf{H}}^\dagger \xrightarrow{\sim} \mathbf{B}$ restricts to a bijection $u_\lambda : \hat{\mathbf{H}}_\lambda \xrightarrow{\sim} {}_\lambda\mathbf{B}$.*

4.5.   There is a unique vector $\xi \in \underline{\mathbf{B}}_\lambda$ such that $f_i\xi = 0$ for all i and $t\xi = (w_0\lambda)(t)$ for all $t \in T$. (We use the natural action of W on X.)

4.6.   Note that $-w_0\lambda \in X^+$. From [3, 21.1.2] we see that $b \mapsto b^+\xi$ is a bijection of $_{-w_0\lambda}\mathbf{B}$ onto $\underline{\mathbf{B}}_\lambda$. Hence we have a unique bijection $\kappa : {}_\lambda\mathbf{B} \xrightarrow{\sim} {}_{-w_0\lambda}\mathbf{B}$ such that $\kappa(b)^+\xi = b^-\eta$ for all $b \in {}_\lambda\mathbf{B}$.

Lemma 4.7. *Let $f^* : I \to \mathbf{N}$ be defined by $f^*(i) = f(i^*)$.*
(a)   *We have $u(h_{f^*}) \in {}_\lambda\mathbf{B}$ and $u(h_{f^*})^-\eta = \xi$; hence $\kappa(u(h_{f^*})) = 1$.*
(b)   *We have $u(h_f) \in {}_{-w_0\lambda}\mathbf{B}$ and $u(h_f)^-\xi = \eta$; hence $\kappa(1) = u(h_f)$.*

The subset $\hat{\mathbf{H}}^\dagger_\lambda$ can be described combinatorially (see [2, 8.12, 8.13]); moreover, the weight of an element $u(\zeta)^-\eta$ for $\zeta \in \hat{\mathbf{H}}^\dagger_\lambda$ can be described purely combinatorially (see [2, 2.8, 2.9]). In terms of these decriptions it is easy to verify that $h_{f^*} \in \hat{\mathbf{H}}^\dagger_\lambda$ and that the weight of $u(h_{f^*})^-\eta$ is the same as the weight of ξ; this forces the first equality in (a). The proof of (b) is similar.

4.8. For $i \in I$, let $\tilde{F}_i, \tilde{E}_i : \mathbf{B}_\lambda \to \mathbf{B}_\lambda \cup \{0\}$ be the operators defined by Kashiwara [1] (here we follow the notation of [3, §18]). (Strictly speaking, these operators are defined on the limit as $q = \infty$ of the q-analogue of $\mathbf{B}_\lambda$; but this limit is in natural bijection with $\mathbf{B}_\lambda$.)

We shall use the following known properties of these operators [1]:

(a) If $\beta \in \mathbf{B}_\lambda$, then there exists a sequence $i_1, i_2, \ldots, i_k$ in I such that $\beta = \tilde{F}_{i_1} \tilde{F}_{i_2} \ldots \tilde{F}_{i_k} \eta$.

(b) Let $i \in I$ and let β', β'' be elements of $\mathbf{B}_\lambda$ such that $\beta'' = \tilde{F}_i \beta'$. Then:

   (b1) $\beta' = \tilde{E}_i \beta''$;

   (b2) if $b_1', b_1'' \in {}_\lambda\mathbf{B}$ are defined by $b_1'{}^-\eta = \beta', b_1''{}^-\eta = \beta''$, then $b_1'' = \tilde{\phi}_i b_1'$;

   (b3) if $b_2', b_2'' \in {}_{-w_0\lambda}\mathbf{B}$ are defined by $b_2'{}^+\xi = \beta', b_2''{}^+\xi = \beta''$, then $b_2'' = \tilde{\epsilon}_{i*} b_2'$.

Proposition 4.9. *We have a commutative diagram of bijections*

$$
\begin{array}{ccc}
\hat{\mathbf{H}}_\lambda & \xrightarrow{u_\lambda} & {}_\lambda\mathbf{B} \\
\downarrow & & \kappa \downarrow \\
\hat{\mathbf{H}}_{-w_0\lambda} & \xrightarrow{u_{-w_0\lambda}} & {}_{-w_0\lambda}\mathbf{B}
\end{array}
$$

where κ is as in 4.6 and the left vertical arrow is given by the restriction of $S_f \circ \Phi : \hat{\mathbf{H}}^\dagger \to \hat{\mathbf{H}}^\dagger$ (see 2.4, 2.9).

Using 4.7(b), we have

$$\kappa(u(h_0)) = \kappa(1) = u(h_f) = u(S_f(h_0)) = u(S_f\Phi(h_0)).$$

Now let ζ', ζ'' be two elements of $\hat{\mathbf{H}}_\lambda$ such that $\beta' = u(\zeta'), \beta'' = u(\zeta'')$ satisfy $\beta'' = \tilde{F}_i \beta'$ for some $i \in I$. We will show that:

(a) *if $\kappa(u(\zeta')) = u(S_f\Phi(\zeta'))$, then $\kappa(u(\zeta'')) = u(S_f\Phi(\zeta''))$.*

First note that

(b) $\kappa(\tilde{F}_i \beta') = \tilde{\epsilon}_{i*} \kappa(\beta')$

(by 4.8(b3)). Next we show that

(c) $T_{i,1} \zeta' = \zeta''$.

Indeed, it suffices to check this after applying u to both sides. We have $u(T_{i,1}\zeta') = \tilde{\phi}_i u(\zeta') = \tilde{\phi}_i \beta' = \beta'' = u(\zeta'')$ (we have used 4.3(c) and 4.8(b2)). Thus, (c) is verified.

With the assumption of (a), we have

$$\kappa(u(\zeta'')) = \kappa(\beta'') = \kappa(\tilde{F}_i \beta') \overset{(b)}{=} \tilde{\epsilon}_{i*}\kappa(\beta') = \tilde{\epsilon}_{i*}\kappa u(\zeta') = \tilde{\epsilon}_{i*} u(S_f\Phi(\zeta'))$$

$$\overset{4.3(d)}{=} u(T_{i*,-1} S_f\Phi(\zeta')) = u(S_f\Phi(T_{i,1}\zeta')) \overset{(c)}{=} u(S_f\Phi(\zeta''))$$

(the last but one equality follows from 2.6 and 2.9(b)). Thus, (a) is verified.

By 4.8(a), for any $\zeta \in \hat{\mathbf{H}}_\lambda$ there exists a sequence $\zeta_0, \zeta_1, \ldots, \zeta_k$ in $\hat{\mathbf{H}}_\lambda$ such that $\zeta_0 = h_0, \zeta_k = \zeta$ and such that the argument above is applicable to any two consecutive terms ζ', ζ'' of the sequence. We then deduce by induction on k that $\kappa u(\zeta) = u(S_f\Phi(\zeta))$. The proposition is proved.

Corollary 4.10. *We have* $\Phi(h_{f^*}) = h_{-f}$.

We apply the commutativity of the diagram in 4.9 to the vector $h_{f^*} \in \hat{\mathbf{H}}_\lambda$. Using 4.7(a), we have

$$u(S_f\Phi(h_{f^*})) = \kappa(u(h_{f^*})) = 1 = u(h_0).$$

Hence $S_f\Phi(h_{f^*}) = h_0$. Applying S_{-f} to both sides, we obtain $\Phi(h_{f^*}) = S_{-f}(h_0) = h_{-f}$. The corollary is proved.

4.11. Proof of 2.10. We may clearly assume that G is simply connected (the statement of 2.10 depends only on the Coxeter diagram). By 4.10, the statement 2.10(a) is true for any $f : I \to \mathbf{Z}$ that is attached to some $\lambda \in X^+$ by $f(i) = \langle \check{\alpha}_i, \lambda \rangle$ for all i. The set $\mathscr{G}$ of such f is a set of generators of the additive group of all $f : I \to \mathbf{Z}$ (since G is simply connected).

Let $f \in \mathscr{G}$. Since 2.10(a) holds for f, it follows that the operator $S_f \circ \Phi \circ S_{f^*}$ takes h_0 to itself. Since this operator has the same commutation formula with $T_{i,p}$ as Φ (by 2.6), it follows that this operator coincides with Φ (by the uniqueness of Φ). Thus, 2.10(b) holds for $f \in \mathscr{G}$.

Next we note that the set of all $f : I \to \mathbf{Z}$ such that 2.10(b) holds for f is a subgroup of the group of all $f : I \to \mathbf{Z}$. (This is easily verified, using 2.4(a).) Since this set contains $\mathscr{G}$, which generates the whole group, it follows that 2.10(b) holds for any f. Finally, if 2.10(b) holds for f, then 2.10(a) holds for f. (Apply both sides of 2.10(a) to h_0.) This completes the proof of 2.10.

5. Inequalities defining $G_{\geq 0}, G_{>0}$

5.1. *In this section we assume that* $K = \mathbf{R}$.

As in [4], we define $\mathscr{B}_{>0}$ to be the subset $\{uB^-u^{-1}|u \in U^+_{>0}\} = \{\tilde{u}B^+\tilde{u}^{-1}|\tilde{u} \in U^-_{>0}\}$ of $\mathscr{B}$ and $\mathscr{B}_{\geq 0}$ to be the closure of $\mathscr{B}_{>0}$ in the manifold $\mathscr{B}$.

Let us fix $\lambda^1, \lambda^2, \ldots, \lambda^s$ in X^+. For $r = 1, \ldots, s$, let $\mathbf{P}^r$ be the set of lines in the vector space Λ_{λ^r}. Let $\mathbf{P}^r_{\geq 0}$ (resp. $\mathbf{P}^r_{>0}$) be the set of all lines in Λ_{λ^r} which are spanned by a non-zero vector whose coordinates with respect to the canonical basis $\underline{\mathbf{B}}_{\lambda^r}$ are all in $\mathbf{R}_{\geq 0}$ (resp. all in $\mathbf{R}_{>0}$).

For $B \in \mathscr{B}$ let L^r_B be the unique B-stable line in $\mathbf{P}^r$.

The vectors η, ξ in Λ_λ (see 4.4, 4.5) for $\lambda = \lambda^r$ will be denoted by η^r, ξ^r.

Proposition 5.2. *Assume that* $\langle \check{\alpha}_i, \lambda_1 + \cdots + \lambda_s \rangle > 0$ *for any* $i \in I$. *Let* $B \in \mathscr{B}$.

(a) $B \in \mathscr{B}_{>0}$ *if and only if, for* $r = 1, \ldots, s$, *we have* $L^r_B \in \mathbf{P}^r_{>0}$.
(b) $B \in \mathscr{B}_{\geq 0}$ *if and only if, for* $r = 1, \ldots, s$, *we have* $L^r_B \in \mathbf{P}^r_{\geq 0}$.

In the case where $s = 1$, this is just [4, 8.17]. The proof in the general case requires only minor modifications. We will omit it.

Proposition 5.3. *Assume that* $\langle \check{\alpha}_i, \lambda_1 + \cdots + \lambda_s \rangle > 0$ *for any* $i \in I$. *Let* $u \in U^+$.

(a) $u \in U^+_{>0}$ *if and only if for any* r, $u\xi^r$ *is a linear combination with coefficients* > 0 *of elements in the canonical basis* $\underline{\mathbf{B}}_{\lambda^r}$.

(b) $u \in U_{\geq 0}^+$ if and only if for any r, $u\xi^r$ is a linear combination with coefficients ≥ 0 of elements in the canonical basis $\mathbf{B}_{\lambda^r}$.

We prove (a). We consider the following conditions:

(c) $u \in U_{>0}^+$;
(d) $uB^-u^{-1} \in \mathscr{B}_{>0}$;
(e) $L^r_{uB^-u^{-1}} \in \mathbf{P}^r_{>0}$ for any r;
(f) $\mathbf{R}u\xi^r \in \mathbf{P}^r_{>0}$ for any r;
(g) $u\xi^r$ is a > 0 linear combination of elements in $\mathbf{B}_{\lambda^r}$.

Then (c),(d) are equivalent by the definition of $\mathscr{B}_{>0}$. (d),(e) are equivalent by 5.2(a); (e),(f) are obviously equivalent; (f),(g) are equivalent since ξ^r appears with coefficient 1 in $u\xi$ (expressed in terms of the basis $\mathbf{B}_{\lambda^r}$). (a) follows.

We prove (b). We consider the following conditions:

(h) $u \in U_{\geq 0}^+$;
(i) $uB^-u^{-1} \in \mathscr{B}_{\geq 0}$;
(j) $L^r_{uB^-u^{-1}} \in \mathbf{P}^r_{\geq 0}$ for any r;
(k) $\mathbf{R}u\xi^r \in \mathbf{P}^r_{\geq 0}$ for any r;
(l) $u\xi^r$ is a ≥ 0 linear combination of elements in the basis $\mathbf{B}_{\lambda^r}$.

Then (h),(i) are equivalent (the proof is as in [4, 8.4]); (i),(j) are equivalent by 5.2(b); (j),(k) are obviously equivalent; (k),(l) are equivalent since ξ^r appears with coefficient 1 in $u\xi$ (expressed in terms of the basis $\mathbf{B}_{\lambda^r}$). (b) follows.

Proposition 5.4. *Assume that* $\langle \check{\alpha}_i, \lambda_1 + \cdots + \lambda_s \rangle > 0$ *for any* $i \in I$. *Let* $u \in U^-$.

(a) $u \in U_{>0}^-$ *if and only if for any* r, $u\eta^r$ *is a linear combination with coefficients* > 0 *of elements in the canonical basis* $\mathbf{B}_{\lambda^r}$.
(b) $u \in U_{\geq 0}^-$ *if and only if for any* r, $u\eta^r$ *is a linear combination with coefficients* ≥ 0 *of elements in the canonical basis* $\mathbf{B}_{\lambda^r}$.

The proof is just like that of 5.3.

Theorem 5.5. *Assume that* $\lambda^1, \lambda^2, \ldots, \lambda^s$ *generate the abelian group* X. *Let* $g \in G$.

(a) $g \in G_{>0}$ *if and only if for any* r, $g\xi^r$ *and* $\Pi(g)\eta^r$ *are linear combinations with coefficients* > 0 *of elements in the canonical basis* $\mathbf{B}_{\lambda^r}$.
(b) $g \in G_{\geq 0}$ *if and only if for any* r, $g\xi^r$ *is a linear combination with coefficients* ≥ 0 *of elements in the canonical basis* $\mathbf{B}_{\lambda^r}$ *(and the coefficient of* ξ^r *is* > 0) *and* $\Pi(g)\eta^r$ *is a linear combination with coefficients* ≥ 0 *of elements in the canonical basis* $\mathbf{B}_{\lambda^r}$ *(and the coefficient of* η^r *is* > 0).

First note that our assumption implies that $\langle \check{\alpha}_i, \lambda_1 + \cdots + \lambda_s \rangle \neq 0$ for any $i \in I$. Since $\langle \check{\alpha}_i, \lambda_1 + \cdots + \lambda_s \rangle \geq 0$ for any $i \in I$, it follows that $\langle \check{\alpha}_i, \lambda_1 + \cdots + \lambda_s \rangle > 0$ for any $i \in I$. Hence 5.3, 5.4 are applicable.

We prove (a). Assume that $g\xi^r, \Pi(g)\eta^r$ are as in (a). Then, for any r, ξ^r appears with a non-zero coefficient in $g\xi^r$. It follows that $g \in U^+TU^-$. Thus $g = utu'$ where $u \in U^+, u' \in U^-, t \in T$. We have $g\xi^r = ut\xi^r = a_r u\xi^r$ where $t\xi^r = a_r\xi^r, a_r \in \mathbf{R}$. The coefficient of ξ^r in $a_r u\xi^r$ is a_r, while the coefficient of ξ^r

in $g\xi^r$ is > 0 by assumption. Hence $a_r > 0$ and we see that $u\xi^r = (1/a_r)g\xi^r$ is a linear combination with > 0 coefficients of elements in the basis $\underline{\mathbf{B}}_{\lambda^r}$. Using 5.3(a), we see that $u \in U^+_{>0}$. Now $\Pi(g) = \Pi(u')t^{-1}\Pi(u)$ and $\Pi(u') \in U^-, \Pi(u) \in U^+$. We have $\Pi(g)\eta^r = \Pi(u')t^{-1}\eta^r = a'_r\Pi(u')\eta^r$. As before we see that $a'_r > 0$. As in the previous argument we see, using 5.4(a), that $\Pi(u') \in U^-_{>0}$ hence $u' \in U^-_{>0}$. As we have seen, we have $a'_r > 0$ for any r. Since (λ^r) generate X, it follows that $t \in T_{>0}$. Thus $g \in G_{>0}$. The converse is clear. This proves (a).
The proof of (b) is similar.

Remark 5.6. (a) If G is simply connected, we may take $\lambda^1, \lambda^2, \ldots, \lambda^s$ to be the fundamental weights.

(b) Assume that $G = GL_s(\mathbf{R})$ with the standard épinglage. Let us take $\lambda^1, \ldots, \lambda^s$ to be the exterior powers of the standard representation of G. Then 5.5 becomes a classical criterion for total positivity of a matrix in terms of minors.

REFERENCES

1. M. Kashiwara, *On crystal bases of the q-analogue of universal enveloping algebras*, Duke Math. J. **133** (1991), 465–516.
2. G. Lusztig, *Canonical bases arising from quantized enveloping algebras*, J. Amer. Math. Soc. **3** (1990), 447–498.
3. _____, *Introduction to quantum groups*, Progress in Math., no. 110, Birkhauser, Boston, 1993.
4. _____, *Total positivity in reductive groups*, Lie Theory and Geometry: in honor of Bertram Kostant, Progress in Math., no. 123, Birkhauser, Boston, 1994, pp. 531–568.

DEPARTMENT OF MATHEMATICS, M.I.T., CAMBRIDGE, MA 02139

ON THE NUMBER OF ORBITS OF A PARABOLIC
SUBGROUP ON ITS UNIPOTENT RADICAL

VLADIMIR POPOV* AND GERHARD RÖHRLE**

Dedicated to the memory of Roger Richardson

ABSTRACT. In this paper we prove some general results concerning algebraic group actions on an algebraic variety and apply these to the action of a parabolic subgroup of a reductive algebraic group on its unipotent radical. We assume that the ground field is algebraically closed and of characteristic zero. As an application we determine each minimal parabolic subgroup that admits a finite number of orbits under this action. We also exhibit other canonical families of parabolic subgroups with this property.

1. INTRODUCTION

Let G be a reductive algebraic group and $P = LP_u$ a parabolic subgroup of G with P_u the unipotent radical of P and L a Levi complement in P. In 1974 R. W. Richardson proved that P admits a dense orbit on P_u [11]. The proof relies on the fact that the number of unipotent classes of G is finite. This latter fact was first proved by Richardson under some mild restrictions on the characteristic of the ground field [10] which were later removed by G. Lusztig [6].

In view of these results it is natural to ask when the number of P-orbits on P_u is finite. In general, one certainly expects the number of orbits to be infinite. An indication for this behavior is illustrated by Theorem 1.1 where the answer is given for the case of a Borel subgroup. This note is a first step in solving the following general problem:

Problem 1. *Determine each parabolic subgroup $P = LP_u$ of a reductive algebraic group G that has a finite number of orbits on P_u.*

We emphasize that in case there are infinitely many orbits, the geometry of orbits is somewhat complicated, because then, by Richardson's Dense Orbit Theorem, infinitely many orbits must occur in a proper invariant subvariety of P_u, while the complement to this subvariety is the dense orbit in P_u. Also, it follows from another result of R. W. Richardson that L, and hence P, only has a finite number of orbits on consecutive quotients of the descending central series of P_u [12, Theorem E].

In particular, this result implies that the number of P-orbits on P_u is finite if P_u is Abelian. Thus parabolic subgroups with an Abelian unipotent radical

---

*Research supported in part by Grant # MQZ000 from the International Science Foundation.
**Research supported by ARC Grant # A69030627 (principal investigator: Prof. G. Lehrer).

provide a natural family with a finite number of orbits. A detailed analysis of this case can be found in [13]. In addition, there exist other natural families of parabolic subgroups which share this property. See Section 5 of this paper.

To avoid technical considerations we assume that the ground field K is algebraically closed of characteristic zero. Therefore, the exponential mapping is a P-equivariant isomorphism between P_u and its Lie algebra $\text{Lie}(P_u)$.

In Section 2 we prove some general results concerning the structure of orbits of algebraic groups and apply these to the action of a parabolic subgroup on its unipotent radical.

In a forthcoming manuscript by the second author a close connection between the number of orbits of P on P_u and the length of the descending central series of P_u is investigated [14].

We observe that our results can be interpreted in terms of the general concept of the *modality* of an algebraic group acting morphically on an algebraic variety. Recall from [18] (or [9, 5.2]) that for an algebraic group R acting morphically on an algebraic variety X the *modality* of the action, $\text{mod}(R : X)$, is defined as

$$\text{mod}(R : X) = \max_Z \min_{z \in Z} \text{codim}_Z R^0 \cdot z,$$

where Z runs through all irreducible R^0-invariant subvarieties of X. Informally, $\text{mod}(R : X)$ is the maximal number of parameters on which a family of R-orbits on X may depend. For a parabolic subgroup P of an algebraic group we call $\text{mod}(P : P_u)$ the *modality* of P. The modality of P is 0 precisely when P admits only a finite number of orbits on P_u. Thus the problem of determining these parabolics can be extended to the classification of all parabolics of a given modality. This leads to the following natural generalization of Problem 1:

Problem 2. *(1) Determine all parabolic subgroups of a reductive algebraic group G of a given modality. (2) Determine the modality of each parabolic subgroup of G.*

Some of our general results at the end of Section 2 are formulated in this more general setting.

In a forthcoming manuscript by the first author a general finiteness theorem for parabolic subgroups of a fixed modality is proved [7].

Because a reductive algebraic group is the direct product of simple ones over its center, and similarly for parabolic subgroups, we can reduce to the case of a simple algebraic group.

The following result of V. Kashin [5] is the starting point for our study:

Theorem 1.1. *Let G be a simple algebraic group and B a Borel subgroup of G. The number of orbits of B on its unipotent radical B_u is finite if and only if G is of type A_n for $n \leq 4$ or B_2.*

By a minimal parabolic subgroup of G we mean a parabolic subgroup of semi-simple rank one. Our main theorem on these subgroups is

Theorem 1.2. *Let G be a simple algebraic group, $P = LP_u$ a minimal parabolic subgroup, and B a Borel subgroup of G. Assume that the number of orbits of B on its unipotent radical B_u is infinite. Then the number of P-orbits on P_u is finite if and only if G is of type A_5, B_3, C_3, D_4, or G_2.*

We give a proof of Theorem 1.2 in Sections 3 and 4. An elementary observation in Section 4 leads to the following consequences of Theorems 1.1 and 1.2:

Theorem 1.3. *If G is as in Theorem 1.2, then any non-Borel parabolic subgroup of G admits a finite number of orbits on its unipotent radical. If G is as in Theorem 1.1, then any parabolic subgroup of G has this property.*

In particular, we obtain a classification for the minimal parabolic case:

Corollary 1.4. *Let G be a simple algebraic group and let $P = LP_u$ be a minimal parabolic subgroup of G. Then the number of P-orbits on P_u is finite if and only if G is of type A_n for $n \leq 5$, B_n for $n \leq 3$, C_n for $n \leq 3$, D_4, or G_2.*

Remark 1.5. It is worth noting that in Theorem 1.2 and in Corollary 1.4 in each of the cases the number of P-orbits on P_u is finite for *any* minimal parabolic subgroup P of G.

For an algebraic group R we often denote its Lie algebra by $\mathrm{Lie}(R)$ or simply by the corresponding lower case gothic letter $\mathfrak{r}$. The identity component of R is denoted by R^0 and its unipotent radical by R_u. The Lie algebra of R_u is denoted by $\mathfrak{r}_u$. Throughout the identity element of an algebraic group is denoted by e.

Let X be an algebraic variety. For a point $x \in X$ we write $\mathrm{T}_x(X)$ for the tangent space of X at x. We say that a given property holds for a point in general position of X, if the property is fulfilled for each point of a dense open subset of X (this subset depends on the property under consideration). For a subvariety Z of X we denote the Zariski closure of Z in X by $\overline{Z}$.

For an irreducible algebraic variety X we denote by $K(X)$ the field of rational functions on X. If an algebraic group R acts rationally on X, then $K(X)^R$ denotes the field of R-invariant rational functions on X. If R acts morphically on an algebraic variety X, we denote by $R \cdot x$ the orbit of x in X under the action of R and if Z is a subset of X, then $R \cdot Z$ is the union of all orbits $R \cdot z$ in X, where $z \in Z$.

The tangent space of a vector space V at any point is canonically identified (as a vector space) with V itself. Accordingly, we identify the tangent space of any algebraic variety $Y \subset V$ with a (linear) subspace of V.

For a subset S of a vector space V we write $\langle S \rangle$ for the linear span of S in V.

Let G be a connected reductive algebraic group, T a maximal torus in G and Ψ the set of roots of G with respect to T. As usual, for a root $\alpha \in \Psi$ we denote the corresponding root space in $\mathfrak{g}$ by $\mathfrak{g}_\alpha$. Fix a set of simple roots Π of Ψ and let B be the Borel subgroup of G corresponding to Π. A parabolic subgroup of G containing B is called a standard parabolic subgroup. For a subset J of Π we denote by P_J the standard parabolic subgroup corresponding to J. We also write P_α instead of $P_{\{\alpha\}}$.

For a root $\alpha \in \Psi$ the corresponding one-parameter unipotent subgroup is denoted by U_α. We denote by $U_\alpha(\xi)$ the image of $\xi \in K$ with respect to a fixed isomorphism $K \longrightarrow U_\alpha$. If C is a subset of K, we denote $\{U_\alpha(\xi) \mid \xi \in C\}$ by $U_\alpha(C)$. In particular, $U_\alpha(0) = e$ and $U_\alpha(K) = U_\alpha$. The *support* of a subvariety Y of B_u, $\mathrm{supp}(Y)$, is the set of all roots $\alpha \in \Psi$ such that the projection of Y into U_α is non-trivial.

As a general reference for algebraic groups we cite Borel's book [1] and for information on root systems we refer the reader to Bourbaki [2]. In particular, we index the simple roots in a base of a root system of a simple algebraic group in accordance with *loc. cit.* (Planches I - IX).

2. Orbits of Algebraic Groups on Algebraic Varieties

The following result is proved in [9, Lemma 1.10]. In a somewhat weaker form it appeared for the first time in [10].

Theorem 2.1. *Let R be an algebraic group acting on an algebraic variety X. Further, let S be a closed subgroup of R and Y a locally closed S-invariant subset of X. Assume that*

$$\mathrm{T}_y(R \cdot y) \cap \mathrm{T}_y(Y) = \mathrm{T}_y(S \cdot y) \quad \text{for each point } y \in Y. \qquad (2.1.1)$$

Then for any R-orbit $\mathscr{O}$ intersecting Y the intersection $\mathscr{O} \cap Y$ is a union of finitely many S-orbits each of which is open and closed in $\mathscr{O} \cap Y$. More precisely, each irreducible component of $\mathscr{O} \cap Y$ is a single S^0-orbit.

Proof. Without loss we may assume that $S = S^0$. Let $y \in \mathscr{O} \cap Y$, i.e., $R \cdot y = \mathscr{O}$. Since Y is S-invariant, $S \cdot y \subseteq R \cdot y \cap Y$. Thus $\mathrm{T}_y(S \cdot y) \subseteq \mathrm{T}_y(R \cdot y \cap Y)$, and since $\mathrm{T}_y(R \cdot y \cap Y) \subseteq \mathrm{T}_y(R \cdot y) \cap \mathrm{T}_y(Y)$, it follows from 2.1.1 that

$$\mathrm{T}_y(S \cdot y) = \mathrm{T}_y(R \cdot y \cap Y). \qquad (2.1.2)$$

Let Z be an irreducible component of $R \cdot y \cap Y$ containing $S \cdot y$. Then $S \cdot y \subseteq Z \subseteq R \cdot y \cap Y$ implies

$$\dim S \cdot y \leq \dim Z \leq \dim \mathrm{T}_y(R \cdot y \cap Y). \qquad (2.1.3)$$

But as $S \cdot y$ is smooth, we also have $\dim S \cdot y = \dim \mathrm{T}_y(S \cdot y)$. It then follows from 2.1.2 and 2.1.3 that $\dim S \cdot y = \dim Z$. Whence the closure of $S \cdot y$ in $R \cdot y \cap Y$ coincides with Z. Thus, as $S \cdot y$ is open in its closure, $S \cdot y$ is open in Z. Further, $Z \setminus S \cdot y$ is a union of S-orbits each of which is also open in Z by the same argument. Therefore, $S \cdot y$ is also closed in Z and we conclude that $S \cdot y = Z$. The result now follows.

Here is an immediate consequence of Theorem 2.1:

Corollary 2.2. *Assume the same hypothesis as in Theorem 2.1. If the number of S-orbits on Y is infinite, then the number of R-orbits on X is infinite as well. Equivalently, if the number of R-orbits on X is finite, then so is the number of S-orbits on Y.*

In Propositions 2.4 and 2.7 we describe configurations that ensure hypothesis 2.1.1 of Theorem 2.1.

Let G be a connected reductive algebraic group, T a maximal torus in G and Ψ the set of roots of G with respect to T. For subsets of roots $A, B \subseteq \Psi$ we denote by $A + B$ the set of all roots of the form $\alpha + \beta$, where $\alpha \in A$ and $\beta \in B$.

Hypothesis 2.3. *Suppose* $A \subseteq B \subseteq \Gamma \subseteq \Psi$ *satisfy*

(H1) B *and* Γ *are closed with respect to addition in* Ψ.

(H2) $A + B \subseteq A$.

(H3) $(A + (\Gamma \setminus B)) \cap A = \varnothing$.

Then define $\mathfrak{a} = \bigoplus_{\alpha \in A} \mathfrak{g}_\alpha$ and let C be the connected subgroup of G such that $\mathrm{Lie}(C) = \bigoplus_{\gamma \in \Gamma} \mathfrak{g}_\gamma \oplus \mathfrak{t}$. Let $\mathfrak{t}_B$ be the linear span in $\mathfrak{t}$ of the coroots $h_\beta \in \mathfrak{t}$, where $\beta \in B$. Let D be the connected subgroup of G such that $\mathrm{Lie}(D) = \bigoplus_{\beta \in B} \mathfrak{g}_\beta \oplus \mathfrak{t}_B$. Note that D acts on $\mathfrak{a}$, because of H2.

The situation we have in mind in which closed subsets of roots $A \subseteq B \subseteq \Gamma \subseteq \Psi$ satisfying (H1) - (H3) arise naturally, is the setting of Proposition 2.5 below.

Proposition 2.4. *Assume that* $A \subseteq B \subseteq \Gamma \subseteq \Psi$ *satisfy Hypothesis 2.3 and let* $\mathfrak{a}$, $\mathfrak{t}_B$, C, *and* D *be as above. Then condition 2.1.1 of Theorem 2.1 holds for* $R = C$, $S = D$, $Y = \mathfrak{a}$ *and for* X *any* C-*invariant algebraic subvariety of* $\mathrm{Lie}(C)$ *containing* $\mathfrak{a}$. *More explicitly,*

$$T_x(C \cdot x) \cap T_x(\mathfrak{a}) = T_x(D \cdot x) \quad \text{for every } x \in \mathfrak{a}.$$

Proof. Since $T_x(\mathfrak{a}) = \mathfrak{a}$ for any $x \in \mathfrak{a}$, the result to be proved becomes

$$[\mathrm{Lie}(C), x] \cap \mathfrak{a} = [\mathrm{Lie}(D), x].$$

We have a decomposition of $\mathrm{Lie}(C)$ as follows

$$\mathrm{Lie}(C) = (\bigoplus_{\beta \in B} \mathfrak{g}_\beta \oplus \mathfrak{t}_B) \oplus (\bigoplus_{\gamma \in \Gamma \setminus B} \mathfrak{g}_\gamma \oplus \mathfrak{t}_B^\perp),$$

where $\mathfrak{t}_B^\perp$ is the orthogonal complement to $\mathfrak{t}_B$ in $\mathfrak{t}$ with respect to the Killing form. Observe that by H3 we have $[\bigoplus_{\gamma \in \Gamma \setminus B} \mathfrak{g}_\gamma, x] \cap \mathfrak{a} = \{0\}$ and, since $A \subseteq B$, we have $[\mathfrak{t}_B^\perp, x] = \{0\}$ for each $x \in \mathfrak{a}$. Whence

$$[\mathrm{Lie}(C), x] \cap \mathfrak{a} = [\bigoplus_{\beta \in B} \mathfrak{g}_\beta \oplus \mathfrak{t}_B, x] \cap \mathfrak{a} = [\mathrm{Lie}(D), x] \cap \mathfrak{a} = [\mathrm{Lie}(D), x],$$

as claimed.

We use the same terminology as Dynkin [4] by calling a closed connected subgroup H of an algebraic group G *regular* if the normalizer of H in G contains a maximal torus T of G. In that case the root spaces of $\mathfrak{h}$ relative to T are also root spaces of $\mathfrak{g}$ relative to T, and the set of roots of H with respect to T, $\Psi(H)$, is a subset of Ψ, the root system of G relative to T, and is closed under addition in Ψ. Additionally, if H is a reductive regular subgroup of G, then $\Psi(H)$ is a subsystem of Ψ.

Proposition 2.5. *Let G be a reductive algebraic group and H a reductive regular subgroup of G relative to a maximal torus T of G. Let Ψ be the set of roots of G with respect to T. Assume that P is a parabolic subgroup of G containing T. Let $Q = P \cap H$ and set*

$$A = \Psi(Q_u), \quad B = \Psi(Q), \quad and \quad \Gamma = \Psi(P).$$

Then $A \subseteq B \subseteq \Gamma \subseteq \Psi$ *satisfy Hypothesis 2.3.*

Proof. Clearly, H1 holds, as both Q and P are regular subgroups of G. Further, H2 holds as Q_u is a normal subgroup of Q. Finally, suppose that $\alpha, \delta \in A$ and $\gamma \in \Gamma$ such that $\alpha = \delta + \gamma$. Thus

$$\alpha - \delta = \gamma \in \Psi(H) \cap \Psi(P) = \Psi(Q) = B.$$

But this implies H3.

Corollary 2.6. *Let G be a reductive algebraic group and P a parabolic subgroup of G relative a maximal torus T of G. Suppose that H is a reductive regular subgroup of G with respect to T. Set $Q = H \cap P$ which is a parabolic subgroup of H. If the number of Q-orbits on Q_u is infinite, then so is the number of P-orbits on P_u.*

Proof. The result follows readily from Propositions 2.4 and 2.5 and Corollary 2.2.

Let Θ be a semisimple automorphism of G and let θ denote the corresponding (semisimple) automorphism of $\mathfrak{g}$, i.e., $d\Theta_e = \theta$. Suppose H is a closed connected subgroup of G and N is a closed connected normal subgroup of H such that both H and N are Θ-stable. Then $\mathfrak{h}$ and $\mathfrak{n}$ are θ-stable as well. Let G^Θ be the Θ-fixed point subgroup of G and likewise $\mathfrak{g}^\theta$ the θ-fixed point subalgebra of $\mathfrak{g}$. It is known that G^Θ is reductive and that $\mathrm{Lie}(G^\Theta) = \mathfrak{g}^\theta$. Moreover, if G is simply connected, then G^Θ is connected [17]. Let $H^\Theta = H \cap G^\Theta$ be the Θ-fixed point subgroup of H. We have $\mathrm{Lie}(H^\Theta) = \mathfrak{h}^\theta := \mathfrak{g}^\theta \cap \mathfrak{h}$, and $\mathrm{Lie}(N \cap H^\Theta) = \mathfrak{n}^\theta := \mathfrak{n} \cap \mathfrak{h}^\theta$.

Proposition 2.7. *Let Θ be a semisimple automorphism of G. Suppose H is a closed connected subgroup of G and N is a closed connected normal subgroup of H. Also assume that both H and N are Θ-stable. Then condition 2.1.1 of Theorem 2.1 holds for $R = H$, $S = H^\Theta$, $X = \mathfrak{n}$, and $Y = \mathfrak{n}^\theta$. More explicitly,*

$$T_x(H \cdot x) \cap T_x(\mathfrak{n}^\theta) = T_x(H^\Theta \cdot x) \quad for \; every \;\; x \in \mathfrak{n}^\theta.$$

Proof. The result to be proved is

$$[\mathfrak{h}, x] \cap \mathfrak{n}^\theta = [\mathfrak{h}^\theta, x] \quad \text{for every } x \in \mathfrak{n}^\theta.$$

Let Δ be the set of eigenvalues of θ on $\mathfrak{h}$ and let $\mathfrak{h}_\lambda$ denote the eigenspace of θ on $\mathfrak{h}$ with respect to $\lambda \in \Delta$. In particular, $\mathfrak{h}_1 = \mathfrak{h}^\theta$. Then $\mathfrak{h} = \bigoplus_{\lambda \in \Delta} \mathfrak{h}_\lambda$, since θ is semisimple. Further, since θ is a Lie algebra automorphism, for $\lambda, \mu \in \Delta$ we have $[\mathfrak{h}_\lambda, \mathfrak{h}_\mu] \subseteq \mathfrak{h}_{\lambda\mu}$ provided $\lambda\mu \in \Delta$ and 0 otherwise. Let $\mathfrak{h}^- = \bigoplus_{\lambda \neq 1} \mathfrak{h}_\lambda$. Then $\mathfrak{h} = \mathfrak{h}^\theta \oplus \mathfrak{h}^-$ and $[\mathfrak{h}^\theta, \mathfrak{h}^-] \subseteq \mathfrak{h}^-$.

Observe that $[\mathfrak{h}^\theta, x] \subseteq \mathfrak{n}^\theta$ and $[\mathfrak{h}^-, x] \subseteq \mathfrak{n}^-$. Therefore, $[\mathfrak{h}^-, x] \cap \mathfrak{n}^\theta = \{0\}$. Whence

$$[\mathfrak{h}, x] \cap \mathfrak{n}^\theta = [\mathfrak{h}^\theta \oplus \mathfrak{h}^-, x] \cap \mathfrak{n}^\theta = [\mathfrak{h}^\theta, x] \cap \mathfrak{n}^\theta + [\mathfrak{h}^-, x] \cap \mathfrak{n}^\theta = [\mathfrak{h}^\theta, x],$$

as desired.

We now apply Proposition 2.7 to parabolic subgroups of G.

Corollary 2.8. *Let Θ be a semisimple automorphism of G. Assume that P is a Θ-stable parabolic subgroup of G. Then Proposition 2.7 holds for the case $H = P$ and $N = P_u$. Consequently, if the number of P^Θ-orbits on $\mathfrak{p}_u^\theta$ is infinite, then so is the number of P-orbits on $\mathfrak{p}_u$. Equivalently, if P has a finite number of orbits on $\mathfrak{p}_u$, then so does P^Θ on $\mathfrak{p}_u^\theta$.*

Proof. Since P_u is a characteristic subgroup of P and P is Θ-stable, $\Theta P_u = P_u$. The remaining statements follow from Proposition 2.7 and Corollary 2.2.

In sections 3 and 4 we apply Corollary 2.8 in the case when Θ is an automorphism of G which is induced from a graph automorphism of the Dynkin-diagram of G.

Next we consider a specific situation for the action of P on $\mathfrak{p}_u$. Let S be a linear subvariety in $\mathfrak{p}_u$. Let $\mathcal{O}$ be a P-orbit in $\mathfrak{p}_u$ such that $S \cap \mathcal{O} \neq \varnothing$. Since $\mathcal{O}$ is open dense in its closure, the intersection $S \cap \mathcal{O}$ is open dense in $\overline{S \cap \mathcal{O}}$.

Suppose P acts on $\mathfrak{p}_u$ with a finite number of orbits. Then, since S is the disjoint union of its intersection with the P-orbits on $\mathfrak{p}_u$, there is a P-orbit $\mathcal{O}$ such that $\overline{S \cap \mathcal{O}} = S$. Let v be in $S \cap \mathcal{O}$ and consider the orbit map $\pi : P \to \mathcal{O}$, i.e., $p \mapsto p \cdot v$. Since char $K = 0$, the map $d\pi_e$ is surjective. Since $S \cap \mathcal{O}$ is a subvariety of $\mathcal{O}$, it follows that $T_v(S \cap \mathcal{O}) \subseteq T_v(\mathcal{O})$ and because $S \cap \mathcal{O}$ is open in S, we have $T_v(S \cap \mathcal{O}) = T_v(S)$. According to our identification, $T_v(S)$ is a linear subspace of $\mathfrak{p}_u$ and since S is a linear subvariety, $S = v + T_v(S)$. Since $d\pi_e$ is given by the formula $d\pi_e(X) = [X, v]$, for $X \in \mathfrak{p}$, and since $d\pi_e$ is surjective, we obtain

Lemma 2.9. *Let S be a linear subvariety in $\mathfrak{p}_u$ and let $\mathfrak{s}$ be the linear subspace of $\mathfrak{p}_u$ such that $S = v + \mathfrak{s}$ for any v in S. If P has a finite number of orbits on $\mathfrak{p}_u$, then there exists an element $v \in S$ such that $[\mathfrak{p}, v] = \mathfrak{s}$.*

Here is an immediate consequence of Lemma 2.9:

Corollary 2.10. *Let S be a linear subvariety in $\mathfrak{p}_u$ of positive dimension and let $\mathfrak{s}$ be the linear subspace of $\mathfrak{p}_u$ such that $S = v + \mathfrak{s}$ for any v in S. If $[\mathfrak{p}, v] \cap \mathfrak{s} = \{0\}$ for every $v \in S$, then P has an infinite number of orbits on $\mathfrak{p}_u$.*

We now discuss the concept of modality and prove a "monotonicity" result for this more general setting.

Theorem 2.11. *Let R be an algebraic group acting morphically on an irreducible algebraic variety X, S a subgroup of R and Z an irreducible S-invariant subvariety of X. Assume that*

(i) $X = \overline{R \cdot Z}$,

(ii) $R \cdot z \cap Z$ is a union of finitely many S-orbits for a point $z \in Z$ in general position.

Then

$$\operatorname{tr} \deg_K K(X)^R = \operatorname{tr} \deg_K K(Z)^S.$$

Proof. Changing X and Z by appropriate invariant open subsets, we may assume, because of Rosenlicht's Theorem [15, Theorem 2] (see also [9, Theorem 4.4]), that there exist geometrical quotients

$$\pi_{R,X} : X \to X/R \quad \text{and} \quad \pi_{S,Z} : Z \to Z/S.$$

For any function $f \in K(X)^R$ the set of points in X where f is not regular is R-invariant and closed. Therefore, it follows from hypothesis 2.11(i) that the restriction of f to Z, $f|_Z$, is well-defined. The mapping $f \to f|_Z$ is the embedding of the fields $K(X)^R \hookrightarrow K(Z)^S$. Since

$$\pi^*_{R,X}(K(X/R)) = K(X)^R \quad \text{and} \quad \pi^*_{S,Z}(K(Z/S)) = K(Z)^S, \quad (2.11.1)$$

this embedding defines a dominant rational mapping $\eta : Z/S \to X/R$ such that the following diagram is commutative

$$
\begin{array}{ccc}
X & \xleftarrow{\ id\ } & Z \\
{\scriptstyle \pi_{R,X}}\big\downarrow & & \big\downarrow{\scriptstyle \pi_{S,Z}} \\
X/R & \xleftarrow{\ \eta\ } & Z/S
\end{array}
$$

Since the fibers of $\pi_{R,X}$ are R-orbits, it follows from this diagram and hypothesis 2.11(ii) that, for a point $a \in X/R$ in general position, the set

$$(\eta \circ \pi_{S,Z})^{-1}(a) = \pi^{-1}_{R,X}(a) \cap Z$$

is a union of finitely many S-orbits. Since the fibers of $\pi_{S,Z}$ are S-orbits, this shows that $\eta^{-1}(a)$ is a finite set. Further, as η is dominant, it follows from here and from the theorem on the dimension of the fibers of a morphism [1, Theorem 10.1] that $\dim X/R = \dim Z/S$. The claim now follows from (2.11.1) and the definition of dimension.

Corollary 2.12. *Assume the notation and hypothesis of Theorem 2.11. Then*

$$\min_{x \in X} \operatorname{codim}_X R \cdot x = \min_{z \in Z} \operatorname{codim}_Z S \cdot z.$$

Proof. This follows from Theorem 2.11 and the general facts that the transcendence degree of the field of invariant rational functions equals the minimum of the codimensions of orbits (e.g. see Corollary to Lemma 2.4 in [9]), and that this minimum is attained on the orbits of points in general position [9, sec. 1.4].

Theorem 2.13. *Assume the notation and hypothesis of Theorem 2.1. Then*

$$\operatorname{mod}(R : X) \geq \operatorname{mod}(S : Y).$$

Proof. Let Z be an irreducible S^0-invariant subvariety of Y such that

$$\mathrm{mod}(S:Y) = \min_{z \in Z} \mathrm{codim}_Z S^0 \cdot z.$$

It follows from Corollary 2.12 that

$$\min_{z \in Z} \mathrm{codim}_Z S^0 \cdot z = \min_{x \in \overline{R^0 \cdot Z}} \mathrm{codim}_{\overline{R^0 \cdot Z}} R^0 \cdot x.$$

From the definition of modality we have

$$\mathrm{mod}(R:X) \geq \min_{x \in \overline{R^0 \cdot Z}} \mathrm{codim}_{\overline{R^0 \cdot Z}} R^0 \cdot x.$$

Finally, combining these three formulas yields the claim.

Remark 2.14. In view of Theorem 2.13 one can now formulate results analogous to Corollaries 2.6 and 2.8 in terms of monotonicity of the modality of the parabolics involved.

3. Minimal Parabolic Subgroups with an infinite Number of Orbits

In this section we determine those minimal parabolic subgroups P of a simple algebraic group G with an infinite number of orbits on $\mathrm{Lie}(P_u)$. Our principal results are Theorems 3.1 and 3.2.

Theorem 3.1. *If G is a simple algebraic group of type A_6, B_4, C_4, D_5, or F_4 and P is any minimal parabolic subgroup of G, then the number of P-orbits on $\mathrm{Lie}(P_u)$ is infinite.*

Granting Theorem 3.1 we can prove our main result of this section:

Theorem 3.2. *If G is a simple algebraic group of type A_n for $n \geq 6$, B_n for $n \geq 4$, C_n for $n \geq 4$, D_n for $n \geq 5$, F_4, E_6, E_7, or E_8 and P is any minimal parabolic subgroup of G, then the number of P-orbits on $\mathrm{Lie}(P_u)$ is infinite.*

Proof. This follows from Theorem 3.1 and Corollary 2.6. For, if G is of type A_6, B_4, C_4, D_5, or F_4, then this is just Theorem 3.1. Otherwise there is a simple regular subgroup H of G of smaller rank than that of G (more precisely, $\Psi(H)$ is the subsystem of Ψ generated by a subset of simple roots of G and hence the Dynkin-diagram of H is a subdiagram of the Dynkin-diagram of G) such that the corresponding parabolic subgroup $Q = P \cap H$ of H is known to have an infinite number of orbits on its unipotent radical, according to either Theorem 1.1 or Theorem 3.1. It then follows from Corollary 2.6 that P has the same property. Specifically, in the cases when G is of type E_6, E_7, or E_8 the corresponding regular subgroup can always be chosen of type D_5 or D_4. We illustrate this process by the following example, but leave the remaining cases to the reader.

Example 3.3. We illustrate the procedure described above in terms of Dynkin-diagrams. Let G be of type E_6. We indicate the minimal standard parabolic subgroup $P = P_\alpha$ of G in the Dynkin-diagram of G by coloring the simple root α. We also list the diagram for H and label the corresponding parabolic subgroup $Q = P \cap H$ likewise. Here and elsewhere, we attach a 'prime' to the simple roots of H in order to distinguish them from those of G.

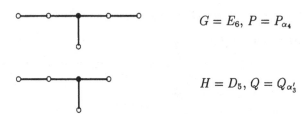

$$G = E_6, \; P = P_{\alpha_4}$$

$$H = D_5, \; Q = Q_{\alpha_3'}$$

FIGURE 1.

The proof of Theorem 3.1 follows from a series of lemmas.

The following elementary but nevertheless useful observation originated in [3] and was used by V. Kashin in [5] to show that a Borel subgroup of an algebraic group of type A_5, B_3, C_3, D_4, or G_2 has an infinite number of orbits on its unipotent radical.

Let P be a parabolic subgroup of G. Let $\mathfrak{r}$ be a subalgebra of $\mathfrak{p}_u$ which is normalized by $\mathfrak{p}$, i.e., an ideal of $\mathfrak{p}$ contained in $\mathfrak{p}_u$. Then the adjoint action of $\mathfrak{p}$ on $\mathfrak{r}/[\mathfrak{r}, \mathfrak{r}]$ factors through $\mathfrak{p}/\mathfrak{r}$. Whence the following is clear:

Lemma 3.4. *Let* $\mathfrak{r} \subseteq \mathfrak{p}_u \subseteq \mathfrak{p}$ *be as above. If*

$$\dim \mathfrak{p}/\mathfrak{r} < \dim \mathfrak{r}/[\mathfrak{r}, \mathfrak{r}], \qquad (3.4.1)$$

then $\min_{x \in \mathfrak{r}} \operatorname{codim}_{\mathfrak{r}} P \cdot x > 0$, *and hence the number of P-orbits on $\mathfrak{p}_u$ is infinite.*

Remarks 3.5. Observe that any $\mathfrak{r}$ satisfying 3.4.1 is properly contained in $\mathfrak{p}_u$, because of Richardson's Dense Orbit Theorem. Therefore, in such a case there are infinitely many P-orbits in a proper *linear* subspace of $\mathfrak{p}_u$.

However, if the number of P-orbits on $\mathfrak{p}_u$ is infinite, then there need not be an ideal $\mathfrak{r}$ of $\mathfrak{p}$ contained in $\mathfrak{p}_u$ satisfying 3.4.1 in general. For instance, if G is of type A_6 and $P = P_{\alpha_2}$, then it follows from Lemma 3.9 below that P has an infinite number of orbits on $\mathfrak{p}_u$, but $\mathfrak{p}$ does not have an ideal $\mathfrak{r}$ obeying 3.4.1, as can easily be checked. Another example is $P = P_{\alpha_3}$ in B_4.

Table 1 below contains two cases of minimal standard parabolic subalgebras $\mathfrak{p}$ together with subalgebras $\mathfrak{r}$ of $\mathfrak{p}_u$ which are ideals in $\mathfrak{p}$ that satisfy the dimension condition 3.4.1 above. In particular, in each of these cases the number of P-orbits on $\mathfrak{p}_u$ is infinite. We list the simple root α corresponding to $P = P_\alpha$ in column 2. In column 5 we list the roots β whose root spaces $\mathfrak{g}_\beta$ generate $\mathfrak{r}$ as an ideal in $\mathfrak{p}$. The details are left for the reader to check. Further examples of this kind can be found in Table 3 at the end of this section.

Type of G	α	$\dim \mathfrak{p}$	$\dim \mathfrak{p}_u$	$\mathfrak{r}$	$\dim \mathfrak{r}$	$\dim[\mathfrak{r}, \mathfrak{r}]$	$\dim \mathfrak{p}/\mathfrak{r}$	$\dim \mathfrak{r}/[\mathfrak{r}, \mathfrak{r}]$
B_4	α_2	21	15	α_1, α_3	14	6	7	8
C_4	α_2	21	15	α_1, α_4	13	4	8	9

TABLE 1.

We collect the results from Table 1 and the previous discussion in our next lemma.

Lemma 3.6. *If G is of type B_4 or C_4 and P is the minimal standard parabolic subgroup of G corresponding to α_2, then the number of P-orbits on $\mathrm{Lie}(P_u)$ is infinite.*

Proof. This is clear from Lemma 3.4 and the information provided in Table 1.

Lemma 3.7. *If G is of type B_4 and P is the minimal standard parabolic subgroup of G corresponding to α_3, then the number of P-orbits on $\mathrm{Lie}(P_u)$ is infinite.*

Proof. We apply Corollary 2.10. Let $\{e_\beta, h_i \mid \beta \in \Psi, 1 \le i \le 4\}$ be a Chevalley basis of $\mathfrak{g}$. Set $\gamma = \alpha_3 + 2\alpha_4$, $\delta = \alpha_2 + 2\alpha_3 + 2\alpha_4$, and $v_0 = e_{\alpha_1} + e_{\alpha_2} + e_{\alpha_4} + e_\gamma$. Consider the line $\ell = v_0 + \langle e_\delta \rangle$ in $\mathfrak{p}_u$. We claim that $[\mathfrak{p}, v] \cap \langle e_\delta \rangle = \{0\}$ for each $v \in \ell$, which is precisely the condition of Corollary 2.10. Suppose otherwise that there are elements $X \in \mathfrak{p}$ and $\lambda, \mu \in K$ with $\mu \ne 0$ satisfying

$$[X, v_0 + \lambda e_\delta] = \mu e_\delta.$$

We write $X = h + \sum_{\beta \in \Psi(P)} \lambda_\beta e_\beta$, where $h \in \mathfrak{t}$ and $\lambda_\beta \in K$. Whence we have

$$\mu e_\delta = [X, v_0 + \lambda e_\delta]$$
$$= \alpha_1(h)e_{\alpha_1} + \alpha_2(h)e_{\alpha_2} + \alpha_4(h)e_{\alpha_4} + (\pm\lambda_{\alpha_2+\alpha_3} \pm \lambda_{\alpha_3+\alpha_4})e_{\alpha_2+\alpha_3+\alpha_4}$$
$$+ (\pm\lambda_{\alpha_2+\alpha_3})e_{\alpha_1+\alpha_2+\alpha_3} + (\gamma(h) \pm 2\lambda_{\alpha_3+\alpha_4})e_\gamma + (\lambda\delta(h) \pm \lambda_{\alpha_2+\alpha_3})e_\delta + \dots .$$

A careful comparison of the coefficients yields $\mu = 0$, which is a contradiction. Whence, the claim and the lemma follow.

In the proof of Lemma 3.7 it is not apparent why we chose this particular line $\ell = v_0 + \langle e_\delta \rangle$. This will be motivated in Section 6.

We now apply Corollary 2.6 to some minimal parabolic cases in a simple algebraic group G which allows us to reduce to the case of either a Borel or a minimal parabolic case of a simple subgroup H of a maximal-rank subgroup of G. In each of the latter cases the Borel or parabolic obtained this way is known to admit an infinite number of orbits on its unipotent radical, either by Theorem 1.1, Lemma 3.6, or else by Lemma 3.7. As before we employ Corollary 2.6 to conclude that then the initial minimal parabolic subgroup of G has an infinite number of orbits on its unipotent radical as well.

In each case the subgroup H is carefully chosen as a simple regular subgroup of G (with respect to T) such that the negative of the highest root in Ψ is a simple root in $\Psi(H)$, i.e., H is the derived subgroup of a Levi subgroup of a maximal-rank subgroup of G, and moreover, the additional node of the extended Dynkin-diagram for G is also a node of the Dynkin-diagram for H.

As above we present this information in form of a list. In Table 2 below we give the types of both G and H. Further, P_α is a minimal parabolic subgroup of G and $Q = H \cap P_\alpha$ is a standard parabolic subgroup of H of semisimple rank at most one, i.e., either a minimal parabolic or a Borel subgroup of H. In column 3 of Table 2 we list each simple root α for which this argument applies. In the last column we list the type of the parabolic Q of H. If Q is a minimal parabolic in H, then we indicate the corresponding simple root $\alpha' \in \Psi(H)$.

Type of G	Type of H	P_α	Q
A_6	A_5	α_2, α_3	Borel
B_4	D_4	α_4	Borel
C_4	C_3	α_3, α_4	Borel
D_5	D_4	α_4, α_5	Borel
F_4	B_4	α_2	α'_3
F_4	B_4	α_3	α'_4

TABLE 2.

The cases $\alpha = \alpha_4$ or α_5 for G of type A_6 are omitted, as they are dual to the ones already listed in Table 2. As an example we illustrate two cases from Table 2 in terms of extended Dynkin-diagrams:

Examples 3.8. In our first example G is of type A_6 and of type F_4 in our second. The same notation as in Example 3.3 applies. The diagrams for H and Q are also given.

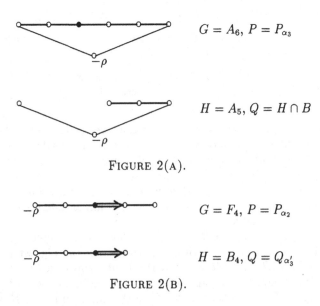

FIGURE 2(A).

FIGURE 2(B).

Collecting the results from Table 2 yields

Lemma 3.9. *If G is as in Theorem 3.1 and $P = P_\alpha$ is one of the minimal standard parabolic subgroups of G as in Table 2, then the number of P-orbits on $\text{Lie}(P_u)$ is infinite.*

Proof. Apart from the F_4 cases this is a consequence of Corollary 2.6 and Theorem 1.1. The first F_4 case follows from Corollary 2.6 and Lemma 3.7 and the last one from Corollary 2.6 and the B_4 case in Table 2.

We continue the proof of Theorem 3.1 with an application of Corollary 2.8.

Lemma 3.10. *If G is of type D_5 and P is the minimal standard parabolic subgroup of G corresponding to α_2 or α_3, then the number of P-orbits on $\mathrm{Lie}(P_u)$ is infinite.*

Proof. Let Θ be the outer automorphism of G induced by the non-trivial graph automorphism of D_5. Then G^Θ is of type B_4 and both parabolic subgroups of G given are Θ-invariant. By Lemmas 3.6 and 3.7 both of the corresponding parabolic subgroups of B_4 already have an infinite number of orbits on the Lie algebra of their unipotent radicals. It then follows from Corollary 2.8 that this also holds for both parabolic subgroups of G.

Lemma 3.11. *Let G be as in Theorem 3.1 and let $P = P_\alpha$ be the minimal standard parabolic subgroup of G corresponding to α. If $\alpha = \alpha_1$, or if G is of type A_6 and $\alpha = \alpha_6$, or if G is of type F_4 and $\alpha = \alpha_4$, then the number of P-orbits on $\mathrm{Lie}(P_u)$ is infinite.*

Proof. In each of the cases above let H be the derived subgroup of the standard Levi subgroup of G corresponding to the set $\Pi \setminus \{\alpha\}$ of simple roots. Note that $H \cap P$ is a Borel subgroup of H and by Theorem 1.1 it admits an infinite number of orbits on its unipotent radical. We can now apply Corollary 2.6 and conclude the desired result for P.

Finally, Theorem 3.1 now follows from Lemmas 3.6 through 3.11.

We close this section with a list of numerous examples of minimal parabolic subalgebras $\mathfrak{p}$ that satisfy 3.4.1. The same notation as in Table 1 applies. The details of the information provided in Table 3 are again left for the reader to check.

Type of G	α	$\dim \mathfrak{p}$	$\dim \mathfrak{p}_u$	$\mathfrak{r}$	$\dim \mathfrak{r}$	$\dim[\mathfrak{r},\mathfrak{r}]$	$\dim \mathfrak{p}/\mathfrak{r}$	$\dim \mathfrak{r}/[\mathfrak{r},\mathfrak{r}]$
A_7	α_3, α_4	36	27	$\alpha_2, \alpha_5, \alpha_7$	23	9	13	14
B_5	α_4	31	24	α_1, α_3	20	8	11	12
B_4	α_4	21	15	α_1, α_3	14	6	7	8
C_4	α_3	21	15	α_1, α_4	13	4	8	9
D_5	α_2	26	19	α_3	15	3	11	12
D_5	α_3	26	19	$\alpha_2, \alpha_4, \alpha_5$	18	9	8	9
D_5	α_4, α_5	26	19	α_3	15	3	11	12
F_4	α_3	29	23	α_2, α_4	22	14	7	8

TABLE 3.

Remark 3.12. Apart from one, the D_5 cases in Theorem 3.1 were dealt with in Lemmas 3.9 and 3.10. It is apparent from Table 3 that we could have applied Lemma 3.4 to each of these cases as well. However, instead of using the *ad hoc* method via ideals of Lemma 3.4, our treatment, apart from being an application of our results from Section 2, provides a more systematic approach.

4. Minimal Parabolic Subgroups with a finite Number of Orbits

Our main results of this section are Theorems 4.1 and 4.2.

Theorem 4.1. *Let G be a simple algebraic group of type A_5, or D_4, and let P be any minimal parabolic subgroup of G. Then P acts with a finite number of orbits on P_u.*

Theorem 4.2. *Let G be a simple algebraic group of type B_3, C_3, or G_2, and let P be any minimal parabolic subgroup of G. Then P acts with a finite number of orbits on P_u.*

We need an elementary but useful observation.

Lemma 4.3. *Let G be a reductive algebraic group and $Q \subseteq P$ (standard) parabolic subgroups of G. If the number of orbits of Q on Q_u is finite, then so is the number of P-orbits on P_u. Equivalently, if P has an infinite number of orbits on P_u, then so does Q on Q_u.*

Proof. Since $Q \subseteq P$, we have $P_u \subseteq Q_u$ and, since P_u is normal in P, it is normal in Q. Therefore, if Q acts with a finite number of orbits on Q_u, then it acts likewise on P_u, and clearly P does then as well.

The following restatement of Theorem 1.3 is an immediate consequence of Lemma 4.3:

Corollary 4.4. *If G is as in Theorem 1.2, then any non-Borel parabolic subgroup of G admits a finite number of orbits on its unipotent radical. If G is as in Theorem 1.1, then any parabolic subgroup of G has this property.*

Granting Theorem 4.1 we can now prove Theorem 4.2.

Proof of Theorem 4.2. Let G be of type A_5 and let Θ denote the outer automorphism of G, preserving the maximal torus T and the set of simple roots, which is induced from the non-trivial graph automorphism of the A_5 Dynkin-diagram. Then G^Θ is of type C_3. It follows from Theorem 4.1 and Lemma 4.3 that each of $P_{\{\alpha_1,\alpha_5\}}$, $P_{\{\alpha_2,\alpha_4\}}$, and P_{α_3} has a finite number of orbits on its unipotent radical. Moreover, each of these subgroups is Θ-stable and the corresponding fixed point subgroup is a minimal parabolic subgroup of C_3. We can now apply Corollary 2.8 and conclude that each minimal parabolic subgroup of the group of type C_3 has a finite number of orbits on its unipotent radical.

Next let G be of type D_4 and let Θ and τ denote the outer automorphisms of G, preserving the maximal torus T and the set of simple roots, which are induced respectively from the non-trivial graph automorphism of the D_4 Dynkin-diagram by interchanging α_3 and α_4 and from the triality graph automorphism. Then G^Θ is of type B_3, while G^τ is of type G_2. It follows from Theorem 4.1 and Lemma 4.3 that each of P_{α_1}, P_{α_2}, and $P_{\{\alpha_3,\alpha_4\}}$ has a finite number of orbits on its unipotent radical. In addition, each of these subgroups is Θ-stable and the corresponding fixed point subgroups are minimal parabolic subgroups of B_3. As above it follows from Corollary 2.8 that each minimal parabolic subgroup of B_3 has a finite number of orbits on its unipotent radical. Finally, the same argument applied to P_{α_2}, and $P_{\{\alpha_1,\alpha_3,\alpha_4\}}$, both of which are τ-stable, yields the desired result for G_2. This completes the proof of Theorem 4.2.

Examples 4.5. We illustrate two examples from the proof of Theorem 4.2. Let G be of type D_4 with Θ and τ the automorphisms of G defined above. We also use Θ and τ to denote the corresponding automorphisms of the Dynkin-diagram of G. The information for G^Θ and G^τ is presented as well.

$$\tau \qquad G = D_4,\ P = P_{\alpha_2}$$

$$G^\tau = G_2,\ Q = P^\tau = Q_{\alpha_2'}$$

FIGURE 3(A).

$$\Theta \qquad G = D_4,\ P = P_{\alpha_1}$$

$$G^\Theta = B_3,\ Q = P^\Theta = Q_{\alpha_1'}$$

FIGURE 3(B).

The proof of Theorem 4.1 follows from an analysis of each of the relevant cases. Since our arguments for these are based on some explicit computations which are quite lengthy at times, we postpone them until Section 6. We find it more appealing to continue with some additional general results in our next section.

5. MORE PARABOLIC SUBGROUPS WITH A FINITE NUMBER OF ORBITS

As we have already pointed out in the Introduction, parabolic subgroups with an Abelian unipotent radical form a natural class of parabolic subgroups admitting a finite number of orbits on the unipotent radical. In this section we exhibit additional natural families of such subgroups that share this finiteness property.

Let Π be a set of fundamental roots of G. Let ρ be the highest root of Ψ. Define a subset J of Π by

$$J := \{\alpha \in \Pi \mid \langle \alpha, \rho \rangle = 0\}$$

and let $P = P_J$ be the standard parabolic subgroup of G corresponding to J. Then $\Psi(P_u) = \{\beta \in \Psi^+ \mid \langle \beta, \rho \rangle = 1\} \cup \{\rho\}$, and it follows from the commutator relations that $\Psi(P'_u) = \{\rho\}$, i.e., $P'_u = U_\rho$.

Proposition 5.1. *Let $P = P_J = LP_u$ be as above. Then the number of P-orbits on P_u equals the number of L-orbits on P_u/P'_u plus 1; in particular, the number of orbits is finite.*

Proof. It follows from [12, Theorem E] that L and thus P operate on P_u/P'_u with a finite number of orbits. Thus it suffices to show the first assertion. We may represent an element x in P_u as a product of root group elements relative to a fixed ordering of $\Psi(P_u)$. Suppose that $x \notin P'_u$. Then there exists a root $\beta \in \Psi \setminus \{\rho\}$ in supp(x). Note, $\rho - \beta \in \Psi(P_u)$. Since $P'_u = U_\rho$, it follows from the commutator relations that $U_{\rho-\beta} \cdot x = xP'_u$. Thus in particular, $xP'_u \subset P \cdot x$. Further, no two L-orbits on P_u/P'_u are P-conjugate, as P_u acts trivially on P_u/P'_u. Finally, there are two P-orbits on $P'_u = U_\rho$, namely $U_\rho \setminus \{e\}$ and $\{e\}$. The desired result now follows.

Remark 5.2. The parabolic subgroups P_J from Proposition 5.1 are sometimes called *extraspecial*. If G is of type A_n, then $J = \Pi \setminus \{\alpha_1, \alpha_n\}$, while in all other cases P_J is a maximal parabolic subgroup of G.

Proposition 5.3. *Let G be of type B_n and let $P = P_{\Pi \setminus \{\alpha_n\}} = LP_u$ be the maximal standard parabolic subgroup of G corresponding to the short simple root α_n in Π. Then the number of P-orbits on P_u is finite.*

Proof. For $1 \leq j \leq n$ let $\beta_j = \sum_{i=j}^n \alpha_i$. In particular, $\beta_n = \alpha_n$. Note that $\Psi(P_u) = \{\beta_j, \beta_j + \beta_k \mid 1 \leq j < k \leq n\}$, and $\Psi(P'_u) = \{\beta_j + \beta_k \mid 1 \leq j < k \leq n\}$. By [12, Theorem E] there are only finitely many P-orbits in P'_u. Let $y \in P_u \setminus P'_u$. Then there exists a β_j in supp(y). By applying elements from W_L, the Weyl group of L, we may assume that $\beta_n = \alpha_n$ is in the support of y. Then by conjugating y by suitable elements from $U_{\alpha_{n-1}}, U_{\alpha_{n-2}+\alpha_{n-1}}, \ldots, U_{\alpha_1+\ldots+\alpha_{n-1}}$ we can remove all roots β_j with $j < n$ from supp(y). In addition, by conjugating with suitable elements from $U_{\beta_{n-1}}, U_{\beta_{n-2}}, \ldots, U_{\beta_1}$ we can also remove all roots $\beta_j + \beta_n$ with $j < n$ from the support of y. So, using the action of T, we may assume that

$$x = U_{\alpha_n}(1)y'$$

is a representative of $P \cdot y$ with supp$(y') \subseteq \{\beta_j + \beta_k \mid 1 \leq j < k \leq n-1\}$. If supp$(y') = \varnothing$, then $x = U_{\alpha_n}(1)$ is a representative for $P \cdot y$. So we may suppose that supp$(y') \neq \varnothing$. Then, by applying reflections from W_L relative to roots which are orthogonal to α_n, we can obtain $\beta_{n-2} + \beta_{n-1}$ in the support of y'.

Next we conjugate x by suitable elements from $U_{\alpha_{n-3}}, U_{\alpha_{n-4}+\alpha_{n-3}}, \ldots,$ and $U_{\alpha_1+\ldots+\alpha_{n-3}}$ in order to remove all roots $\beta_j + \beta_{n-1}$ with $j < n-2$ from the support of y'. Then conjugating the resulting element further by appropriate elements from $U_{\alpha_{n-3}+\alpha_{n-2}}, U_{\alpha_{n-4}+\alpha_{n-3}+\alpha_{n-2}}, \ldots,$ and $U_{\alpha_1+\ldots+\alpha_{n-2}}$ allows us to remove all roots $\beta_j + \beta_{n-2}$ with $j < n-2$ from supp(y'). Observe that this does not reintroduce roots into supp(y') which have been removed previously, as α_n is orthogonal to each of $\alpha_{n-3} + \alpha_{n-2}, \alpha_{n-4} + \alpha_{n-3} + \alpha_{n-2}, \ldots$, and $\alpha_1 + \ldots + \alpha_{n-2}$. So we may assume that

$$x' = U_{\alpha_n}(1)U_{\beta_{n-2}+\beta_{n-1}}(1)y''$$

is a representative of $P \cdot y$ with $\text{supp}(y'') \subseteq \{\beta_j + \beta_k \mid 1 \leq j < k \leq n - 3\}$. If $\text{supp}(y'') = \varnothing$, then $x' = U_{\alpha_n}(1)U_{\beta_{n-2}+\beta_{n-1}}(1)$ is a representative for $P \cdot y$. If not, then we continue this process in the same fashion as above, by first obtaining $\beta_{n-4} + \beta_{n-3}$ in $\text{supp}(y'')$ and then by removing all roots $\beta_j + \beta_{n-3}$ for $j < n - 4$ and also all roots $\beta_j + \beta_{n-4}$ for $j < n - 4$ from $\text{supp}(y'')$, etc. This process ends, and once it does, we will have arrived at a representative of $P \cdot y$ of the form

$$\prod_{j=0}^{s} U_{\beta_{n-2j}+\beta_{n-2j+1}}(1),$$

where $0 \leq s \leq [\frac{n+1}{2}] - 1$ and $\beta_{n+1} := 0$. As shown, all the parameters can be chosen to be 1, using the action of T, as $\{\beta_{n-2j} + \beta_{n-2j+1} \mid 0 \leq j \leq [\frac{n+1}{2}] - 1\}$ is a set of linearly independent roots in $\Psi(P_u)$. In particular, there are only finitely many P-orbits in $P_u \setminus P_u'$. The result now follows.

Theorem 5.4. *Let G be classical and let $P = P_{\Pi \setminus \{\alpha\}} = LP_u$ be a maximal standard parabolic subgroup of G, where α is a simple root corresponding to an end-node in the Dynkin-diagram of G. Then the number of P-orbits on P_u is finite.*

Proof. If P is as in the statement of the theorem, then one easily checks that, unless G is of type B_n and $P = P_{\Pi \setminus \{\alpha_n\}}$, either P_u is Abelian or P is extraspecial. Whence the result follows from [12, Theorem E] and Propositions 5.1 and 5.3.

Remark 5.5. It is known that the statement of Theorem 5.4 also holds for exceptional groups of type G_2, F_4, and E_6. In the first instance this follows from Theorem 4.2. In E_6 an end-node maximal parabolic subgroup either has an Abelian unipotent radical, or else is extraspecial. One of the F_4 cases is extraspecial, while the other is easily analyzed using the methods outlined in Section 6. We certainly expect that the same result is true for groups of type E_7 and E_8. In other words, we expect that for any *maximal* parabolic subgroup $P = LP_u$ of a simple algebraic group G such that L' is *simple* the number of P-orbits on P_u is finite.

We already mentioned in the Introduction that a parabolic subgroup P of G has a finite number of orbits on P_u provided P_u is Abelian. It is known, that each of these is a maximal parabolic subgroup of G [13, Lemma 2.2]. Taking the preceding, together with Lemma 4.3, Remark 5.2, and Theorem 5.4, into account, we would like to ask at this point the following general question.

Problem 5.6. *Is it true that if P is a maximal parabolic subgroup of a reductive algebraic group G, then the number of orbits of P on P_u is finite?*

6. A Proof of Theorem 4.1

In this section we provide proofs for the relevant cases of Theorem 4.1. Some of the subsequent ideas are inspired by similar concepts introduced in [3].

We first outline the general procedure. Suppose that $P = LP_u$ is a given parabolic subgroup of G and that we aim to show that the number of P-orbits on P_u is finite. The idea is roughly to decompose P_u into a finite union of suitable subvarieties and to show that each of the latter is contained in only a finite union of P-orbits.

To describe these subvarieties, we first fix a total ordering on $\Psi(P_u)$ which may be arbitrary, say $\Psi(P_u) = \{\beta_1, \ldots, \beta_r\}$, where $r = \dim P_u$.

Set $M := K \setminus \{0\}$. Let $C_1, \ldots, C_r$ be any sequence, where $C_i = C_{\beta_i} \in \{0, K, M\}$, and define

$$P_u(C_1, \ldots, C_r) := \prod U_{\beta_i}(C_i).$$

These are the subvarieties to be considered. In particular, $P_u(K, \ldots, K) = P_u$ and $P_u(0, \ldots, 0) = e$.

Our goal is to select a collection of subvarieties $P_u(C_1, \ldots, C_r)$ such that

(1) each P-orbit in P_u intersects at least one subvariety of this collection, and

(2) each subvariety of this collection is contained in a finite union of P-orbits.

If it is possible to select such a collection, then we may conclude that P acts with a finite number of orbits on P_u. To achieve this, we start with the collection consisting of just one subvariety, namely $P_u(K, \ldots, K) = P_u$, and then apply to it a sequence of 'splitting' and 'reduction' operations, each of which produces a new collection of subvarieties and preserves property (1). The idea is that in applying suitable combinations of these operations we aim at decreasing the support of these subvarieties and hope to eventually arrive at a collection of subvarieties $P_u(C_1, \ldots, C_r)$, each of which is supported by at most rank $G = \dim T$ linearly independent roots in $\Psi(P_u)$. In that case each of these final subvarieties of this procedure is contained in a finite union of T-orbits on P_u, whence each is contained in a finite union of P-orbits. Thus, these final subvarieties fulfil both properties (1) and (2).

Mainly we use two operations which are taken from [3]. The first one consists in 'splitting' one of the varieties $P_u(C_1, \ldots, C_r)$ 'along a root', say β_k, with $C_k = K$. By that we mean that the element $P_u(C_1, \ldots, C_r)$ of a given collection is replaced by the pair $P_u(C_1, \ldots, C_k', \ldots, C_r)$ and $P_u(C_1, \ldots, C_k'', \ldots, C_r)$, where $C_k' = 0$ and $C_k'' = M$. That is, the first subvariety of the pair contains all elements of $P_u(C_1, \ldots, C_r)$ whose support does not contain β_k, while the second consists of those which do have β_k in their support. In particular, $P_u(C_1, \ldots, C_k', \ldots, C_r)$ has smaller support in $\Psi(P_u)$ than $P_u(C_1, \ldots, C_r)$. Since $P_u(C_1, \ldots, C_r)$ is the disjoint union of these two subvarieties, property (1) is preserved, in passing from the given to the new collection of subvarieties, where $P_u(C_1, \ldots, C_r)$ is replaced by the pair $P_u(C_1, \ldots, C_k', \ldots, C_r)$ and $P_u(C_1, \ldots, C_k'', \ldots, C_r)$.

The second operation also aims at reducing the support of a given $P_u(C_1, \ldots, C_r)$ in $\Psi(P_u)$. Namely, consider a collection containing a subvariety $P_u(C_1, \ldots, C_r)$ such that $C_\alpha = M$ for some $\alpha \in \Psi(P_u)$. Suppose also that $\alpha + \beta \in$

$\Psi(P_u)$ for some $\beta \in \Psi(P)$. Define a subvariety $P_u(C_1', \ldots, C_r')$ of P_u as follows. Let $C_{\alpha+\beta}' = 0$, $C_\gamma' = K$, whenever $\gamma = \delta + d\beta$ for some $\delta \in \Psi(P_u)$, $d \in \{1, 2, \ldots\}$, with $(\delta, d) \neq (\alpha, 1)$, and $C_\delta \neq 0$, and finally $C_\gamma' = C_\gamma$ otherwise. Then every element of $P_u(C_1, \ldots, C_r)$ can be obtained by conjugating an element from $P_u(C_1', \ldots, C_r')$ by a suitable element in $U_\beta \subseteq P$. Thus every P-orbit passing through $P_u(C_1, \ldots, C_r)$ has a representative in $P_u(C_1', \ldots, C_r')$. Therefore, in passing to a new collection, by replacing $P_u(C_1, \ldots, C_r)$ by $P_u(C_1', \ldots, C_r')$, condition (1) above is preserved. We refer to this procedure as a 'reduction' operation, since it removes the element $\alpha + \beta$ from the support of the subvariety under consideration. However, at the same time, this may, and generally does, introduce new roots of the form $\delta + d\beta$ in the support of the resulting subvariety, as shown. Thus, this method is only effective in reducing the support of these varieties, if one can avoid new roots being introduced. In that case $P_u(C_1', \ldots, C_r')$ has smaller support in $\Psi(P_u)$ than $P_u(C_1, \ldots, C_r)$. For instance, in the context above, this is the case if $C_\delta = 0$, whenever $\delta + d\beta$ is a root, where $\delta \neq \alpha$, and $\alpha + d\beta$ is not a root for $d \geq 2$. In order to achieve this, one has to perform suitable splittings which then allow for the desired reductions without introducing new roots.

Suppose that one obtains a subvariety $P_u(C_1, \ldots, C_r)$ along this process, where $C_{\beta_k} = C_k = M$ for a set of roots $\{\beta_{k_1}, \ldots, \beta_{k_t}\}$ with $t > \dim T$, and moreover, assume that the reduction technique above does not decrease the size of the support of $P_u(C_1, \ldots, C_r)$. Then we can not conclude that the subvariety $P_u(C_1, \ldots, C_r)$ is contained in a finite union of P-orbits, as desired. In such a case, we call the subvariety $P_u(C_1, \ldots, C_r)$ *unresolved*. For instance, if P has an infinite number of orbits on P_u, then one necessarily encounters unresolved subvarieties. Such an unresolved subvariety is the basis for the argument in the proof of Lemma 3.7 above. Namely, if one applies the methods described above to the case in Lemma 3.7 in the attempt to show that the number of orbits is finite, then one encounters a one-dimensional subvariety in P_u which is unresolved. The line ℓ used in the proof of Lemma 3.7 is the precise counterpart in the Lie algebra of this subvariety.

We would like to stress that in the cases of Theorem 4.1 the procedure of applying suitable combinations and repetitions of these splitting and reduction techniques can be continued until one arrives at a complete set of orbit representatives. Thus, our proof of Theorem 4.1 provides much more information apart from the finiteness of the number of orbits. However, as we are interested here only in the question of finiteness, we did not carry out a full classification of the orbits. Nevertheless, an upper bound for the number of orbits is easily obtained at this stage, see Remark 6.1.

After having outlined the general concept for the proof of Theorem 4.1 above, we now introduce some notation that will allow us to handle the computations more efficiently. We fix a total order on $\Psi(P_u)$, say $\Psi(P_u) = \{\beta_1, \ldots, \beta_r\}$. Relative to this fixed order we abbreviate the variety $P_u(C_1, \ldots, C_r)$ by the *string* $C_1 \ldots C_r$. That is, at position β_k we simply write the parameter set C_k, which equals one of 0, M, or K. Sometimes we abbreviate this notation even further by writing C^j for $C \ldots C$ (j times). This notational simplification is taken from [3].

We keep track of the various strings and their splittings by numbering them. A splitting, or branching, of a string along a root γ is indicated by the symbol $\overset{\gamma}{\Rightarrow} N$. Then in N.(a) and N.(b) we continue dealing with the separate subvarieties, where the (a)-string is always the one with γ eliminated from its support, while in the (b)-string the K corresponding to γ is replaced by an M etc.

For example, an elimination of roots in passing from a given subvariety, say $\cdots U_\alpha(M) \cdots U_{\alpha+\beta} \cdots U_{\alpha+\delta} \cdots$ to $\cdots U_\alpha(M) \cdots U_{\alpha+\beta+\delta} \cdots$, by conjugation with elements from U_β and U_δ, is indicated by the diagram

$$\ldots M \ldots K \ldots K \ldots 0 \ldots \overset{\beta,\delta}{\longrightarrow} \ldots M \ldots 0 \ldots 0 \ldots K' \ldots$$

(only the positions α, $\alpha + \beta$, $\alpha + \delta$, and $\alpha + \beta + \delta$ are shown). The order in which these conjugations are applied may be significant. Also, if $\alpha + \beta + \delta$ is a root, then by this conjugation process, we may introduce elements in $U_{\alpha+\beta+\delta}$, so the resulting variety may have $\alpha + \beta + \delta$ in its support. In the diagram above, the 0 in the initial string indicates that $\alpha + \beta + \delta$ is not in the support of the corresponding subvariety, while in the second string, the symbol K' in position for the same root, indicates that $\alpha + \beta + \delta$ is in the support of the resulting subvariety. Any occurrence of such a newly introduced root is indicated by the symbol K' in the corresponding position.

For $\alpha \in \Psi(L) \subseteq \Psi(P)$ let s_α be the reflection relative to α in the Weyl group of L. Note that s_α acts on each string by interchanging the parameter sets corresponding to γ and $s_\alpha(\gamma)$. In our analysis this is indicated by $\overset{s_\alpha}{\longrightarrow}$. An application of $\overset{s_\alpha}{\longrightarrow}$ allows us to reduce the study of a given subvariety $P_u(C_1, \ldots, C_r)$ to a previously analyzed case $P_u(C'_1, \ldots, C'_r)$. To indicate this, we end the corresponding string with the symbol "$\bullet p$".

If, after a number of splittings along roots and eliminations of roots, we arrive at a subvariety whose support in $\Psi(P_u)$ consists of at most rank G linearly independent roots, then this subvariety is contained in a finite union of T-orbits on P_u, and thus it is contained in a finite union of P-orbits. We call such a subvariety *terminal*. To indicate that we have arrived at such a terminal subvariety, we simply end the corresponding string by a "$\bullet$". If a string corresponds to a subvariety whose support consists of rank G linearly dependent roots in $\Psi(P_u)$, then we indicate this by writing "(lin. dep.)" after it.

Remark 6.1. Since a terminal subvariety is supported by at most rank G linearly independent roots, it is contained in at most 2^t P-orbits in P_u, where t is the number of occurrences of K in the corresponding terminal string consisting of 0's, M's and K's. Thus we can easily calculate an upper bound for the number of P-orbits on P_u by simply adding up these numbers over all terminal subvarieties. This of course is a very crude estimate and can easily be improved by further analyzing the terminal subvarieties.

We now proceed with our proof of Theorem 4.1.

Let G be of type D_4. Here and later on we abbreviate a root of G simply by its index set relative to the usual labeling of the simple roots of D_4. For instance, we write 123 for $\alpha_1 + \alpha_2 + \alpha_3$ etc., and for the highest root in D_4 we write $1\bar{2}34$, where $\bar{2}$ indicates that α_2 occurs twice. Whence, if G is of type D_4, then $\Psi^+ = \{1, 2, 3, 4, 12, 23, 24, 123, 124, 234, 1234, \rho = 1\bar{2}34\}$ denotes the set of positive roots of G.

Lemma 6.2. *Let G be of type D_4 and consider the subvariety*

$$Y = U_{12}(M)U_{23}(M)U_{24}(M)U_{234}(M)U_{1234}$$

of B_u. Then Y is contained in the union of the two B-orbits passing through $U_{12}(1)U_{23}(1)U_{24}(1)$ and $U_{12}(1)U_{23}(1)U_{24}(1)U_{234}(1)$.

Proof. For our purpose we need to have precise information on the signs of the structure constants of various commutators in G. We may assume that we have chosen a realization of G which is consistent with [16, 2.12]. Thus, relative to the basis $\{1, 2, 3, 4\}$, we can explicitly compute the structure constants $c_{\alpha,\beta}$ of the commutator relations in G as outlined in *loc. cit.* One readily checks that

$$
\begin{aligned}
c_{1,23} &= c_{1,24} = c_{1,234} = -1, \quad \text{and} \\
c_{3,12} &= c_{3,24} = c_{3,124} = c_{4,23} = c_{4,12} = c_{4,123} = 1.
\end{aligned}
\tag{6.2.1}
$$

We omit the details.

For a fixed root α we denote by T_α the one-dimensional torus of G whose Lie algebra contains the coroot h_α. Thus, T_α is the maximal torus of the rank one simple regular subgroup of G with root system $\{-\alpha, \alpha\}$. Note that for G of type D_4, the torus T is the subgroup of G generated by T_{12}, T_{23}, T_{24}, and T_{1234}, as the roots $12, 23, 24, 1234$ are linearly independent.

Since $\{12, 23, 24\}$ is a set of linearly independent roots, we can conjugate an arbitrary element from Y by suitable elements from T_{12}, T_{23}, and T_{24} into the form

$$x = U_{12}(1)U_{23}(1)U_{24}(1)U_{234}(\zeta)U_{1234}(\xi),$$

where $\zeta, \xi \in K$ and $\zeta \neq 0$. If $\xi = 0$, we can apply a suitable element from T_{1234} to conjugate x into $U_{12}(1)U_{23}(1)U_{24}(1)U_{234}(1)$, as 1234 is orthogonal to $12, 23$ and 24. Thus we may assume that $\xi \neq 0$. Let $\eta \in K$ be a solution of the equation

$$-t^2 - \zeta t + \xi = 0. \tag{6.2.2}$$

We now conjugate x above by $U_1(\eta)$. Using (6.2.1) we obtain

$$
\begin{aligned}
x' &= U_{12}(1)U_{23}(1)U_{123}(c_{1,23}\eta)U_{24}(1)U_{124}(c_{1,24}\eta)U_{234}(\zeta)U_{1234}(\xi + c_{1,234}\zeta\eta) \\
&= U_{12}(1)U_{23}(1)U_{123}(-\eta)U_{24}(1)U_{124}(-\eta)U_{234}(\zeta)U_{1234}(\xi - \zeta\eta).
\end{aligned}
$$

Next we conjugate x' by $U_3(\eta)$. Using (6.2.1) again and (6.2.2) we obtain

$$y = U_{12}(1)U_{23}(1)U_{123}(c_{3,12}\eta - \eta)U_{24}(1)U_{124}(-\eta)U_{234}(\zeta + c_{3,24}\eta)$$
$$U_{1234}(\xi - \zeta\eta - c_{3,124}\eta^2)$$
$$= U_{12}(1)U_{23}(1)U_{123}(\eta - \eta)U_{24}(1)U_{124}(-\eta)U_{234}(\zeta + \eta)U_{1234}(\xi - \zeta\eta - \eta^2)$$
$$= U_{12}(1)U_{23}(1)U_{24}(1)U_{124}(-\eta)U_{234}(\zeta + \eta).$$

We continue by conjugating y by $U_4(\eta)$. Using (6.2.1) we obtain

$$y' = U_{12}(1)U_{124}(c_{4,12}\eta)U_{23}(1)U_{24}(1)U_{124}(-\eta)U_{234}(\zeta + \eta + c_{4,23}\eta)$$
$$= U_{12}(1)\big(U_{124}(\eta)U_{23}(1)U_{124}(\eta)^{-1}\big)U_{24}(1)U_{234}(\zeta + 2\eta)$$
$$= U_{12}(1)U_{23}(1)U_{24}(1)U_{234}(\zeta + 2\eta)U_\rho(c_{124,23}\eta).$$

Then by applying a suitable element from U_{123} we can remove ρ from the support of y' and obtain an element of the form

$$z = U_{12}(1)U_{23}(1)U_{24}(1)U_{234}(\zeta + 2\eta).$$

If $\zeta + 2\eta \neq 0$, we argue as in the case $\xi = 0$ above. Otherwise $z = U_{12}(1)U_{23}(1)U_{24}(1)$, as desired. The result now follows.

Proposition 6.3. *Let G be of type D_4 and $P = P_{\alpha_1}$. Then the number of P-orbits on P_u is finite. There are at most 68 orbits.*

Proof. Note that $\dim P_u = 11$. We choose the following ordering on $\Psi(P_u)$: $\Psi(P_u) = \{2, 12, 3, 4, 23, 123, 24, 124, 234, 1234, 1\bar{2}34\}$. We proceed to analyze the various subvarieties $P_u(C_1, \ldots, C_r)$:

1. $0^7 K^4 \bullet$.
2. $0^6 M K^4 \xrightarrow{1,3,123} 0^6 M 00 K 0 \bullet$.
3. $0^5 M K^5 \xrightarrow{4,24} 0^5 M K^3 00 \bullet$.
4. $0^4 M K^6 \xrightarrow{1} 0^4 M 0 K^5 \xrightarrow{s_1} 0^5 M K^5 \bullet p$.
5. $0^3 M K^7 \xrightarrow{2,12,23,123} 0^3 M K^2 0^4 K \bullet$.
6. $0^2 M K^8 \xrightarrow{2,12,24,124} 0^2 M K 00 K' K' 00 K \xRightarrow{24} 7$.
7.(a) $0^2 M K 000 K 00 K \bullet$.
7.(b) $0^2 M K 00 M K 00 K \xrightarrow{1} 0^2 M K 00 M 000 K \bullet$.
8. $0 M K^9 \xrightarrow{3,4,234} 0 M K^3 0 K 0 K K 0 \xRightarrow{3} 10$.
9. $M K^{10} \xrightarrow{1} M 0 K^9 \xrightarrow{s_1} 0 M K^9 \bullet p$.
10.(a) $0 M 0 K^2 0 K 0 K K 0 \xRightarrow{4} 11$.
10.(b) $0 M M K^2 0 K 0 K K 0 \xrightarrow{2,24,124} 0 M M K 00 K' 000 K' \xrightarrow{234} 0 M M K 00 K 0^4 \bullet$.
11.(a) $0 M 00 K 0 K 0 K K 0 \xRightarrow{23} 12$.
11.(b) $0 M 0 M K 0 K 0 K K 0 \xrightarrow{2,23,123} 0 M 0 M K 0^5 K'$ (lin. dep.) $\xrightarrow{234} 0 M 0 M K 0^6 \bullet$.
12.(a) $0 M 0^4 K 0 K K 0 \bullet$.
12.(b) $0 M 0^2 M 0 K 0 K K 0 \xRightarrow{24} 13$.
13.(a) $0 M 0^2 M 0^3 K K 0 \bullet$.
13.(b) $0 M 0^2 M 0 M 0 K K 0 \xrightarrow{234} 14$.
14.(a) $0 M 0^2 M 0 M 0^2 K 0 \bullet$.

14.(b) $0M0^2M0M0MK0 \implies$ Lemma 6.2 •.

Finally, the given upper bound on the number of P-orbits on P_u follows in view of Remark 6.1 and Lemma 6.2.

Proposition 6.4. *Let G be of type D_4 and $P = P_{\alpha_2}$. Then the number of P-orbits on P_u is finite. There are at most 56 orbits.*

Proof. Note that $\dim P_u = 11$. We choose the following ordering on $\Psi(P_u)$: $\Psi(P_u) = \{1, 12, 3, 23, 4, 24, 123, 124, 234, 1234, 1\bar{2}34\}$. We proceed to analyze the various subvarieties $P_u(C_1, \dots, C_r)$:

1. $0^7 K^4$•.
2. $0^6 M K^4 \xrightarrow{4,24} 0^6 M K^2 00$•.
3. $0^5 M K^5 \xrightarrow{1,3,123} 0^5 M K 0 0 K 0$•.
4. $0^4 M K^6 \xrightarrow{2} 0^4 M 0 K^5 \xrightarrow{s2} 0^5 M K^5 \bullet p$.
5. $0^3 M K^7 \xrightarrow{1,4,124} 0^3 M K^2 0 K 0 K 0 \overset{4}{\implies} 6$.
6.(a) $0^3 M 0 K 0 K 0 K 0$•.
6.(b) $0^3 M M K 0 K 0 K 0 \xrightarrow{2,12,123} 0^3 M M 0^5 K'$•.
7. $0^2 M K^8 \xrightarrow{2} 0^2 M 0 K^7 \xrightarrow{s2} 0^3 M K^7 \bullet p$.
8. $0 M K^9 \xrightarrow{3,4,234} 0 M K^4 0 0 K' K' 0 \overset{3}{\implies} 10$.
9. $M K^{10} \xrightarrow{1} M 0 K^9 \xrightarrow{s2} 0 M K^9 \bullet p$.
10.(a) $0 M 0 K^3 0 0 K K 0 \overset{23}{\implies} 11$.
10.(b) $0 M M K^3 0 0 K K 0 \xrightarrow{2,24,124} 0 M M 0 K K 0 0 0 0 K' \xrightarrow{234} 0 M M 0 K K 0^5$•.
11.(a) $0 M 0 0 K K 0 0 K K 0 \overset{4}{\implies} 12$.
11.(b) $0 M 0 M K K 0 0 K K 0 \overset{4}{\implies} 13$.
12.(a) $0 M 0^3 K 0 0 K K 0$•.
12.(b) $0 M 0^2 M K 0 0 K K 0 \xrightarrow{2,23,123} 0 M 0^2 M 0^5 K' \xrightarrow{234} 0 M 0^2 M 0^6$•.
13.(a) $0 M 0 M 0 K 0 0 K K 0 \overset{24}{\implies} 14$.
13.(b) $0 M 0 M M K 0 0 K K 0 \xrightarrow{2,123,23} 0 M 0 M M 0^5 K' \xrightarrow{234} 0 M 0 M M 0^6$•.
14.(a) $0 M 0 M 0^4 K K 0$•.
14.(b) $0 M 0 M 0 M 0^2 K K 0 \overset{234}{\implies} 15$.
15.(a) $0 M 0 M 0 M 0^3 K 0$•.
15.(b) $0 M 0 M 0 M 0^2 M K 0 \implies$ Lemma 6.2 •.

Finally, the given upper bound on the number of P-orbits on P_u follows in view of Remark 6.1 and Lemma 6.2.

Corollary 6.5. *Let G be of type D_4 and P a minimal parabolic subgroup of G. Then the number of P-orbits on P_u is finite.*

Proof. This follows from Propositions 6.3 and 6.4 and the fact that the cases for P_{α_3} and P_{α_4} are equivalent to the one for P_{α_1}.

Remark 6.6. The A_5 cases relevant for Theorem 4.1 can be analyzed in the same fashion as the ones for D_4. We omit them here, as the computations involved are somewhat lengthy. They can be found in [8].

320   VLADIMIR POPOV AND GERHARD RÖHRLE

Both authors began work together on the subject of this paper while they were visitors of the University of Sydney, Australia. We would like to express our gratitude to the members of the School of Mathematics and Statistics of the University of Sydney for their hospitality. We would also like to thank the referee who made numerous suggestions improving the exposition of the manuscript.

REFERENCES

1. A. Borel, *Linear algebraic groups*, W. A. Benjamin, New York, 1969.
2. N. Bourbaki, *Groupes et algèbres de Lie*, Hermann, Paris, 1975, Chapitres 4,5 et 6.
3. H. Bürgstein and W. H. Hesselink, *Algorithmic orbit classification for some Borel group actions*, Comp. Math. **61** (1987), 3–41.
4. E. B. Dynkin, *Semisimple subalgebras of semisimple Lie algebras*, Amer. Math. Soc. Transl. Ser. 2 **6** (1957), 111–244.
5. V. V. Kashin, *Orbits of adjoint and coadjoint actions of Borel subgroups of semisimple algebraic groups*, Questions of Group Theory and Homological algebra, Yaroslavl' (1990), 141–159, (in Russian).
6. G. Lusztig, *On the finiteness of the number of unipotent classes*, Inv. Math. **34** (1976), 201–213.
7. V. Popov, *A finiteness theorem for the parabolics of a given modality*, Indag. Math., (to appear).
8. V. Popov and G. Röhrle, *On the number of orbits of a parabolic subgroup on its unipotent radical*, Tech. Report Report 94-24, The University of Sydney, 1994, (preprint).
9. V. L. Popov and E. B. Vinberg, *Invariant theory*, Encyclopaedia of Math. Sci.: Algebraic Geometry IV., vol. 55, Springer Verlag, 1994, (Translated from Russian Series: Itogi Nauki i Tekhniki, Sovr. Probl. Mathem., Fund. Napravl., **55** (1989), 137–314), pp. 123–284.
10. R. W. Richardson, *Conjugacy classes in Lie algebras and algebraic groups*, Ann. Math. **86** (1967), 1–15.
11. ———, *Conjugacy classes in parabolic subgroups of semisimple algebraic groups*, Bull. London Math. Soc. **6** (1974), 21–24.
12. ———, *Finiteness Theorems for Orbits of Algebraic Groups*, Indag. Math. **88** (1985), 337–344.
13. R. Richardson, G. Röhrle, and R. Steinberg, *Parabolic subgroups with Abelian unipotent radical*, Inv. Math. **110** (1992), 649–671.
14. G. Röhrle, *Parabolic subgroups of positive modality*, Geom. Ded., (to appear).
15. M. Rosenlicht, *Some basic theorems on algebraic groups*, Amer. J. Math. **78** (1956), 401–443.
16. T. Springer, *Linear Algebraic Groups*, Birkhäuser, Boston, 1981.
17. R. Steinberg, *Endomorphisms of linear algebraic groups*, Mem. Am. Math. Soc. **80** (1968).
18. E. B. Vinberg, *Complexity of actions of reductive groups*, Functional Anal. Appl. **20** (1986).

DEPARTMENT OF MATHEMATICS, MOSCOW STATE UNIVERSITY, MGIEM, BOL'SHOI, TREKHSVYATITEL'SKIĬ PER., 3/12, 109028 MOSCOW, RUSSIA
E-mail address: popov@ppc.msk.ru

FAKULTÄT FÜR MATHEMATIK, UNIVERSITÄT BIELEFELD, 33615 BIELEFELD, GERMANY
E-mail address: roehrle@mathematik.uni-bielefeld.de

ON A HOMOMORPHISM OF HARISH-CHANDRA

GERALD W. SCHWARZ

Dedicated to the memory of Roger W. Richardson

0. INTRODUCTION

Let $\mathfrak{g}$ be a reductive Lie algebra with adjoint group G, Cartan subalgebra $\mathfrak{t}$ and Weyl group $\mathcal{W}$. In [2] and [3], Harish-Chandra defined an exact sequence

$$0 \to I \to \mathscr{D}(\mathfrak{g})^G \xrightarrow{\delta} \mathscr{D}(\mathfrak{t})^{\mathcal{W}}$$

where $\mathscr{D}(\mathfrak{g})$ and $\mathscr{D}(\mathfrak{t})$ denote the algebras of algebraic differential operators on $\mathfrak{g}$ and $\mathfrak{t}$, respectively. The kernel I of δ consists of the elements of $\mathscr{D}(\mathfrak{g})^G$ vanishing on $\mathcal{O}(\mathfrak{g})^G$. Harish-Chandra's proof of the existence of δ is long and analytic in nature, and one of the aims of this paper is to give a conceptual and short algebraic proof. We also give a short proof of the recent theorem of Levasseur and Stafford [4] which establishes that δ is surjective. Previously, Wallach [12] had established (graded) surjectivity for all simple Lie algebras except for those of type E. He and Hunziker [13] have also given an algebraic proof of the existence of δ using methods completely different from those used here. A fact which we rely upon is the existence of natural actions of SL_2 on $\mathscr{D}(\mathfrak{g})^G$ and $\mathscr{D}(\mathfrak{t})^{\mathcal{W}}$ such that δ is equivariant.

I thank the Department of Mathematics of the University of Bochum and the ICTP Trieste for their hospitality while this paper was prepared, and I thank Toby Stafford for illuminating discussions.

1. DIFFERENTIAL OPERATORS AND SL_2-ACTIONS

All varieties we consider will be algebraic and defined over our base field $\mathbb{C}$. Let Z be an affine variety, and set $A := \mathcal{O}(Z)$. Then we define the algebra of (algebraic) differential operators on A and Z as follows: If $P \in \mathrm{End}_{\mathbb{C}}(A)$ and $a \in A$, then $[P, a]$ denotes the usual commutator: $[P, a](b) = P(ab) - a(P(b))$, $b \in A$. Define $D^n(A) = 0$ for $n < 0$, and for $n \geq 0$ inductively define:

$$D^n(A) = \{P \in \mathrm{End}_{\mathbb{C}}(A) : [P, a] \in D^{n-1}(A) \text{ for all } a \in A\}.$$

Clearly, $D^0(A) \simeq A$ acting on itself by multiplication. Note that $D^n(A) \subseteq D^{n+1}(A)$ for all n, and we define $D(A) := \bigcup D^n(A)$. Now we set $\mathscr{D}^n(Z) := D^n(A)$, and similarly for $\mathscr{D}(Z)$. We call $\mathscr{D}^n(Z)$ (resp. $D^n(A)$) the *differential operators on Z (resp. A) of order at most n*, and $\mathscr{D}(Z)$ (resp. $D(A)$) the *algebra of differential operators on Z (resp. A)*. Note that $\mathcal{O}(Z)$ acts on $\mathscr{D}^n(Z)$, etc. by left multiplication, making $\mathscr{D}^n(Z)$, etc. into left $\mathcal{O}(Z)$-modules.

Research partially supported by the NSF

Proposition 1. (see [10, §3]) *Let Z be as above. Then*

(1) $\mathscr{D}^n(Z)$ *is finitely generated for all n.*

(2) *If $P \in \mathscr{D}^n(Z)$ and $Q \in \mathscr{D}^m(Z)$, then $Q \circ P \in \mathscr{D}^{n+m}(Z)$ and $[Q, P] = Q \circ P - P \circ Q \in \mathscr{D}^{n+m-1}(Z)$.*

Suppose that $Z = \mathbb{C}^k$, so that $A := \mathscr{O}(\mathbb{C}^k) = \mathbb{C}[x_1, \ldots, x_k]$. Then $D(A)$ is the kth *Weyl algebra*, i.e., the noncommutative algebra $\mathbb{C}\langle x_1, \ldots, x_k, \partial_1, \ldots, \partial_k \rangle$ generated by the x_i and the $\partial_j := \partial/\partial x_j$ with their usual commutation relations. Note that $\operatorname{gr} D(A) \simeq \mathbb{C}[x_1, \ldots, x_k, y_1, \ldots, y_k]$ is a polynomial ring. If $\rho = (\rho_1, \ldots, \rho_k) \in \mathbb{N}^k$, let $|\rho|$ denote $\sum_i \rho_i$, let x^ρ denote $x_1^{\rho_1} \cdots x_k^{\rho_k}$ and let ∂^ρ denote $\partial_1^{\rho_1} \cdots \partial_k^{\rho_k}$. Then every element $Q \in D^n(A)$ is a sum $\sum_{|\rho| \leq n} a_\rho \partial^\rho$ where the a_ρ are in A. The (nth order) *symbol of Q* is $\sum_{|\rho|=n} a_\rho \partial^\rho$. If U is a Zariski open subset of $\mathbb{C}^k$, then $\mathscr{D}(U) = \{\sum_\rho a_\rho \partial^\rho : a_\rho \in \mathscr{O}(U)\} \simeq \mathscr{O}(U) \otimes_{\mathscr{O}(\mathbb{C}^k)} \mathscr{D}(\mathbb{C}^k)$.

We define a degree function on $\mathscr{D}(\mathbb{C})^k$ as follows: A differential operator P has degree m if it sends the homogeneous functions $\mathscr{O}(\mathbb{C}^k)_s$ of degree s to $\mathscr{O}(\mathbb{C}^k)_{s+m}$ for all s. Equivalently, $P = \sum_{\rho, \eta} c_{\rho, \eta} x^\eta \partial^\rho$ has degree m iff for each constant $c_{\rho, \eta} \neq 0$, $|\eta| - |\rho| = m$.

Suppose that we have a complex vector space V of dimension k, and we are given a nondegenerate symmetric bilinear form q on V. Then we can choose coordinates x_i so that $q = 1/2 \sum_{i=1}^k x_i^2$. There is a natural isomorphism (given q), between V and V^*, so $\operatorname{gr} \mathscr{D}(V) \simeq \mathscr{O}(V \oplus V^*) \simeq \mathscr{O}(V \oplus V) \simeq \mathscr{O}(V \otimes \mathbb{C}^2)$. There is a natural action of SL_2 on $\mathscr{O}(V \otimes \mathbb{C}^2)$, hence an action of SL_2 on $\operatorname{gr} \mathscr{D}(V)$. We lift this action to $\mathscr{D}(V)$:

Set $Q = -1/2 \sum_i \partial_i^2 \in S^2(V)$, and let E denote the Euler operator $\sum_i x_i \partial_i$. Noting that $\operatorname{ad} q$ and $\operatorname{ad} Q$ act locally nilpotently on $\mathscr{D}(V)$ one can easily establish the following:

Proposition 2. *Let q, etc. be as above. Then*

(1) $e := q$, $f := Q$ *and* $h := [q, Q] = E + k/2$ *form a simple Lie algebra of type $\mathfrak{sl}_2$.*

(2) *The $\mathfrak{sl}_2$-action integrates to SL_2.*

(3) $x^\eta \partial^\rho$ *has weight $|\eta| - |\rho|$ with respect to the action of the standard maximal torus $T \subset \operatorname{SL}_2$.*

(4) *The element $w := \begin{pmatrix} 0 & -i \\ -i & 0 \end{pmatrix} \in \operatorname{SL}_2$ generates the Weyl group and acts as follows: $w(x_j) = -i\partial_j$ and $w(\partial_j) = -i x_j$ for all j.*

Remark 3. Consider the real span $V_\mathbb{R}$ of the x_j and the space of complex valued Schwartz functions $\mathscr{S}(V_\mathbb{R})$. Note that $\mathscr{D}(V)$ acts in a natural way on $\mathscr{S}(V_\mathbb{R})$. Let $\tilde{f}$ denote the Fourier transform of $f \in \mathscr{S}(V_\mathbb{R})$. For $P \in \mathscr{D}(V)$ we define $\tilde{P}$ by: $\widetilde{P(f)} = \tilde{P}(\tilde{f})$, $f \in \mathscr{S}(V_\mathbb{R})$ (see [3]). It is easy to see that $\tilde{x}_j = w(x_j)$ and $\tilde{\partial}_j = w(\partial_j)$ for all j. Since $P \mapsto w(P)$ and $P \mapsto \tilde{P}$ are automorphisms, they are identical, hence w "is" Fourier transform.

Suppose that X is an affine G-variety, where G is complex reductive. Then $\mathscr{O}(X)^G$ is finitely generated, and there is a canonical morphism $\pi_{X,G}$ (or just π_X) : $X \to X /\!\!/ G$ where $X /\!\!/ G$ is the affine variety corresponding to $\mathscr{O}(X)^G$, and $(\pi_X)^*$ is the inclusion $\mathscr{O}(X)^G \subset \mathscr{O}(X)$. The morphism π_X is surjective and induces a one-to-one correspondence between the closed G-orbits in X and the points of $X /\!/ G$. Moreover, G acts rationally on $\mathscr{O}(X)$ and $\mathscr{D}(X)$ ([10, §3]). There is a canonical morphism $(\pi_X)_* : \mathscr{D}(X)^G \to \mathscr{D}(X /\!/ G)$, where $(\pi_X)_*(P)$ is the restriction of $P \in \mathscr{D}(X)^G$ to $\mathscr{O}(X)^G \simeq \mathscr{O}(X /\!/ G)$. It follows from the definitions that $\operatorname{order}(\pi_X)_*(P) \le \operatorname{order} P$. We let $\mathscr{K}^n(X)$ denote the elements of $\mathscr{D}^n(X)$ which annihilate $\mathscr{O}(X)^G$. Then, by definition, $\mathscr{K}^n(X)^G$ is the kernel of $(\pi_X)_*$ restricted to $\mathscr{D}^n(X)^G$, and $\mathscr{K}(X)^G := \bigcup_n \mathscr{K}^n(X)^G$ is the kernel of $(\pi_X)_*$. Note that $\mathscr{D}^{n-1}(X)\tau(\mathfrak{g}) \subset \mathscr{K}^n(X)$, where $\tau(A)$ denotes the action of $A \in \mathfrak{g}$ on $\mathscr{O}(X)$ as a derivation. Multiplication by $f \in \mathscr{O}(X)$, considered as an element in $\mathscr{D}^0(X)$, is denoted by m_f.

2. The Homomorphism δ'

Let G be a reductive complex algebraic group and V a G-module. If U is a subspace of V, then $C_G(U)$ denotes the subgroup of G fixing U pointwise, while $N_G(U)$ denotes the subgroup stabilizing U. Suppose that we can find a vector subspace $U \subset V$ such that

(1) $H := N_G(U)/C_G(U)$ is a finite group.
(2) The restriction $\rho : \mathscr{O}(V) \to \mathscr{O}(U)$ induces an isomorphism of $\mathscr{O}(V)^G$ and $\mathscr{O}(U)^H$.
(3) (V, G) and (U, H) are coregular, i.e., $\mathscr{O}(V)^G \simeq \mathscr{O}(U)^H$ is a polynomial algebra.

The subspace $U \subset V$ is called a *Cartan subspace of* (V, G) ([1]). Examples are any Cartan subalgebra of a reductive Lie algebra (adjoint representation). If G is connected semisimple, then a result of Panyushev [8] shows that (1) and (2) imply (3).

Let U_{pr} denote the principal orbits in U, i.e., the points with trivial isotropy group. Then U_{pr} is open and dense in U. Since $\pi_U : U_{\mathrm{pr}} \to U_{\mathrm{pr}} /\!/ H$ is a covering map, $(\pi_U)_*$ induces an isomorphism $\mathscr{D}^n(U_{\mathrm{pr}})^H \xrightarrow{\sim} \mathscr{D}^n(U_{\mathrm{pr}} /\!/ H)$ for all n [10, 4.5].

Definition 4. Define δ' by the following commutative diagram:

$$
\begin{array}{ccc}
\mathscr{D}(V)^G & \xrightarrow{\ \delta'\ } & \mathscr{D}(U_{\mathrm{pr}})^H \\
\downarrow{\scriptstyle (\pi_V)_*} & & \uparrow{\scriptstyle (\pi_U)_*^{-1}} \\
\mathscr{D}(V /\!/ G) & \xrightarrow{\ \mathrm{incl.}\ } & \mathscr{D}(U_{\mathrm{pr}} /\!/ H)
\end{array}
$$

Remark 5. If $f \in \mathscr{O}(V)^G$ and $P \in \mathscr{D}(V)^G$, then

$$\delta'(P)(f|_U) = P(f)|_U \in \mathscr{O}(U)^W. \qquad (*)$$

In fact, this property determines δ' uniquely. The δ' that Harish-Chandra constructs satisfies $(*)$, so his construction and ours are equivalent. Note that $\operatorname{Ker} \delta' = \mathscr{K}(V)^G$.

By the remark above, $\delta'(P)$ is a differential operator on $\mathscr{O}(U)^H$, but we do not know if $\delta'(P)$ extends to a differential operator on $\mathscr{O}(U)$. In fact, except in trivial cases, this is false. Harish-Chandra's idea was to modify δ' to obtain a homomorphism $\delta : \mathcal{D}(V)^G \to \mathcal{D}(U)^H$ with kernel $\mathcal{K}(V)^G$. In the next few paragraphs we explore this idea.

Let $G = \mathrm{SO}(k)$, $k \geq 3$, or $\mathrm{O}(k)$, $k \geq 2$, act on $V = \mathbb{C}^k$ in the usual way. Let $e_1, \ldots, e_k$ be an orthonormal basis with dual basis $x_1, \ldots x_k$. The generators of $\mathcal{D}(V)^G$ are q, Q and E where $q = 1/2 \sum_i x_i^2$, etc. as before. We choose a coordinate function y on $V /\!/ G \simeq \mathbb{C}$ so that $(\pi_V)_* y = q$. Since $(\pi_V)_*(Q)$, etc. have order at most 2, applying Q, etc. to 1, q and q^2 and expressing the input and output in terms of powers of y we obtain:

(1) $(\pi_V)_*(Q) = -y\partial_y^2 - (k/2)\partial_y$
(2) $(\pi_V)_*(q) = y$,
(3) $(\pi_V)_*([q, Q]) = (\pi_V)_*(E + k/2) = 2y\partial_y + k/2$.

Note that E induces twice the usual Euler operator (plus a constant) on $\mathcal{O}(V)^G = \mathbb{C}[y]$ since q (hence y) has degree 2.

Now $\mathcal{D}(V)^G$ is a copy of $\mathfrak{U}(\mathfrak{sl}_2)$, with $X := q$, $Y := Q$ and $H := E + k/2$ a basis of $\mathfrak{sl}_2$. The center of $\mathfrak{U}(\mathfrak{sl}_2)$ is generated by the Casimir operator $\mathscr{C} = (H^2 + 4YX + 2H)$. The formulas above show that the image of $\mathscr{C}$ in $(\pi_V)_*(\mathcal{D}(V)^G) \subset \mathcal{D}(V /\!/ G)$ is the constant $(k^2/4 - k)$. For λ, $\mu \in \mathbb{C}$, $n \in \mathbb{N}$, define $S_\lambda := \mathfrak{U}(\mathfrak{sl}_2)/(\mathscr{C} - \lambda)$, $R_\mu = S_{(\mu^2 + 2\mu)}$ and $B_n = R_{-n/2} = S_{(n^2/4 - n)}$. Then we have

Theorem 6. *(see J. T. Stafford [11])*
(1) $S_\lambda \simeq S_{\lambda'}$ *if and only if* $\lambda = \lambda'$.
(2) S_{-1} *is not Morita equivalent to* S_λ *for* $\lambda \neq -1$.
(3) R_μ *is simple if and only if* $-1 \neq \mu \in \mathbb{Z}$.
(4) R_μ *is Morita equivalent to* $R_{\mu+1}$ *if* $\mu \neq -1, -2$.
(5) $R_n = R_{-n-2}$, $n = \{0, 1, \ldots\}$ *has a unique nontrivial quotient induced from the standard action of* $\mathfrak{U}(\mathfrak{sl}_2)$ *on* $\mathrm{S}^n(\mathbb{C}^2)$.

Corollary 7. *Let* n, $n' \in \mathbb{Z}$, n, $n' \geq 1$. *Then*
(1) $B_n \simeq B_{n'}$ *if and only* $n = n'$ *or* $\{n, n'\} = \{1, 3\}$.
(2) B_n *is Morita equivalent to* B_1 *if* n *is odd, and* B_n *is Morita equivalent to* B_4 *if* n *is even*, $n \geq 4$.
(3) *No two of* B_1, B_2, B_4 *are Morita equivalent.*

We make no use of the results on Morita equivalence, but they are interesting nonetheless. Note that $(\pi_V)_*(\mathcal{D}(V)^G)$ is an infinite dimensional quotient of B_k, hence $(\pi_V)_*(\mathcal{D}(V)^G) \simeq B_k$.

Let x denote x_k, and let $U := \mathrm{span}\{x\} \simeq \mathbb{C}^1$. Then U is a Cartan subspace of V, with $H = \mathrm{O}(1)$. (This is a special case of the Luna-Richardson Theorem [6].) As generators of $\mathcal{D}(U)^H$ we choose $\bar{q} = (1/2)x^2$, etc. as in Proposition 2. By the results above, $\mathcal{D}(U)^H \simeq (\pi_U)_*(\mathcal{D}(U)^H) \simeq B_1$.

Proposition 8. *There is a surjective* $\mathbb{C}$-*algebra homomorphism* $\delta : \mathcal{D}(V)^G \to \mathcal{D}(U)^H$ *with kernel* $\mathscr{K}(V)^G$ *if and only if* $k = 3$, *in which case* δ *can be chosen to be* $m_x \circ \delta' \circ m_{x^{-1}}$.

Proof. If δ exists, then it induces an isomorphism of B_k with B_1, and this implies that $k = 3$ (since $k \geq 2$). Note that the case $k = 3$ can be considered as the adjoint action of $\mathrm{SL}_2 / \{\pm I\} \simeq \mathrm{SO}(3)$ on $\mathfrak{sl}_2$. We assume now that $k = 3$.

Clearly $\delta'(q) = \bar{q}$ and $\delta'(E) = \bar{E}$. A priori, $\delta'(Q)$ must have order 2 and degree -2, so it must be a linear combination of $\bar{Q}$, $\bar{E}/\bar{q}$ and $1/\bar{q}$. In fact, the formulas for $(\pi_V)_*Q$ and $(\pi_U)_*Q$, etc., show that $\delta'(Q) = \bar{Q} - \bar{E}/\bar{q}$. One computes easily that $m_{x^{-1}} \circ \bar{Q} \circ m_x = \bar{Q} - \bar{E}/\bar{q}$. Thus $\delta := m_x \circ \delta' \circ m_{x^{-1}}$ sends Q to $\bar{Q}$. We have:

(1) $\delta(Q) = \bar{Q}$.

(2) $\delta(q) = \bar{q}$,

(3) $\delta(E + 3/2 := [q, Q]) = [\bar{q}, \bar{Q}] = \bar{E} + 1/2$.

Remark 9. We translate the result above into the usual coordinates for $\mathfrak{sl}_2$: Let X, Y, $H \in \mathfrak{sl}_2$ be the usual basis and x, y, and h the dual basis. We can choose $q := (1/2)(xy + h^2)$ and $Q := -2\partial_x\partial_y - (1/2)\partial_h^2$. Our Cartan subspace is $\mathfrak{t} = \mathrm{span}\{H\}$ with the action of $\mathcal{W} = \{\pm 1\}$. Set $\bar{q} := (1/2)h^2$ and $\bar{Q} := -(1/2)\partial_h^2$. Then $\delta'(Q) = \bar{Q} - (1/h)\partial_h = m_{h^{-1}} \circ \bar{Q} \circ m_h$, hence $\delta(Q) = m_h \circ \delta'(Q) \circ m_{h^{-1}} = \bar{Q}$, and δ maps $\mathscr{D}(\mathfrak{sl}_2)^{\mathrm{SL}_2}$ to $\mathscr{D}(\mathfrak{t})^{\mathcal{W}}$.

3. THE HOMOMORPHISM δ

Let G be connected reductive with maximal torus T. Let $\mathfrak{t}$ denote the Lie algebra of T and $\mathcal{W}$ the Weyl group. Choose a system of positive roots Φ^+, and set $\sigma := \prod_{\alpha \in \Phi^+} \alpha$. Then σ transforms by the sign representation of $\mathcal{W}$, and σ vanishes to order 1 on each root hyperplane $\mathscr{H}_\alpha := \{\alpha = 0\}$. Define δ to be the composition $m_\sigma \circ \delta' \circ m_{\sigma^{-1}}$. (Note that in case $\mathfrak{g} = \mathfrak{sl}_2$, this is how we defined δ above.) Harish-Chandra showed that δ has the first six of the following properties.

($\delta 1$) $\delta : \mathscr{D}(\mathfrak{g})^G \to \mathscr{D}(\mathfrak{t})^{\mathcal{W}}$ is an algebra homomorphism.

($\delta 2$) The restriction of δ to $\mathscr{O}(\mathfrak{g})^G$ is the restriction map $\mathscr{O}(\mathfrak{g})^G \to \mathscr{O}(\mathfrak{t})^{\mathcal{W}}$.

($\delta 3$) The restriction of δ to $\mathrm{S}(\mathfrak{g})^G$ (considered as the invariant constant coefficient differential operators on $\mathfrak{g}$) is the canonical projection $\mathrm{S}(\mathfrak{g})^G \to \mathrm{S}(\mathfrak{t})^{\mathcal{W}}$.

($\delta 4$) The kernel of δ is $\mathscr{K}(\mathfrak{g})^G$.

($\delta 5$) δ commutes with Fourier transform (see §4).

($\delta 6$) δ commutes with formal adjoint of differential operators (see §5).

($\delta 7$) δ is surjective.

Levasseur and Stafford [4] proved a theorem (see §5) which implies ($\delta 7$). Recently [5], they have improved ($\delta 4$) by showing that $\mathscr{K}(\mathfrak{g}) = \mathscr{D}(\mathfrak{g})\tau(\mathfrak{g})$, hence that $\mathscr{K}(\mathfrak{g})^G = (\mathscr{D}(\mathfrak{g})\tau(\mathfrak{g}))^G$. The graded version of this result is still unknown, i.e., whether or not $\mathscr{K}^n(\mathfrak{g})^G = (\mathscr{D}^{n-1}(\mathfrak{g})\tau(\mathfrak{g}))^G$ for all n.

To establish the properties ($\delta 1$)–($\delta 7$), we may clearly reduce to the case that $\mathfrak{g}$ is simple, which we assume from now on. Note that δ and δ' satisfy ($\delta 2$) and ($\delta 4$).

We first show that $\mathrm{Im}\,\delta \subset \mathscr{D}(\mathfrak{t})^{\mathcal{W}}$. The proof is conceptually simple. Let $P \in \mathscr{D}(\mathfrak{g})^G$. Then $\delta(P) = \sum_\rho a_\rho \partial^\rho$ where the a_ρ are rational functions with no poles on $\mathfrak{t}_{\mathrm{pr}}$. Thus $\delta(P)$ lies in $\mathscr{D}(\mathfrak{t})^{\mathcal{W}}$ iff the a_ρ are polynomial. If one of the

a_ρ is not polynomial, then it must have poles along some reflection hyperplane $\mathcal{H}_\gamma$, $\gamma \in \Phi^+$. However, $\mathcal{H}_\gamma$ corresponds to a copy $\mathrm{SL}_{2,\gamma}$ of $\mathrm{SL}_2 \subset G$, and we can reduce to the case of SL_2, where we already know that $\mathrm{Im}\,\delta \subset \mathscr{D}(\mathfrak{t})^{\mathscr{W}}$.

Let $x_0 \in \mathcal{H}_\gamma \setminus (\cup_{\alpha \neq \gamma} \mathcal{H}_\alpha)$. Then $\mathscr{W}' := \mathscr{W}_{x_0} \simeq \mathscr{W}(\mathrm{SL}_{2,\gamma}) \simeq \{\pm 1\}$. Modulo a finite central subgroup, $G_{x_0} = \mathrm{SL}_{2,\gamma} \times T'$ where $T' \subset T$ denotes a torus of corank 1 with Lie algebra $\mathfrak{t}' = \mathcal{H}_\gamma$. We have $\mathfrak{t} = \mathfrak{t}_\gamma \oplus \mathfrak{t}'$ where $\mathfrak{t}_\gamma = \mathfrak{t} \cap \mathfrak{sl}_{2,\gamma}$. Choose $X_1, \ldots, X_s \in \mathfrak{g}$ such that the $\tau(X_i)(x_0)$ are a basis of $T_{x_0}(Gx_0)$. We consider $N := \mathfrak{g}_{x_0} = \mathfrak{sl}_{2,\gamma} \oplus \mathfrak{t}'$ as a subspace of $T_{x_0}(\mathfrak{g})$ (note that $x_0 \in N$). There is a G_{x_0}-stable decomposition

$$T_{x_0}(\mathfrak{g}) = T_{x_0}(Gx_0) \oplus N = \mathrm{span}\{\tau(X_i)(x_0)\}_{i=1}^s \oplus N.$$

Let x, y and h be coordinate functions on $\mathfrak{sl}_{2,\gamma}$ as in Remark 9. Let $t_1, \ldots, t_r$ be coordinates on $\mathfrak{t}'$, so that h, $t_1, \ldots, t_r$ are coordinates on $\mathfrak{t}$. Set $U := \{u \in \mathfrak{g} \mid \mathfrak{g} = N \oplus \mathrm{span}\{\tau(X_i)(u)\}\}$. Then U is a G_{x_0}-stable neighborhood of x_0. Set $\tilde{N} := N \cap U$, $\tilde{\mathfrak{t}} := \mathfrak{t} \cap U$ and $\tilde{\mathfrak{t}}' := \mathfrak{t}' \cap U$.

On U we may write any element $P \in \mathscr{D}(\mathfrak{g})^G$ uniquely as a sum

$$P = \sum_{i,j,k \in \mathrm{N}, \rho \in \mathrm{N}^r, \eta \in \mathrm{N}^s} a_{i,j,k,\rho,\eta} \partial_x^i \partial_y^j \partial_h^k \partial_t^\rho \tau(X)^\eta, \quad a_{i,j,k,\rho,\eta} \in \mathcal{O}(U),$$

where $\partial_t^\rho = \partial_{t_1}^{\rho_1} \cdots \partial_{t_r}^{\rho_r}$ and $\tau(X)^\eta = \tau(X_1)^{\eta_1} \cdots \tau(X_s)^{\eta_s}$. When we apply P to any $f \in \mathcal{O}(\mathfrak{g})^G$, the terms involving the $\tau(X_i)$ give zero, so

$$P(f)|_{\tilde{N}} = (\tilde{P} := \sum_{i,j,k \in \mathrm{N}, \rho \in \mathrm{N}^r} (a_{i,j,k,\rho,0}|_{\tilde{N}}) \partial_x^i \partial_y^j \partial_h^k \partial_t^\rho)(f|_{\tilde{N}}).$$

Note that $\tilde{P}$ is $\mathrm{SL}_{2,\gamma}$-invariant.

Let $\tilde{\delta}'$, $\tilde{\delta} : \mathscr{D}(N)^{\mathrm{SL}_{2,\gamma}} \to \mathscr{D}(\mathfrak{t} \setminus \mathfrak{t}')^{\mathscr{W}'}$ be the "δ'" and "δ" homomorphisms for the G_{x_0} (equivalently, $\mathrm{SL}_{2,\gamma}$)-action on N. The formulas above show that $\delta(P)$ is smooth at x_0 if and only if $m_\sigma \circ \tilde{\delta}'(\tilde{P}) \circ m_{\sigma^{-1}}$ is smooth at x_0. But $\sigma = h\gamma$ where $h = \prod_{\alpha \in \Phi^+, \alpha \neq \gamma} \alpha$ does not vanish at x_0. Hence $\delta(P)$ is smooth at x_0 if and only if $\tilde{\delta}(\tilde{P})$ is smooth at x_0. But the latter fact follows immediately from the case for $\mathrm{SL}_{2,\gamma}$, i.e., from Proposition 8.

4. SL_2-ACTIONS AND δ

We show that δ has properties $(\delta 3)$ and $(\delta 5)$: Since $\mathfrak{g}$ is simple, there are unique up to scalars $q \in \mathcal{O}(\mathfrak{g})_2^G$ and $Q \in S^2(\mathfrak{g})^G$ such that $[q, Q] = E+$ constant. Let $\bar{q}$ and $\bar{Q}$ be analogous invariants for the action of $\mathscr{W}$ on $\mathfrak{t}$ with $\delta(q) = \bar{q}$ and $[\bar{q}, \bar{Q}] = \bar{E}+$ constant. Since $(\mathrm{ad}\,E)f = mf$ if $f \in \mathcal{O}(\mathfrak{g})_m^G$, it follows that $\delta(E) = \bar{E}+$ constant. Hence $\delta(Q) = \bar{Q}+$ lower order invariant operators. But no invariant operator can be of lower order and degree -2 (and have polynomial coefficients), so $\delta(Q) = \bar{Q}$. It follows that δ is equivariant with respect to the SL_2-actions on $\mathscr{D}(\mathfrak{g})$ and $\mathscr{D}(\mathfrak{t})$.

Proof of $(\delta 3)$. Let $P \in S^n(\mathfrak{g})^G$, and let $w \in \mathrm{SL}_2$ be as in Proposition 2. Then $w(P)$ is in $\mathcal{O}(\mathfrak{g})^G$, hence $\delta(P) = w^{-1}(\delta(w(P))) = w^{-1}(w(P)|_{\mathfrak{t}})$ is the projection of P onto $S^n(\mathfrak{t})^{\mathscr{W}}$.

Proof of ($\delta 5$). Choose a compact real form $\mathfrak{g}_{\mathbb{R}}$ of $\mathfrak{g}$. Then one can apply Fourier transform to $\mathscr{S}(\mathfrak{g}_{\mathbb{R}})$ and $\mathscr{S}(\mathfrak{t}_R = \mathfrak{t} \cap \mathfrak{g}_{\mathbb{R}})$, and by Remark 3, Fourier transform is just the action of w. Since δ commutes with the SL$_2$-actions, it commutes with Fourier transform.

5. THE THEOREM OF LEVASSEUR AND STAFFORD

We give a short proof of the following result of Levasseur and Stafford [4].

Theorem 10. *Let H be a finite group and W a finite dimensional H-module. Then $\mathscr{D}(W)^H$ is generated by $\mathscr{O}(W)^H$ and $S(W)^H$, i.e., by the invariant functions and invariant constant coefficient differential operators.*

Let A denote the subalgebra of $B := \mathscr{D}(W)^H$ generated by $\mathscr{O}(W)^H$ and $S(W)^H$. We may assume that H acts faithfully on W.

Lemma 11. *There are differential operators $P_1, \ldots, P_k \in B$ which generate B as both a left and right A-module.*

Proof. Since H is finite, $\mathscr{O}(W)$ is a finite $\mathscr{O}(W)^H$-module and $S(W)$ is a finite $S(W)^H$-module. Thus, since $\mathscr{O}(W)^H \otimes S(W)^H \subset \operatorname{gr} A \subset \operatorname{gr} B \subset \mathscr{O}(W) \otimes S(W)$, $\operatorname{gr} B$ is a finite $\operatorname{gr} A$-module. Hence B is finite as both a left and right A-module.

Lemma 12. *Let H be a finite group and W a faithful H-module. Let $Hw \subset W$ be a principal orbit, i.e., $H_w = \{e\}$. Let $\chi(A)$ denote the vector fields in A. Then*

(1) *The elements of $\chi(A)$ evaluated at w span the tangent space $T_w(W)$.*
(2) *There is an $f \in \mathscr{O}(W)^H$ such that for every $P \in B$ there is a $k \in \mathbb{Z}^+$ such that $f^k P \in A$.*

Proof. Since Hw is principal, there are $f_1, \ldots, f_n \in \mathscr{O}(W)^H$, $n = \dim W$, such that $T_w^*(W) = \operatorname{span}\{df_i(w)\}$ ([9]). Let $x_1, \ldots, x_n$ be coordinate functions on W. We may assume that $f_i(w) = 0$ and that $df_i(w) = dx_i(w)$, $i = 1, \ldots, n$. If $\chi(A)(w)$ fails to span $T_w(W)$, then without loss of generality we may assume that $X(f_1)(w) = 0$ for all $X \in \chi(A)$. We derive a contradiction:

If $Q = \sum_{|\rho| \leq r} a_\rho \partial^\rho \in \mathscr{D}(W)$, let $\sigma_w^r(Q)$ denote $\sum_{|\rho| = r} a_\rho(w) \partial^\rho$ (the symbol of Q evaluated at w). Since $S(W)$ is finite over $S(W)^H$, there is an $r \geq 1$ such that

$$\sigma_w^r(\partial_1^r + \sum_{i=1}^r Q_i \partial_1^{r-i}) = 0, \qquad (*)$$

where the $Q_i \in A$ are of order at most i (of course, the statement is true with $Q_i \in S^i(W)^H$ and σ_w omitted). If $r > 1$, bracketing with f_1 we obtain

$$\sigma_w^{r-1}(r(\partial f_1/\partial x_1)\partial_1^{r-1} + \sum_{i=1}^r (r-i)(\partial f_1/\partial x_1)Q_i\partial_1^{r-i-1} + [Q_i, f_1]\partial_1^{r-i}) = 0.$$

Since $(\partial f_1/\partial x_1)(w) = 1$ and $[Q_1, f_1](w) = 0$, we obtain a version of $(*)$ with r replaced by $r - 1$. Eventually we arrive at the case $r = 1$, which implies that $\sigma_w^1(\partial_1 - Q_1) = 0$, contradicting our assumption about $\chi(A)$. We have (1).

Let $P = \sum_{|\rho|\le r} a_\rho \partial^\rho$, $P \in B$. Let $X_1, \ldots, X_n$ be elements of $\chi(A)$ such that $\mathrm{span}\{X_i(w)\} = T_w(W)$. The determinant $f(w') := \det(X_1(w'), \ldots, X_n(w'))$ does not vanish on an H-stable open subset W' of W containing w, and replacing f by a power of f, we may assume that f is H-invariant. On W' we may write

$$P = \sum_{|\rho|\le r} a'_\rho X_1^{\rho_1} \cdots X_n^{\rho_n}$$

for some rational functions a'_ρ whose denominators are a power of f. Averaging over H, we may arrange that the a'_ρ are H-invariant.

Proof of Theorem 10. Let $P_1, \ldots, P_k$ generate B as both a left and right A-module, and let $f \in \mathscr{O}(W)^H$ be as in the lemma above. Choose $k \in \mathbb{N}$ so that $k \ge$ order P_i and $f^k P_i \in A$, $i = 1, \ldots, k$. Set $I := \{Q \in B : BQB \subset A\}$, a two-sided ideal in B. Note that $f^{3k} \in I$, since $f^k B = \sum_i f^k P_i A \subset A$ and $B f^{2k} = \sum_i A P_i f^{2k} \subset A f^k B \subset A$. Since B is a simple algebra [7], I must equal B, hence $A = B$.

Corollary 13. *Harish-Chandra's homomorphism is surjective (property (δ7)).*

Corollary 14. *Let $\mathfrak{g}$ be simple, and let $\tilde\delta \colon \mathscr{D}(\mathfrak{g})^G \to \mathscr{D}(\mathfrak{t}_{pr})^{\mathscr{W}}$ be an algebra homomorphism which satisfies (δ2) and (δ4). Suppose that*

$$\tilde\delta(S^2(\mathfrak{g})^G) \subset \mathscr{D}(\mathfrak{t})^{\mathscr{W}}. \tag{$\delta 3'$}$$

Then $\tilde\delta = \delta$.

Proof. Let q, $Q \in \mathscr{D}(\mathfrak{g})^G$ and $\bar{q} := \tilde\delta(q)$, $\bar{Q} \in \mathscr{D}(\mathfrak{t})^{\mathscr{W}}$ be chosen as in §4 so that $E = [q, Q]$ and $\bar{E} = [\bar{q}, \bar{Q}]$ up to addition of constants. Since $\tilde\delta$ satisfies (δ2), we have that order $\tilde\delta(Q) \le 2$, degree $\tilde\delta(Q) = -2$ and $\tilde\delta(E) = \bar{E} + f$ where $f \in \mathscr{O}(\mathfrak{t}_{pr})^{\mathscr{W}}$. The argument in §4 now goes through to show that $\tilde\delta(Q) = \bar{Q}$ and that $\tilde\delta = \delta$ when restricted to $S(\mathfrak{g})^G$. By Theorem 10, $\tilde\delta = \delta$.

Let H be reductive, and let W be an H-module with H-invariant nondegenerate quadratic form q. Let $x_1, \ldots, x_k$ be coordinate functions dual to an orthonormal basis of W. Then $P \mapsto P^t$ (formal adjoint) denotes the anti-automorphism of $\mathscr{D}(W)$ (preserving $\mathscr{D}(W)^H$) which sends x_i to x_i and ∂_i to $-\partial_i$.

Proof of (δ6). The homomorphism δ commutes with formal adjoint on $\mathscr{D}(\mathfrak{g})^G$ and $\mathscr{D}(\mathfrak{t})^{\mathscr{W}}$ iff $\delta(P) = \bar\delta(P) := \delta(P^t)^t$ for all $P \in \mathscr{D}(\mathfrak{g})^G$. Since $\bar\delta$ satisfies (δ1) through (δ3), we need only show that formal adjoint preserves $\mathscr{K}(\mathfrak{g})^G$. Rather than working with formal adjoint, we consider the composition μ of formal adjoint with the automorphism given by $w = \left(\begin{smallmatrix} 0 & 1 \\ -1 & 0 \end{smallmatrix}\right) \in SL_2$. Then μ is the antiautomorphism of $\mathscr{D}(\mathfrak{g})$ which sends $x^\rho \partial^\eta$ to $x^\eta \partial^\rho$. Let $P \in \mathscr{D}(\mathfrak{g})^G$. If $(QP(f))(0) = 0$ for any $Q \in S(\mathfrak{g})^G$ and $f \in \mathscr{O}(\mathfrak{g})^G$, then $P \in \mathscr{K}(\mathfrak{g})^G$. But $(QP(f))(0) = 0$ iff $(\mu(f)\mu(P)(\mu(Q)))(0) = 0$. Switching the roles of f and Q, we see that $(Q\mu(P)(f))(0) = 0$ for all $Q \in S(\mathfrak{g})^G$ and $f \in \mathscr{O}(\mathfrak{g})^G$, hence $\mu(P)$ and P^t are in $\mathscr{K}(\mathfrak{g})^G$.

Remark 15. In [2], Harish-Chandra showed that $\delta \colon \mathscr{D}(\mathfrak{g})^G \to \mathscr{D}(\mathfrak{t}_{\mathrm{pr}})^{\mathscr{W}}$ was a homomorphism satisfying $(\delta 2)$–$(\delta 4)$. In [3] he showed that $\mathrm{Im}(\delta) \subset \mathscr{D}(\mathfrak{t})^{\mathscr{W}}$ using the line of argument given below. The interested reader can supply proofs, using the techniques above. Recall the definition of $P \mapsto \tilde{P}$ (Remark 2). Let $D \in \mathscr{D}(\mathfrak{g})^G$.

1) If $\delta(D) \in \mathscr{D}(\mathfrak{t})^{\mathscr{W}}$, then $\delta(\tilde{D}) = \widetilde{\delta(D)} \in \mathscr{D}(\mathfrak{t})^{\mathscr{W}}$. Set $n := \deg \sigma^2$, let $\omega \in \mathscr{O}(\mathfrak{g})^G$ restrict to $\sigma^2 \in \mathscr{O}(\mathfrak{t})^{\mathscr{W}}$, and set $\Omega := \delta(\tilde{\omega}) \in S^n(\mathfrak{t})^{\mathscr{W}}$.

2) There is an $r \in \mathbb{N}$ such that $\delta(D)m_\sigma^{2r} \in \mathscr{D}(\mathfrak{t})^{\mathscr{W}}$. Set $D_1 := Dm_\omega^r$. Then $\delta(D_1) \in \mathscr{D}(\mathfrak{t})^{\mathscr{W}}$ and $\delta(\tilde{D}_1) = \delta(\tilde{D})\Omega^r \in \mathscr{D}(\mathfrak{t})^{\mathscr{W}}$.

3) Write $\delta(\tilde{D})$ as a sum $\sum_i a_i P_i$ where the a_i are in $\mathbb{C}(\mathfrak{t})$ and the $P_i \in S(\mathfrak{t})$ are linearly independent over $\mathbb{C}$. Then

$$\sum_i a_i P_i \Omega^r = \delta(\widetilde{D_1}) \in \mathscr{D}(\mathfrak{t})^{\mathscr{W}},$$

which implies that the a_i lie in $\mathscr{O}(\mathfrak{t})$. Hence $\delta(\tilde{D})$ and $\delta(D)$ are in $\mathscr{D}(\mathfrak{t})^{\mathscr{W}}$.

Nolan Wallach (private communication) has devised a short analytic proof which also establishes that $\mathrm{Im}\,\delta \subset \mathscr{D}(\mathfrak{t})^{\mathscr{W}}$, using the hypotheses $(\delta 2)$–$(\delta 4)$.

REFERENCES

1. V. Gatti and E. Viniberghi, *Spinors of 13-dimensional space*, Adv. in Math. **30** (1978), 137–155.
2. Harish-Chandra, *Differential operators on a semisimple Lie algebra*, Amer. J. Math. **79** (1957), 87–120.
3. _____, *Invariant differential operators and distributions on a semisimple Lie algebra*, Amer. J. Math. **86** (1964), 534–564.
4. T. Levasseur and J. T. Stafford, *Invariant differential operators and an homomorphism of Harish-Chandra*, J. Amer. Math. Soc. **8** (1995), 365–372.
5. _____, *The kernel of an homomorphism of Harish-Chandra*, Ann. Sci. École Norm. Sup., to appear.
6. D. Luna and R. W. Richardson, *A generalization of the Chevalley restriction theorem*, Duke Math. J. **46** (1979), 487–496.
7. Susan Montgomery, *Fixed rings of finite automorphism groups of associative rings*, Lecture Notes in Mathematics, no. 818, Springer-Verlag, Berlin, New York, 1980.
8. D. I. Panyushev, *On orbit spaces of finite and connected linear groups*, Math. USSR-Izv. **20** (1983), 97–101.
9. C. Procesi and G. Schwarz, *Inequalities defining orbit spaces*, Invent. Math. **81** (1985), 539–554.
10. G. Schwarz, *Lifting differential operators from orbit spaces*, Ann. Sci. École Norm. Sup. **28** (1995), 253–306.
11. J. T. Stafford, *Homological properties of the enveloping algebra $U(\mathfrak{sl}_2)$*, Mat. Proc. Camb. Phil. Soc. **91** (1982), 29–37.
12. N. R. Wallach, *Invariant differential operators on a reductive Lie algebra and Weyl group representations*, J. Amer. Math. Soc. **6** (1993), 779–816.
13. N. R. Wallach and M. Hunziker, *On the Harish-Chandra homomorphism of invariant differential operators on a reductive Lie algebra*, preprint.

DEPARTMENT OF MATHEMATICS, BRANDEIS UNIVERSITY, PO BOX 9110, WALTHAM, MA 02254-9110

TWO NOTES ON A FINITENESS PROBLEM IN THE REPRESENTATION THEORY OF FINITE GROUPS

PETER SLODOWY

Dedicated to the memory of Roger Richardson

INTRODUCTION

The following notes address two questions, recently raised by B. Külshammer, concerning the representations of a finite group Γ into a linear algebraic group G over an algebraically closed field K:

(1) Let char(K) be prime to the order of Γ. Are there only finitely many representations $\rho : \Gamma \to G$ up to conjugation by G?

(2) Let $p = \text{char}(K)$ and $\Gamma_p \subset \Gamma$ a Sylow p–subgroup. Fix a conjugacy class of representations $\bar{\rho} : \Gamma_p \to G$. Are there, up to conjugation by G, only finitely many representations $\rho : \Gamma \to G$ whose restrictions to Γ_p belong to the given class?

As it turns out, and was observed by several people, the answer to (1) is positive and essentially contained in a paper by A. Weil of 1964.

We are able to give a positive answer to (2) provided G is reductive (not necessarily connected) and the characteristic of K is good for G. We also give a sufficient criterion for a positive answer to (2).

Both notes may be read independently. We have also repeated a number of basic definitions as well as given reports on the contributions of other authors. Our results may not be serious advances, but their presentation here might nonetheless be of some general interest.

Work on these notes was started while visiting the School of Mathematics at the University of Sydney, Nov. – Dec. 92, and completed at RIMS, Kyoto University, Jan. – Feb. 93. We would like to thank both institutions, and especially our respective hosts, Gus Lehrer and Kyoji Saito, for their hospitality and support. In December 92 we also visited Roger in Canberra and discussed the matters in this article. His letter [7] was our last exchange.

I. A VARIATION ON A THEME OF R. W. RICHARDSON

Let G be a linear algebraic group over an algebraically closed field K and H a closed subgroup. Following Richardson [8] we call (G, H) a *reductive pair* if the Lie algebra $\mathbf{g}$ of G decomposes as a direct sum

$$\mathbf{g} = \mathbf{h} \oplus \mathbf{m}$$

where $\mathbf{h}$ is the Lie algebra of H and where $\mathbf{m}$ is a vector space supplement to $\mathbf{h}$ which is stable under the natural adjoint representation of H on $\mathbf{g}$.

This notion is crucial in [8] for the derivation of finite–decomposition–results for conjugacy classes of objects in G when restricted to corresponding objects in H. For example, if $\mathrm{char}(K) = 0$ and (G, H) is a reductive pair, then the intersection $C \cap H$ of the G-conjugacy class of an element $h \in H$ with H decomposes into a finite number of H–conjugacy classes (loc. cit. Theorem 3.1).

Here, we first want to prove a similar result for simultaneous conjugation of G and H on n–tuples of elements. This has then some bearing on Külshammer's questions [6].

I.1. Orbit maps. Let G be as before acting morphically on a K–variety X. Let $x \in X$. Then the morphism

$$\eta : G \to X$$
$$g \mapsto g \cdot x$$

is called the *orbit map* attached to x. This map is called *separable* if its differential at e (neutral element)

$$D_e \eta : T_e G \to T_x X$$

maps surjectively onto the tangent space $T_x C$ of the orbit $C = \{g \cdot x \mid g \in G\}$ of x (a locally closed submanifold of X). If this is the case, η induces an isomorphism of algebraic varieties

$$G/G_x \overset{\sim}{\to} C,$$

where $G_x = \{g \in G \mid g \cdot x = x\}$ is the stabilizer of x in G. Obviously, separability of an orbit map and its implications do not depend on the point x chosen in the orbit. Let us recall that in the cases of interest to us, $D_e \eta$ attains simple forms.

Let us first look at conjugation of G on itself:

$$\eta : G \to G$$
$$g \mapsto gxg^{-1}$$
$$D_e \eta : T_e G \to T_x G.$$

If we identify both tangent spaces with $\mathbf{g}$, considering $\mathbf{g}$ as the Lie algebra of right–invariant vector fields on G, this differential reads

$$D_e \eta : \mathbf{g} \to \mathbf{g}$$
$$X \mapsto (1 - \mathrm{Ad}\, x)X.$$

Here 1 is the identity on $\mathbf{g}$ and $\mathrm{Ad}\, x$ the adjoint representation of the element x.

This example immediately generalizes to simultaneous conjugation, G acting on G^n by

$$g.(x_1, \cdots, x_n) = (gx_1 g^{-1}, \cdots, gx_n g^{-1}).$$

Let $\underline{x} := (x_1, \cdots, x_n)$, then $T_{\underline{x}} G^n = \mathbf{g}^n$ and the differential of the orbit map $\eta : g \mapsto g \cdot \underline{x}$ takes the form

$$D_e \eta : \mathbf{g} \to \mathbf{g}^n = \bigoplus_{i=1}^{n} \mathbf{g}$$

$$X \mapsto \bigoplus_{i=1}^{n} (1 - \operatorname{Ad} x_i)(X).$$

We see that η is separable if and only if the centralizer $Z_G(\underline{x}) = \{g \in G \mid g x_i g^{-1} = x_i \text{ for all } i\}$ has the same dimension as the infinitesimal centralizer $\mathbf{z}_\mathbf{g}(\underline{x}) = \{X \in \mathbf{g} \mid (\operatorname{Ad} x_i)(X) = X \text{ for all } i\}$ (cf. [8] §2 or [2] $III.9$, for example). In particular, if G is the general linear group, $G = GL(V)$, then $Z_G(\underline{x})$ forms the open subset of invertible elements in the associative subalgebra $\mathbf{z}_\mathbf{g}(\underline{x})$ of $\operatorname{End}(V)$, and thus $dim Z_G(\underline{x}) = dim\, \mathbf{z}_\mathbf{g}(\underline{x})$, i.e. η is separable.

Assume now that (G, H) is a reductive pair with H–stable decomposition $\mathbf{g} = \mathbf{h} \oplus \mathbf{m}$, and let $\underline{x} = (x_1, \cdots, x_n) \in H^n$. Since $\operatorname{Ad} x_i(\mathbf{m}) \subset \mathbf{m}$ for all $x_i \in H$ the differential $D_e \eta : \mathbf{g} \to \mathbf{g}^n$ of the orbit map $g \mapsto g \cdot \underline{x}$ "respects" this decomposition, i.e.

$$D_e \eta(\mathbf{h}) \subset \mathbf{h}^n, \quad D_e \eta(\mathbf{m}) \subset \mathbf{m}^n.$$

I.2. Finite decomposition for simultaneous conjugation. We have assembled the relevant facts to follow [8] §3 almost verbatim in the derivation of:

Theorem 1. *Let (G, H) be a reductive pair and $C \subset G^n$ the simultaneous G–conjugacy class of an element $\underline{x} \in H^n$. Assume that the associated orbit map $\eta : G \to C$ is separable. Then the intersection $C \cap H^n$ decomposes into finitely many simultaneous H–conjugacy classes. Each corresponding orbit map is again separable.*

Proof. Let $\underline{y} = (y_1, \cdots, y_n)$ be an arbitrary element of $C \cap H^n$ and let $T(\underline{y}) := T_{\underline{y}} C \cap T_{\underline{y}} H^n = T_{\underline{y}} C \cap \mathbf{h}^n$ denote the Zariski tangent space of $C \cap H^n$ in $\underline{y}$. Let $\underline{Y} = (Y_1, \cdots, Y_n)$ be an element of $T(\underline{y})$. Since $D_e \eta$ is surjective onto $T_{\underline{y}} C$ there is an $X \in \mathbf{g}$ such that $\underline{Y} = D_e \eta(X)$. Split $X = X_0 + X_1$ with $X_0 \in \mathbf{h}, X_1 \in \mathbf{m}$. Then $\underline{Y}_0 = D_e \eta(X_0) \in \mathbf{h}^n$ and $\underline{Y}_1 = D_e \eta(X_1) \in \mathbf{m}^n$. Since $\underline{Y} = \underline{Y}_0 + \underline{Y}_1 \in \mathbf{h}^n$ we must have $\underline{Y}_1 = 0$, i.e. $\underline{Y} = \underline{Y}_0 = D_e \eta(X_0)$. Thus, if Z denotes the H–orbit through $\underline{y}$ we have that

$$T_{\underline{y}} Z \subset T(\underline{y}) \subset D_e \eta(\mathbf{h}) = T_{\underline{y}} Z,$$

or

$$T_{\underline{y}} Z = T(\underline{y}).$$

This implies that Z is open in $C \cap H^n$. Since this holds for an arbitrary point of $C \cap H^n$ and since $C \cap H^n$ is a noetherian space we deduce that $C \cap H^n$ decomposes into a finite number of H–orbits. Since $T_{\underline{y}} Z = D_e \eta(\mathbf{h})$ for all $\underline{y} \in C \cap H^n$, we see that the orbit maps of H onto the orbits in $C \cap H^n$ are separable (cf. also [8] Lemma 6.6).

Remark. Note that Richardson's assumption of connectedness of G and H plays no role in the proof.

Corollary. *Let $(GL(V), H)$ be a reductive pair. Then the intersection $C \cap H^n$ of any simultaneous $GL(V)$-conjugacy class in $GL(V)^n$ with H^n is either empty or decomposes into finitely many simultaneous H-conjugacy classes.*

I.3. Examples of reductive pairs. For applications, it is useful to know sufficiently many examples of reductive pairs. Richardson [8] provides us already with the following list:

(1) (G, H) where char$(K) = 0$ and H is reductive,
(2) $(GL(V), H)$ where char$(K) \neq 2$ and H is the automorphism group of some nondegenerate (anti-) symmetric bilinear form B on V:

$$H = \{g \in GL(V) \mid B(gv, gw) = B(v, w) \text{ for all } v, w \in V\},$$

(3) $(GL(\mathbf{h}), H)$ where $H = \mathrm{Aut}(\mathbf{h})$ for a simple Lie algebra of type D_4, E_6, F_7, F_4, G_2 or E_8 and where char$(K) \neq 2, 3$ in the first cases and char$(K) \neq 2, 3, 5$ for E_8.

Richardson, in [8], considers only connected H. However, his arguments pertain to the non–connected H in the above list as well, the case of D_4 in (3) only requiring the easy computation of the discriminant of a $\mathbb{Z}$-Killing form on D_4 (cf. [10] I 4.8, for example).

Further reductive pairs can be built by using the following easy observations:

(4) If (G, H) and (H, K) are reductive pairs, then (G, K) is one.
(5) If (G, H) and (G', H') are reductive pairs, then $(G \times G', H \times H')$ is one.
(6) Let (G, H) be such that either $G^0 \subset H \subset G$ or $\mid H \mid < \infty$, i.e. H finite. Then (G, H) is a reductive pair.
(7) Let V be a direct sum $V = \bigoplus_{i=1}^{n} W_i$. Then $(GL(V), H)$ with $H = \prod_{i=1}^{n} GL(W_i)$ embedded naturally is a reductive pair (the supplement $\mathbf{m}$ to $\mathbf{h}$ being $\bigoplus_{i \neq j}$ Hom(W_i, W_j)).
(8) Let V be a direct sum $V = W^n = W \oplus \cdots \oplus W$ and $H = GL(W)^n \rtimes S_n$, where the symmetric group S_n acts by permutations of summands of V and factors of $GL(W)^n$. Then $(GL(V), H)$ is a reductive pair (same $\mathbf{m}$ as in 7)).

I.4. Külshammer's first problem. In [6] Külshammer introduces "algebraic representations" of finite groups Γ. These are homomorphisms $\rho : \Gamma \to G$ of Γ into a linear algebraic group G. Two such representations ρ, ρ' are called *equivalent* if there is a $g \in G$ such that

$$\rho'(\gamma) = g\,\rho(\gamma)g^{-1} \quad \text{for all } \gamma \in \Gamma.$$

The first problem he raised is answered by the following

Theorem 2. *Assume that the order $|\Gamma|$ of Γ is prime to the characteristic of K. Then there are only finitely many equivalence classes of homomorphisms of Γ into G.*

Apart from a number of reductions and solutions in special cases by Külshammer [6], the result had essentially been proved already by A. Weil [11], as was pointed out by A. Borel [1]. Weil proved in fact more general rigidity theorems for representations of finitely generated groups Γ into Lie or algebraic groups. His rigidity results depend on the vanishing of the cohomology group $H^1(\Gamma, \mathbf{g})$ where Γ acts on the Lie algebra $\mathbf{g}$ of G by means of the composed adjoint action

$$\Gamma \xrightarrow{\rho} G \xrightarrow{\mathrm{Ad}} GL(\mathbf{g})$$

(one needs $H^1(\Gamma, \mathbf{g}) = 0$ for all ρ). The same proof has also been proposed by T. A. Springer [9] and R. Richardson [7]. Moreover, S. Donkin [4] has given an independent proof using Külshammer's reductions (to connected semisimple G), the finite decomposition of simultaneous conjugacy classes (for reductive pairs in char$(K) = 0$) and subtle reduction techniques from $char = 0$ to $char = p > 0$.

Since there is no interest in reproving the result over and over, let us just indicate, as a warm-up for later developments, how one can derive it for representations $\rho : \Gamma \to G$ into groups G which admit a realization $(GL(V), G)$ in some reductive pair. For that we consider a representation $\rho : \Gamma \to G$ as an element $\rho = (\rho(\gamma), \gamma \in \Gamma) \in G^\Gamma = Map(\Gamma, G)$. Two representations ρ and ρ' are equivalent if and only if they lie in the same simultaneous conjugacy class of G on G^Γ. Assume $\iota : G \hookrightarrow GL(V)$ is an embedding such that $(GL(V), G)$ is a reductive pair. For any representation $\rho : \Gamma \to G$, we get a representation $\iota \circ g : \Gamma \to GL(V)$. By Maschke's theorem, there are only finitely many classes of representations $\Gamma \to GL(V)$ $((\mid \Gamma \mid, char(K)) = 1\,!)$. By our finiteness result in section 2), any such class, considered as a $GL(V)$-orbit on $GL(V)^\Gamma$, decomposes into finitely many G-orbits after intersection with G^Γ. Thus, there can only be finitely many representations $\Gamma \to G$, up to equivalence.

In case the characteristic of K divides $\mid \Gamma \mid$, the statement of the theorem need not hold any longer, already for $G = GL(V)$. Külshammer has asked a modified question in this case, to which we turn now.

I.5. Külshammer's second problem. Let G be an arbitrary linear algebraic group over an algebraically closed field K of characteristic p, and let Γ_p be a Sylow p-subgroup of the finite group Γ.

Question. ([6]): Fix an equivalence class of representations of Γ_p into G. Are there only finitely many equivalence classes of representations $\rho : \Gamma \to G$ such that the restriction $\rho|_{\Gamma_p}$ belongs to that fixed class.

If $G = GL(V)$ this question is known to have a positive answer [5] or [1]. Our aim in this section is to give a positive answer for the case when G is reductive (not necessarily connected) and char(K) is good for G (for the definition of "good", cf. below; it deviates slightly from the definition for connected groups).

It has been pointed out by A. Borel [1], that A. Weil's arguments of [11] can be adapted to the new situation, giving a positive answer provided the centralizer of Γ_p in G is reduced. As a by-product of our proof, we obtain the reducedness of this group. However, this condition might not really be necessary, as already surmised by Borel (loc. cit.). In our following note, we shall analyse Weil's argument in more detail and replace Borel's condition by a more general one.

So assume now that G is reductive, i.e. its unipotent radical is trivial. Let R be its solvable radical, i.e. the connected center of the identity component G^0, and $\pi : G \to G/R$ the natural projection. We follow Külshammer ([6] p. 3) in showing that it is sufficient to consider representations into G/R.

Lemma 1. *Let $\bar{\rho} : \Gamma \to G/R$ be a homomorphism. Then, up to G–equivalence, there are only finitely many homomorphisms (possibly none) $\rho : \Gamma \to G$ lifting $\bar{\rho}$, i.e. such that $\pi \circ \rho = \bar{\rho}$.*

Proof. Assume that there exists one lift ρ at least. Then Γ acts via ρ on the normal abelian subgroup R of G. The assertion now follows from two facts:

(1) There is a bijection between the set of R-equivalence classes of lifts of $\bar{\rho}$ to G and the first cohomology group $H^1(\Gamma, R)$.
(2) The group $H^1(\Gamma, R)$ is finite.

Both facts are standard, cf. e.g. [3] $IV, 2, III, 10,$[6].

Definition. Let $\pi : G \to G'$ be a morphism of algebraic groups. It is called an *isogeny* if it is surjective with finite kernel.

The following lemma is obvious.

Lemma 2. *Let $\pi : G \to G'$ be an isogeny and $\bar{\rho} : \Gamma \to G'$ a homomorphism. Then there are only finitely many homomorphisms $\rho : \Gamma \to G$ lifting $\bar{\rho}$ (possibly none).*

With Lemmata 1 and 2 we may from now on assume that G^0 is semisimple of adjoint type and that G itself is a subgroup of the automorphism group $\mathrm{Aut}(G^0)$ of G^0, the last group being itself a semidirect product of G^0 and the finite group $\mathrm{Out}(G^0) = \mathrm{Aut}(G^0)/\mathrm{Aut}(G^0)^0 = \mathrm{Aut}(G^0)/G^0$.

Avoiding the finite prime numbers p for which the Killing form on the Lie algebra $\mathbf{g}$ of G is degenerate, we might embed G into $\mathrm{Aut}(\mathbf{g})$ to obtain a reductive pair $(\mathrm{Aut}(\mathbf{g}), G)$ and to apply our results of section 2. However, if G^0 contains simple factors of classical type (A_n, B_n, C_n, D_n) this would impose unnecessary restrictions on $\mathrm{char}(K)$, which we are able to circumvent by means of the following construction. Let $\mathrm{char}(K)$ be *good* for G, that means:

(i)   if G^0 contains a factor not of type A_n, $\mathrm{char}(K) \neq 2$,
(ii) if G^0 contains a factor of type E_6, E_7, E_8, F_4, G_2 or a factor of type D_4 on which G induces a triality, i.e. an outer automorphism of order 3, $\mathrm{char}(K) \neq 3$,
(iii) if G^0 contains a factor of type E_8, $\mathrm{char}(K) \neq 5$.

We now build a new group $\tilde{G}$ from G in the following way, starting with $\tilde{G}^0$:

(i)   replace each factor of type A_n in G^0 by a factor GL_{n+1} in $\tilde{G}^0$,
(ii) replace each factor of type C_n by Sp_{2n},
(iii) replace each factor of type D_n by SO_{2n} provided $n \neq 4$ or G induces no triality on it.

Due to our care with the case D_4, it is easily seen that the action of the subgroup $G/G^0 \cong F \subset \mathrm{Out}(G^0)$ lifts to this new group $\tilde{G}^0$. We then define $\tilde{G} := \tilde{G}^0 \rtimes F$ and obtain a natural surjective morphism of algebraic groups.

$$\pi : \tilde{G} = \tilde{G}^0 \rtimes F \to G^0 \rtimes F = G$$

with kernel of the form

$$G_m^s \times (\mathbb{Z}/2\mathbb{Z})^t$$

where s is the cardinality of type A-factors and t the cardinality of type C and non–triality D-factors in G^0. Starting now with the almost simple factors of $\tilde{G}^0$ and observing that the automorphism group of the n–fold cartesian product H^n of a simple group H is $\mathrm{Aut}(H)^n \rtimes S_n$, we can build up, by means of our construction kit for reductive pairs in section 3, a faithful representation

$$r : \tilde{G} \to GL(V)$$

such that $(GL(V), \tilde{G})$ forms a reductive pair.

Here, we have only assumed that $\mathrm{char}(K)$ is good for G. The link with representations of Γ into G is provided by the following result, which is well known.

Lemma 3. *There is an extension*

$$1 \to A \to \tilde{\Gamma} \xrightarrow{\varepsilon} \Gamma \to 1$$

of Γ by a finite abelian group A of order prime to p such that any representation $\rho : \Gamma \to G$ lifts to a representation $\tilde{\rho} : \tilde{\Gamma} \to \tilde{G}$, i.e.

$$
\begin{array}{ccc}
\tilde{\Gamma} & \xrightarrow{\tilde{\rho}} & \tilde{G} \\
\varepsilon \downarrow & & \downarrow \pi \\
\Gamma & \xrightarrow[\rho]{} & G
\end{array}
$$

is commutative.

Proof. We sketch how to obtain such a $\tilde{\Gamma}$ in an elementary way. Let $S\tilde{G} \lhd \tilde{G}$ denote the "special" subgroup obtained by replacing the GL–factors by SL–factors, i.e. $(S\tilde{G})^0 = (\tilde{G}^0, \tilde{G}^0)$. Then the restriction $S\pi$ of π to $S\tilde{G}$ induces an isogeny onto G with finite abelian kernel A_0 of order prime to p. The pull back of $S\pi$ by any representation $\rho : \Gamma \to G$ induces an extension $1 \to A_0 \to \Gamma_\rho \to \Gamma \to 1$ of Γ. Let $\Gamma_1, \cdots, \Gamma_N$ be representatives for the non–isomorphic extensions arising that way and denote their fibre product $\Gamma_1 \times_\Gamma \Gamma_2 \times_\Gamma \cdots \times_\Gamma \Gamma_N$ by $\tilde{\Gamma}$. Then $\tilde{\Gamma}$ has all the properties asked for (the kernel A is A_0^N, now).

Lemma 4. *Let $\tilde{\Gamma}_p$ denote a Sylow p–subgroup of Γ_p. Then $\varepsilon|_{\tilde{\Gamma}_p}$ induces an isomorphism $\tilde{\Gamma}_p \xrightarrow{\sim} \Gamma_p$. Moreover, for any homomorphism $\rho_p : \Gamma_p \to G$ there is, up to conjugation in $\tilde{G}$, a unique lift $\tilde{\rho}_p : \Gamma_p \cong \tilde{\Gamma}_p \to \tilde{G}$.*

Proof. The first assertion is trivial, and the second follows from $H^1(\tilde{\Gamma}_p, A) = 0$.

In view of Lemmata 1 and 2, there are, up to $\tilde{G}$-equivalence only finitely many lifts $\tilde{\rho} : \tilde{\Gamma} \to \tilde{G}$ for any given $\rho : \Gamma \to G$. Combining with Lemma 4, we see that, with respect to Külshammer's question, we may restrict our attention to representations of $\tilde{\Gamma}$ into $\tilde{G}$.

After these preparations, we can prove:

Theorem 3. *Let G be a reductive group over an algebraically closed field K of characteristic $p > 0$ such that p is good for G. Let Γ be a finite group with p-Sylow subgroup Γ_p and fix a G-equivalence class of representations $\Gamma_p \to G$. Then there are only finitely many G-equivalence classes of representations $\rho : \Gamma \to G$ such that the restriction $\rho|_{\Gamma_p}$ belongs to that fixed class.*

Proof. After our reductions made above, we may assume without loss of generality that G admits a faithful representation $r : G \to GL(V)$ such that $(GL(V), G)$ forms a reductive pair. By composition with r, all representations $\rho : \Gamma \to G$ as in the statement above provide representations $r \circ \rho : \Gamma \to GL(V)$ whose restriction to Γ_p belongs to a given class. Thus, by the positive answer to Külshammer's question for $G = GL(V)$, there are only finitely many $GL(V)$-equivalence classes of such representations. Since $(GL(V), G)$ forms a reductive pair, we can invoke Theorem 1 (on simultaneous conjugation) to deduce that the intersection of such a $GL(V)$-class with the set of representations whose images lie in G decomposes into finitely many G-classes. This implies the assertion of the theorem.

Remark. If G allows a direct embedding into a reductive pair $(GL(V), G)$ we can directly deduce that the orbit map

$$\eta : G \to G^{\Gamma} = \text{Map}(\Gamma, G)$$

for any given representation $\rho : \Gamma \to G$ is separable, i.e. the Lie algebra of

$$C_G(\Gamma) = \{g \in G \mid g\rho(\gamma) = \rho(\gamma)g \quad \text{for all } \gamma \in \Gamma\}$$

equals the infinitesimal centralizer $\mathbf{z}_{\mathbf{g}}(\Gamma) = \{X \in \mathbf{g} \mid \text{Ad}\,\rho(\gamma)X = X \text{ for all } \gamma \in \Gamma\}$ (since this is true for $G = GL(V)$).

If G contains no factor of type A_n, all our isogenies used are separable, due to the restrictions on the characteristic. One easily sees that this implies the separability for the orbit map η of the original group G. It is also well-known that the inseparability of orbit maps for conjugacy classes of groups of type A_n is an ephemeral phenomenon, which disappears if we pass to the group PGL_{n+1}, which is the one acting effectively, or to GL_{n+1}. As mentioned before, we are going to attack Külshammer's problem along the lines of A. Weil in the following note. The behaviour of tangent maps will be crucial, there.

II. A VARIATION ON A THEME OF A. WEIL

This note is a sequel to our preceding note, I, describing a different approach to Külshammer's finiteness problems for representations of finite groups Γ into linear algebraic groups G over algebraically closed fields K (cf. [6]). The origin of this approach lies in a paper by A. Weil [11], which also lies at the heart of Richardson's work [8] (personal communication of R.W.R.). Its relevance to

Külshammer's questions has been pointed out in detail in letters by A. Borel [1], T. A. Springer [9] and R. W. Richardson [7].

Here, we first want to reproduce Weil's arguments proving the finiteness of algebraic representations of finite groups up to equivalence, provided the group order $|\Gamma|$ is prime to the characteristic p of K.

Then, in the same geometric framework, we establish a sufficient criterion for the finiteness of classes of representations $\rho : \Gamma \to G$ restricting to a given class of representations $\bar{\rho} : \Gamma_p \to G$ for a Sylow p–subgroup, $p = \mathrm{char}(K)$.

At the end, we discuss an example in a related but different context where the desired finiteness property gets lost.

II.1. The tangent bundle of an algebraic group. Let G be a linear algebraic group over an algebraically closed field K with Lie algebra $\mathbf{g}$. Let $K[\varepsilon] = K \oplus K\varepsilon$ denote the K–algebra of "dual numbers", $\varepsilon^2 = 0$, with obvious K–algebra homomorphisms

$$K \xrightarrow{i} K[\varepsilon] \xrightarrow{p} K,$$

and induced homomorphisms of groups (cf. [2] AG 16)

$$G = G(K) \xrightarrow{i} G(K[\varepsilon]) \xrightarrow{p} G(K) = G.$$

Via p, we may identify $G(K[\varepsilon])$ with the tangent bundle $T(G)$ of G, the map i corresponding to the zero–section. And we obtain a split exact sequence (cf. [2] 3.20):

$$0 \to \mathbf{g} \to T(G) \to G \to 1.$$

Writing $T(G)$ as a semidirect product $\mathbf{g} \rtimes G$, where G acts on $\mathbf{g}$ by the adjoint representation, we will regard any $X \in \mathbf{g}$ as a right–invariant vector field by means of the map

$$X* : G \to T(G) = \mathbf{g} \rtimes G$$
$$g \mapsto (X, g).$$

II.2. Varieties of representations. Let G be as before and Γ a finite group of order $|\Gamma| = N$. Let $G^\Gamma = \mathrm{Map}(\Gamma, G)$ denote the algebraic group of maps of Γ into G (it is the N–fold cartesian product of G). We shall view the set $R(\Gamma, G)$ of all representations ρ of Γ into G as the subvariety (possibly non–reduced) of G^Γ given by the N^2 algebraic conditions

$$\rho(\gamma\delta) = \rho(\gamma)\rho(\delta), \quad \gamma, \delta \in \Gamma.$$

Obviously, this subvariety ist stable under simultaneous conjugation by G, the G–orbits corresponding to "algebraic" equivalence classes in the sense of [6]. If H is another algebraic group and $\varphi : G \to H$ a morphism of algebraic groups, we obtain a morphism of algebraic varieties

$$R(\Gamma, \varphi) : R(\Gamma, G) \to R(\Gamma, H),$$
$$\rho \mapsto \varphi \circ \rho$$

equivariant with respect to the actions of G and H related by φ.

Similarly, if $\psi : \Gamma \to \Gamma'$ is a homomorphism of finite groups, we obtain a natural G–equivariant morphism

$$R(\psi, G) : R(\Gamma, G) \to R(\Gamma', G).$$

The tangent "bundle" $TR(\Gamma, G)$ of Zariski tangent spaces to $R(\Gamma, G)$ is easily identified with $R(\Gamma, T(G)) = R(\Gamma, \mathbf{g} \rtimes G)$ (looking at schemes from the functorial point of view, we have $R(\Gamma, G)(A) = R(\Gamma, G(A))$ for any K-algebra A). In particular, the tangent space $T_\rho R(\Gamma, G)$ at the point $\rho \in R(\Gamma, G)$ is the fibre over ρ of the "bundle" map

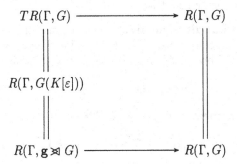

i.e. $T_\rho R(\Gamma, G) = \{\tilde{\rho} : \Gamma \to \mathbf{g} \rtimes G \mid \tilde{\rho}(\gamma) = (X_\gamma, \rho(\gamma))$ and $\tilde{\rho}(\gamma\delta) = \tilde{\rho}(\gamma)\tilde{\rho}(\delta)$ for all $\gamma, \delta \in \Gamma\}$.

However, the condition $\tilde{\rho}(\gamma, \delta) = \tilde{\rho}(\gamma)\tilde{\rho}(\delta)$ is equivalent to saying that $X_{\gamma\delta} = X_\gamma + {}^\gamma X_\delta$, where $\gamma \in \Gamma$ acts on $X_\delta \in \mathbf{g}$ via the adjoint representation, ${}^\gamma X_\delta = (\mathrm{Ad}\,\rho(\gamma))(X_\delta)$. In other words

$$T_\rho R(\Gamma, G) = Z^1(\Gamma, \mathbf{g}),$$

the group of 1–cocycles of Γ in the Lie algebra $\mathbf{g}$ of G (Γ acting on $\mathbf{g}$ via $\Gamma \xrightarrow{\rho} G \xrightarrow{\mathrm{Ad}} GL(\mathbf{g})$).

Let $G \cdot \rho$ denote the orbit of ρ under conjugation by G and $\eta : G \to G \cdot \rho, \eta(g) = g \cdot \rho$, the orbit map. We want to compute the image of its differential

$$d_e\eta : T_e G = \mathbf{g} \to T_\rho R(\Gamma, G).$$

The result, i.e. the explicit form of $d_e\eta$, was already discussed in our previous note. In our language used above, the image of $d_e\eta$ concides with the orbit of $\rho \in R(\Gamma, G) \subset TR(\Gamma, G) = R(\Gamma, \mathbf{g} \rtimes G)$ under the action of the "infinitesimal group" $\mathbf{g} \cong \mathbf{g} \rtimes \{e\} \subset TG$, i.e.

$$\mathrm{Im}\, d_e\eta = \{((X, e)(0, \rho(\gamma))(-X, e)), \gamma \in \Gamma) \in R(\Gamma, \mathbf{g} \rtimes G) | X \in \mathbf{g}\}$$
$$= \{((X - {}^\gamma X, \rho(\gamma)), \gamma \in \Gamma) \in R(\Gamma, \mathbf{g} \rtimes G) | X \in \mathbf{g}\}.$$

Under our identification of $T_\rho R(\Gamma, G)$ with $Z^1(\Gamma, \mathbf{g})$ this means

$$\mathrm{Im}\, d_e\eta = B^1(\Gamma, \mathbf{g}),$$

the 1–coboundaries of Γ with values in $\mathbf{g}$.

Remarks. 1) Note that we can identify $\operatorname{Im} d_e \eta$ with the tangent space $T_\rho(G \cdot \rho)$ if and only if η is separable.

2) We have embarked on the above well known computations to point out more clearly the connection to the theory of group extensions, cf. for instance [3] Chap. 4.2. This is of course also done in [11], however in a slightly different language.

II.3. Külshammer's first problem. The following is a positive answer to the first problem posed by Külshammer in [6]. It is a special case of a more general result of A. Weil [11].

Theorem 1. *Let G be a linear algebraic group over an algebraic closed field K and Γ a finite group of order prime to the characteristic of K. Then, up to G-equivalence, there are only finitely many representations of Γ into G.*

Proof. Because of our assumption on $|\Gamma|$ we have $H^1(\Gamma, \mathbf{g}) = 0$ for all $\rho \in R(\Gamma, G)$ (recall that Γ acts on $\mathbf{g}$ via $\operatorname{Ad} \circ \rho$!), cf. e.g. [3] III, 10.2. Thus for any ρ we have

$$Z^1(\Gamma, \mathbf{g}) = B^1(\Gamma, \mathbf{g}) \subset T_\rho(G \cdot \rho) \subset T_\rho R(\Gamma, G) = Z^1(\Gamma, \mathbf{g}),$$

thus $T_\rho(G \cdot \rho) = T_\rho R(\Gamma, G)$.

This implies that the G-orbit of any ρ is open in $R(\Gamma, G)$ (cf. e.g. [11] Lemma 1, or [8]§3). Since $R(\Gamma, G)$ is a Noetherian topological space, there can be only finitely many such orbits.

Remarks. 1) This proof has also been given by A. Borel [1], T. A. Springer [9], and R. W. Richardson [7].

2) As a by-product of the proof, we get that $R(\Gamma, G)$ is a smooth variety and that all orbit maps $G \to G \cdot \rho$, $\rho \in R(\Gamma, G)$, are separable.

II.4. Külshammer's second problem. Again, let G be a linear algebraic group over K, but assume now that Γ is an arbitrary finite group. Let $\Gamma_p \subset \Gamma$ be a Sylow p-subgroup of Γ, where $p = \operatorname{char}(K)$. It is well known that Γ may already have infinitely many equivalence classes of linear representations. However, the following question appears reasonable.

Question. ([6]): Given an equivalence class of representations of Γ_p into G, are there only finitely many equivalence classes of representations $\rho : \Gamma \to G$ such that the restriction $\rho_{|\Gamma_p}$ belongs to that given class?

Building on a positive answer for $G = GL(V)$, we have derived in part I a positive answer for all reductive groups G provided that the characteristic of K is good for G. Here, we want to approach this question from Weil's point of view.

Let $i : \Gamma_p \hookrightarrow \Gamma$ denote the inclusion. Then we have a G-equivariant "restriction" morphism

$$r = R(i, G) : R(\Gamma, G) \to R(\Gamma_p, G),$$
$$r(\rho) = \rho \circ i =: \bar{\rho}.$$

In terms of this map r, we may rephrase the above question in the form:

Question. Let $\bar{\rho} \in R(\Gamma_p, G)$ and $\mathscr{R}(\bar{\rho}) = r^{-1}(G \cdot \bar{\rho})$. Are there only finitely many G–orbits in $\mathscr{R}(\bar{\rho})$?

Whereas in the previous section our argument depended on the vanishing of the cohomology group $H^1(\Gamma, \mathbf{g})$, we now will make essential use of the following cohomological result.

Lemma. *Let* $\rho : \Gamma \to G$ *be representation of* Γ *into* G, *thus providing* $\mathbf{g}$ *with a* Γ-*module structure. Then the restriction map in cohomology*

$$H^1(\Gamma, \mathbf{g}) \to H^1(\Gamma_p, \mathbf{g})$$

is injective.

Proof. This is a standard fact in group cohomology, see for example [3] III, Proposition 10.4.

To apply this result, we have to analyse the differential structure of the restriction map r. In doing that, we will also take into account inseparability of orbit maps.

Let us consider $\rho \in R(\Gamma, G)$ and $\bar{\rho} = r(\rho) \in R(\Gamma_p, G)$, and let

$$\eta : G \to G \cdot \rho \subset R(\Gamma, G),$$
$$\bar{\eta} : G \to G \cdot \bar{\rho} \subset R(\Gamma_p, G)$$

denote the corresponding orbit maps with differentials (at $e \in G$)

$$D\eta : \mathbf{g} \to T_\rho(G \cdot \rho) \subset T_\rho R(\Gamma, G)$$

and

$$D\bar{\eta} : \mathbf{g} \to T_{\bar{\rho}}(G \cdot \bar{\rho}) \subset T_{\bar{\rho}} R(\Gamma_p, G).$$

We shall call

$$\mathrm{ind}(\rho) := \dim(G \cdot \rho) - \dim(\mathrm{Im}\, D\eta),$$

resp.

$$\mathrm{ind}(\bar{\rho}) := \dim(G \cdot \bar{\rho}) - \dim(\mathrm{Im}\, D\bar{\eta})$$

the *inseparability–defect* of ρ resp. $\bar{\rho}$.
We have

$$\mathrm{ind}(\rho) = \dim z_{\mathbf{g}}(\Gamma) - \dim Z_G(\Gamma)$$

resp.

$$\mathrm{ind}(\bar{\rho}) = \dim z_{\mathbf{g}}(\Gamma_p) - \dim Z_G(\Gamma_p),$$

where

$$Z_G(\Gamma) = \{g \in G | g\rho(\gamma) = \rho(\gamma)g \text{ for all } \gamma \in \Gamma\}$$

is the global centralizer of Γ in G, and

$$z_{\mathbf{g}}(\Gamma) = \{X \in \mathbf{g} | \operatorname{Ad}\rho(\gamma)(X) = X \text{ for all } \gamma \in \Gamma\}$$

is the infinitesimal centralizer of Γ in $\mathbf{g}$; similar definitions apply for Γ_p.

Theorem 2. *Let $\bar{\rho} \in R(\Gamma_p, G)$.*

(i) *For all $\rho \in \mathscr{R}(\bar{\rho})$ we have $\operatorname{ind}(\rho) \leq \underline{\operatorname{ind}}(\bar{\rho})$.*

(ii) *If $\operatorname{ind}(\rho) = \operatorname{ind}(\bar{\rho})$ for all $\rho \in r^{-1}(\bar{\rho})$, then there are only finitely many G-orbits in $\mathscr{R}(\bar{\rho})$.*

Proof. The assertion i) is an immediate consequence of the preceding lemma, once we have rephrased the position of certain tangent spaces in cohomological terms. Note that by G-equivariance, we may restrict our attention to a single $\rho \in r^{-1}(\bar{\rho})$. Fix such a ρ now, and let

$$Dr : T_\rho R(\Gamma, G) \to T_{\bar{\rho}} R(\Gamma_p, G)$$

denote the derivative at ρ of the restriction map r.

Using the identifications

$$T_\rho R(\Gamma, G) = Z^1(\Gamma, \mathbf{g}),$$
$$T_{\bar{\rho}} R(\Gamma_p, G) = Z^1(\Gamma_p, \mathbf{g}),$$

we may easily identify Dr with the restriction map for cocycles. Recall that the coboundaries are obtained as the images of the derivatives of the orbit maps $\eta : G \to G \cdot \rho$ and $\bar{\eta} : G \to G \cdot \bar{\rho}$:

$$\operatorname{Im} D\eta = B^1(\Gamma, \mathbf{g}) \quad,$$
$$\operatorname{Im} D\bar{\eta} = B^1(\Gamma_p, \mathbf{g}) \quad.$$

We have the following commutative diagram of maps between vector spaces, where all vertical arrows are natural inclusions and where all horizontal maps are induced by Dr:

$$
\begin{array}{ccc}
Z^1(\Gamma, \mathbf{g}) & & Z^1(\Gamma_p, \mathbf{g}) \\
\| & & \| \\
T_\rho R(\Gamma, G) & \longrightarrow & T_{\bar\rho} R(\Gamma_p, G) \\
\cup & & \cup \\
T_\rho \mathscr{R}(\bar\rho) & \longrightarrow & T_{\bar\rho}(G \cdot \bar\rho) \\
\cup & & \| \\
T_\rho(G \cdot \rho) & & \| \\
\cup & & \| \\
\operatorname{Im} D\eta & \longrightarrow & \operatorname{Im} D\bar\eta \\
\| & & \| \\
B^1(\Gamma, \mathbf{g}) & & B^1(\Gamma_p, \mathbf{g})
\end{array}
$$

Since, by the preceding lemma, the induced map $T_\rho R(\Gamma, G)/\operatorname{Im} D\eta \to T_{\bar\rho} R(\Gamma_p, G)/\operatorname{Im} D\bar\eta$ is injective, we have

$$
\operatorname{ind}(\rho) = \dim(T_\rho(G \cdot \rho)/\operatorname{Im} D\eta) \le \dim(T_\rho \mathscr{R}(\bar\rho)/\operatorname{Im} D\eta) \le
$$
$$
\le \dim(T_{\bar\rho}(G \cdot \rho)/\operatorname{Im} D\bar\eta) = \operatorname{ind}(\bar\rho).
$$

This proves assertion i).

If we have $\operatorname{ind}(\rho) = \operatorname{ind}(\bar\rho)$ for all $\rho \in r^{-1}(\bar\rho)$, then we obtain $T_\rho(G \cdot \rho) = T_\rho \mathscr{R}(\bar\rho)$ for all $\rho \in \mathscr{R}(\bar\rho)$ from the above (and G–equivariance). Employing the same geometric arguments as in the proof of theorem 1, this gives assertion ii).

Corollary 1. ([1]): *Let $\bar\rho \in R(\Gamma_p, G)$ and assume that the centralizer of $\bar\rho(\Gamma_p)$ in G is reduced. Then there are only finitely many G–orbits in $\mathscr{R}(\bar\rho)$.*

Proof. In this case $\operatorname{ind}(\bar\rho) = 0$, thus $\operatorname{ind}(\rho) = 0$, and the theorem applies.

Corollary 2. *Assume $G = GL(V)$ or that G is reductive and $\operatorname{char}(K)$ is good for G (as in our previous note I). Then Külshammer's question has a positive answer.*

Proof. If $G = GL(V)$ all centralizers of subgroups are reduced (cf. note I, section 1 or [8]§6). If G is reductive and $\operatorname{char}(K)$ is good, this fact was proved in note I, Theorem 3, Remark.

Remarks. 1) If one wants to exhibit examples of representations $\rho : \Gamma \to G$ with G reductive and $Z_G(\Gamma_p)$ non–reduced, one has to resort to bad characteristics. For example, if Γ_p is cyclic of order p, and if ρ maps the nontrivial elements of Γ_p to subregular unipotent elements in G, then $Z_G(\Gamma_p)$ may be non–reduced (at least, it is so for G of type B_n, C_n, F_4 in characteristic 2 or G_2 in characteristic 3).

2) It is definitely harder to find examples of representations $\rho : \Gamma \to G$ where the inseparability defects $\mathrm{ind}(\rho)$ and $\mathrm{ind}(\bar{\rho})$ differ. We haven't found an example, and would like to leave this as an open question. Of course, this question does not replace Külshammer's second question, since we don't know the necessity of the defect relation.

II.5. An example in the theory of transformation groups. The condition on the inseparability defects required in Theorem 2, ii) may not really be necessary for the result deduced. However, the following example, albeit in a different setting, shows that the finiteness property might get lost if that condition is dropped.

Let K be an algebraically closed field of characteristic $p > 0$ and $G = G_a$ the additive group over K. Let $X = \mathbf{A}^2, Y = \mathbf{A}^1$ and $\varphi : X \to Y$ the second projection $\varphi(x,y) = y$. We let act G on X by

$$G \times X \to X, \quad t \cdot (x,y) = (x+t, y+t^p)$$

and on Y by

$$G \times Y \to Y, \quad t \cdot y = y + t^p.$$

Then φ is obviously G–equivariant, mapping X onto a single orbit Y. The plane X itself decomposes into infinitely many orbits

$$X_a = \{(x,y) \in \mathbf{A}^2 | y = (x+a)^p = x^p + a^p\}.$$

Let us now look at the differential properties of this situation. The orbit maps in X are separable whereas those in Y are inseparable (corresponding to inseparability defects 0 and 1), and for any points $\rho = (x,y) \in X, \bar{\rho} = \varphi(\rho) = y \in Y$, we have a commutative diagram:

$$
\begin{array}{ccc}
T_\rho X & \xrightarrow{\;\;D\varphi\;\;} & T_{\bar{\rho}}Y \\[4pt]
\cup & & \cup \\[4pt]
T_\rho(G \cdot \rho) & & 0 \\[4pt]
\| & & \| \\[4pt]
\mathrm{Im}\, D\eta & \longrightarrow & \mathrm{Im}\, D\bar{\eta}
\end{array}
$$

($D\eta$ and $D\bar{\eta}$ denote the derivatives at $e \in G$ of the orbit maps $G \to G \cdot \rho, G \to G \cdot \bar{\rho}$). The induced map

$$T_\rho X / \mathrm{Im}\, D\eta \to T_{\bar{\rho}}Y / \mathrm{Im}\, D\bar{\eta}$$

is injective as in the situation of Theorem 2. However, the inseparability defects differ here!

Of course, we can preserve these two properties and get finitely many G–orbits in the source, if we simply replace X by a single G–orbit, say $X_0 = \{(x,y) \in \mathbf{A}^2 | y = x^p\}$.

In some sense, this shows a limit to the quite general methods which we have employed above.

Remarks added in proof. 1) After the submission of this paper we learned of the book V. Platonov, A. Rapinchuk: "Algebraic Groups and Number Theory", Academic Press, 1994, where in sections 2.4.6 and 2.4.7 theorems 1 and 2 of our note I are proved along the same lines.

2) In a recent note (Dec. 94, appended to this article) Martin Cram has constructed a counterexample to Külshammer's second question by considering representations of S_3 into the semidirect product $Q \rtimes S_3$ of S_3 with a Heisenberg-type group Q over a field of characteristic 2. A closer study of the orbit geometry involved in his example reveals exactly the situation described above (apart from a trivial factor).

REFERENCES

1. A. Borel, *Letter to B. Külshammer*, 23 September 1992.
2. _____, *Linear Algebraic Groups*, 2nd ed., Springer Verlag, 1991.
3. K. S. Brown, *Cohomology of groups*, GTM, no. 87, Springer Verlag, 1982.
4. S. Donkin, *On a question of Külshammer*, manuscript, Queen Mary College, London, 1992.
5. B. Külshammer, *Letter to the author*, 22 December 1992.
6. _____, *Algebraic representations of finite groups*, manuscript, Universität Augsburg, 1992.
7. R. W. Richardson, *Letter to the author*, 13 January 1993.
8. _____, *Conjugacy classes in Lie algebras and algebraic groups*, Ann. of Math. **86** (1967), 1–15.
9. T. A. Springer, *Letter to B. Külshammer*, 14 August 1992.
10. T. A. Springer and R. Steinberg, *Conjugacy classes*, Seminar on Algebraic Groups and Related Finite Groups (A. Borel, ed.), Lecture Notes in Math., no. 131, Springer, 1970, pp. 167–266.
11. A. Weil, *Remarks on the cohomology of groups*, Ann. of Math. **80** (1964), 149–157.

MATHEMATISCHES SEMINAR DER UNIVERSITÄT HAMBURG, D-20146 HAMBURG, GERMANY

E-mail address: slodowy@math.uni-hamburg.de

APPENDIX: ON A QUESTION OF KÜLSHAMMER ABOUT ALGEBRAIC GROUP ACTIONS: AN EXAMPLE

G.-MARTIN CRAM

In [1], Burkhart Külshammer asked the following Question:

Let K be an algebraically closed field of prime characteristic p, let G be a linear algebraic group over K, let Γ be a finite group, let Γ_p be a Sylow-p-group of Γ, and let $\rho : \Gamma_p \to G$ be a representation of Γ_p into the (K-rational points of)G. Then, are there only finitely many equivalence classes (i.e. up to conjugation by G) of representations of Γ in G which contain a representation extending ρ?

The answer is "yes" if the order of the finite group Γ is prime to the characteristic p of the field K (a proof is sketched in [1]), or if the algebraic group G is reductive and the characteristic is good for G, see [2].

But in general, the answer is "no". I will give a counterexample for a field of characteristic $p = 2$. The finite group is $\Gamma := SL(2,2)$, which is isomorphic to the symmetric group on three elements. The algebraic group G has dimension 3 over K. It is a semidirect product of Γ with a Heisenberg-type algebraic group Q defined over the prime field $\mathbf{F}_2$.

As an algebraic variety Q is isomorphic to the affine space $\mathbf{A}^3$ over $\mathbf{F}_2$. For any field $K \supset \mathbf{F}_2$ the group multiplication in $Q(K)$ is given by

$$\mu\left(\begin{pmatrix}\alpha\\\beta\\\gamma\end{pmatrix},\begin{pmatrix}\alpha'\\\beta'\\\gamma'\end{pmatrix}\right):=\begin{pmatrix}\alpha+\alpha'\\\beta+\beta'\\\gamma+\gamma'+\alpha\alpha'+\beta\beta'+\alpha\beta'\end{pmatrix}$$

Straightforward computation yields:

Lemma 1. *μ induces on Q the structure of an algebraic group defined over $\mathbf{F}_2$. The group of $\mathbf{F}_2$-rational points $Q(\mathbf{F}_2)$ is the quaternion group.*

We may define Γ by generators and relations:

$$\Gamma := \langle \sigma, \tau / \sigma^3 = 1, \tau^2 = 1, \tau\sigma\tau = \sigma^2 \rangle$$

We shall fix the Sylow-2-subgroup $\Gamma_2 := \langle \tau \rangle \leq \Gamma$.

Lemma 2. *By defining*

$$\tau\left(\begin{pmatrix}\alpha\\\beta\\\gamma\end{pmatrix}\right):=\begin{pmatrix}\beta\\\alpha\\\gamma+\alpha^2+\beta^2+\alpha\beta\end{pmatrix} \quad and \quad \sigma\begin{pmatrix}\alpha\\\beta\\\gamma\end{pmatrix}:=\begin{pmatrix}\beta\\\alpha+\beta\\\gamma\end{pmatrix}$$

we obtain an action of Γ on Q by algebraic automorphisms of Q.

We let G denote the semidirect product

$$G = Q \rtimes \Gamma$$

where the action of Γ on Q is as above.

Elements of G will be written $\begin{pmatrix}\alpha\\\beta\\\gamma\end{pmatrix}\cdot\xi$, with $\xi \in \Gamma$.

We shall fix the representation $\rho : \Gamma_2 \to G$ given by the natural inclusion

$$\Gamma_2 \hookrightarrow \Gamma \hookrightarrow Q \rtimes \Gamma$$

Now we have to look at extensions of ρ to representations of Γ. There is of course the trivial extension, sending σ to the neutral element of G. All other extensions are given by the following Theorem, which can be proved by elementary computations:

Theorem. *Let $K \supset \mathbf{F}_2$ be a field. Apart from the trivial extension, all equivalence classes of extensions of ρ to representations of Γ into $G(K)$ are provided by the family $\{\rho_\beta : \Gamma \to G(K) | \beta \in K\}$ defined by*

$$\rho_\beta(\sigma) := \begin{pmatrix} 0 \\ \beta \\ 0 \end{pmatrix} \cdot \sigma.$$

In particular, ρ_β and ρ'_β are conjugate by $G(K)$ if and only if $\beta = \beta'$.

Remark. To compare these results with the cases considered in [1] and [2], one should compute the ideals defining the centralizers of Γ_2 and Γ in G:

$$C_G(\Gamma_2) = V(\langle \alpha^2, \beta^2, \alpha - \beta \rangle)$$

and

$$C_G(\Gamma) = V(\langle \alpha, \beta \rangle).$$

Thus, $C_G(\Gamma)$ is reduced, but $C_G(\Gamma_2)$ is not.

REFERENCES

1. B. Külshammer, *Donovan's Conjecture, Crossed Products and Algebraic group actions.*, Israel J. Math. (1995), 1–12.
2. P. Slodowy, *Two Notes on a Finiteness Problem in the Representation Theory of Finite Groups.*, these proceedings.

INSTITUT FÜR MATHEMATIK DER UNIVERSITÄT AUGSBURG, 86135 AUGSBURG, GERMANY
E-mail address: cram@uni-augsburg.de

A DESCRIPTION OF B-ORBITS ON SYMMETRIC VARIETIES.

T. A. SPRINGER

In memory of Roger Richardson

INTRODUCTION

This paper had its origin in an attempt to understand the results of Vogan's paper [12] in another framework. The main result of [12] is a duality for character multiplicities involving two reductive groups. With some oversimplification one can say that it can be formulated in terms of the Lusztig-Vogan polynomials introduced in [3]. These polynomials come from the intersection cohomology complexes of local systems on certain algebraic varieties, which we now describe.

Let G be a connected reductive linear algebraic group over $\mathbb{C}$ (or more generally over an algebraically closed field of characteristic $\neq 2$) with an involutive automorphism θ. Denote by K the fixed point group of θ, a reductive subgroup of G and let B be a Borel group of G. Then B acts with finitely many orbits on the symmetric variety G/K, let v be one. The varieties in question are the Zariski closures $\bar{v}$ and the intersection cohomology complexes are the ones defined by certain local systems on v, for example by the B-equivariant ones.

These facts suggest that Vogan's results should have counterparts in the framework of the preceding paragraph, which is also the framework of [5]. We shall study such counterparts in the present paper.

An important technical tool in [12] are $\mathbb{Z}/2\mathbb{Z}$-gradings of root systems. They appear in our set-up in the following manner. An orbit v as before defines a K-conjugacy class of θ-stable maximal tori of G. Let T be such a torus. The involution θ acts on the root system R of (G,T). Let R_i be the subsystem of imaginary roots, i.e. those fixed by θ. By considering the action of θ on root vectors of the Lie algebra one sees that the involution θ of G defines a $\mathbb{Z}/2\mathbb{Z}$-grading ϵ of R_i. So T determines a triple (R,θ,ϵ), consisting of a root system with involution and a grading of the imaginary roots for θ. An analogous construction, in the context of real Lie groups, occurs in [12], see [loc.cit., 4.11].

The main result of section 2 of the present paper is the characterization 2.4 of the gradings ϵ which can occur for a given θ. Such a result for real Lie groups is given in [12, 10.8]. Our formulation of 2.4 uses the even cohomology group $H_+(\theta, P(R^\vee))$ of the group of order 2, acting in the dual weight lattice $P(R^\vee)$ via θ. This group parametrizes the gradings which can occur. The introductory section 1 contains a discussion of properties of such groups.

Using the parametrization of the set V of orbits v used in [5] we attach to each v a *model*, this being a triple $\mathcal{M}_v = (\Psi_0, \theta_v, \xi)$, where Ψ_0 is a based graded root datum (the root datum of G together with a basis of its root system), θ_v

is an involution of the root datum, and $\xi \in H_+(\theta_v, P(R^\vee))$. The triple is a combinatorial model of an orbit $v \in V$.

In [5] we have introduced an action of the Weyl group W on V. This defines a Weyl group action on models, described in 4.15. (The W-action is similar to a cross-action of [12, 4.1].) Basic for the study of V made in [5] is an operation $(s, v) \mapsto m(s).v$ where s is a reflection in a simple root α. Except for the case that α is odd imaginary (for the grading of the imaginary roots associated to our model) we have $m(s).v = v$ or $s.v$. If α is odd imaginary we call $m(s).v$ a *Cayley transform* (following the terminology of [12]). Cayley transforms are discussed in section 3. Their description in terms of models is given in 4.18.

Section 4 discusses the question to what extent an orbit $v \in V$ (parametrized as in [5]) is determined by its model $\mathscr{M}_v$. One result (see 4.13) is that this is so if G is adjoint. In the discussion of the general case groups like the isotropy group W_v of v in W enter the picture. The material of this section is related to material about $\mathscr{R}$-groups in [12].

In section 5 we discuss one-dimensional B-equivariant local systems on the orbits v. If G is semi-simple and simply connected it turns out that their isomorphism classes are parametrized by the group $H_+(-\theta_v, P(R))$, which by the results of previous sections defines an orbit $v^\vee$ for an involution of the adjoint group with the dual root system. This suggests the existence of a duality between pairs $(v, \mathscr{L})$ of an orbit $v \in V$ and a local system $\mathscr{L}$ on it and similar pairs for a dual group. If G is semi-simple and simply connected with a center of odd order the results of no. 5 do give such a duality. In the general case a further study is required. It is best to carry this out in the context of a geometric study of more general local systems on v (having a possibly non-zero weight for the T-action). In this study the Hecke algebra representations of [3] and [13] will also enter the picture. We hope to take up these matters in another paper.

Work on the material of this paper was started during a stay at the Australian National University in the autumn of 1991. I would like to thank the ANU for its hospitality.

I recall with gratitude the many stimulating talks which I had with the late Roger Richardson during that stay.

1. PRELIMINARIES

1.1. Involutions. Let R be a reduced root system in the real vector space V. Our reference for the theory of root systems is [1]. It will be convenient also to admit the empty root system (in contrast to loc.cit.). An *involution of R* is a linear map θ of order ≤ 2 of V which stabilizes R, such that the contragredient linear map of the dual V^*, which we also denote by θ, stabilizes the dual root system. For $\alpha \in R$ we denote by $s_\alpha \in W$ the corresponding reflection. Then $s_{\theta\alpha} = \theta s_\alpha \theta$. It follows that θ defines an automorphism θ of order ≤ 2 of the Weyl group $W = W(R)$. Its subgroup of fixed points is denoted by W^θ.

We put

$$R_r = R_r(\theta) = \{\alpha \in R \mid \theta\alpha = -\alpha\}, \quad R_i = R_i(\theta) = \{\alpha \in R \mid \theta\alpha = \alpha\}.$$

These are the *real*, respectively *imaginary* roots of R (relative to θ). It is clear that R_r and R_i are closed subsystems of R (which may be empty). A root which is neither real nor imaginary is *complex*.

For $w \in W$, $w\theta w^{-1}$ is an involution of R. We have $R_r(w\theta w^{-1}) = w.R_r(\theta)$, $R_i(w\theta w^{-1}) = w.R_i(\theta)$.

1.2. Let D be a basis of R. There is a unique $w \in W$ and a permutation $\iota = \iota_D$ of D such that for $\alpha \in D$

$$w(\theta\alpha) = \iota\alpha.$$

Then ι acts on the Coxeter system (W, S), where S is the set of reflections defined by the roots in D. Moreover, w is a twisted involution relative to ι, i.e. $\iota(w) = w^{-1}$.

For $x \in W$ we have $\iota_{x.D} = x\iota_D x^{-1}$.

If $I \subset D$ we denote by R_I the subsystem of R with basis I and by W_I its Weyl group. This is the subgroup of W generated by the reflections s_α, $\alpha \in I$. Its longest element (relative to these generators) is w_I. So w_D is the longest element of W. There is a permutation ι_0 of R which stabilizes D (the opposition involution), such that $w_D\alpha = -\iota_0\alpha$ ($\alpha \in D$).

Lemma 1.3. (i) *There exist $x \in W$ and an ι-stable subset I of D such that (a) $\theta = x(w_I\iota)x^{-1}$, (b) $w_I\alpha = -\iota\alpha$ for all $\alpha \in I$;*
(ii) *We have $R_r(\theta) = x.R_I$, where x and I are as in (i);*
(iii) *There exist $y \in W$ and an $\iota_0\iota$-stable subset J of D such that $R_i(\theta) = y.R_J$.*

(i) is a reformulation of part of [8, 3.3] and (ii) follows from (i). Applying (ii) to $-\theta$ we obtain (iii) (observe that the permutation of D associated to $-\theta$ is $\iota_0\iota = \iota\iota_0$).

Lemma 1.4. (i) *If $R_r(\theta) = \emptyset$ then θ fixes a system of positive roots;*
(ii) *Two systems of positive roots as in (i) are conjugate by a unique element of W^θ;*
(iii) *If $R_r(\theta) = \emptyset$, $R_i(\theta) = \emptyset$ there is a sub-rootsystem R_1 of R such that R is the direct sum of R_1 and θR_1.*

If there are no real roots for θ the set I of 1.3 must be empty by 1.3 (i), whence (i). Moreover, if θ stabilizes an irreducible component of R its highest root (relative to a system of positive roots as in (i)) is an imaginary root. This fact implies (iii). The proof of (ii) is easy.

1.5. Gradings. A *grading* of the root system R is a map $\epsilon : R \to \mathbb{Z}/2\mathbb{Z} = \{0,1\}$ such that for $\alpha, \beta, \alpha + \beta \in R$

$$\epsilon(-\alpha) = \epsilon(\alpha), \quad \epsilon(\alpha + \beta) = \epsilon(\alpha) + \epsilon(\beta).$$

Then $\alpha \in R$ is *even* or *odd* if $\epsilon(\alpha) = 0$, respectively 1. The set R_0 of even roots is a closed subsystem. The set of odd roots is denoted by R_1.

Denote by $\mathscr{G}(R)$ the set of gradings of R. We view it as an abelian group, the group structure being the obvious one. The Weyl group W acts on $\mathscr{G}(R)$.

We denote by $P(R)$, respectively $Q(R)$, the weight and root lattices of R.

Lemma 1.6. *There are isomorphisms*

$$\mathscr{G}(R) \simeq \mathrm{Hom}(Q(R), \mathbb{Z}/2\mathbb{Z}) \simeq P(R^\vee)/2P(R^\vee).$$

The lemma follows from [1, p. 159, Cor. 2]. Notice that $P(R^\vee)$ is the dual of $Q(R)$ ($R^\vee$ denoting the dual root system).

1.7. Let D be a basis of R. There is a unique grading $\epsilon = \epsilon_D$ of R with $\epsilon(\alpha) = 1$ for $\alpha \in D$. Such a grading is called *principal*. The principal gradings form a W-orbit in $\mathscr{G}(R)$.

1.8. Graded root data. Let $\Psi = (X, R, X^\vee, R^\vee)$ be a root datum (see [7, p. 190]). Recall that X is a free abelian group with dual $X^\vee$, and that $R \subset X$ is a root system (in the appropriate vector space). $R^\vee$ can be viewed as the dual root system. A *based root datum* Ψ_0 is a root datum Ψ as before, together with a basis D of R (or equivalently, with a system of positive roots in R). We shall view Ψ_0 as a quadruple $(X, D, X^\vee, D^\vee)$, where $D^\vee$ is the basis of $R^\vee$ determined by D.

An *involution* of Ψ is an automorphism θ of X of order ≤ 2 which stabilizes R and whose contragredient stabilizes $R^\vee$. It is clear that θ induces an involution of R, also denoted by θ.

A *graded root datum* is a triple $\Gamma = (\Psi, \theta, \epsilon)$, where Ψ is a root datum, θ is an involution of Ψ and ϵ is a grading of the subsystem R_i of imaginary roots (relative to θ). The roots of R_i which are even (odd) are also called *compact imaginary roots* (respectively, *non-compact imaginary roots*) of R.

A *based graded root datum* is a triple $\Gamma_0 = (\Psi_0, \theta, \epsilon)$, where Ψ_0 is a based root datum and θ, ϵ are as before, relative to the root datum Ψ defined by Ψ_0.

The notion of graded root datum is a slight generalization of the notion of graded root system, introduced in [12].

1.9. It is clear how to define isomorphisms of (based) graded root data. We mention a particular case. Let $\Gamma = (\Psi, \theta, \epsilon)$ be as before and let $w \in W = W(R)$. We define the root datum $w.\Gamma$ by

$$w.\Gamma = (\Psi, w\theta w^{-1}, \epsilon \circ w^{-1}).$$

This is a root datum isomorphic to Γ. We thus have an action of W on the set of graded root data with underlying root datum Ψ. The elements of W^θ permute the graded root data with underlying root datum with involution (Ψ, θ).

1.10. Let A be an abelian group and let σ be an automorphism of A of order ≤ 2. We write

$$H_\pm(\sigma, A) = H_\pm(A) = \mathrm{Ker}(\sigma \mp 1)/\mathrm{Im}(\sigma \pm 1).$$

If A is of finite type these are finite elementary abelian 2-groups. Clearly, $H_-(\sigma, A) = H_+(-\sigma, A)$.

Let $\mathbb{Z}/2\mathbb{Z}$ act on A via σ. Then $H_+(\sigma, A)$ ($H_-(\sigma, A)$) is an even (odd) reduced cohomology group $\hat{H}^{2i}(\mathbb{Z}/2\mathbb{Z}, A)$ (respectively, $\hat{H}^{2i+1}(\mathbb{Z}/2\mathbb{Z}, A)$) of [2, Ch. XII]. If A is free of finite type write $A^\vee = \mathrm{Hom}(A, \mathbb{Z})$, if A is finite write $A^* =$

$\text{Hom}(A, \mathbb{Q}/\mathbb{Z})$. We denote the pairings between A and $A^\vee$ (respectively A^*) by $\langle , \rangle$.

Lemma 1.11. *Assume that A is free of finite type. The pairing $A \times A^\vee \to \mathbb{Z}$ induces a functorial isomorphism $H_\pm(\sigma, A)^* \simeq H_\pm(\sigma, A^\vee)$.*

Here σ acts on $A^\vee$ in the obvious way. 1.11 can be deduced from [loc.cit., 6.5] (or can be proved directly).

Let C be a divisible abelian group, with trivial σ-action.

Lemma 1.12. *Assume that A is free of finite type. There is a functorial isomorphism (relative to A)*

$$H_\pm(\sigma, A \otimes C) \simeq H_\mp(\sigma, A) \otimes H_-(\sigma, C).$$

The σ-action on $A \otimes C$ is the obvious one. We have $A \otimes C \simeq \text{Hom}(A^\vee, C)$. The lemma is a consequence of [loc.cit., 6.4] and 1.11.

Let A and σ be as before, with A free of finite type. Let s be a reflection in A. I. e., there exist $a \in A$ and $a^\vee \in A^\vee$ such that

$$s(x) = x - \langle x, a^\vee \rangle a \quad (x \in A),$$

with $\langle a, a^\vee \rangle = 2$. We assume that a and $a^\vee$ are fixed by σ. Then s commutes with σ. Let $a^\flat$ and $a^\sharp$ denote the image of a in $H_+(\sigma, A)$, respectively $H_-(s\sigma, A)$. Similarly, we have $(a^\vee)^\flat \in H_+(\sigma, A^\vee)$ and $(a^\vee)^\sharp \in H_-(s\sigma, A^\vee)$.

In the next two lemmas we assume that $a^\vee$ takes the value 1 on A.

Lemma 1.13.   (i) *We have exact sequences*

$$\mathbb{Z}/2\mathbb{Z} \xrightarrow{\delta_+} H_-(\sigma, \text{Ker } a^\vee) \to H_-(\sigma, A) \to 0,$$

$$0 \to H_-(\sigma, \text{Ker } a^\vee) \to H_-(s\sigma, A) \xrightarrow{\gamma_-} \mathbb{Z}/2\mathbb{Z};$$

(ii) *δ_+ is injective if and only if $(a^\vee)^\flat = 0$;*
(iii) *γ_- is surjective if and only if $(a^\vee)^\sharp \neq 0$;*
(iv) *If δ_+ is injective the image in $H_-(s\sigma, A)$ of the non-trivial element of $\text{Im } \delta_+$ is $a^\sharp$.*

Lemma 1.14.   (i) *We have exact sequences*

$$0 \to H_+(\sigma, \text{Ker } a^\vee) \to H_+(\sigma, A) \xrightarrow{\gamma_+} \mathbb{Z}/2\mathbb{Z},$$

$$\mathbb{Z}/2\mathbb{Z} \xrightarrow{\delta_-} H_+(\sigma, \text{Ker } a^\vee) \to H_+(s\sigma, A) \to 0;$$

(ii) *δ_- is injective if and only if $(a^\vee)^\sharp = 0$;*
(iii) *γ_+ is surjective if and only if $(a^\vee)^\flat \neq 0$;*
(iv) *If δ_- is injective the image in $H_+(\sigma, A)$ of the non-trivial element of $\text{Im } \delta_-$ is $a^\flat$.*

Applying 1.13 with σ replaced by $-s\sigma$ we obtain 1.14. So it suffices to prove 1.13.

Ker $a^\vee$ is a σ-stable submodule of A, on which s acts trivially. We have an exact sequence

$$0 \to \operatorname{Ker} a^\vee \to A \xrightarrow{a^\vee} \mathbb{Z} \to 0,$$

of groups with σ-action, σ acting trivially on $\mathbb{Z}$. It is also an exact sequence of groups with $s\sigma$-action, $s\sigma$ acting on $\mathbb{Z}$ as -1. We obtain two long exact cohomology sequences. Observing that $H_-(\operatorname{id}, \mathbb{Z}) = H_+(-\operatorname{id}, \mathbb{Z}) = \{0\}$ we obtain the assertions (i), (ii), (iii).

$a^\natural = 0$ if and only if there is $b \in A$ with $\sigma b = b$, $\langle b, a^\vee \rangle = 1$. From 1.11 we see that then $(a^\vee)^b \neq 0$. By 1.13 (ii), $a^\natural \neq 0$ if δ_+ is injective. Choose $c \in A$ with $\langle c, a^\vee \rangle = 1$. Then $a - c + s\sigma(c) = -c + \sigma c$ lies in Ker $a^\vee$. It represents $a^\natural$ and has image zero in $H_-(\sigma, A)$. This proves 1.13 (iv).

Remark 1.15. In 1.13 and 1.14 we had assumed that $a^\vee$ takes the value 1 on A. If this is not the case then A is the direct sum of $\mathbb{Z}a$ and Ker $a^\vee$. We then have $H_-(\sigma, A) = H_-(\sigma, \operatorname{Ker} a^\vee)$, $H_-(s\sigma, A) = \mathbb{Z}/2\mathbb{Z} \oplus H_-(\sigma, \operatorname{Ker} a^\vee)$ (and similarly for H_+).

Let again θ be an involution of the root system R. If $\lambda \in \operatorname{Ker}(\theta - 1, P(R^\vee))$ then $\alpha \mapsto \langle \alpha, \lambda \rangle \pmod{2}$ defines a grading of R_i, which is trivial if $\lambda \in \operatorname{Im}(\theta + 1)$. We obtain a homomorphism $\gamma_i : H_+(\theta, P(R^\vee)) \to \mathscr{G}(R_i)$. In a similar manner, we obtain a homomorphism $\gamma_r : H_-(\theta, P(R^\vee)) \to \mathscr{G}(R_r)$.

Lemma 1.16. γ_i and γ_r are injective.

It suffices to deal with γ_r. By 1.6 and 1.11 the injectivity of γ_r is equivalent to the surjectivity of the canonical homomorphism

$$Q(R_r)/2Q(R_r) = H_-(\theta, Q(R_r)) \to H_-(\theta, Q(R)).$$

Put $A = Q(R)/Q(R_r)$. The desired surjectivity will follow if we show that $H_-(\theta, A) = 0$. Now it follows from 1.3 (i) that A is a free abelian group which has a basis whose elements are permuted by θ (namely the image of $w.(D - I)$, with the notations of loc.cit.). An easy argument gives that indeed $H_-(\theta, A) = 0$.

2. GROUPS WITH INVOLUTION AND GRADED ROOT DATA

2.1. In this section we denote by k an algebraically closed field of characterictic $\neq 2$ and by G a connected reductive linear algebraic group over k. Assume that θ is an automorphism of G (as a linear algebraic group) of order ≤ 2. We say that θ is an *involution* of G.

Let T be maximal torus of G which is θ-stable and denote by $\Psi = (X, R, X^\vee, R^\vee)$ the root datum defined by (G, T) (see [7, p. 191]. So X is the character group of T and R is the root system of (G, T). It is immediate from the definitions that θ induces an involution θ of the root datum Ψ.

For $\alpha \in R$ denote by $x_\alpha : \mathbf{G}_a \to G$ an additive one-parameter subgroup defined by it and put $U_\alpha = \operatorname{Im} x_\alpha$ [loc.cit., p.199]. Put $X_\alpha = dx_\alpha(1)$, this is a root vector for α in the Lie algebra of G. Let G_α be the subgroup of G generated by U_α and

$U_{-\alpha}$. Then G_α is isomorphic to either $\mathbf{SL}_2$ or $\mathbf{PSL}_2$. If α is an imaginary root then the automorphism θ of G stabilizes G_α and the induced automorphism is trivial or not according as $\theta(X_\alpha) = X_\alpha$, respectively $-X_\alpha$.

For $\alpha \in R_i$ we define $\epsilon(\alpha) = 0$ if the restriction of θ to G_α is trivial and $\epsilon(\alpha) = 1$ otherwise. Notice that ϵ can also be defined by

$$\theta(x_\alpha(\xi)) = x_\alpha((-1)^{\epsilon(\alpha)}\xi) \quad (\alpha \in R_i, \ \xi \in k).$$

We claim that ϵ is a grading of R_i.

It is clear that $\epsilon(-\alpha) = \epsilon(\alpha)$. Assume that $\alpha, \beta, \alpha + \beta \in R_i$. If $X_{\alpha+\beta}$ is a non-zero multiple of $[X_\alpha, X_\beta]$ then $\epsilon(\alpha + \beta) = \epsilon(\alpha) + \epsilon(\beta)$. Since $p = \operatorname{char} k \neq 2$, this condition is always fulfilled, except perhaps when $p = 3$ and α and β lie in a subsystem R' of R_i of type G_2. In that case let $\{\gamma, \delta\}$ be a basis of R'. Using [7, 11.2.7 (i)] one shows that if $\zeta = m\gamma + n\delta \in R'$ we have $\epsilon(\zeta) = m\epsilon(\alpha) + n\epsilon(\beta)$, which shows that ϵ is a grading.

It follows that $\Gamma = \Gamma(G, \theta, T) = (\Psi, \theta, \epsilon)$ is a graded root datum.

2.2. In the situation of 2.1 a Borel group B containing T defines a system of positive roots in the root system R of (G, T), whence a based graded root datum. Since any Borel group contains a θ-stable maximal torus, one could attach to such a group a based graded root datum (which turns out to be unique up to isomorphism).

We shall proceed somewhat differently, putting things in the framework of [5]. We recall the relevant results.

Fix a θ-stable maximal torus T of G which is contained in a θ-stable Borel subgroup B. Denote by $W = N(T)/T$ the Weyl group and by R the root system. These are acted upon by θ.

Put $\mathscr{V} = \{x \in G \mid x\theta(x)^{-1} \in N(T)\}$. Let K be the fixed point group of θ, it is a reductive closed subgroup of G. The group $T \times K$ acts on $\mathscr{V}$ and the set $V = V_\theta$ of $T \times K$-orbits in $\mathscr{V}$ is finite. For $v \in V$ denote by $x(v)$ a representative on $\mathscr{V}$. Then G is the disjoint union of the double cosets $Bx(v)K$, $v \in V$ [loc.cit., 1.3]. We may identify V with the set of double cosets BgK, with the set of K-orbits on the flag manifold $\mathscr{B} = B\backslash G$ or with the set of B-orbits on G/K.

Define $\phi : V \to W$ by $\phi(v) = w = \theta(x(v))^{-1}T$. Then $\theta(w) = w^{-1}$, so w is a twisted involution.

The set $\mathscr{V}$ is stable under left multiplication by elements of $N = N(T)$. We obtain an action of W on V [loc.cit., p.397]. If $x \in \mathscr{V}$ the maximal torus $T_x = x^{-1}Tx$ is θ-stable and clearly $T_{n.x} = T_x$ ($n \in N$). The map $x \mapsto T_x$ induces a bijection of the set of orbits V/W onto the set of K-conjugacy classes of θ-stable maximal tori of G [loc.cit., 2.7].

Let $x \in \mathscr{V}$. According to 2.1 the θ-stable maximal torus T_x defines a graded root datum Γ_x. We prefer to describe it in another manner.

Transport the ingredients of the graded root datum Γ_x to T by the inner automorphism $\operatorname{Int}(x)$, obtaining a graded root datum $\Gamma_v = (\Psi, \theta_v, \epsilon_v)$, depending only on the image v of x in V. The root datum Ψ is the one defined by (G, T) and $\theta_v = \phi(v)\theta$.

We denote by $R_i(v)$ and $R_r(v)$ the sets of imaginary, respectively, real roots. Then $\alpha \in R$ is real (imaginary) according as $\phi(v)\theta(\alpha) = -\alpha$ (respectively, α).

Put $n = x\theta(x)^{-1}$. If $\alpha \in R$ is an imaginary root for θ_v then $\psi_x = \text{Int}(n) \circ \theta$ induces an automorphism of the subgroup U_α of 2.1. Then α is even (odd) imaginary according as this automorphism is trivial (respectively, non-trivial).

The Borel group $B \supset T$ defines a system of positive roots R^+ in R and a basis D. We denote by Ψ_0 the corresponding based root datum and by $\Gamma_{v,0}$ the based graded root datum $(\Psi_0, \theta_v, \epsilon_v)$.

2.3. The root datum Γ of 1.8 is said to be *representable* if it is isomorphic to a graded root datum $\Gamma(G, \theta, T)$ of 2.1 (or, equivalently, to a Γ_v of 2.2). Then the triple (G, θ, T) is said to represent Γ.

Let (Ψ, θ) be a root datum with involution and let $\mathscr{G}(R_i)^\sharp$ be the set of $\epsilon \in \mathscr{G}(R_i)$ such that (Ψ, θ, ϵ) is representable.

The following theorem characterizes representable graded root data. It is similar to part (a) of [12, 10.8].

Theorem 2.4.   (i) $\mathscr{G}(R_i)^\sharp$ *contains the principal gradings of* R_i;
(ii) $\mathscr{G}(R_i)^\sharp$ *is a coset of* $\gamma_i(H_+(\theta, P(R^\vee)))$.

In (ii), the homomorphism $\gamma_i : H_+(\theta, P(R^\vee)) \to \mathscr{G}(R_i)$ is as in 1.16. Recall that it is injective.

We first prove (ii). Let G and T be as in 2.1 and assume that ϑ and ϑ' are two involutions of G stabilizing T, which induce the same involution θ of the root datum Ψ of (G, T). Let ϵ and ϵ' be the respective gradings of R_i.

By the isomorphism theorem for reductive groups [7, p. 265] there exists $t \in T$ such that $\vartheta' = \text{Int}(t) \circ \vartheta$. The right-hand side is an involution if and only if $t\vartheta(t)$ lies in the center C of G.

Let $\alpha \in R_i$. Then $\alpha(t) = \pm 1$. If there is $t' \in T$ with $t = t'(\vartheta t')^{-1}C$ then $\alpha(t) = 1$. We see from the description of the gradings given in 2.1 that

$$\alpha(t) = (-1)^{\epsilon(\alpha)+\epsilon'(\alpha)}.$$

We conclude that $\mathscr{G}(R_i)^\sharp$ is acted upon by the group $F = H_-(\vartheta, T/C)$. One knows that

$$T/C \simeq P(R^\vee) \otimes k^*.$$

Applying 1.12 with $C = k^*$ (observing that $H_-(\text{id}, k^*) = \mathbb{Z}/2\mathbb{Z}$) we see that $F \simeq H_+(\theta, P(R^\vee))$. The action of F on $\mathscr{G}(R_i)^\sharp$ corresponds via this isomorphism to the action by translations of $\text{Im}\,\gamma_i$. Now (ii) follows.

The next uniqueness result is an offshoot of the proof of (ii).

Proposition 2.5. *A triple* (G, θ, T) *representing the graded root datum* Γ *is unique up to isomorphism.*

We use the notations introduced above. To prove 2.5 it suffices to show that if $\epsilon = \epsilon'$ there is t' in T such that $t \in t'(\vartheta t')^{-1}C$. For then $\vartheta' = \text{Int}(t') \circ \vartheta \circ \text{Int}(t')^{-1}$, from which 2.5 follows.

Let τ be the image of t in $H_-(\vartheta, T/C) = H_+(\vartheta, P(R^\vee))$. Since $\epsilon = \epsilon'$ we have that $\gamma_i(\tau)$ is the trivial grading of R_i. As γ_i is injective, $\tau = 0$, which means that there exists t' with the desired property.

2.6. We now come to the proof of (i). There is a connected, reductive, linear algebraic group G over k with a maximal torus T such that the root datum of (G, T) is Ψ. Let G' be the derived group of G (a semi-simple group) and C^0 its connected center. Then $G = G'.C^0$, and $G' \cap C^0$ is finite. Moreover, there is a maximal torus T' of G' such that $T = T'.C^0$. The character group X' of T' is $X/(R^\vee)^\perp$, where $(R^\vee)^\perp$ is the annihilator of $R^\vee$ in X. Its dual $(X')^\vee$ is the rational closure $Q(R^\vee)_{rat}$ of $Q(R^\vee)$ in $X^\vee$. The character group of C^0 is $X_0 = X/Q(R)_{rat}$ and its cocharacter group is $X_0^\vee = R^\perp$. The root system of (G', T') and its dual can be identified with R, respectively $R^\vee$.

We now forget about G and T. Define $\Psi' = (X', R, (X')^\vee, R^\vee)$. An involution θ of the root datum Ψ induces an involution θ' of Ψ'. The imaginary roots for θ and θ' coincide. Assume that we have proved the theorem for (Ψ', θ') and let (G', T', ϑ') realize the corresponding graded root datum. Then G' is semi-simple.

Let C^0 be a torus with character group X_0. It is obvious that θ defines an automorphism ϑ^0 of C^0. There is an injective homomorphism $X \to X' \oplus X_0$ (use the obvious map $X \to X/(R^\vee)^\perp \oplus X/Q(R)_{rat}$). This homomorphism defines a central isogeny of reductive groups $G' \times C^0 \to G$, mapping $T' \times C^0$ onto a maximal torus T with character group X. The involution $(\vartheta', \vartheta^0)$ stabilizes the kernel of the isogeny, hence induces an involution ϑ of G which stabilizes T. The graded root datum $\Gamma(G, T, \vartheta)$ then has the properties required in (i).

We conclude that it suffices to prove (i) in the case that Ψ is the root datum of a semi-simple group. It follows from [10, 9.16] that an involution of an adjoint group can be lifted to an involution of its simply connected covering. Hence we may and shall assume that Ψ is the root datum of an adjoint group, i.e. that $X = Q(R)$. We then say that Ψ, or any graded root datum with underlying root datum Ψ, is *adjoint*.

2.7. An involution θ of a reductive group G is *quasi-split* if G contains a Borel subgroup B such that B and θB are opposite Borel groups. Then $T = B \cap (\theta B)$ is a maximal torus. With the notations of 2.1, let R^+ be the system of positive roots defined by B. Then $\theta R^+ = -R^+$, from which we see that the graded root datum $\Gamma(G, \theta, T)$ has no imaginary roots. Call *quasi-split* a graded root datum without imaginary roots. We define an automorphism ι of order ≤ 2 of R stabilizing R^+ by $\iota\alpha = w(\theta\alpha)$, where w is the longest element of the Weyl group (relative to R^+).

Proposition 2.8. *Let Γ be an adjoint quasi-split graded root datum. Then Γ is representable.*

The underlying root datum of Γ comes from a semi-simple adjoint group G with maximal torus T. By the isomorphism theorem [7, p. 265] there exists an automorphism θ of G which stabilizes T and induces on the root system R of (G, T) the involution of Γ, which we denote by ϑ. We shall show that θ can be taken to be an involution. Let R^+ be a system of positive roots in R with $\vartheta R^+ = -R^+$ and let D be the basis defined by R^+. Since the groups $U_{\pm\alpha}$ ($\alpha \in D$)

generate G (see [loc.cit., 10.1.11]) it suffices to prove that we can arrange matters
such that θ^2 fixes the elements of U_α, $\alpha \in D$.

For $\alpha \in R$ let x_α be as in 2.1. Let N be the normalizer of T. We assume the
one-parameter subgroups normalized such that for $\xi \in k^*$

$$x_\alpha(\xi)x_{-\alpha}(-\xi^{-1})x_\alpha(\xi) \in N \tag{1}$$

(see [loc.cit., 11.2.1 (ii)]). This fixes $x_{-\alpha}$ uniquely, once x_α is given.
We have $\theta(x_\alpha(\xi)) = x_{\vartheta\alpha}(c_\alpha\xi)$, where $c_\alpha \in k^*$. Using (1) it follows that $c_{-\alpha} = c_\alpha^{-1}$.
Then

$$\theta^2(x_\alpha(\xi)) = x_\alpha(c_{-\vartheta\alpha}^{-1}c_\alpha\xi).$$

Choose $t \in T$ such that $\alpha(t) = c_\alpha^{-1}$ for all $\alpha \in D$. Then the automorphism
$\text{Int}(t) \circ \theta$ is an involution with the desired properties. This proves 2.8.

Corollary 2.9. *There is a bijection of the set of isomorphism classes of quasi-split adjoint groups with involution (G, θ) onto the set of automorphisms ι of the corresponding ordered root system (R, R^+) of order ≤ 2.*

This follows from 2.8 and 2.5.

We call a graded root datum $\Gamma = (\Psi, \theta, \epsilon)$ *special* if (a) θ stabilizes a system
of positive roots R^+ and (b) ϵ is the principal grading of R_i defined by the basis
determined by $R_i \cap R^+$. (Principal gradings were defined in 1.7.)

Proposition 2.10. *Let G be a connected, adjoint semi-simple group. An involution θ of G is quasi-split if and only if there is a θ-stable maximal torus T of G such that $\Gamma = \Gamma(G, \theta, T)$ is special.*

G is a direct product of simple groups, which are permuted by an involution
θ. There is an obvious reduction to the case that either G is simple, or that G
is a product of two copies of a simple group H, the involution θ permuting the
factors. In the last case θ is quasi-split, and any θ-stable maximal torus has the
property of 2.10 (we skip the easy proofs). Hence the lemma is true in that case.
So we may assume G to be simple.

Let T be a maximal torus of G and denote by D a basis of the root system R
of (G, T), for the system of positive roots R^+. It follows from 2.9 that in order to
prove 2.10, it suffices for any ι as in 2.9, to exhibit an involution θ of G stabilizing
T, such that $\Gamma(G, \theta, T)$ is special and that (G, θ) is quasi-split, corresponding to
ι. Given such a group, an intrinsic way of describing ι is as follows. Let T' be
any θ-stable maximal torus. The root system of (G, T'), together with a system
of positive roots, can be canonically identified with (R, R^+). Then ι is as in 1.2,
relative to the involution of R defined by the restriction of θ to T'.

First assume that $\iota = \text{id}$. Let $a \in T$ be the element with $\alpha(a) = -1$ for
$\alpha \in D$. By [9, 6.1] the inner involution $\theta = \text{Int}(a)$ of G is quasi-split. θ clearly
has the property of the lemma. Let T' be a maximal torus of G which is an
intersection $B \cap (\theta B)$, where B is a Borel group. The involution θ' of the root
system R' of (G, T') defined by θ is that defined by the longest element of its
Weyl group, relative to B (see 2.7). In fact, this property characterizes inner
quasi-split involutions. It is clear that $\Gamma(G, \theta, T)$ is special.

There remains the case that ι is non-trivial (G being simple). It follows from [loc.cit, 7.1] that then a corresponding quasi-split group (G, θ) has the following property. There is a maximal torus T of G stabilized by θ such that the induced involution of the root system R of (G, T) is ι, hence fixes a basis D of R. All imaginary roots in D (i.e. those fixed by ι) are odd. We claim that these facts imply that $\Gamma = \Gamma(G, \theta, T)$ is special.

The simple groups for which we can have a non-trivial ι are those of types A_l, D_l and E_6. We shall deal with these cases separately. We use the description of root systems given in [1, Planches].

Type A_l. The root system is $\{\epsilon_i - \epsilon_j \mid 1 \leq i, j \leq n, \ i \neq j\} \subset \mathbb{R}^{l+1}$. The basis D is $\{\alpha_i = \epsilon_i - \epsilon_{i+1} \mid 1 \leq i \leq l\}$ and ι is induced by $\epsilon_i \mapsto \epsilon_{l+2-i}$. The imaginary roots are of the form $\epsilon_i - \epsilon_{l+2-i}$. The positive ones are mutually orthogonal.

If l is even an imaginary root α can be written as $\beta + \iota\beta$, for some complex root β. By [8, 2.6] α is odd, which shows that Γ is special in this case.

If $l = 2m - 1$ is odd no imaginary root is of the form $\beta + \iota\beta$. There is only one imaginary root in D, namely α_m. If $\alpha \in D - \{\alpha_m\}$ then α and $\iota\alpha$ are orthogonal. The positive imaginary roots are

$$\beta_i = \sum_{i \leq h \leq 2m-i} \alpha_i \quad (1 \leq i \leq m).$$

Now in type A we can take the one-parameter subgroups x_α of 2.1 such that for positive roots α, β with $\alpha + \beta \in R$ we have a commutator formula

$$x_{\alpha+\beta}(\xi) = (x_\alpha(1), x_\beta(\xi)) \quad (\xi \in k),$$

as a check in $\mathbf{SL}_{l+1}$ shows. Put $x_i = x_{\alpha_i}$ $(1 \leq i \leq 2m - 1)$. We may assume that $\theta(x_i(\xi)) = x_{2m-i}(\xi)$ for $i \neq m$ (see [9, no. 7]). We have $\theta(x_m(\xi)) = x_m(-\xi)$. If $1 \leq i \leq m - 1$ then

$$x_{\beta_i}(\xi) = (x_i(1), (x_{2m-i}(1), x_{\beta_{i+1}}(\xi)) = (x_{2m-i}(1), (x_i(1), x_{\beta_{i+1}}(\xi)),$$

from which we see that all β_i are odd. This settles the case of type A_l with l odd. We reduce the proof of 2.10 for the types D_l and E_6 to the case of type A.

Type D_l ($l \geq 4$). With the notations of [1, p. 256] take ι such that it permutes α_{l-1} and α_l and fixes the other roots of the basis D given there. If $l \geq 5$ this is the only possibility for ι. If $l = 4$ there are two other possibilities. But in this case it also suffices to consider only ι (use an outer automorphism of G of order 3). Now the root system R_i has the basis $(\alpha_1, \ldots, \alpha_{l-2}, \alpha_{l-2} + \alpha_{l-1} + \alpha_l)$. The first $l - 2$ roots are odd. Working in the θ-stable semi-simple subgroup of type A_3 of G defined by the subset $\{\alpha_{l-2}, \alpha_{l-1}, \alpha_l\}$ we conclude that $\alpha_{l-2} + \alpha_{l-1} + \alpha_l$ is also odd. This settles type D.

In type E_6 (the basis D being the one of [1, p. 260]) we have that ι permutes α_1, α_6 and α_3, α_5, and fixes α_2 and α_4. One checks (for instance by using the description of positive roots given in loc.cit.) that $\{\alpha_2, \alpha_4, \alpha_3 + \alpha_4 + \alpha_5, \alpha_1 + \alpha_3 + \alpha_4 + \alpha_5 + \alpha_6\}$ is a basis of R_i. The first two roots are odd. Working in the group of type A_5 defined by $D - \{\alpha_2\}$ we see that the other two roots are also odd. This finishes the proof of 2.10.

We return to the proof of 2.4 (i). Let $\Gamma = (\Psi, \vartheta, \epsilon)$ be an adjoint graded root datum, with ϵ a principal grading of $R_i(\vartheta)$. Let w and ι be as in 1.2 for ϑ, where D is the basis of a system of positive roots R^+ inducing the given principal grading ϵ. There exist G, θ and T as in 2.10 such that $\Gamma(G, T) = (\Psi, \iota, \epsilon')$, where ϵ' is the principal grading of $R_i(\iota)$ defined by R^+.

Let $B \supset T$ be the Borel group defined by R^+. It is fixed by θ. We use the notations of 2.2. By [5, 8.3] the map ϕ of 2.2 is a surjection of V onto the set of twisted involutions in W (relative to ι). Take $v \in V$ with $\phi(v) = w$ and let $x \in \mathcal{V}$ represent x. The torus $T' = x^{-1}Tx$ is θ-stable. Let Ψ' be its root datum and θ' the involution of Ψ' defined by θ. Then (Ψ', θ') is isomorphic to (Ψ, ϑ). To finish the proof of 2.4 (i) we show that the induced grading of $(R')_i$ is principal. Let T'_- be the identity component of the group $\{t \in T' \mid \theta(t) = t^{-1}\}$ and let H be the centralizer of T'_-. This is a connected, θ-stable, reductive group with maximal torus T'. Let κ be the involution induced on H. The root system R_1 of (H, T) contains no real roots. By 1.4 (i) κ fixes a system of positive roots of R_1. We claim that κ is quasi-split. Once this has been established, 2.4 (i) follows from 2.10. The claim is a consequence of the following lemma.

Lemma 2.11. *Let G be a group with a quasi-split involution θ. If S is a split subtorus of G then its centralizer $Z(S)$ is quasi-split for the involution induced by θ.*

Recall that a torus S is split if $\theta(s) = s^{-1}$ for all $s \in S$.

Let B a Borel group of G such that $T = B \cap (\theta B)$ is a maximal torus and let T_- be the identity component of $\{t \in T \mid \theta(t) = t^{-1}\}$. Then T_- is a maximal split subtorus, and S is conjugate under K^0 to a subtorus of T_- (see [4, 2.6]). Hence we may assume that S is a subtorus of T_-. But then the assertion of the lemma follows from the fact that $B \cap Z(S)$ is a Borel group of $Z(S)$.

This finishes the proof of 2.4.

2.12. Theorem 2.4 leads to another description of representable based graded root data.

Let $\Gamma_0 = (\Psi_0, \theta, \epsilon)$ be a representable based graded root datum. Denote by R^+ the system of positive roots of Ψ_0. Then $R^+ \cap R_i$ is a system of positive roots in R_i. Let $\rho_i^\vee$ be half the sum of the positive roots of R_i. If ϵ_0 is the principal grading of R_i defined by the basis of $R^+ \cap R_i$ (1.7), then

$$\epsilon_0(\alpha) \equiv \langle \alpha, \rho_i^\vee \rangle \pmod 2 \quad (\alpha \in R_i).$$

By 2.4 and 1.16 there exists a unique $\xi \in H_+(\theta_v, P(R^\vee))$ such that $\epsilon = \epsilon_0 + \gamma_i(\xi)$. Then Γ_0 is characterized by the triple $M = (\Psi_0, \theta, \xi)$. We call such a triple, with Ψ_0, θ as before and $\xi \in H_+(\theta, P(R^\vee))$ a *model*. It is a combinatorial model of a K-orbit on the flag manifold $\mathcal{B} = B \backslash G$.

We denote by $M_v = (\Psi_0, \theta_v, \xi)$ the model associated to the representable based graded root data $\Gamma_{v,0}$ $(v \in V)$ of 2.2.

Let $\alpha \in R$ be a root and put $s = s_\alpha$. Recall that the Weyl group W acts on the set V (see 2.2).

Proposition 2.13. *Let α be a simple root.*

(i)   If α is complex for v then $M_{s.v} = (\Psi_0, s\theta_v s^{-1}, s.\xi)$;
(ii) If α is odd imaginary for v then $M_{s.v} = (\Psi_0, \theta_v, \xi + (\alpha^\vee)^b)$;
(iii) If α is real or even imaginary for v then $s.v = v$ and $M_{s.v} = M_v$;

In (i) $s.\xi$ is the obvious element of $H_+(s\theta_v s^{-1}, P(R^\vee))$ and in (iii) $(\alpha^\vee)^b$ is the image of $\alpha^\vee$, as in 1.14.

The proof of (i) is straightforward from the definitions, observing that because α is not imaginary, we have $R_i(s.v)^+ = s.R_i(v)^+$.

Assume that α is imaginary. We have $\theta_{s.v} = \theta_v$ and $R_i(s.v) = R_i(v)$. Put $M_{s.v} = (\Psi_0, \theta_v, \eta)$. For $\beta \in R_i(v)$ we have

$$\langle \beta, \alpha^\vee \rangle \epsilon_\alpha = \epsilon_v(\beta) + \epsilon_{s.v}(\beta) = \langle \beta^b, \xi + \eta \rangle,$$

from which we see (using 1.16) that $\eta = \xi + \epsilon_v(\alpha)(\alpha^\vee)^b$. The assertions about an imaginary α follow. The case that α is real is easy. The proposition is proved.

We also record the following fact, which is a consequence of the definitions.

Lemma 2.14. $\alpha \in R_i(v) \cap D$ is odd if and only if $\langle \alpha^b, \xi \rangle = 0$.

On V we also have an operation $m(s)$ for a reflection s in a simple imaginary root. The effect of this operation on models will be discussed in the next section.

3. Cayley transforms

3.1. A basic construction. We use the notations of 2.1 and 2.2. Let D be the basis of the system of positive roots R^+ defined by B. Let $S = \{s_\alpha \mid \alpha \in D\}$ be the set of simple reflections. If $s = s_\alpha \in S$ denote by P_s the parabolic subgroup of G generated by B and $U_{-\alpha}$. Let $v \in V$ and view it as a double coset $Bx(v)K$, where $n = x(v)(\theta x(v))^{-1}$ lies in $N(T)$. Then $\psi = \psi_{x(v)} = \mathrm{Int}(n) \circ \theta$ is an involution of G.

We recall some facts which were established in [5, no. 4] (see also [6]). $P_s v$ contains a unique double coset $m(s).v$ of maximal dimension. Assume that $m(s).v \neq v$. Then $\dim m(s).v = \dim v + 1$ and either $m(s).v = s.v$ or $\alpha \in R_i(v)_1$ is odd.

Assume that we are in the latter case. The involution ψ fixes the group G_α of 2.1 and $\psi(x_{\pm\alpha}(\xi)) = x_{\pm\alpha}(-\xi)$ ($\xi \in k$). Let $n_\alpha \in G_\alpha$ be a representative in $N(T)$ of s_α. There is $z = z_{x(v)} \in G_\alpha$ such that $z(\psi(z))^{-1} = n_\alpha$. Then $zx(v) \in \mathscr{V}$. Denote by $m(s).v$ its image in V. We have $\phi(m(s).v) = s\phi(v) = \phi(v)(\theta s)$.

We have defined $m(s)$ only for a reflection s defined by a *simple* odd root. But the definition can be extended. Let $s = s_\alpha$ be a reflection, with $\alpha \in R_i(v)_1$. We say that s is odd for v. The parabolic subgroup P_s is only defined for $\alpha \in \pm D$. But $z \in G_\alpha$ exists for all odd imaginary roots α, and there is a well-defined element $m(s).v \in V$, represented by $zx(v)$. We call $m(s).v$ the *Cayley transform* of v through α or s. The name is taken from [12]. In fact, 3.2 shows that $(R, \epsilon_{m(s).v})$ is the Cayley transform of (R, ϵ_v) through α, in the sense of [loc. cit., 5.2].

Let $\Gamma_{m(s).v} = (\Psi, s\theta_v, \epsilon_{m(s).v})$ be the graded root datum associated to $m(s).v$ (see 2.2). The grading $\epsilon_{m(s).v}$ is described in the following lemma. Recall that two linearly independent roots $\beta, \gamma \in R$ are *strongly orthogonal* if R contains no root of the form $\pm\beta \pm \gamma$.

Lemma 3.2. (i) $R_i(m(s).v) = \{\beta \in R_i(v) \mid \langle \beta, \alpha^\vee \rangle = 0\}$;
(ii) *If* $\beta \in R_i(m(s).v)$ *then* $\epsilon_{m(s).v}(\beta)$ *equals* $\epsilon_v(\beta)$ *if and only if* α *and* β *are strongly orthogonal.*

Since $\theta_{m(s).v} = s\theta_v$ we have that $\beta \in R$ is imaginary for $m(s).v$ if and only if $\theta_v(\beta) = \beta - \langle \beta, \alpha^\vee \rangle \alpha$. If $\langle \beta, \alpha^\vee \rangle$ were non-zero, this would imply $\theta_v\alpha = -\alpha$, which is absurd. (i) follows.

With the notations of 3.1 we can take $\psi_{x(m(s).v)} = \text{Int}(n_\alpha) \circ \psi_{x(v)}$. We then have for $\beta \in R_i(m(s).v)$

$$\psi_{x(m(s).v)}(x_\beta(\xi)) = n_\alpha x_\beta((-1)^{\epsilon_v(\beta)}\xi)n_\alpha^{-1} \quad (\xi \in k).$$

If α and β are strongly orthogonal this equals $x_\beta((-1)^{\epsilon_v(\beta)}\xi)$. Otherwise it equals $x_\beta(-(-1)^{\epsilon_v(\beta)}\xi)$. The first point follows from the fact that if α and β are strongly orthogonal U_β commutes with $U_{\pm\alpha}$. To prove the second point it suffices to work in type B_2. One can take α to be a short simple root and β the sum of α and the other simple root. Then the asserted formula can be read off from the results in [7, p. 245].

Remark 3.3. One can define the Cayley transform of any graded root datum through an odd imaginary root, using the prescription of 3.2 (ii) as a definition, cf. [12, 5.2].

The next result is a version of [12, 6.12]). We say that the based graded root datum $\Gamma_{v,0}$ of 2.2 is *principal* if ϵ_v is the principal grading defined by the basis of the system of positive roots $R_i(v) \cap R^+$ in R_i.

Proposition 3.4. *Let* s *be a simple reflection which is odd for* v. *Assume that* $\Gamma_{v,0}$ *is principal. Then* $\Gamma_{m(s).v,0}$ *is principal.*

It follows from 3.2 that the proposition is a consequence of the next lemma.

Let R be a root system and let R^+ be a system of positive roots, with basis D. Let $\alpha \in D$ and denote by R_α the set of roots which are orthogonal to α, it is a subsystem of R. Let D_α be the basis of R_α defined by the system of positive roots $R^+ \cap R_\alpha$. The notation ϵ_D is as in 1.7.

Lemma 3.5. *For* $\beta \in R_\alpha$ *we have* $\epsilon_{D_\alpha}(\beta) = \epsilon_D(\beta)$ *if and only if* α *and* β *are strongly orthogonal.*

We may assume that R is irreducible. The case that R is of type G_2 is easily disposed of. So assume that R is not of type G_2. We have (cf. 2.12)

$$\epsilon_D(\beta) \equiv \frac{1}{2} \sum_{\gamma \in R^+} \langle \beta, \gamma^\vee \rangle \pmod 2.$$

We conclude that the lemma follows if we prove the following claim:

Let $\langle \beta, \alpha^\vee \rangle = 0$. Then the integer

$$a = \frac{1}{2} \sum_{\gamma \in R^+, \langle \gamma, \alpha^\vee \rangle \neq 0} \langle \beta, \gamma^\vee \rangle \tag{2}$$

is even if and only if α and β are strongly orthogonal.

Clearly, the summation can be restricted to the γ which are not orthogonal to β. Then $s_\beta \gamma \neq \gamma$ and $\langle \beta, \gamma^\vee \rangle + \langle \beta, (s_\beta \gamma)^\vee \rangle = 0$. It follows that we may restrict in (2) the summation to the set A of $\gamma \in R^+$ with

$$\langle \gamma, \alpha^\vee \rangle \neq 0, \quad \langle \beta, \gamma^\vee \rangle \neq 0, \quad s_\beta(\gamma) \in -R^+.$$

If $\gamma \in A$ then the same is true for $s_\alpha \gamma, -s_\beta \gamma, -s_\alpha s_\beta \gamma$ and the scalar products of each of their coroots with β equals $\langle \beta, \gamma^\vee \rangle$. Hence if for all $\gamma \in A$ the four roots $\gamma, s_\alpha \gamma, -s_\beta \gamma, -s_\alpha s_\beta \gamma$ are distinct, the integer a is even. If these four roots are not distinct we must have $\gamma = -s_\alpha s_\beta \gamma$, whence

$$2\gamma = \langle \gamma, \alpha^\vee \rangle \alpha + \langle \gamma, \beta^\vee \rangle \beta. \tag{3}$$

The claim now follows from the following lemma. (Recall that R is not of type G_2.)

Lemma 3.6. *Let $\gamma \in A$ be such that (3) holds, with $\beta \in R^+$. Then R is not simply laced. α and β are either orthogonal short roots with $\gamma = \pm\alpha + \beta$ or orthogonal long roots with $2\gamma = \pm\alpha + \beta$.*

If $\delta \in R$ put $m(\gamma, \delta) = \langle \gamma, \delta^\vee \rangle \langle \delta, \gamma^\vee \rangle$. This integer can take the values $0, 1, 2$. If (3) holds we have $m(\gamma, \alpha) + m(\gamma, \beta) = 4$, which can only be if $m(\gamma, \alpha) = m(\gamma, \beta) = 2$. Hence R is not simply laced. If α is a short root then γ is long and β is short. We have $\gamma = \pm\alpha \pm \beta$. Since α is simple, and β, γ are positive, we must have $\gamma = \pm\alpha + \beta$. This gives the first possibility. If α is long then γ is short and β is long, which leads to the other possibility.

This finishes the proof of 3.4.

3.7. Let $M_v = (\Psi_0, \theta_v, \xi)$ and $M_{m(s).v} = (\Psi_0, s\theta_v, m(s).\xi)$ be the models associated to v and $m(s).v$ (see 2.12). s is an odd reflection and $\xi \in H_+(\theta_v, P(R^\vee))$. By 2.14 we have $\langle \alpha^\flat, \xi \rangle = 0$. Apply 1.14 (i), with $A = P(R^\vee)$, $\sigma = \theta_v$, $a = \alpha^\vee$. Then $\xi \in \mathrm{Ker}\, \gamma_+$. The following description of $m(s).\xi$ now follows from 1.14 (i), using 3.2 (ii) and 3.5.

Proposition 3.8. *The image in $H_+(\theta_{m(s).v}, P(R^\vee))$ of $\xi \in H_+(\theta_v, \mathrm{Ker}\,\alpha)$ is $m(s).\xi$.*

4. Some subgroups of the Weyl group

We keep the notations of 2.1 and 2.1. Let T be a θ-stable torus in G, let $\Gamma = \Gamma(G, \theta, T)$ be as in 2.1. Denote by C the center of G. For $w \in W$ we have defined in 1.9 the graded root datum $w.\Gamma$.

Lemma 4.1. *We have $w.\Gamma = \Gamma$ if and only w can be represented by an element n of the normalizer $N(T)$ such that $(\theta n)n^{-1} \in C$.*

Assume that $\theta w = w$, let $n \in N(T)$ represent w and put $n^{-1}(\theta n) = t$. Then $t \in T$. If x_α is as in 2.1 then $t x_\alpha(\xi) t^{-1} = x_\alpha(\alpha(t)\xi)$ $(\xi \in k)$. For $\alpha \in R_i$ we have

$$\theta(x_\alpha(\xi)) = x_\alpha((-1)^{\epsilon(\alpha)}\xi), \quad n x_\alpha(\xi) n^{-1} = x_{w.\alpha}(d\xi),$$

with $d \in k^*$. By a straightforward computation we find from these formulas that

$$\alpha(t) = (-1)^{\epsilon(\alpha)+\epsilon(w.\alpha)}, \tag{4}$$

for $\alpha \in R_i$. If $t \in C$ then $\alpha(t) = 1$ for all roots α. It follows from (4) that then $w.\Gamma = \Gamma$.

Now assume that $w.\Gamma = \Gamma$. Then (4) gives that $\alpha(t) = 1$ for $\alpha \in R_i$. Using 1.3 (i) (for $-\theta$) we see that we may assume that there is a basis D of R, a permutation ι of D of order ≤ 2 and a subset I of D stabilized by ι, such that $\theta\alpha = \alpha$ for $\alpha \in I$ and that $\theta\alpha + \iota\alpha$ is an integral linear combination of roots in I for $\alpha \in D - I$. Since $I \subset R_i$ we have $\alpha(t) = 1$ for $\alpha \in I$. Using that $\theta t = t^{-1}$ it follows that $\iota\alpha(t) = \alpha(t)$ for $\alpha \in D - I$. There exists $t' \in T$ such that $\alpha(t) = \alpha(t')(\iota\alpha)$ for all $\alpha \in D$. Now $n(t')^{-1}$ is a representative of w with the required property. The lemma is proved.

If $w.\Gamma = \Gamma$ then w commutes with θ, hence lies in the group W^θ of 1.1. The next result describes the structure of W^θ. We write $W_i = W(R_i)$, $W_r = W(R_r)$, these are commuting subgroups of W^θ. Fix a system of positive roots R^+. Put $\rho_i = \frac{1}{2}(\sum_{\alpha \in R^+ \cap R_i} \alpha)$ and define ρ_r similarly. Put

$$R_c = \{\alpha \in R \mid \langle \rho_i, \alpha^\vee \rangle = \langle \rho_r, \alpha^\vee \rangle = 0\}.$$

This is a θ-stable root system without real or imaginary roots. Put $W_c = W(R_c)$. Using 1.4 (iii) we see that the fixed point group W_c^θ is generated by products $s_\alpha s_{\theta\alpha}$, where $\alpha \in R$ is a complex root which is strongly orthogonal to $\theta\alpha$.

Proposition 4.2. W^θ *is the semi-direct product of* W_c *and the normal subgroup* $W_i \times W_r$.

See [12, 3.12].

4.3. Assume that we are in the situation of 2.2. Let $v \in V$ and let $x \in \mathscr{V}$ be a representative. Put

$$W_v = \{w \in W \mid w.v = v\}, \quad W_v^* = \{w \in W \mid w.\Gamma_v = \Gamma_v\}.$$

Then W_v is isomorphic to the image in W of $x N_K(x^{-1}Tx)x^{-1} = N(T) \cap xKx^{-1}$ (see [5, p. 398 below]). We denote by W_v^0 the image of $N(T) \cap xK^0x^{-1}$.

Let $\bar{G}$ be the adjoint semi-simple group $\bar{G} = G/C$. Let $\bar{\theta}$ be the involution of $\bar{G}$ induced by θ. Denote by $\tilde{G}$ the simply connected covering of $\bar{G}$ and by $\tilde{\theta}$ the unique lift of $\bar{\theta}$ to $\tilde{G}$ (see [10, 9.16]). Then $\tilde{\theta}$ is an involution. By [loc.cit.,8.1] the fixed point group $\tilde{K}$ of $\tilde{\theta}$ is connected. It follows that the image of $\tilde{K}$ in the fixed point group $\bar{K}$ of $\bar{\theta}$ is the identity component $(\bar{K})^0$. We conclude that W_v^0 is a group like W_v for $\tilde{G}$.

Let K^* be the inverse image in G of $\bar{K}$, i.e.

$$K^* = \{g \in G \mid g(\theta g)^{-1} \in C\}.$$

Then K^* is the normalizer of K in G (we shall not need this fact).

Lemma 4.4. W_v^*/W_v^0 is an elementary abelian 2-group.

The commutator subgroup G' of G is connected, semi-simple and θ-stable. We have $G = G'.C^0$. Representatives of the elements of W may be taken in G'. It follows that we may assume G to be semi-simple and adjoint.

By 4.1 the elements of W_v^* are represented by elements of $K^* = K$. It follows that W_v^*/W_v^0 is a subgroup of K^*/K^0, which is an elementary abelian 2-group (as a consequence of [4, 8.1 (a)]). The lemma follows.

4.5. We see from 4.4 that W_v is a normal subgroup of W_v^*. Denote by P_v the quotient W_v^*/W_v.

The exact sequence

$$\{1\} \to C \to T \to T/C \to \{1\}$$

of groups with θ_v-action gives rise to a long exact sequence, part of which is

$$H_+(\theta_v, T/C) \xrightarrow{\delta} H_-(\theta_v, C) \to H_-(\theta_v, T) \to H_-(\theta_v, T/C).$$

It follows from 4.1 that P_v is isomorphic to a subgroup of $H_-(\theta_v, C)/\operatorname{Im}\delta$, hence to a subgroup of $H_-(\theta_v, T)$. Since $T \simeq X^\vee \otimes k^*$ we have by 1.12

$$H_-(\theta_v, T) \simeq H_+(\theta_v, X^\vee),$$

whence an injective homomorphism $P_v \to H_+(\theta_v, X)$. We identify P_v with its image in the latter group.

We have $H_-(\theta_v, T/C) \simeq H_+(\theta_v, P(R^\vee))$. It follows that P_v is contained in the kernel of the homomorphism $H_+(\theta_v, X^\vee) \to H_+(\theta_v, P(R^\vee))$, induced by the homomorphism $X^\vee \to P(R^\vee)$ which is the contragredient of the inclusion $Q(R) \to X$.

Let $w \in W$. We have an isomorphism

$$w. : H_+(\theta, X^\vee) \to H_+(w\theta w^{-1}, X^\vee).$$

Then $P_{w.v} = w.P_v$ $(v \in V)$.

We now discuss a more explicit description of the subgroups of W introduced in 4.3. The groups of 4.2 for the graded root datum Γ_v will be denoted by $W_{v,i}, \ldots$. Let $W_{v,i}^0$ be the subgroup of $W_{v,i}$ generated by the reflections s_α with α even imaginary for v.

We say that a root α of a root system R is *isolated* if it lies in an irreducible component of R of type A_1. Then α or $-\alpha$ is a simple root, for any system of positive roots in R.

Theorem 4.7. (i) W_v^0 *is the semi-direct product of* $W_{v,c}$ *and the normal subgroup* $W_{v,r} \times W_{v,i}^0$;

(ii) W_v *and* W_v^* *are generated by* W_v^0 *and reflections* s_α, *where* α *is an isolated odd root of* $R_i(v)$.

If α and β are imaginary roots then

$$\epsilon_v(s_\alpha\beta) = \epsilon_v(\beta) + \langle\beta, \alpha^\vee\rangle\epsilon_v(\alpha), \tag{5}$$

which shows that $s_\alpha \in W_v^*$ if α is even imaginary. Then the group G_α of 3.1 is contained in K^0, whence $s_\alpha \in W_v^0$. Hence $W_{v,i}^0 \subset W_v^0$. Also notice that it follows from (5) that the s_α of (ii) lie in W_v^*.

If α is a real root for v then $s_\alpha.v = v$ (see 2.13 (iii)), whence $s_\alpha \in W_v$. Applying this for $\tilde{G}$, using the connectedness of $\tilde{K}$, we conclude that $s_\alpha \in W_v^0$ and $W_{v,r} \subset W_v^0$.

Put $\phi(v) = a$. If α is complex and strongly orthogonal to $a\theta(\alpha)$, we have $s_\alpha s_{a\theta(\alpha)}.v = v$, as a straightforward check shows. Using again $\tilde{G}$ we see that $s_\alpha s_{a\theta(\alpha)} \in W_v^0$, whence $W_{v,c} \subset W_v^0$. We have proved that the three groups of (i) lie in W_v^0. By 4.2, (i) will follow if we show that $W_v^0 \cap W_{v,i} \subset W_{v,i}^0$. As in the end of the proof of 2.4 (i) we reduce to the case that $R_r(v) = \emptyset$. From 1.4 we see that there is $w \in W$ such that $w.v$ is minimal. We may replace v by $w.v$. Then the element x of 4.4 lies in $N(T)$, T is θ-stable and is contained in a θ-stable Borel group. If $G = \tilde{G}$ then the Weyl group of $(K, (K \cap T)^0)$ is generated by $W_{v,c}$ and $W_{v,i}^0$, by results of Steinberg ([10, no. 8]). This implies what we wanted and finishes the proof of (i).

We next claim that W_v and W_v^* are generated by W_v^0 and certain s_α, with α odd imaginary. It suffices to prove this for W_v. The case of W_v^* then follows, working in the adjoint group $\overline{G}$. Put $A = \mathbb{R} \otimes Q(R)$. If $w \in W_v$ has an eigenvalue 1 in A there is a non-central θ_v-stable subtorus S of T whose elements are fixed by w. Replacing G by the centralizer $Z(S)$, which is θ_v-stable and has smaller dimension, we can use induction. So we are reduced to dealing with the case that w has no eigenvalue 1 in A. Let α be a root. Then there is $x \in V$ with $(w - 1)x = \alpha$ and a familiar argument (see e.g. [1, p. 193]) shows that $s_\alpha w$ fixes x. Using (i) we see that we can proceed by induction if there are real or even imaginary roots. So we may assume that there are no such roots. By 1.4 (i) there is a basis D whose elements are permuted by θ_v. We may assume that G is semi-simple and simply connected, and is either quasi-simple or the product of two quasi-simple groups which are permuted by θ. In the latter case there are no imaginary roots, and the claim is obvious. So assume that G is quasi-simple. If θ_v acts trivially on D, the fact that there are no compact imaginary roots implies that R must be of type A_1, and the claim readily follows.

So assume that θ_v acts non-trivially on D. Then G is of one of the types A_l, D_l ($l \geq 4$), E_6. The last two possibilities are ruled by the assumption of the absence of even imaginary roots. So we remain with type A_l. We may assume that $G = SL_{l+1}$. If l is even we also may assume that $\theta_v(g) = ({}^t g)^{-1}$, so that $K = SO_{l+1}$. In that case $K^* = K^0$, and the claim follows. If l is odd the absence of compact imaginary roots implies that we may assume θ_v to be the familiar involution with fixed point group Sp_{l+1}. In that case R_i is of type $(A_1)^{\frac{1}{2}(l+1)}$ and all imaginary roots are non-compact. A direct check shows that $W_v^* = W_v^0$. The claim follows.

To finish the proof of the theorem we have to show that if α is an odd imaginary root with $s_\alpha \in W_v^*$ it is an isolated root of $R_i(v)$. By (5) we have for such an α that $\langle \beta, \alpha^\vee \rangle$ is even for all $\beta \in R_i$. Taking for granted the next lemma, we see that R is of type A_1 or of type B_l with $l \geq 2$. In the latter case an argument as in the first part of the proof, using $\tilde{K}$, shows that $s_\alpha \in W_{v,i}^0$, whence the theorem in this case. In type A_1 the theorem is obvious.

Lemma 4.8. *Let R be an irreducible root system. Let $\alpha \in R$ be such that $\langle \beta, \alpha^\vee \rangle$ is even for all roots β. Then either R is of type A_1 or of type B_l ($l \geq 2$) and α is a short root.*

Assume that R is not of type A_1. We may assume α to be simple. R cannot be simply laced and α must be a short root which is orthogonal to all other short roots. This forces R to be of type B.

Put $W_{v,i}^* = W_v^* \cap W(R_i)$.

Corollary 4.9.   (i) $W_v^*/W_v \simeq W_{v,i}^*/W_{v,i}$;
(ii) $W_{v,i}^*$ *is the semi-direct product of $W_{v,i}^0$ and the normal elementary abelian 2-group generated by the s_α with $\alpha \in R_i(v)$ odd and isolated;*
(iii) $P_v \simeq W_{v,i}^*/W_{v,i}$.

(i) follows from 4.2 and 4.7 (i) and (ii) is a consequence of the proof of 4.7 (ii). Then (iii) is clear.

Corollary 4.10. *The subgroup P_v of $H_+(\theta_v, X^\vee)$ is generated by the elements $(\alpha^\vee)^\flat$, where α runs through the set of isolated odd roots of R_i.*

As in no. 1, $(\alpha^\vee)^\flat$ is the image of $\alpha^\vee$. The corollary follows from 4.9 (ii). The proof is straightforward. One uses that a reflection s_α with α odd imaginary has a representative n with $n(\theta n)^{-1} = \alpha^\vee(-1)$.

Corollary 4.11. *P_v depends only on the graded root datum Γ_v.*

This is a consequence of the preceding corollary.

Remarks 4.12. (a) $W_{v,i}$ is an $\mathscr{R}$-group in the sense of [12, 3.15]. 4.7 (ii) is similar to [loc.cit., 4.14].
(b) Let $M_v = (\Psi_0, \theta_v, \xi)$ be the model of $v \in V$ (2.12). It follows from 2.14 that the odd imaginary roots of 4.9 are the imaginary roots α of $R_i(v)$ with $\langle \alpha^\flat, \xi \rangle = 0$ (notations of 2.14). For such a root α we have two cases, according as $(\alpha^\vee)^\flat$ is zero or not. These cases correspond to the type II and type I roots of [11, p. 549].

4.13. With the notations of 2.2, we say that two elements $v, v' \in V$ lie in the same *packet* if $\Gamma_v = \Gamma_{v'}$. It follows from [5, 2.5] that then $v' \in W.v$. Using 4.3 we conclude that P_v acts simply transitively on the packet of v. In particular, if G is adjoint semi-simple a packet consists of one element. In that case v is uniquely determined by the model M_v

4.14. We record a functorial property of P_v. Let $s = s_\alpha$ with α simple and $m(s).v > v$. Then α is either complex or odd imaginary. In the first case the isomorphism s. of 4.6 induces an isomorphism $P_v \to P_{s.v}$.

Let α be odd imaginary and let β be an isolated root of $R_i(v)$, which we may take to be simple. It follows from 3.2 (i) that β is an isolated root of $R_i(m(s).v)$, except when $\beta = \alpha$. Apply 1.14 with $A = X^\vee$, $a = \alpha^\vee$. Observing that $(\beta^\vee)^b \in \operatorname{Ker}\gamma_+$ and using 4.10 we obtain a homomorphism $\phi : P_v \to P_{s.v}$. If α is an isolated odd imaginary root then $\operatorname{Ker}\phi = \{0, (\alpha^\vee)^b\}$, by 1.14 (iv).

If α is odd imaginary the subgroup P_v of $H_+(\theta_v, P(R^\vee))$ lies in $H_+(\theta_v, \operatorname{Ker}\alpha)$ (use 4.13 and 1.14 (i)). It follows that $\phi(\xi) = m(s).\xi$ ($\xi \in P_v$), where $m(s)$ is as in 3.8.

We now give some results about root systems which follow from what was established in the present section. Let R be a root system with basis D and let S be the corresponding set of simple reflections. Denote by Σ the set of pairs (θ, ξ), where θ is an involution of R and $\xi \in H_+(\theta, P(R^\vee))$.

Proposition 4.15. *There is an action of the Weyl group W of R on Σ such that for a simple reflection $s = s_\alpha$ ($\alpha \in D$) we have*

$$s.(\theta, \xi) = (s\theta s^{-1}, s.\xi) \qquad \text{if } \alpha \text{ is complex for } \theta,$$
$$= (s\theta, \xi + (\alpha^\vee)^b) \quad \text{if } \alpha \text{ is imaginary for } \theta \text{ and } \langle \alpha^b, \xi \rangle = 0,$$
$$= (\theta, \xi) \qquad\qquad \text{otherwise.}$$

Let G be an adjoint semi-simple group with root system R, relative to a maximal torus T. Denote by $B \supset T$ the Borel subgroup determined by D and let Ψ_0 be the corresponding based root datum. If $(\theta, \xi) \in \Sigma$ we have the model (Ψ_0, θ, ξ). By 2.4 it comes from a B-orbit on G/K, where K is the fixed point group an involution of G. Changing the notation, we may assume that we are in the situation of 2.2 and that we have $v \in V$ such that the model M_v of 2.12 is (Ψ_0, θ, ξ). In the present situation the elements of V are uniquely determined by their models (see 4.13). We have the action of W on V of 2.2. Now 4.15 follows from 2.13.

4.16. Let $(\theta, \xi) \in \Sigma$ and $s = s_\alpha$ where α is imaginary for θ and $\langle \alpha^b, \xi \rangle = 0$. An argument as that of the proof of 4.15 shows that one has a Cayley transform $m(s).(\theta, \xi) = (s\theta, m(s).\xi)$, where $m(s).\xi$ is as in 3.8.

4.17. With the Coxeter group (W, S) there is associated a monoid $M = M(W)$, generated by elements $\mu(s)$ ($s \in S$), satisfying the relations $\mu(s)^2 = \mu(s)$ ($s \in S$) and braid relations (see [5, 3.10], to avoid confusion we write $\mu(s)$ instead of $m(s)$ as in [loc.cit.]). It is shown in [loc.cit., 4.7] that M acts on the set V of 2.2. It follows that there is an action of M on the set Σ. Using the results of [loc.cit., no.4] we see that this action of M is decribed as follows. Let $\sigma = (\theta, \xi) \in \Sigma$, $s = s_\alpha$ ($\alpha \in D$). Then

$$\mu(s).\sigma = s.\sigma$$

if α is complex for θ and $\theta\alpha$ is positive (relative to D),

$$\mu(s).\sigma = m(s).\sigma$$

if α is imaginary for θ and $\langle \alpha^b, \xi \rangle = 0$ and $\mu(s).\sigma = \sigma$ otherwise.

4.18. Let $\sim$ be the equivalence relation on Σ generated by the elementary equivalences $\sigma \sim \sigma'$ if there is $s \in S$ with $\sigma' = \mu(s).\sigma$. An equivalence classes is isomorphic, as a W-set, to a set V of 2.2, for an involution of an adjoint semisimple group.

The Bruhat order on the sets V, studied in [loc.cit.] can also be transferred to Σ. By [loc.cit., 7.11, 5.2] this Bruhat order on Σ can be completely described in terms of the M-action.

Notice that the elements of Σ which are maximal for the Bruhat order (they correspond to the maximal elements of the various V) are the pairs (θ, ξ) such that (a) $\theta\alpha$ is negative for all $\alpha \in D$ which are complex for θ, (b) if $\alpha \in D$ is imaginary for θ then $\langle \alpha^b, \xi \rangle \neq 0$.

4.19. We return to the situation of 4.3. We shall give another description of the group P_v of 4.5. The group K^* acts on the right on the set $\mathscr{V}$ of 2.2, whence a right action of $\Phi = K^*/K$ on V, commuting with the W-action. Let C the center of G. Let $\tau(x) = x(\theta x)^{-1}$. Then τ induces an isomorphism of Φ onto the subgroup $\tau(G) \cap C$ of $\mathrm{Ker}(\theta + 1, C)$. The action of Φ on V comes from an action of its image in $H_-(\theta, C)$.

Lemma 4.18. *Let A be a maximal θ-split subtorus of G. Then $\tau(G) \cap C = A \cap C$.*

This is a consequence of [4, 6.3, 2.7]. If $\bar{R}$ is the root system of (G, A) (see [loc.cit., no.4]) then $A \cap C$ is the intersection of the kernels of the roots of $\bar{R}$. This gives a description of Φ in terms of A and $\bar{R}$.

Let $v \in V$ and denote by Φ_v its stabilizer in Φ.

Lemma 4.19. $P_v \simeq \Phi/\Phi_v$.

It follows from the definitions that for $f \in \Phi$ we have $\Gamma_{v.f} = \Gamma_v$, whence (see 4.13) $v.f \in W.v$. We conclude that we have an injective map $\Phi/\Phi_v \to W_v^*/W_v$, which in fact is a homomorphism. It follows from 4.1 that the map is surjective.

5. LOCAL SYSTEMS, BLOCKS

5.1. We use the notations of 2.2. We now view V to be the set of B-orbits on $Y = G/K$. Let $v \in V$ and let $\mathscr{L}$ be a B-equivariant local system of rank one on v, in the l-adic sense. For $k = \mathbb{C}$ one could also work with local systems in the sense of classical topology. The isomorphism classes of such local systems form an abelian group $L(v)$. In both cases it can be described in the following manner.

Fix $y \in v$ and let B_y be its isotropy group in B. Then $L(v)$ is isomorphic to the character group of the finite group $A_y = B_y/B_y^0$, with values in a suitable field E. We shall identify this character group with the dual $A_y^* = \mathrm{Hom}(A_y, \mathbb{Q}/\mathbb{Z})$.

Lemma 5.2. *There is an isomorphism* $\phi_y : A_y \rightarrow H_-(\theta_v, X^\vee)$ *such that for* $b \in B$ *the isomorphism* $\phi_{b,y}$ *is the composite of* ϕ_y *and the map induced by translation by* b.

Let $x \in \mathscr{V}$ represent v. The isotropy group B_y is isomorphic to the group $B \cap xKx^{-1}$, which is isomorphic to the direct product of $\mathrm{Ker}(\theta_v - 1, T)$ and a connected unipotent group (see [8, 4.8]). Since the identity component of the first group is $\mathrm{Im}(\theta_v + 1, T)$, we obtain an isomorphism $A_y \simeq H_+(\theta_v, T)$. Now $T \simeq X^\vee \otimes k^*$. Applying 1.12 we obtain the isomorphism of the lemma. The argument also gives its equivariance.

It follows from 5.2 and 1.11 that $L(v) \simeq H_-(\theta_v, X)$. In the sequel we shall identify these groups.

The following lemma is immediate from the definitions.

Lemma 5.3. *If* v *and* v' *lie in the same packet (4.13) then* $L(v) = L(v')$.

5.4. Bimodels. A *bimodel* is a quadruple $\mathscr{M} = (\Psi_0, \theta, \xi, \eta)$, where (Ψ_0, θ, ξ) is a model (2.12) and $\eta \in H_-(\theta, P(R))$. Bimodels can be viewed as weak versions of the bigradings associated to regular characters of [12, 4.12].

A pair $(v, \mathscr{L})$ as in 5.1 defines a bimodel $\mathscr{M}_{v,\mathscr{L}}$, in the following manner. Let $M_v = (\Psi_0, \theta_v, \xi)$ be the model of v (2.12) and let $\zeta \in L(v) = H_-(\theta_v, X)$ be the class of $\mathscr{L}$. We have an injective homomorphism $Q(R^\vee) \rightarrow X^\vee$. The contragredient homomorphism $X \rightarrow P(R)$ defines a homomorphism

$$\phi : H_-(\theta_v, X) \rightarrow H_-(\theta_v, P(R)).$$

Let $\eta = \phi\zeta$. Then $\mathscr{M}_{v,\mathscr{L}} = (\Psi_0, \theta_v, \xi, \eta)$.

A bimodel $\mathscr{M}$ as above is *representable* if it is isomorphic to an $\mathscr{M}_{v,\mathscr{L}}$.

Lemma 5.5. *If* Ψ_0 *is the based graded root datum of a simply connected, semi-simple group then* $\mathscr{M}$ *is representable.*

There is v such that $M_v = (\Psi_0, \theta, \xi)$ (see 2.12). The lemma follows by observing that $\phi = \mathrm{id}$.

5.6. Let $\mathscr{M}$ be as before and put $\Psi_0 = (X, D, X^\vee, D^\vee)$ (see 1.8). The *dual* of the based graded root datum is $\Psi_0^\vee = (X^\vee, D^\vee, X, D)$. The *dual* of the bimodel $\mathscr{M}$ is $\mathscr{M}^\vee = (\Psi_0^\vee, -\theta, \eta, \xi)$. (Notice that in the situation of 1.10 we have $H_\pm(\sigma, A) = H_\mp(-\sigma, A)$.)

Let $\overline{\Psi}_0 = (Q(R), D, P(R^\vee), D^\vee)$, this the based root datum of an adjoint semi-simple group. Put $\overline{\mathscr{M}} = (\overline{\Psi}_0, \theta, \xi, \eta)$.

Lemma 5.7. $\overline{\mathscr{M}}^\vee$ *is representable.*

Since $\overline{\mathscr{M}}^\vee$ is the based graded root datum of a simply connected, semi-simple group this follows from 5.5.

The lemma shows that for any pair $(v, \mathscr{L})$ as in 5.1 with G adjoint, semi-simple, there exists a similar pair $(v^\vee, \mathscr{L}^\vee)$, relative to a simply connected, semi-simple group whose root system is the dual of that of G, such that $(\overline{\mathscr{M}_{v,\mathscr{L}}})^\vee = \mathscr{M}_{v^\vee, \mathscr{L}^\vee}$.

5.8. Let Σ be as in 4.15 and denote by $\Sigma^\vee$ the corresponding set, for the dual root system $R^\vee$ with basis $D^\vee$. The bimodel $\mathscr{M} = (\Psi_0, \theta, \xi, \eta)$ of 5.4 defines elements $\pi\mathscr{M} = (\theta, \xi)$ and $\pi^\vee\mathscr{M} = (-\theta, \eta)$ of Σ, respectively $\Sigma^\vee$ (where now θ is viewed as an involution of $R^\vee$).

We define an equivalence relation $\sim$ on the set of bimodels with the same underlying based root datum as follows. Let $\mathscr{M}' = (\Psi_0, \theta', \xi', \eta')$. Then $\sim$ is generated by the following elementary equivalences: $\mathscr{M} \sim \mathscr{M}'$ if there is a simple reflection such that $\pi\mathscr{M}' = \mu(s).\pi\mathscr{M}$, $\pi^\vee\mathscr{M} = \mu(s).\pi^\vee\mathscr{M}'$. Here $\mu(s)$ is as in 4.18 (both for Σ and Σ'). Notice that if $s = s_\alpha$ ($\alpha \in D$) we have $\mathscr{M} = \mathscr{M}'$ unless $\theta\alpha$ is positive and α is either complex or imaginary for θ. In the first case we have (with obvious notation) $\mathscr{M}' = s.\mathscr{M}$. In the second case we must have $\langle \alpha^\flat, \xi \rangle = \langle \eta, (\alpha^\vee)^\sharp \rangle = 0$ (see 4.18, the notations are as in no. 1). We call the equivalence classes of bimodels *blocks*.

Lemma 5.9. *If $\sigma \in \Sigma$ is equivalent with $\pi\mathscr{M}$ in Σ there is a bimodel $\mathscr{M}'$ with $\mathscr{M}' \sim \mathscr{M}$ and $\pi\mathscr{M}' = \sigma$.*

It suffices to prove this in the case that the equivalence in Σ is elementary. Let $s = s_\alpha$ ($\alpha \in D$) be the simple reflection which is involved. Assume that α is imaginary for θ, with $\langle \alpha^\flat, \xi \rangle = 0$. An $\mathscr{M}' = (\Psi_0, \theta', \xi', \eta')$ as required must have $\theta' = s\theta$, $\xi' = m(s).\xi$ (as in 3.9). By 1.13 (i), $\eta \in H_-(\theta, P(R))$ is the image of at most two elements of a subgroup of $H_-(s\theta, P(R))$. If we take η' to be one of these elements, $\mathscr{M}'$ is as required. We have dealt with one possibility for α. The other cases are easier and are omitted.

It follows from the definitions that we have a unique action of the Weyl group W on the set of bimodels, such that $\pi(w.\mathscr{M}) = w.\pi\mathscr{M}$, $\pi^\vee(w.\mathscr{M}) = w.\pi^\vee\mathscr{M}$ ($w \in W$).

Lemma 5.10. *Blocks are W-stable.*

Since W acts on the sets V of 2.2, the equivalence classes in Σ of 4.18 are W-stable. This implies the lemma.

5.11. Denote by $\overline{G}$, respectively $\overline{G^\vee}$ an adjoint semi-simple group with root system R, respectively $R^\vee$.

Fix a block $\mathbf{B}$ of bimodels. By the observations made in 4.18, using 5.9, it follows that there is an involution θ of $\overline{G}$ such that $\pi\mathbf{B}$ is isomorphic to a set V as in 2.2, relative to a Borel group and maximal torus of $\overline{G}$ (which we fix). Dually, $\pi^\vee\mathbf{B}$ is isomorphic to a set $V^\vee$, relative to an involution $\theta^\vee$ of $\overline{G^\vee}$ (and similar data).

We say that $\mathscr{M} \in \mathbf{B}$ is *maximal* if $\pi\mathscr{M}$ is maximal (see 4.18) and *minimal* if $\pi^\vee\mathscr{M}$ is maximal.

We have introduced in 4.15 an action of W of Σ. If θ is an involution of R, the group W^θ (1.1) acts on the set of (θ, ξ) in Σ, i.e. it acts on $H_+(\theta, P(R^\vee))$. The structure of W^θ is described in 4.3. The groups W_c and W_r of loc.cit. act trivially on $H_+(\theta, P(R^\vee))$, as follows from the proof of 4.7. By 4.3 the action is really one of the group $W_i = W_i(\theta)$ of loc.cit. Dually, $W_r(\theta)$ acts on $H_-(\theta, P(R))$.

Proposition 5.12.   (i) $\mathbf{B}$ *contains maximal and minimal elements;*

(ii) *If $\mathscr{M} = (\Psi_0, \theta, \xi, \eta)$ is maximal (minimal) then the element (θ, ξ) (respectively, $(-\theta, \eta)$) of Σ is uniquely determined.*

This follows from the fact that the sets V and $V^{\vee}$ of 5.11 have unique maximal elements, using 5.9.

5.13. If θ is an involution of R there exist a unique factorization $\theta = w\iota$, where ι is an involution stabilizing D and $w \in W$. Let $\mathscr{I} = \mathscr{I}_{\iota} = \{a \in W \; \iota(a) = a^{-1}\}$, the set of twisted involutions for ι. Then $w \in \mathscr{I}$. We have $-\theta = (ww_D)(\iota_0 \iota)$, where ι_0 is the opposition involution and w_D is the longest element of W (see 1.2).

On $\mathscr{I}$ we have the *weak order* $\vdash$, generated by the elementary relations $a \vdash b$, where $a < b$ and either $b = sa(\theta s)$ or $b = sa = a(\theta s)$ (see [5, 3.17]).

Lemma 5.14. *$a \mapsto aw_D$ defines a bijection of $\mathscr{I}_{\iota}$ onto $\mathscr{I}_{\iota_0 \iota}$ which reverses the weak order.*

This is a weak version of [loc.cit., 8.20]. The proof is straightforward.

Let $\mathscr{M} = (\Psi_0, \theta, \xi \eta) \in \mathbf{B}$. The involution ι defined by θ is an invariant of $\mathbf{B}$ (as one sees by passing to a set V). We define $\phi(\mathscr{M}) = w$ where w is as above. Let w_{max} ($w^{\vee}_{max}$) be $\phi\mathscr{M}$ (respectively, $\phi\mathscr{M}^{\vee}$), where $\mathscr{M}$ is maximal (respectively, minimal). Put $w_{min} = w^{\vee}_{max} w_D$.

The following result contains a characterization of the elements of the block $\mathbf{B}$.

Theorem 5.15.   (i) *For $\mathscr{M} \in \mathbf{B}$ we have $w_{min} \vdash \phi\mathscr{M} \vdash w_{max}$. In particular,*
$w_{min}w_D \vdash w_{max}$;
(ii) *Let $a \in \mathscr{I}$. There exists $\mathscr{M} \in \mathbf{B}$ with $\phi\mathscr{M} = a$ if and only if $w_{min} \vdash a \vdash$*
w_{max};
(iii) *If θ is given, the $(\xi, \eta) \in H_+(\theta, P(R^{\vee})) \times H_-(\theta, P(R))$ such that $\mathscr{M} = (\Psi_0, \theta, \xi, \eta) \in \mathbf{B}$ form a $W_i(\theta) \times W_r(\theta)$-orbit.*

Let $\mathscr{M} \in \mathbf{B}$. By [5, 7.13], applied to the sets V and $V^{\vee}$ of 5.10, we have $\phi\mathscr{M} \vdash w_{max}$, $\phi\mathscr{M}^{\vee} \vdash w^{\vee}_{max}$. The latter relation is equivalent to $w_{min} \vdash \phi\mathscr{M}$. This proves (i).

Let $a \in \mathscr{I}$ and put $\theta = a\iota$, an involution of R. It follows from [loc.cit.] that there exist $\xi \in H_+(\theta, P(R^{\vee}))$ and $\eta \in H_-(\theta, P(R))$ such that $(\theta, \xi) \in \pi\mathbf{B}$ and $(-\theta, \eta) \in \pi^{\vee}\mathbf{B}$. Then $(\Psi_0, \theta, \xi, \eta) \in \mathscr{M}$. This proves (ii).

Let $(\Psi_0, \theta, \xi, \eta)$ and $(\Psi_0, \theta, \xi', \eta')$ lie in $\mathbf{B}$. It follows that the elements (θ, ξ) and (θ, ξ') of Σ are equivalent. Let v, v' be the corresponding elements of V. They satisfy the assumption of [5, 2.5]. Hence there is $w \in W$ with $v' = w.v$. It follows that there is $w \in W$ such that $(\theta, \xi') = w.(\theta, \xi)$. Then $w \in W^{\theta}$. Using 4.3 and observing that the groups W_r and W_c of loc.cit. act trivially (see the proof of 4.7) we conclude that $\xi' \in W_i.\xi$. Similarly, $\eta' \in W_r.\eta$. Now (iii) follows from 5.10.

The results of 5.15 are weak analogues of some of the results of [12, no. 8]. There is a strong version of 5.15, dealing with blocks of pairs $(v, \mathscr{L})$ as in 5.1. The two versions would coincide in the case of a simply connected semi-simple group whose center is of odd order. We shall not go into these matters here.

REFERENCES

1. N. Bourbaki, *Groupes et algèbres de Lie*, Hermann, 1968, Chap. 4,5,6.
2. H. Cartan and S. Eilenberg, *Homological algebra*, Princeton, 1956.
3. G. Lusztig and D. A. Vogan, *Singularities of closures of K-orbits on flag manifolds*, Inv. Math. **71** (1983), 365–379.
4. R. W. Richardson, *Orbits, invariants and representations associated to involutions of reductive groups*, Inv. Math. **66** (1982), 287–312.
5. R. W. Richardson and T. A. Springer, *The Bruhat order on symmetric varieties*, Geom. Dedic. **35** (1990), 389–436.
6. _____, *Complements to "The Bruhat order on symmetric varieties"*, Geom. Dedic. **49** (1994), 231–238.
7. T. A. Springer, *Linear algebraic groups*, Birkhäuser, 1981, 2nd ed.
8. _____, *Some results on algebraic groups with involutions*, Adv. Studies in Pure Math., vol. 6, Kinokuniya/North-Holland, 1985, pp. 525–543.
9. _____, *The classification of involutions of simple algebraic groups*, J. Fac. Sci. Univ. Tokyo (sect.IA) **34** (1987), 655–670.
10. R. Steinberg, *Endomorphisms of linear algebraic groups*, Mem. Amer. Math. Soc. **80** (1968).
11. D. A. Vogan, *Representations of real reductive Lie groups*, Birkhäuser, 1981.
12. _____, *Irreducible characters of semisimple Lie groups IV. Character-multiplicity duality*, Duke Math. J. **49** (1982), 943–1073.
13. _____, *Irreducible characters of semisimple Lie groups III. Proof of the Kazhdan-Lusztig conjectures in the integral case*, Inv. Math. **71** (1983), 381–417.

MATHEMATISCH INSTITUUT, UNIVERSITEIT UTRECHT, BUDAPESTLAAN 6, 3584 CD UTRECHT, NETHERLANDS

NAGATA'S EXAMPLE

ROBERT STEINBERG

To the memory of my friend Roger Richardson

1.

At the International Congress of Mathematicians held in Edinburgh in 1958 M. Nagata [12] presented an example of a group acting linearly on a finite-dimensional vector space such that algebra of polynomials on the space invariant under the group is not finitely generated. He thereby answered in the negative Problem 14 of Hilbert's famous list presented to the Congress of 1900. Our object in this article is to present other examples which are simpler and easier to establish than Nagata's and also yield a better result. We hasten to add that our development is close to his, with one twist which produces the improved examples.

Here is our first example.

Theorem 1.1. *Let k be a field of characteristic 0 and $V_2 = k^2$, viewed as a vector space over k. Let a_i $(1 \le i \le 9)$ be distinct elements of k such that $\Sigma a_i \ne 0$. Let G_a denote the additive group of k. Let G_a^9 act on $V = V_2^9$, given the coordinates (x_i, t_i) $(1 \le i \le 9)$, thus: $t_i \to t_i$ and $x_i \to x_i + c_i t_i$ $(1 \le i \le 9)$ for each $(c_1, c_2, \ldots, c_9)$ in $k^9 \simeq G_a^9$. Finally, let $G \simeq G_a^6$ be the subgroup of G_a^9 consisting of all $(c_1, c_2, \ldots, c_9)$ such that*

(1) $\Sigma c_i = 0$,
(2) $\Sigma a_i c_i = 0$,
(3) $\Sigma a_i^3 c_i = 0$.

Then $k[V]^G = k[x_1, t_1, x_2, t_2, \ldots]^G$, the algebra of polynomials on V invariant under G, is not finitely generated.

In Nagata's example (his first one, given in [12]) 16 copies of V_2 figure rather than 9, and accordingly the group acting is G_a^{13} rather than G_a^6.

Following Nagata, we prove instead of Theorem 1.1 the following result.

Theorem 1.2. *Enlarge G in Theorem 1.1 to $H = GT$ with T the torus $\{(d_1, d_2, \ldots, d_9) \in k^{*9} \mid \Pi d_i = 1\}$ acting on V via $x_i \to d_i x_i$, $t_i \to d_i t_i$ $(1 \le i \le 9)$. Then $k[V]^H$ is not finitely generated.*

As will be seen below the proof of 1.2 to be given becomes one of 1.1 once some obvious changes are made. Alternatively, one can note that 1.1 follows from 1.2 and the fact that if a torus acts on an algebra which is finitely generated, then the algebra of invariants is also finitely generated (see, e.g., Ch. 2 of [17] for this).

To get an example in char $k \ne 0$ we choose the a_i's in 1.1 so that Πa_i is not 0 or any nth root of 1 and replace a_i^3 in (3) by $a_i^2 - a_i^{-1}$. As is easily seen, such a_i's exist

375

if k is any infinite field which is not locally finite, i.e., not an algebraic extension of a finite field, hence, in particular, if char $k = 0$. In this sense our second example is more general than our first. However, both examples are contained in a more general example which we now explain.

As is known (for this, see Ch. 5 of [6]), the nonsingular points of an irreducible cubic curve C in the projective plane form an Abelian group A in which every line not containing any singular point meets C in 3 points (counting multiplicities) whose sum in A is 0. The last condition requires only that one of the flexes of C (whose coordinates need not lie in k) is chosen as the neutral element of A. We choose one of them and call A "the" group of C. For example, if C is $y - x^3$, in affine coordinates, then the only singular point is the point at infinity in the direction of the y-axis, and every line not through this point has the form $y - (ax + b)$. Since such a line always meets C in 3 points for which the sum of the x-coordinates is 0 (since they satisfy $x^3 - (ax + b)$), it follows that A is, in this case, the additive group of k. Similarly, for $y - (x^2 - x^{-1})$, or, equivalently, $xy - x^3 + 1$, the product of the x-coordinates is 1, so that A is the multiplicative group of k. Thus our two examples are contained in the following result.

Theorem 1.3. *In the construction of G in Theorem 1.1 let the 9 points $(1 : a_i : a_i^3)$ used in the conditions (1), (2) and (3) be replaced by 9 nonsingular points on an arbitrary irreducible cubic curve. If the sum of these points has infinite order in the group of the curve, then the conclusions of Theorem 1.1 and 1.2 remain true.*

Other examples have been given (see [1], [4], [10, 12, 13], [14], [15], [16]), some of them for other versions of Hilbert's problem. The one by A' Campo [1], based on Roberts [16], is quite simple. The group there is G_a^{12} and the field must be of characteristic 0. Also, many positive results have been obtained; for example, for finite or reductive groups the algebra of invariants is always finitely generated. For discussions of some of these matters we refer the reader to the survey articles [8] and [9].

In our main development we establish our first two examples because this can be done using only elementary algebra, and then we indicate the changes needed for Theorem 1.3. This is done in the next three sections. In Section 5 we present yet other examples which involve reducible cubic curves and figure in a converse to our main lemma. In a final section we consider some problems suggested by our development.

2.

In this section the two main lemmas are presented and Theorem 1.2 is deduced from them. The lemmas themselves are proved in the following two sections.

Lemma 2.1. (Nagata) *With the notations as in Theorems 1.1 and 1.2, set $t = \Pi t_i$, $z_1 = \Sigma x_i t/t_i$, $z_2 = \Sigma a_i x_i t/t_i$, $z_3 = \Sigma a_i^3 x_i t/t_i$. Then t, z_1, z_2 and z_3 are algebraically independent over k and each of them is invariant under H. Further $k[V]^H$ consists of the elements of $k[z_1, z_2, z_3, t, t^{-1}]$ that are expressible as sums of* *elements of the form* $f(z_1, z_2, z_3)/t^m$ *in which* f *is a nonzero homogeneous polynomial such that the corresponding plane projective curve has multiplicity at least* m *at each of the points* $P_i = (1 : a_i : a_i^3)$ $(1 \leq i \leq 9)$.

This elementary lemma, which relates the structure of $k[V]^H$ to an interpolation problem in the projective plane, is the heart of Nagata's method. Actually, he proved a more general result which will be formulated and proved in the next section.

Lemma 2.2. *Let* k *be a field of characteristic* 0 *and* $a_1, a_2, \ldots, a_9$ *distinct elements of* k *such that* $\Sigma a_i \neq 0$. *For each* i *let* P_i *be the point* (a_i, a_i^3) *on the plane affine cubic curve* $f_0 : y - x^3$.

(a) *For each* $m \geq 0$ *there exists, up to scalar multiplication, a unique curve (or polynomial) of degree* $\leq 3m$ *with multiplicity* $\geq m$ *at every* P_i, *namely,* f_0^m, *the cubic counted* m *times.*

(b) *For every* $d \geq 3m$ *the multiplicity conditions of* (a) *are linearly independent on the space of all polynomials of degree* $\leq d$.

(c) *There exists a polynomial of degree* $3m + 1$ *which has multiplicity* $\geq m$ *at every* P_i *and is not divisible by* f_0.

It is here that we have deviated from Nagata in his development in [10] and [12]. He uses instead of 2.2 at this point the result that if k has a large enough transcendence degree over some subfield and if $r \geq 4$, then, for each $m \geq 1$, there does not exist a curve of degree rm with multiplicity $\geq m$ at each of r^2 general points of the plane. Nagata's ingenious proof of this is a tour de force but the results from algebraic geometry that he uses are by no means elementary. Since his result is false for $r = 3$, his minimal example has $r = 4$ and $G = G_a^{16-3} = G_a^{13}$.

In order to be able to combine these two lemmas we imbed $\mathbb{A}^2$ in $\mathbb{P}^2$ and match up the coordinates so that $x = z_2/z_1$ and $y = z_3/z_1$, and accordingly match up the nonzero polynomials in x and y of degree $\leq d$ with the homogeneous ones in z_1, z_2 and z_3 of degree exactly d via $f(x, y) \leftrightarrow z_1^d f(z_2/z_1, z_3/z_1)$. The point P_i of 2.2 gets identified with that of 2.1, and the cubic $y - x^3$, viewed as a polynomial of degree ≤ 3, takes on the homogeneous form $z_1^2 z_3 - z_2^3$.

We now give the proof of Theorem 1.2. Assuming that $k[V]^H$ is finitely generated, we shall come to a contradiction. By Lemmas 2.1 and 2.2 there exists a finite generating set of the form $S = \{f_0/t, f_1/t^{m_1}, \ldots, f_r/t^{m_r}\}$, where $f_0 = z_1^2 z_3 - z_2^3$ is our basic cubic and each later generator has the form specified by Lemma 2.1, and where it can be further arranged that no f_j $(j \geq 1)$ is divisible by f_0. Set $d_j = \deg f_j$. Then $d_j > 3m_j$, so that $d_j - 3m_j$ is positive, for every $j \geq 1$. This is so for $d_j > 0$ by Lemma 2.2(a) and also for $d_j = 0$ since then $m_j \leq 0$ and we can rule out the case of equality by dropping every generator that is a constant. Let m be a positive integer larger than every m_j $(j \geq 1)$. By Lemma 2.2(c) there exists a polynomial f of degree $3m + 1$ which has multiplicity $\geq m$ at every P_i and is not divisible by f_0. We show that f/t^m is not expressible as a polynomial in the elements of S and thus reach the desired contradiction. Assume that such a polynomial expression does exist. Then there exists one such that all of the terms appearing in it have the same degree in $\{z_1, z_2, z_3\}$, and also in t, as does f/t^m.

This matching of degrees makes sense because t, z_1, z_2 and z_3 are algebraically independent over k. Since f_0 does not divide f there exists at least one of the terms, $c\Pi(f_j/t^{m_j})^{e_j}$, which does not involve f_0/t. Equating the degrees of this term and of f/t^m, we get

(a) $\Sigma d_j e_j = 3m + 1$ (degree in $\{z_1, z_2, z_3\}$)

(b) $\Sigma m_j e_j = m$ (degree in t)

The combination   (a) - 3(b) yields

(c) $\Sigma(d_j - 3m_j)e_j = 1$.

Since every $d_j - 3m_j$ is positive, as noted above, it follows from (c) that exactly one e_j, say e_{j_0}, is 1 while all the others are 0, and then from (b) that $m = m_{j_0}$. This contradiction to the choice of m shows that $k[V]^H$ is not finitely generated and thus completes the proof that Theorem 1.2 follows from Lemmas 2.1 and 2.2.

3.

In this section we present a proof of Lemma 2.1 which is essentially Nagata's own but is changed in one or two places so that it can be given entirely in the language of elementary algebra. Because of the utility of this result we prove it in full generality. Thus k is replaced by an arbitrary infinite field and the points P_i $(1 : a_i : a_i^3)$ by an arbitrary number n of points P_i $(a_{i1} : a_{i2} : a_{i3})$ not all on one line, a condition which holds for the original points since at most 3 points of the cubic can lie on one line. The conditions (1), (2) and (3) defining G in Theorem 1.1 and the definition of the z_j's given in Lemma 2.1 are then changed accordingly.

First, it is routine to verify that t, z_1, z_2 and z_3 are invariant under H. For example, if the element $(c_1, c_2, \ldots, c_n)$ of G acts on $z_j = \Sigma a_{ij} x_i t/t_i$ the result is $\Sigma a_{ij}(x_i + c_i t_i)t/t_i$, which is again z_j in view of condition (j) of Theorem 1.1.

The rest of the proof is given in three steps.

(1) It is shown first that t, z_1, z_2 and z_3 are algebraically independent over k and that $k[V]^H = k[V] \cap k[z_1, z_2, z_3, t, t^{-1}]$.

Since the P_i's are not all on one line we can, by interchanging P_3 and one of the later P_i's if necessary, arrange that P_1, P_2 and P_3 are not on a line and thus that the matrix of their coordinates, (a_{ij}) $(1 \le i, j \le 3)$, is invertible. It follows that if we set $w_i = x_i t/t_i$ $(1 \le i \le n)$ so that $z_j = \Sigma a_{ij} w_i$ $(1 \le j \le 3)$, then the latter equations can be solved for w_1, w_2 and w_3 as linear functions of the other w_i's and z_1, z_2 and z_3. If we then use the definition of the w_i's we get that $k[V][t^{-1}]$ is the polynomial ring over $k[t^{-1}]$ in the variables z_1, z_2, z_3, $x_4, \ldots, x_n$, $t_1, t_2, \ldots, t_n$. Now for an arbitrary c in k the equations (1), (2) and (3) of 1.1 have a solution with $c_4 = c$ and $c_i = 0$ for every $i > 4$, because, once these values are assigned, (1), (2) and (3) can be solved for c_1, c_2 and c_3, again by the invertibility of the above matrix. If the corresponding element of G acts on an element of $k[V][t^{-1}]$, then only the variable x_4 in the above list changes and it goes to $x_4 + ct_4$. Since c takes on infinitely many values here it follows that if the element is invariant then it does not involve x_4; and similarly for $x_5, \ldots, x_n$. Thus we see that $k[V][t^{-1}]^G = k[z_1, z_2, z_3, t_1, t_2, \ldots, t_n, t^{-1}]$, and, if we then make T act, that

$k[V][t^{-1}]^H = k[z_1, z_2, z_3, t, t^{-1}]$, and finally that $k[V]^H = k[V] \cap k[z_1, z_2, z_3, t, t^{-1}]$. Also, since k is infinite, the x_i's are algebraically independent over $k[t_1, t_2, \ldots, t_n]$, hence so are z_1, z_2, z_3, $x_4, \ldots, x_n$ by the above, and thus t, z_1, z_2 and z_3 are algebraically independent over k.

(2) It is next shown that $k[V]^H$ is spanned by its elements of the form $f(z_1, z_2, z_3)/t^m$, in which f is a homogeneous polynomial in z_1, z_2 and z_3 which is divisible by t^m in $k[V]$.

If F is any element of $k[V]^H$, then by (1) it may be written as a Laurent polynomial in t with coefficients which are polynomials in z_1, z_2 and z_3 : (*) $F = \Sigma f_m(z_1, z_2, z_3)/t^m$. Assume that F is homogeneous in the union of the x_i's and the t_i's. Since, from the definitions, each z_j is homogeneous in the x_i's and also in the t_i's, of degrees 1 and $n-1$, respectively, and t is homogeneous in the t_i's, of degree n, it follows that each term $p z_1^a z_2^b z_3^c$ of f_m in (*) satisfies $d = (a+b+c)n - mn$. This yields $a + b + c = m + d/n$, which depends only on m. Thus f_m is a homogeneous polynomial in the z_j's, of degree $m + d/n$. By the above f_m is also homogeneous of this degree when viewed as a polynomial in the x_i's, with coefficients polynomials in the t_i's. Since also this degree changes as m does it follows that (*) gives the decomposition of F into its x-homogeneous components. Finally, since F is in $k[V]$ so is each of these components, which proves (2).

(3) By (2) it remains to show that a nonzero homogeneous polynomial $f(z_1, z_2, z_3)$ is divisible by t^m in $k[V]$ if and only if (the curve in $\mathbb{P}^2$ represented by) f has multiplicity at least m at all n of the points P_i.

Consider first the point P_1 which by a projective change of coordinates (e.g., $z_1' = z_1/a_{11}$, $z_2' = a_{11}z_2 - a_{12}z_1$, $z_3' = a_{11}z_3 - a_{13}z_1$ in case $a_{11} \neq 0$) may be taken to be $(1 : 0 : 0)$. Then (from the definition of the z_j's) (*) $z_1 = x_1t/t_1 + t_1u_1$, $z_2 = t_1u_2$, $z_3 = t_1u_3$ with $u_j = a_{2j}x_2t/t_1t_2 + a_{3j}x_3t/t_1t_3 + \cdots$ for $j = 1$, 2 and 3. Observe that t_1 is not involved in any u_j. Since P_1, P_2 and P_3 are not all on one line, the sums for u_2 and u_3 above are linearly independent in the terms involving x_2 and x_3, and hence they are algebraically independent over $k(t_2, \ldots, t_n)$. Now if $f(z_1, z_2, z_3)$ is expanded as $f_m(z_2, z_3)z_1^{d-m} + f_{m+1}(z_2, z_3)z_1^{d-m-1} + \cdots$ with $f_m \neq 0$ and each f_k homogeneous of degree k, then $f = t_1^m f_m(u_2, u_3) + t_1^{m+1}(\cdot)$ by (*) above, and $f_m(u_2, u_3) \neq 0$ by the algebraic independence just noted. Thus if m is the multiplicity of f at P_1 then the highest power of t_1 dividing f in $k[V]$ is t_1^m; and similarly for every P_j and t_j since in the above development P_1 can be any of the n given points. Since $t = \Pi t_i$, (3) is proved, and with it Lemma 2.1.

4.

In this section we prove Lemma 2.2, thus establishing our first example, and then consider our second example and Theorem 1.3.

We are again in the affine plane and x and y are the coordinates, and now the multiplicity of a curve, or of a polynomial $f(x, y)$ representing it, at a point P is the degree of the first nonzero term in the (finite) power series expansion of f around P. The following facts will be used repeatedly.

Over an arbitrary infinite field the (linear) space of all polynomials (which we shall also call curves) of degree $\leq d$ has dimension $\binom{d+2}{2}$ (binomial coefficient). The (linear) conditions that specify that a curve of degree $\geq m - 1$ should have multiplicity $\geq m$ at a given point P are independent and in number equal to $\binom{m+1}{2}$.

The first formula comes from counting the monomials in $f = \Sigma c_{ij} x^i y^j$ $(i+j \leq d)$ (one of degree 0, two of degree 1, and so on), while the second counts the number of coefficients in the power series expansion around P which must be set equal to 0 to achieve the required multiplicity (e.g., all c_{ij} $(i + j < m)$ in case $P = (0, 0)$).

We turn now to the proof of Lemma 2.2.

(a) Let $f(x, y) = c_0(x)y^{3m} + c_1(x)y^{3m-1} + \cdots + c_{3m}(x)$ have degree $\leq 3m$ and multiplicity $\geq m$ at every $P_i = (a_i, a_i^3)$. Here c_j is, for each j, a polynomial of degree $\leq j$. The substitution $y \to x^3$ converts f into a polynomial in x of degree $\leq 9m$ which is just the remainder when f is divided by $y - x^3$. This polynomial has multiplicity $\geq m$ at every a_i because f has multiplicity $\geq m$ at P_i and, in terms of the coordinates there, $y - a_i^3$ has been replaced by $x^3 - a_i^3$ which is a multiple of $x - a_i$. Thus it is divisible by $\Pi(x - a_i)^m$, whose degree is exactly $9m$. We thus have the equation

$$c_0 x^{9m} + c_1(x)x^{9m-3} + \cdots + c_{3m}(x) = c_0 \Pi(x - a_i)^m. \qquad (*)$$

The coefficient of x^{9m-1} on the left is 0 since $dg\ c_j \leq j$ for every j, while it is $-c_0 m \Sigma a_i$ on the right, if we assume that $m > 0$. Since char $k = 0$, and $\Sigma a_i \neq 0$ by assumption, we get $c_0 = 0$. Thus the left side of $(*)$, i.e., the remainder when f is divided by $y - x^3$, is 0, so that $f_0 = y - x^3$ divides f. Since f_0 has multiplicity exactly 1 at every P_i, f/f_0 has degree $\leq 3(m - 1)$ and multiplicity $\geq m - 1$ at every P_i, and (a) follows by induction, the case $m = 0$ being obvious.

(b) The polynomials of degree $\leq 3m$ form a space of dimension $\binom{3m+2}{2} = (9m^2 + 9m + 2)/2$, while the multiplicity conditions at the 9 points number $9\binom{m+1}{2} = (9m^2 + 9m)/2$, which is exactly 1 less. Thus the conditions are independent on the space if and only if they define a subspace of dimension 1 exactly, and that is the case by part (a). Since for every $d \geq 3m$ the polynomials of degree $\leq 3m$ are included in those of degree $\leq d$, the conditions remain independent on the space of polynomials of degree $\leq d$. (Linear functions on a vector space are linearly independent if their restrictions to some subspace are so.)

(c) The space of polynomials of degree $\leq 3m+1$ with multiplicity $\geq m$ at every P_i has, by (b), the dimension $\binom{3m+3}{2} - 9\binom{m+1}{2}$, which is equal to $3m + 3$. The subspace of those that are also divisible by f_0 has the dimension $3m$, obtained by replacing m by $m - 1$ in the preceding formula. Therefore there exists a polynomial in the space and not in the subspace (in fact, a punctured 3-dimensional subspace of such polynomials), whence (c).

This completes the proof of Lemma 2.2 and hence also of Theorem 1.2.

Our second example, the one mentioned soon after Theorem 1.2 in Section 1 can be established in essentially the same way. The only proof that has to be changed is that of (a) given above. If we replace y by $x^2 - x^{-1}$ instead of by x^3 and then multiply by x^{3m} to clear away the negative powers of x, we get an equation like $(*)$ which yields $c_0((\Pi a_i)^m - 1) = 0$ and thus that $c_0 = 0$ because of the assumptions made on the a_i's, and that is all that is needed.

We turn now to the proof of Theorem 1.3. Again the only proof that has to be changed is that of Lemma 2.2(a). Let C be the curve and A its group of nonsingular points. We use the following known result: (*) If G is a curve of degree n $(n \geq 1)$ which does not contain C or any of its singular points then the intersection $G.C$ (in which each point is counted with its intersection multiplicity) consists of $3n$ points whose sum in A is 0. Assume the result (*) for a moment. Let F be a curve of degree $3m$ with multiplicity $\geq m$ $(m \geq 1)$ at each of the 9 given points. To prove our required result by induction as before it is enough to show that F contains C as a component. Suppose that it does not. By Bezout's Theorem $F.C$ consists of $3m.3 = 9m$ points, while by the multiplicity assumptions $F.C$ contains each of the 9 points at least m times. Thus $F. C$ consists of the 9 points, each counted exactly m times. By (*) this yields that the sum of these points in A has finite order ($\leq m$), in contradiction to one of our assumptions. It remains to prove (*). Since F is a line if $n = 1$, we assume $n \geq 2$ and start with the case n even. Set $F.C = S$. Partition the $3n$ points of S into $3n/2$ pairs. Let G be the union of the lines through these pairs, a curve of degree $3n/2$, and T the set of third points in their intersections with C. We have $G.C = S \cup T$. By Max Noether's Fundamental Theorem it follows from $G.C \supset F.C$ that there exists a curve E of degree deg $G -$ deg $F = n/2$ such that $E.C = T$, the residual intersection. By induction the sum of the points of T is 0, and because G is a union of lines the sum of those of $S \cup T$ is also; thus the same holds for those of S, as required. If n is odd, the same argument with $(3n - 1)/2$ lines through pairs of points of S and one extra line through the remaining point produces a curve E of degree $(n + 1)/2$. Thus we are done. For an account of the results about plane curves used here we refer the reader to Chapter 5 of Fulton [6].

5.

It does not seem to be widely known that reducible cubic curves also have groups attached to them. This fact leads to other negative examples for Hilbert's problem and to a converse of our main interpolation result, Lemma 2.2. We start with the examples.

Examples 5.1. If each of the sets of 9 points below is used in Lemma 2.2 and Theorem 1.1 then the conclusions there remain true.

(a) (char $k = 0$) 3 points on each of the lines $y - 1$, y and $y + 1$ in the affine plane, $(a_i, 1)$, $(b_j, 0)$ and $(c_k, -1)$, such that $\Sigma a_i + \Sigma c_k - 2\Sigma b_j \neq 0$.

(b) 3 points on each of the coordinate lines z_1, z_2 and z_3 of the projective plane, $(0 : 1 : -a_i)$, $(-b_j : 0 : 1)$ and $(1 : -c_k : 0)$ such that $\Pi a_i \cdot \Pi b_j \cdot \Pi c_k \neq 0$ or any root of 1.

(c) (char $k = 0$) 6 points (a_i, a_i^2) on the parabola $y - x^2$ and 3 points at infinity (on the lines with slopes) m_j such that $\Sigma a_i - \Sigma m_j \neq 0$.

(d) 6 points (a_i, a_i^{-1}) on the hyperbola $xy - 1$ and 3 points m_j at infinity such that $\Pi a_i \cdot \Pi m_j \neq 0$ or any root of 1.

(e) ($k = \mathbb{R}$) 6 points on the unit circle $x^2 + y^2 - 1$ in the Euclidean plane and 3 points at infinity specified by their angular coordinates α_i (mod 2π) and

β_j (mod π), all measured from the positive x-axis, such that $\Sigma\alpha_i - 2\Sigma\beta_j$ is not a rational multiple of π.

The interested reader is invited to establish (a) by restricting the polynomial $f(x,y)$ used in the proof of 2.2(a) in Section 4 to the lines $y-1$, y and $y+1$ and comparing the results; and similarly for (b) and the lines z_1, z_2 and z_3.

Proposition 5.2. *Let C be a cubic curve without any multiple component. Then, with the singular points of C removed, the various irreducible components of C can be made into algebraic groups which can be identified by suitable isomorphisms so that* (*) *every transversal meets C in 3 points whose sum is 0.*

Here and later, "transversal" means a line not through any singular point of C. The group involved in 5.1(a), for example, is the additive group since every transversal meets the cubic $y^3 - y$ in points $(a,1)$, $(b,0)$ and $(c,-1)$ such that the weighted sum $a + c - 2b$ is 0. Similarly in (b) we have the multiplicative group since the points $(0:1:-a)$, $(-b:0:1:1)$ and $(1:-c:0)$ on the cubic $z_1 z_2 z_3$ are in a line if and only if $abc = 1$. (A version of this in Euclidean geometry is known as Menelaus' Theorem.) This actually proves 5.2 in case C is the union of 3 lines since then C can, by appropriate choice of coordinates, be put in the form $y^3 - y$ or $z_1 z_2 z_3$ according as the lines are concurrent or not. To slightly shorten the discussion we have assumed in the first case that char $k \neq 2$. Similarly (c) and (d) take care of the cases in which C is the union of a conic Q and a line L which meets Q in 1 or 2 distinct points. Left, among the reducible cubics, are those for which L fails to meet Q. An example occurs in (e) where the group is the circle group. In general, when Q contains rational points, we get the group of units of a definite quadratic form $x^2 + axy + by^2$ over k.

We now given an outline of a proof of 5.2 in which all cubic curves are treated together. Let C' be the set of nonsingular points of C. Let Z be the additive group of divisors $\Sigma n_i P_i$ ($n_i \in \mathbb{Z}$, $P_i \in C'$) on C' such that the sum of the n_i's for every irreducible component of C' is 0. Let B be the subgroup generated by all divisors $C \cdot L_1 - C \cdot L_2$ with L_1 and L_2 transversals to C. Let $D = Z/B$. The key point of our development is that if P_0 is any point of C' then every element of D has a unique representation $P - P_0$ with P a point of the irreducible component of C' that contains P_0. The uniqueness is proved as follows. Let $P' - P_0$ be another representation. Then, by the definitions, there exist curves F and F', each the union of n transversals, for some n, such that the divisors $F \cdot C + P$ and $F' \cdot C + P'$ are equal. Let L be any transversal through P, meeting C again in Q and R. Then $(F \cup L) \cdot C = F \cdot C + P + Q + R = F' \cdot C + P' + Q + R \supset F' \cdot C$. By Max Noether's Theorem there exists a line L' containing the residual intersection $P' + Q + R$. Since P' is the third point of intersection of QR with C, and P is also, it follows that $P' = P$. We next choose a transversal L_0 which meets each component of C' in a unique point. (If C is irreducible, for example, then L_0 must be the tangent at a flex, while if C is the union of 3 lines then L_0 can be any transversal.) We then make every component A of C' into a group thus: if Q and R are points of A and P_0 is the unique point of L_0 in A, then $Q + R$ is defined to be the point S of A such that $(Q - P_0) + (R - P_0)$ is equivalent to $S - P_0 \mod B$.

This group is seen to be isomorphic to D via the map $Q \to Q - P_0$ ($Q \in A$). We thus get isomorphisms of the groups of the various components of C' with D and hence with each other for which the condition (*) of 5.2 holds because, for any transversal L, the divisor $C \cdot L - C \cdot L_0$ represents O in D.

This brings us to our main interpolation result.

Theorem 5.3. *For the distinct points P_i ($1 \le i \le 9$) in the projective plane, the conditions* (a) *and* (b) *below are equivalent.*

(a) *For every $m \ge 1$ there exists a unique curve of degree $\le 3m$ with multiplicity $\ge m$ at every P_i, i.e., the multiplicity conditions being imposed here are linearly independent.*

(b) *There exists a cubic curve C with the following properties.*

   (1) *C has no multiple component.*
   (2) *C contains every P_i as a nonsingular point.*
   (3) *The number of P_i's on any irreducible component A of C is 3 times the degree of A.*
   (4) *ΣP_i is an element of infinite order of the group of C.*

Because of 5.2 the argument in the proof of Theorem 1.3 given at the end of Section 4, slightly extended, can be used to prove that (b) implies (a). Since (a) implies the conclusions of Theorem 1.1, the five examples of 5.1 are thus established. The proof that (a) implies (b), which proceeds mainly by contradiction, is omitted.

6.

We conclude with some problems. The first one is to decide whether the group G_a^6 of our examples can be made even smaller. In char 0 it is known (Weitzenbock's Theorem, proved in [18] and in [5]) that the group $G_a = G_a^1$ always produces a finitely generated algebra of invariants, but in char $p \ne 0$ not even this is known. There is also the problem of finding a negative example over every infinite locally finite field or showing that one does not exist. One can not expect one of the type presented above since the group of every cubic curve over a locally finite field is itself locally finite.

Finally, the geometric results discussed in our development focus attention on the following fundamental problem, hardly a new one. Find the dimension of the space of all polynomials (or curves) of a given degree with prescribed multiplicities at the points of a given finite set in general position in the plane, thus also determine if there is a curve, i.e., a nonzero polynomial, in the space and if the multiplicity conditions are independent. If the set has 9 points or fewer, the problem has been solved, by Nagata himself (see [11]). His algorithm, which has been rediscovered by others (including ourselves), is short and elementary and it uses a weak version of our main lemma.

For 10 points or more the problem becomes considerably more difficult. Hirschowitz [7] has solved it in case every $m_i \le 3$ and he and Alexander [2] have extended his results for $m_i \le 2$ to higher dimensions, but results of this generality are rare. Chudnovsky [3] has tried to establish Nagata's geometric lemma for $r = \sqrt{10}$

(and thus to bring his group down to G_a^7), but his arguments are inconclusive. Thus our final problem is to prove it, i.e., to show that there is no curve of degree $d \leq m\sqrt{10}$ with multiplicity $\geq m$ at each of 10 general points, or to do related special cases such as $(d,m) = (19,6)$, $(38,12)$, $(177,55)$,... which should just fail to exist because $\binom{d+2}{2} - 10\binom{m+1}{2} = 0$ in each case. Here the first two cases have been done (cf. [7]), with a number of ingenious but ad hoc constructions, but the third case remains undone.

REFERENCES

1. A. A'Campo-Neuen, *Note on a counterexample to Hilbert's fourteenth problem given by P. Roberts*, Indag. Math., N.S. **5** (1994), 253–257.

2. J. Alexander and A. Hirschowitz, *La méthode d'Horace éclatée: application à l'interpolation en degré quatre*, Invent. Math. **107** (1992), 585–602.

3. G. V. Chudnovsky, *Sur la construction de Rees et Nagata pour le 14ᵉ problème de Hilbert*, C. R. Acad. Sci. Paris **286** (1978), A1133–1135.

4. J. Dixmier, *Solution négative du problème des invariants (d'après Nagata)*, Sém. Bourbaki, 1958/59, Exp. 175, also published by Benjamin, 1966.

5. J. Fogarty, *Invariant theory*, Benjamin, 1969.

6. W. Fulton, *Algebraic curves*, Benjamin, 1969.

7. A. Hirschowitz, *La méthode d'Horace pour l'interpolation à plusieurs variables*, Manuscr. Math. **50** (1985), 337–388.

8. J. E. Humphreys, *Hilbert's fourteenth problem*, Amer. Math. Monthly **85** (1978), 341–353.

9. D. Mumford, *Hilbert's fourteenth problem - the finite generation of subrings such as the ring of invariants*, AMS Proc. Symp. Pure Math. **28** (1976), 431–444.

10. M. Nagata, *On the fourteenth problem of Hilbert*, Amer. J. Math. **81** (1959), 766–772.

11. _____, *On rational surfaces II*, Mem. Coll. Sci. Kyoto **33** (1960), 271–293.

12. _____, *On the fourteenth problem of Hilbert*, Proc. I.C.M. 1958, Cambridge University Press, 1960, 459–462.

13. _____, *Lectures on the Fourteenth Problem of Hilbert*, Tech. report, Tata Institute, 1965.

14. M. Nagata and K. Otsuka, *Some remarks on the 14th problem of Hilbert*, J. Math. Kyoto Univ. **5** (1965), 61–66.

15. D. Rees, *On a problem of Zariski*, Illinois J. Math. **2** (1958), 145–149.

16. P. Roberts, *An infinitely generated symbolic blow-up in a power series ring and a new counterexample to Hilbert's fourteenth problem*, J. Algebra **132** (1990), 461–473.

17. T. A. Springer, *Invariant theory*, Springer Lect. Notes in Math., no. 585., Springer, 1977.

18. R. Weitzenbock, *Über die Invarianten von linearen Gruppen*, Acta Math. **58** (1932), 231–293.